3rd International Conference on Design, Analysis, Manufacturing and Simulation (ICDAMS 2018)

MATEC Web of Conferences Volume 172 (2018)

Chennai, India
6-7 April 2018

Editors:

V. Jayakumar
D. Devika
S. Ranganathan
S. Sridevi

ISBN: 978-1-5108-6467-2

Printed from e-media with permission by:

Curran Associates, Inc.
57 Morehouse Lane
Red Hook, NY 12571

Some format issues inherent in the e-media version may also appear in this print version.

Printed by Curran Associates, Inc. (2018)

For additional information, please contact EDP Sciences – Web of Conferences at the address below.

EDP Sciences – Web of Conferences
17, Avenue du Hoggar
Parc d'Activité de Courtabœuf
BP 112
F-91944 Les Ulis Cedex A
France

Phone: +33 (0) 1 69 18 75 75
Fax: +33 (0) 1 69 28 84 91

contact-edps@webofconferences.org

Additional copies of this publication are available from:

Curran Associates, Inc.
57 Morehouse Lane
Red Hook, NY 12571 USA
Phone: 845-758-0400
Fax: 845-758-2633
Email: curran@proceedings.com
Web: www.proceedings.com

TABLE OF CONTENTS

PREFACE

The 3rd International Conference on Design, Analysis, Manufacturing and Simulation, ICDAMS 2018, was organized during 6-7 April 2018 by the Departments of Mechanical and Automobile Engineering, Saveetha School of Engineering, Saveetha Institute of Medical and Technical Sciences (SIMATS), Chennai, INDIA. This conference was intended to provide a common platform to bring together researchers, engineers, scientists, and scholars to exchange and share their experiences, new ideas, and research results about all aspects of Mechanical Engineering. ICDAMS 2018 was successful in establishing a better rapport and cooperation between the academic institutions, the industries and the research establishments.

We are humbled by the overwhelming patronage from delegates across the globe and eventually ICDAMS 2018 had attracted 368 papers. After scrutiny by our Technical Committee, 180 papers (39 papers in Design Engineering; 25 papers in Manufacturing Engineering; 23 papers in Materials & Processing; 59 papers in Thermal Engineering; 11 papers in Industrial Engineering; 13 papers in Interdisciplinary and 10 papers in Bio-Engineering) have been short listed for Oral Presentation and 48 papers have been recommended for publication in MATEC Web of Conferences (Scopus Indexed). From the International arena, ICDAMS 2018 received papers from 11 countries namely New Zealand, Russia, Lithuania, Bahrain, Botswana, Egypt, Malaysia, Nigeria, Ethiopia, Indonesia and India.

The conference facilitated keynote address by eminent people from reputed academia, Industries and research organizations from both within and outside the country. Totally 15 technical sessions were held during this event. The feedback from the participants was encouraging and many of them expressed their willingness to participate in such future conferences to be held at SIMATS. ICDAMS2018 culminated as a successful event in integrating researchers across the globe to bring about an integrated approach towards creating a more comfortable world by establishing better rapport and cooperation between the academic institutions, the industries and the research establishments.

An event of this dimension is the culmination of many individuals. I wish to register my heartiest thanks to all the ICDAMS 2018 organizing committee members for extending their valuable time & effort and to all the International and National Advisory Members, Authors, Reviewers, Sponsors and Contributors for their sparkling efforts and their belief in the success of ICDAMS 2018. Also I am deeply indebted to EDP Sciences for excellently bringing out this conference proceedings through the MATEC Web of Conferences.

Dr. V. Jayakumar
Convener, ICDAMS2018

Statement of Peer review

In submitting conference proceedings to *Web of Conferences*, the editors of the proceedings certify to the Publisher that

1. They adhere to its **Policy on Publishing Integrity** in order to safeguard good scientific practice in publishing.
2. All articles have been subjected to peer review administered by the proceedings editors.
3. Reviews have been conducted by expert referees, who have been requested to provide unbiased and constructive comments aimed, whenever possible, at improving the work.
4. Proceedings editors have taken all reasonable steps to ensure the quality of the materials they publish and their decision to accept or reject a paper for publication has been based only on the merits of the work and the relevance to the journal.

Title, date and place of the conference

Title: 3rd International Conference on Design, Analysis, Manufacturing and Simulation - ICDAMS2018
Date: 6-7 April 2018
Place: Chennai, Tamilnadu, INDIA

Proceedings editor(s):

Dr. V. Jayakumar
Dr. S. Ranganathan
Dr. D. Devika
Dr. S. Sridevi
Professors, Department of Mechanical Engineering, Saveetha School of Engineering, Saveetha Institute of Medical and Technical Sciences, SIMATS, Chennai - 602 105, Tamilnadu, INDIA

Date and editor's signature

Dr. V. Jayakumar
22.05.2018

Recirculating flows analysis and estimation inside channels

Christina G. Georgantopoulou[1*], *Nikolaos S.* Vasilikos[1], *George A.* Georgantopoulos[2]

[1]School of Engineering, Bahrain Polytechnic, Isa Town, PO Box 33349, Kingdom of Bahrain
[2]Aerodynamics and Flight Mechanics Sector, Hellenic Air Force Academy, Dekeleia Attika, Greece

Abstract:-The recirculation which is developed during the flows inside pipes present a high interest in many industrial applications. In the present paper, a Cartesian grid method is presented which can be applied in pipes geometry approximation, even if the solid bounds are not lying on grid lines. A refinement technique using rectangular nested sub-girds is applied in order to avoid the unnecessary grid cells in the areas with no particular flow interest and cluster the grid when is needed. Important and useful for the industries results are extracted by these numerical simulations and estimations regarding the exact position and extend of the recirculation zones and the relating points. The estimation is taking placefor incompressible laminar, viscous flows inside inclined step channelsfor a range of inclination angles and Reynolds numbers values. The Navier – Stokes equations are solved using the artificial compressibility method according to the necessary boundary conditions arrangement. Flow results are presented for several grid sizes and Reynolds numbers focused on the recirculationzones length, in upper and lower channel' walls. Accepted accuracy of the flow results is produced using the aforementioned refinement algorithm, while the flow zones can be located according to the inlet flow rate, in order to avoid possible problems in the industries as corrosion or energy losses.

1 Introduction

The computational models are flexible tools for the industrial flow applications as useful results can be extracted for the optimization of the production line. The case of sudden expansion channels is also quite common in various industrial applications. For this reason, several research attempts can be found regarding the solution of the flows inside this type of channels [1,2] or the validation of their results using experimental methods [3,4].

The grid generation approach is also important in these cases, as a huge number of cells can be createddue to the desired accuracy. Although several grids can be applied providing the necessary accuracy, the Cartesian grids seem to be more attractive due to the simple arrangement, easy specification of the geometry description and not complicated equations discretization [5, 6]. In many channels' flow simulation and estimation cases, the uniform Cartesian grid requires high computational memory without any benefit to the accuracy of the results. In these cases, a refinement technique must be applied, according to the geometrical bound description and flow field requirements [7,8,9].

In the present paper, the flow inside sudden expansion channels are simulated and estimated as well. The main results are not limited only to velocity or pressure distribution, but also to locate the exact position of the recirculation zone extend at the upper and lower wall of the channels. It is expected such a flow pattern due to the sudden step inside the channels. The grid methodology is developed using only Cartesiangrid lines, which is quite acceptable for such geometries providing robust results [10]. This method is based on Tseng approach[11], while it has been already applied successfully for pipe flows prediction in our recent research work [12]. A refinement technique is required to be developed, as grid clustering is required only to specific areas with flow interest. Rectangular nested grids are generated, where the numerical simulation is taking placeafter the necessaryboundary conditions applications along the common edges [13,14]. The flow Navier – Stokes (NS) equations are solved using the flux vector splitting methodology in combination with an artificial compressibility algorithm [15]. The final results

*Corresponding Author: Christina@polytechnic.bh

provide acceptable accuracy and have been validated with other computational techniques and literature corresponded ones.

2 Mathematical and Numerical approach

2.1 Channel geometry and flow equations

The channels' geometry is easy to be projected due to the most bounds' alignment with the cartesian grid lines. The original channel geometry is projected onto the cartesian grid lines and a set of points is set accordingly. These points are lied either along the x- either on y-axis. The rule of minimum distance is used in order to "transfer" these points on the grid nodes (Fig. 1). The new set of points (on the grid nodes) are the approximate geometry points and define a new saw-tooth channel's bound. These points can be connected using straight lines by two ways: using Cartesian grid lines or diagonal lines. In our paper the Cartesian grid lines are chosen as the saw-tooth method is applied.

As we have initially mentioned, the geometry approximation is simple in the channels' cases, as all the bounds are aligned with Cartesian lines apart of one. Only for this part of the geometry, approximated points and bound will be created. The more clustering the grid, the more accurate approximated bound is extracted. This technique has successfully been applied in our previous work and this is the main reason that we choose to develop it for these cases as well [12, 16, 17].

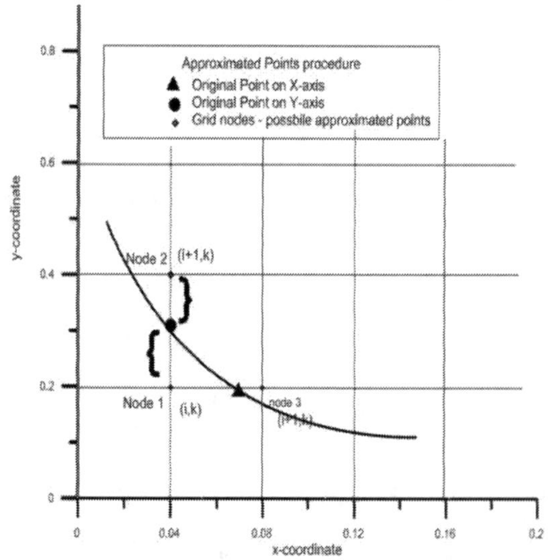

Fig. 1. Approximated points definition and procedure

Although, this flow numerical estimation case, seems not to require ununiform grids, the high number of computational cells increase the required computational memory without offering any benefit to the accuracy of the results. This effect is even worst in the case of unsteady or turbulent flows. There are several refinement techniques that can be applied but according to the simplicity of the channel geometry a rectangular nested technique is developed which produced accurate results despite of the fact that the flow variables values are transferred among the sub-grids. The method is based on using a hierarchical sequence of nested rectangular (block) meshes in which the numerical simulation is taking place (Fig. 2). In the beginning the flow solver procedure is applied to the initial grid, while the estimation is continued to the next level adjacent sub-grids. The integer refinement factor *I* is defined, as follows:

$$I = dx_m / dx_{m+1} = dz_m / dz_{m+1}. \qquad (1)$$

where I is the refinement factor, dx_m *and* dx_{m+1} the grid's resolution in longitude for the m grid level and for the $m+1$ (next) grid level, respectively, while dz_m *and* dz_{m+1} the grid's resolution in latitude for the grid levels as above. This method provides robust and accurate results minimizing the computational time as well [16, 17].

The flow inside the channels is received as laminar, steady, viscus and incompressible. For these reasons the flow is described by the Navier – Stokes equations as per below:

$$\frac{\partial \rho}{\partial t} + \frac{\partial (\rho u)}{\partial x} + \frac{\partial (rw)}{\partial z} = 0 \qquad (2)$$

$$[\Gamma]\frac{\partial q}{\partial t} + \frac{\partial e}{\partial x} + \frac{\partial g}{\partial z} = \frac{1}{\mathrm{Re}}\left(\frac{\partial r}{\partial x} + \frac{\partial s}{\partial z}\right) \qquad (3)$$

$$[\Gamma]\frac{\partial q}{\partial t} + \frac{\partial e}{\partial x} + \frac{\partial g}{\partial z} + \alpha \cdot \frac{g_1}{z} = \frac{1}{\mathrm{Re}}\left(\frac{\partial r}{\partial x} + \frac{\partial s}{\partial z} + \alpha \cdot \frac{s_1}{z}\right) \qquad (4)$$

where,

$$[\Gamma] = \mathrm{diag}(\frac{1}{\beta}, 1, 1)^{\mathrm{T}} \qquad (5)$$

and β is the artificial compressibility term, a is a switch for the activation of the axisymmetric terms ($a=0$ is non-axisymmetric, $a=1$ axisymmetric flow field), *Re* the Reynolds number and Q the unknown solution vector, as follows:

$$Q = (p \quad u \quad w)^T \quad (6)$$

with p being the pressure, and (u, w) the velocity components in physical space, e, g, g_1 and r, s, s_1 are respectively the convective and diffusive flux vectors at the plane (x, z).

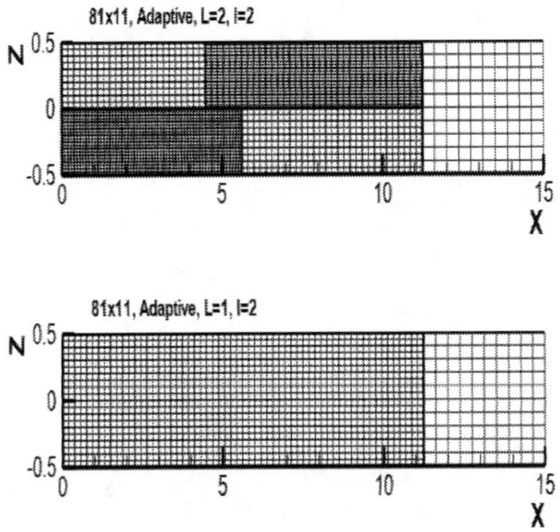

Fig. 2. Refinement technique for the channel geometry and grid generation. Grid size: 81x11, one (1) and two (2) levels of refinement, where I=2.

As it can be easily seen, this type of governing equations is applied for unsteady flows. However, this form of the equations can be used for the solution of steady flow fields as in our case as we know that these are non-linear and iterative method are needed for their solution [17, 18].

2.2 Boundary conditions

In this flow case, two types of boundary conditions must be applied, either at the physical bounds, either at the neighboring grids bounds inside the channel. The relating conditions are presented at the below Table 1 with reference to Fig. 3, where u *and* w are the velocity components, p the pressure and the indicator *ref* shows the corresponded reference values of the flow field.

Because of the sub-grids creation, the velocity and pressure values have to be transferred among the grids in order to ensure the consistent numerical estimation. If we assume that the velocity values have to be transferred from a coarse interface (grid level $= l_i$) to a refine one (grid level $l = l_{i+1}$), we receive:

$$u_{l,m} = u_{l,m+1} = u_{i,j} \quad (7)$$

where $u_{l,m}$ and $u_{l,m+1}$ are the axial velocity components at the refined grid level and $u_{i,j}$ the corresponded one at the coarse level. In the case of the physical bound conditions artificial (ghost) cells are considered, while these are neglected in the case of the sub-grids.

Table 1: Channel flow estimation boundary conditions (Fig. 3)

Geometry Bounds (Fig. 4)	**Boundary Conditions**
AB	Inlet conditions $u = u_{ref}$, $w=0, \frac{\partial p}{\partial x}=0$
AC	No slip condition (wall) $u=w=0, \frac{\partial p}{\partial z}=0$
CD	No slip condition (wall) $u=w=0, \frac{\partial p}{\partial z}=0$
DE	No slip condition (wall) $u=w=0, \frac{\partial p}{\partial z}=0$
EF	Outlet $\frac{\partial u}{\partial x} = 0, w = 0, p = p_{ref}$
BF	No slip condition (wall) $u=w=0, \frac{\partial p}{\partial z}=0$

At the specific flow case study in various sub-grids domains, the vertical velocity component is equal to zero (0) and so this is set at the aforementioned boundaries. However, even if $\neq 0$, the values will be transferred by a similar way with the axial velocity components as above:

$$w_{l,m} = w_{l,m+1} = w_{i,j} \ (8)$$

All the symbols are as these described above.

In the case that the boundary values must be transferred through fine – course interface the below formula is applied:

$$u_{i,j} = \frac{u_{l,m+1}+u_{l,m}}{2} \ (9) \ \text{and} \ w_{i,j} = \frac{w_{l,m+1}+w_{l,m}}{2} (10)$$

where the values are as described above.

In some cases, according for example the specific flow patterns inside the channels, fine grids are adjacent, and the variables values have to be transferred among them. The level of the grids receives the same value in both of the sub-grids and the velocity values are given as it can be seen below:

$$u_{l,m} = u_{k,n} \quad and \quad w_{l,m} = w_{k,n} \ (11)$$

As the aforementioned grid method is applied only for rectangular Cartesian grids and sub-grids, there are some cases that course grids will be adjacent and the related boundary conditions must be set in order the numerical calculation to take place. In that case the boundaries application is developed as at the above case fine-fine interface (Eq. 11).

3 Results

The flow inside sudden expansion channels will be numerically analyzed, simulated and estimated. By presenting the following extended analysis we attempt to enhance previous research work which has been done on theseflow fields [3,19] and extend our research work in incompressible viscous flows inside pipes [17, 18, 20] providing specific, accurate and useful results for the flow development.

The geometry description and the related dimensions can be seen at the above Fig.3. The relevant height H_{ref} has set equal to 1 for dimensionless purposes and it is the maximum hydraulic diameter of the pipe. The inlet area receives $H_{ref}/2$ height as we study a sudden expansion pipe.

In order to investigate the inclination angle effect to the flow, we present various calculations for three different inclination angles values: 30°, 45° and 90° (Fig. 5), for five different *Re* values: 100, 400, 800, 1200 and 1700.

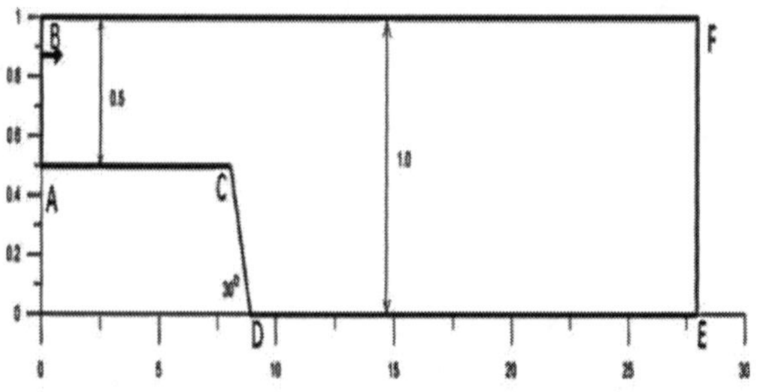

Fig. 3. Channel geometry description. The dimensionless lengths of the channel vary according to the *Re* value.

3.1Recirculation lengths and points

The results that are presented have been produced using grid size 351x26, with two refinement grid levels and *I=2*. In all the cases, the expansion ratio is set to be equal to 2 (Fig. 3). At the following tables the separation zones of the lower and the upper wall of the pipes are presented according to the inlet flow rate for the above *Re* values. Detachment and reattachment points have been located in order to provide even more efficient information concerning the exact location of the separation zones.

For validation purposes, all the numerical cases solved using the commercial software ANSYS and some of these are presented for inclination angle equal to 90 *deg* for all the range of the used *Re*, using uniform grid sized 651x26.

In addition, the present work results are compared with the corresponded ones by the literature where the convergence is accepted.

Table 2: Recirculation lengths and relating points – lower wall: Inclination angle 30 *deg*. Grid, 351x13, *l=2, I=2*.

Re	**Upper Wall**		
	Recirculation Length	**Detachment Point**	**Reattachment Point**
100	No detection	No detection	No detection
400	No detection	No detection	No detection
800	4.74	13.54	18.28
1200	7.54	14.51	22.05
1700	10.07	15.85	25.92

Table 3: Recirculation lengths and relating points – lower wall: Inclination angle 30 *deg*, ANSYS – Fluent, Cartesian uniform grid 651x26.

Re	Lower Wall		
	Recirculation Length	**Detachment Point**	**Reattachment Point**
100	0.44	8.89	9.33
400	4.28	8.28	12.56
800	6.58	8.28	14.86
1200	7.74	8.28	16.02
1700	9.03	8.28	17.31

Table 4: Recirculation lengths and related points – upper wall: Inclination angle 30 *deg*, ANSYS – Fluent, Cartesian uniform grid 651x26.

Re	Upper Wall		
	Recirculation Length	**Detachment Point**	**Reattachment Point**
100	No detection	No detection	No detection
400	No detection	No detection	No detection
800	5.02	13.55	18.57
1200	7.72	14.55	22.27
1700	10.13	15.93	26.06

Table 5: Recirculation lengths and related points – lower wall: Inclination angle 45 deg. Grid, 351x13, l=2, I=2.

Re	Lower Wall		
	Recirculation Length	**Detachment Point**	**Reattachment Point**
100	1.24	8.10	9.34
400	3.82	8.10	11.92
800	6.68	8.10	14.78
1200	8.99	8.10	17.09
1700	10.86	8.10	18.96

Table 6: Recirculation lengths and related points – upper wall: Inclination angle 45 deg. Grid, 351x13, l=2, I=2.

Re	Upper Wall		
	Recirculation Length	**Detachment Point**	**Reattachment Point**
100	No detection	No detection	No detection
400	1.78	11.49	13.27
800	4.20	13.81	18.01
1200	6.19	15.75	21.94
1700	8.72	17.20	25.92

As it is depicted by all the tables, there are mainly two separation zones the most, inside the channel ($Re < 1700$), while the recirculation at the upper wall cannot be detected when the*Re* is equal to 100 and 400 in most of the cases.As the flow rate increases, the recirculation length increases as well ason both channel's walls. However, although the detachment points at the lower wall remain at the same position for all the inlet velocities (around 8.26 except for the case of *Re*=100), the one at the upper wall is moving forward according to the related Re. Despite of this fact the recirculation length at the upper bound is increased as the *Re* increases. For the value of *Re* equal to 100 although the separation zones are not so extended, it seems that there is a low width recirculation at the upper surface of the pipe, which, when the *Re*is equal to 1700, receives a significant value. In Tables 2 – 6 the related information is presented for the inclination angle value equal to 30 deg.

Table 7: Recirculation lengths and related points – lower wall: Inclination angle 90 deg. ANSYS – Fluent, Cartesian uniform grid 651x26.

Re	Lower Wall		
	Recirculation Length	**Detachment Point**	**Reattachment Point**
100	1.30	8.05	9.35
400	3.35	8.05	11.40
800	4.15	8.05	12.20
1200	5.25	8.05	13.30
1700	5.85	8.05	13.90

Table 8: Recirculation lengths and related points – upper wall: Inclination angle 90 deg. ANSYS – Fluent, Cartesian uniform grid 651x26.

Re	Upper Wall		
	Recirculation Length	**Detachment Point**	**Reattachment Point**
100	No detection	No detection	No detection
400	1.20	11.55	12.75
800	4.60	12.00	16.60
1200	6.80	12.60	19.40
1700	8.60	13.10	31.70

The numerical estimation has been developed using both of the computational techniques (block – Cartesian and ANSYS) for validation purposes, where the convergence of the results is acceptable, presenting very low relative error ($4 \times 10^{-3} < rel.error < 5.5 \times 10^{-3}$). It seems that the development of the recirculation in both of the pipe's walls is higher than the previous case (angle equal to 10 *deg)*, the detachment point of the lower wall remains in constant position and the upper wall separation zone is moving forward according to the Re.

The present work numerical data have been retrieved by 351x13 grid size, *l=2* and *I=2*. It seems that the convergence is acceptable since the relative errors receive very low values. However, in some cases, certain differences are depicted. These seem to be developed due to the different grid sizes, the different entrance pipe lengths or the differences to the outflow boundary conditions application.

Table 9: Recirculation lengths and related points lower wall: Inclination angle 90 deg, grid 351x13, *l=2, I=2*.

Re	Lower Wall		
	Recirculation Length	**Detachment Point**	**Reattachment Point**
100	1.17	8.05	9.22
400	3.71	8.05	11.76
800	6.25	8.42	14.67
1200	7.53	8.45	15.98
1700	8.12	8.50	16.62

Table 10: Recirculation lengths and related points upper wall: Inclination angle 90 deg, grid 351x13, *l=2, I=2*.

Re	Upper Wall		
	Recirculation Length	**Detachment Point**	**Reattachment Point**
100	No detection	No detection	No detection
400	1.72	11.55	13.27
800	4.40	12.75	17.15
1200	6.75	15.15	21.90
1700	7.85	16.01	23.86

As we can see by the following figures, no detection of recirculation is presented in very low *Re*, while the appearance of the upper bound recirculation depends on the *Re* values and the inclination angle as well (Fig. 6). Through these results , the influence of the step angle to the flow pattern can be specified. It seems that the higher step angle, the longest recirculation zone at the lower channel wall. By the other hand, the flow detachment on both channel walls is not the same (Fig. 7 and 8).

As we can identify by the Fig. 8, the more the inclination angle is increased the less extension of the upper wall recirculation is developed through the pipe if we assume that the Re remains constant. However, again at the highest Re values, the more extended lengths are

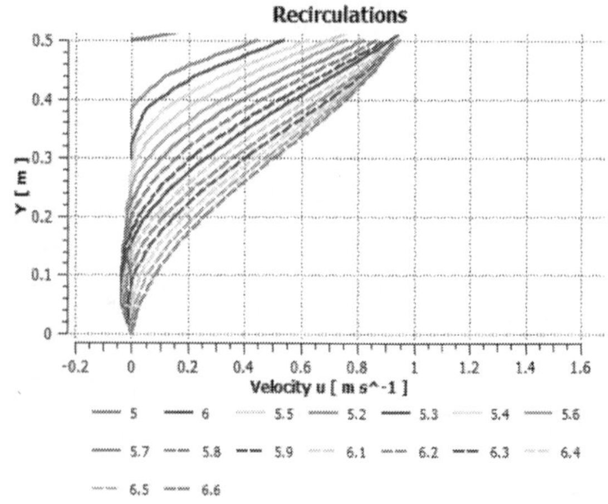

Fig 4. Lower wall recirculation inside the channel. Re=100, inclination angle=30°. ANSYS solution.

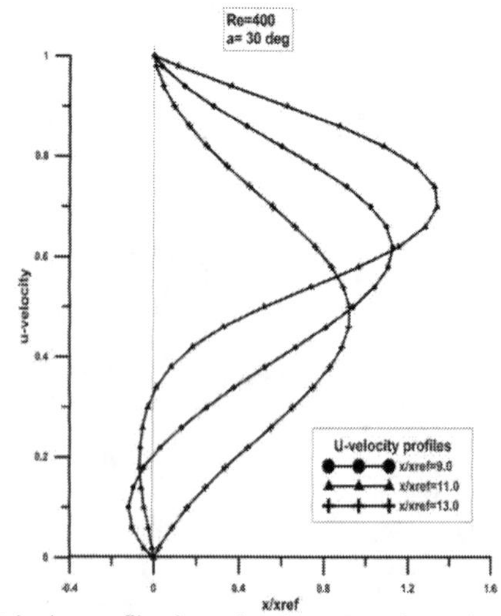

Fig. 5. Velocity profile along the pipe; a=30 deg and Re=400.

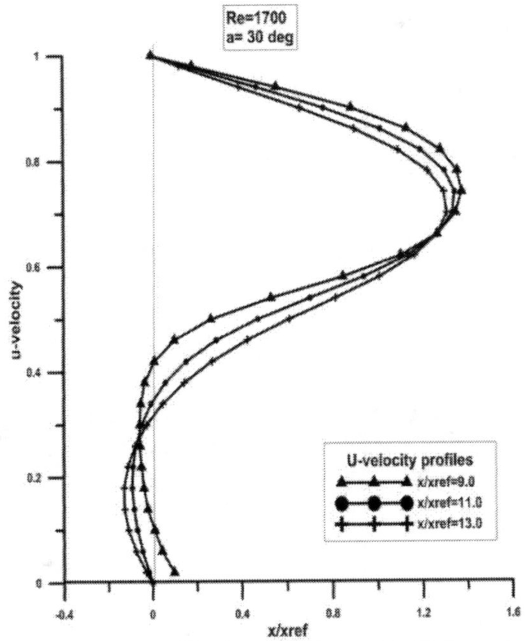

Fig. 6. Velocity profile along the pipe; angle=30 deg and Re=1700.

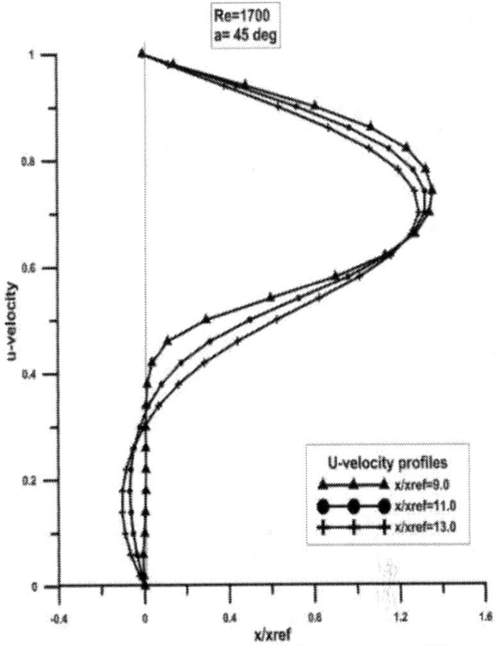

Fig 7. Velocity profiles; a=45 deg and Re=1700.

presented. It is important that this relationship is followed by the flow in all the various Re and by this way we may conclude that this relationship applies for all the range of Reynolds number of laminar flow.

However, as we have concluded after several test cases in this field, the channel length downstream of the step affects the quality and the accuracy of the numerical results as sometimes the flow field is converged without oscillations. In our future research, the appropriate limits for this parameter requires further investigation, ensuring that every solution is length independent [21].

We agree with most of the researchers, who they claim that the entrance length of the pipe affects the numerical solution in low Re [22, 23]. Accordingtothe literature an entrance length more than 6 dimensionless ones should be enough in order not to affect the flow distribution inside the pipe. In order to create an independent solution we have applied eight (8) lengths for the entrance pipe and it seems that there is no influence along the fluid flow field.

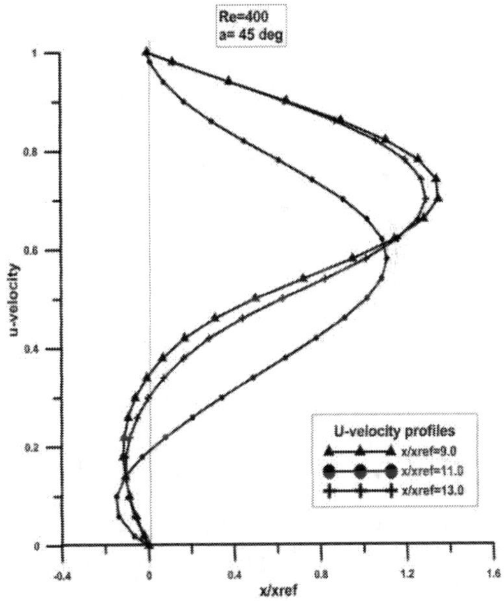

Fig. 8. Velocity profiles; a=45 deg and Re=400.

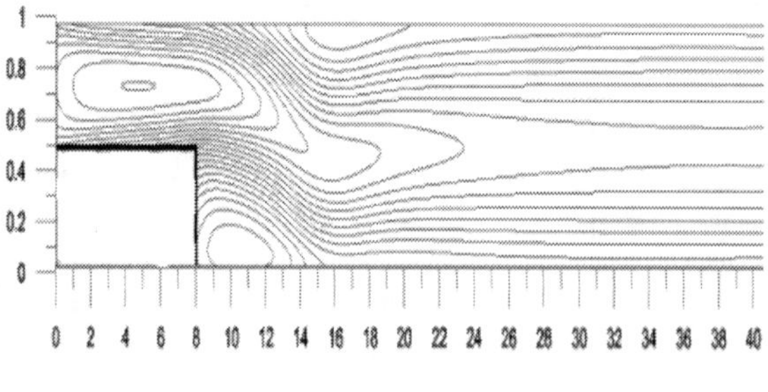

Fig. 9. Velocity contours along the pipe. Block nested grid, I=2, l=2. a=90° and Re=800.

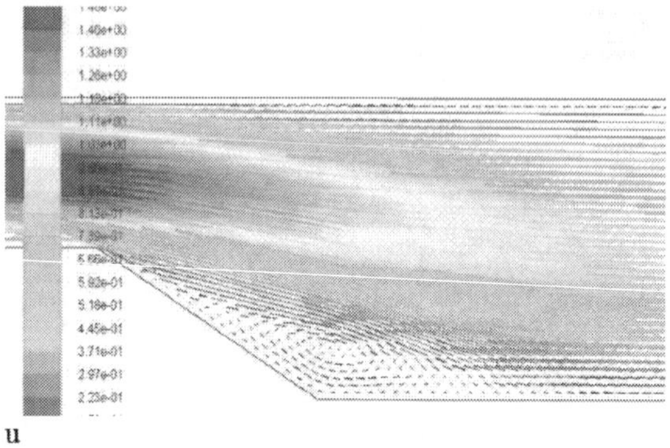

Fig. 11. Velocity contours along the pipe. ANSYS solution, a=30° and Re=100.

The overall flow development inside the channel can be seen through the velocity contours (Fig. 9-11) for various step angles and Re values. It seems that there is a good and acceptable convergence among the current work and ANSYS results.

3 Conclusions

The sudden expansion channel steady, laminar and viscus flows have been numerically analyzed, simulated, estimated and presented for a range of Re values and inclination angles. The flow characteristics, variables distribution, certain data information and points of upper and lower bound separation zones have been analytically presented for various values of *Re*. For the physical domain discretization, we use Cartesian grids, despite of the non-Cartesian bounds existence and in order to

In fig. 5 to 8, the axial velocity profiles are presented for various a and Re values, while velocity profiles and recirculation graphs are presented below with the ANSYS solver as well for Re=100 and a=30o. As it was expected the profile is parabolic inside the pipe, while in the certain x-coordinate location the separation areas are developed. It can be seen that the mass flow rate is conserved as well as the development of the flow is not affected by the downstream and upstream pipe's lengths.

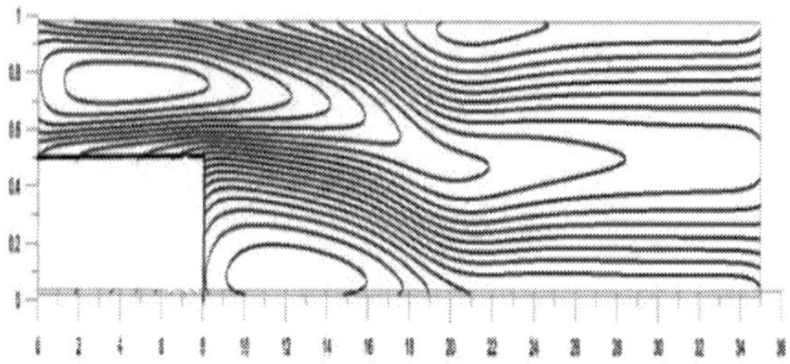

Fig. 10. Velocity contours along the pipe. Block nested grid, I=2, l=2. a=90° and Re=1200.

overcome the high computational memory which is needed due to the grid cells number and reduce the computational time without decreasing the accuracy of the solution, we develop a block-nested refinement algorithm, which can be easily applied to the channel flow domain. It seems that if the *Re* is higher than 400 and less than 1700, two separation zones are developed along the pipe in both bounds. The related detachment and reattachment points are clearly presented for every case and the final results show that in most of the cases the higher *Re* the more extended recirculation area for a certain inclination angle. The most interesting and useful conclusion is the influence of the step angle of the pipe to the separation zones extend and development, which seems to be not the same for both pipe's bounds (upper and lower). More specifically, as the step angle is increased, it seems that the recirculation length at the lower wall is increased as well, for all therange of *Re* values. It's worth mentioning, that our results have been validated for various Re values and grid sizes uniform and ununiformed as well, either by the corresponded ones by the literature either by the commercial ANSYS software ones. Further research is required regarding the experimental investigation of these flows taking into consideration the viscosity effect as well to the flow development and behavior inside the channel.

References

1 Santos R., Oliveira K., Figueiredo J., Mech.Comp, vol. XXIX, pp. 3347-3358, (2010).
2 Lee T. and Mateescu D. J. of fluids and structures, vol. 12, pp. 703–716, (1998).
3 Armaly B., Durst F., Pereira J., and Schounung B. J. of Fluid mech., vol. 127, pp. 473–496, (1983).
4 Coirier, W.J. and Powell, K.G., J. of Comp. Physics, Vol. 117, pp. 121-131, (1995)
5 Georgantopoulou C, Pappou Th., Tsangaris S.G., Proc. of the 4th GRACM conf. on Comp. Mech., pp. 526-534, (2002)
6 Wang, Z.J., Computers and Fluids, Vol. 27, pp.529-549, (1998)
7 Karai E., Kultari A., Haluk M., Hydrology and Ecology, Vol. 1 pp. 15-25
8 Agresar, G., Linderman J.J., Tryggvason, G. and Powell, K.G., J. of Comp. Physics, Vol. 143, pp. 346-380, (1998)
9 Georgantopoulou C., Georgantopoulos G., Vasilikos N., Int. J. of Mathematical Models and Method in Applied Sciences, 9 248-260, (2015).
10 Tseng Y., Ferziger J., J. of Computational Physics, vol.192, pp. 593-623, (2003)
11 Georgantopoulou C., Georgantopoulos G., Vasilikos N., WSEAS Transactions on Fluid mechanics, 10 69-79, (2015) .
12 C. Georgantopoulou and S. Tsangaris, Applied Mathematical Modeling, Vol.31, pp2136-2148.
13 Barton Ie., Int. J.Num.Methods Fluids, vol. 25, 663-644, (2015).
14 Berger M.J. and Collela P., J. of Comp. Physics, Vol. 83, pp.64-84, 1989.
15 Pappou, Th. and Tsangaris, S., Int. J. for Num. Methods in Fluid, Vol. 25, pp.523-545, (1997)
16 Georgantopoulou C., Georgantopoulos G., Vasilikos N. and Tsangaris S., Proc. of Int. Conf. on Applied Mathematics, simulation, modeling, pp.256-263, (2010).
17 Georgantopoulou C., WSEAS Transactions on Fluid Mechanics, vol. 17, (2017).
18 Mavromatides, L., Michel, P., El Mankibi, M., &Santamuris, M., Building Simulation, 3, 279–294. (2010).
19 Biswas G., Breuer M., Durst, Transactions of ASME, vol. 126, pp. 362-374, (2004)
20 C. Georgantopoulou, Feras Ali, N. Vasilikos, G. Georgantopoulos, Applied Mechanics and Materials, vol. 829, pp. 15-20, (2016).
21 Kanfoudi H., Zgoli R., Int. J. of Mod., Simulation and Scientific Comp., vol. 6, (2015).
22 Erturk E.,Computers and Fluids, vol. 37, pp. 633-655, (2008).
23 Gartling D, Int. J. Num. methods, vol. 11, pp. 953-967, (1990).

Slug-bubble regime identification in a square channel using a IR Sensor

Aafrin Sulthana. S[1], *Marieswaran*. T[1], *Braghadesh*. N[1], *Mithran*. N[1], *Venkatesan*. M[1]*

[1]School of Mechanical Engineering, SASTRA Deemed University, Thanjavur 613401, Tamil Nadu, India

Abstract. Design of micro thrusters for nano satellites, require a detailed understanding of multiphase flow phenomena in micro/mini-channels. This work focuses on the experimental and numerical investigation of an Infra-red sensor behavior during two phase flow of a slug-bubble train (air-water two-phase flow). The regime flows inside a square channel of sides 2 mm and 0.5 mm thickness made of borosilicate glass. The interference of the slug-bubble train flow pattern on the IR transceiver characteristics is experimentally studied as current signals corresponding to the number of photons received by the photodiode. A numerical model is developed to analyze the IR transceiver characteristics using COMSOL Multiphysics package. The experimental and numerical results are in good agreement and the developed system with proper calibration can be used to design feedback loops for micro thrusters.

1 Introduction

Capillary gas-liquid two-phase flow is prevalent in many applications for industries including nuclear reactors and modern micro-chemical propulsion systems. The thickness of liquid film surrounding bubble and slug regime play a significant role in deciding two-phase flow characteristics. The film thickness randomly varies for other flow regimes like stratified, churn and annular flows. Many of the liquid propellants, for space applications are mixed with an oxidizing agent before undergoing combustion. Precise knowledge of liquid film-thickness is required to have an efficient design of micro-thrusters capable of producing precise thrust of the order of few Newtons. Previous works reported on two-phase flow characterization is discussed in this section.

Particle image velocimetry (PIV) was employed by Thulasidas et al. [1] to study the bubble-train flow regime inside a capillary tube of circular and square cross-section. The velocity profile, circulating vortices and circulation time were analysed theoretically and experimentally. Mehta and Khandekar [2] measured heat transfer coefficient values during Taylor bubble/slug train flow in mini-channel using IR thermography. The results detailed that the heat transfer was enhanced by 2.5 times during a bubble flow when compared with that of a fully developed laminar single-phase flow. Barreto et al. [3] analyzed flow of air-water in a vertical micro-channel of 1.2 mm diameter. Shi and Wang [4] attempted to measure the water velocity in air-water flow as a function of voltage induced in a inductive flow sensor with electrodes which are arc-shaped.

Greenfield et al. [5] numerically studied multiphase flow inside supersonic micro-nozzles and analyzed the droplet impact on the thrust produced. Liu and Pan [6] studied heat transfer during flow boiling inside a microchannel using infrared thermography. The measured temperatures were observed to be within the theoretical predictions. Air-water flow distribution inside a serpentine mini-channel was analyzed by Malhotra et al. [7] and the pressure drop was studied for different flow regimes. Identification of two-phase flow regimes was done with a IR sensor by Arunkumar et al. [8]. A numerical model was developed using VOF method and validated with experiments. The effect of orientation of IR Transceiver during gas-liquid two-phase flow was analysed by Mithran and Venkatesan [9]. The results showed that keeping the transceiver at 90° orientation gave better results.

The works mentioned above details various studies reported on two-phase flow identification and sensors developed. Also, very few works on two-phase flow behavior characterization using IR sensor were reported in the past. In literature, when compared to other methods which are used to predict the flow regimes, IR sensor could provide more accuracy and is a non-intrusive type. In the current work, along with an experiment, a numerical model is developed using COMSOL Multiphysics package to study the irradiation behavior of infrared rays while passing through a slug-

* Corresponding author: mvenkat@mech.sastra.edu,

bubble two phase flow regime. The experimental and numerical results are compared and the results match reasonably well.

2 Experimental Setup

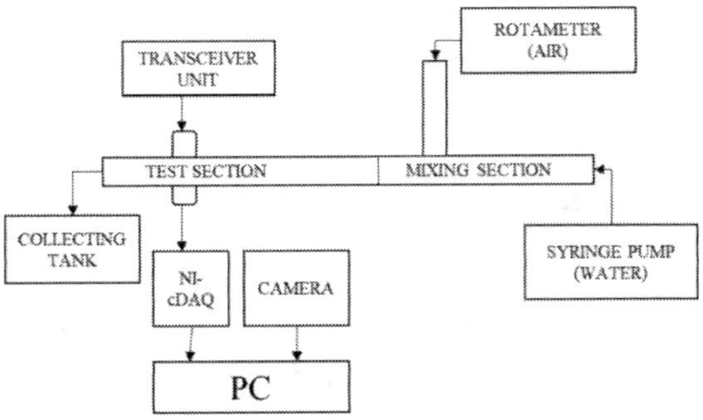

Fig. 1. Experimental setup

The experimental setup shown in Fig. 1 consists of a square channel of 2 mm side and 0.5 mm thickness made of borosilicate glass of length 15 cm. A T-section is used for mixing of two separate fluids. The regimes are formed for combination of various flow rates which are measured using rotameter. Air is pumped using a 240 LPH, 3 W air pump. A syringe pump (E-spin SPK101 model) is used to supply water of TDS concentration of 400 ppm at room temperature. The IR transceiver unit used in this study consists of IR 333-A transmitter and IR PD333-3B photodiode. 1.5 V DC power from AA battery is supplied to the IR LED which emits IR rays. The test section is placed between the IR transmitter and receiver. The IR photon rays pass through the test section and flow regimes having different refractive index which tends to refract some of the photons. The IR transceiver unit is kept vertically perpendicular to the channel, almost touching the outer surface of the tube. The received photon particles are converted into electrical signals by the photodiode and these signals are fed into a NI-cDAQ 9174 chassis with NI 9203 module. The module records the current signals in a computer. The two-phase flow in test section is recorded with the help of a AOS PROMON 501 high-speed camera with 7000 NAVITAR zoom lens. The flow is analyzed frame by frame with AOS studio and the slug-bubble train regime in the two-phase flow is captured. The size of the flow regime is measured with the help of imtool in image processing tool box of MATLAB. The obtained dimensions are used to develop a similar slug-bubble train in COMSOL Multiphysics

3 Numerical modelling

A numerical model is developed as shown in Fig. 2 using COMSOL Multiphysics to analyze the incident radiation behavior of the IR sensor which is same as that of the experimental setup.

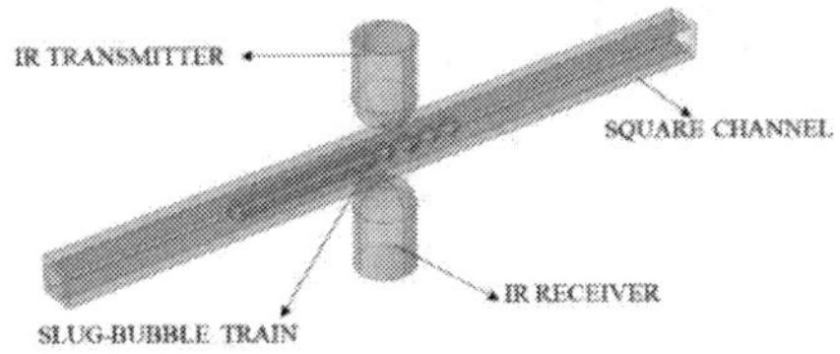

Fig. 2. Representation of numerical model

The refraction of the IR rays follow the Snell's law which postulates that the ratio of the incidence angle to refraction of photon ray is proportional to the ratio of the refractive indices of the two mediums through which the photon rays pass through.

$$\sin\theta_1 / \sin\theta_2 = n_2/n_1 \tag{1}$$

Where, θ_1 - angle of incidence;

θ_2 - angle of refraction;

n_1 - refractive index of the second medium;

n_2 - refractive index of the second medium.

The number of rays emitted from the source point is set as 100 (which can be varied) to predict the percentage of rays received by the receiver directly. A plane corresponding to the photo-diode cross-section is chosen as a freezing plane to collect the photons that enter the receiver after refracting through a different medium. The thickness of liquid film at the top will be thin when compared to the liquid film thickness at bottom due to the effect of gravity. However, in the developed model, an average uniform liquid film thickness around the voids is considered so that the interpretation of the resulting signal is easier. The model is designed such that the water medium remains stationary while the slug-bubble train sweeps an appropriate distance equivalent to the experimental flow rate.

4 Results and discussion

The high speed photograph of slug-bubble train obtained during the experiment is shown in Fig. 3 (a). The corresponding IR transceiver output in terms of current (mA) and numerical results in percentage of photons received is plotted. The position of the slug-bubble train regime is shown in x-axis of Fig. 3 (b).

(a)

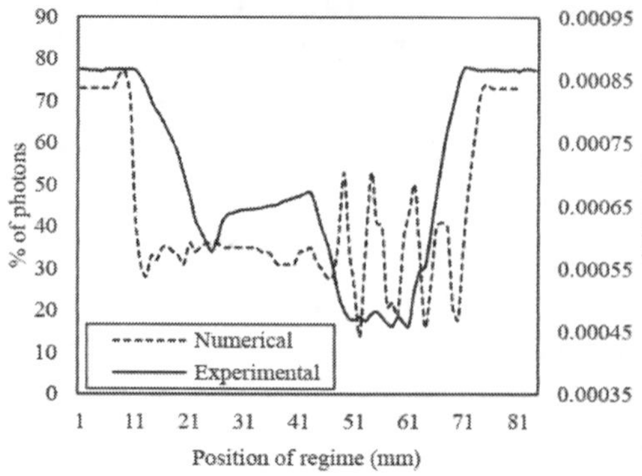

(b)

Fig. 3. Comparison of the experimental and numerical result

The IR rays pass through various mediums like borosilicate glass, water (liquid film) and air (voids), with refractive indices of 1.5, 1.33, and 1 respectively. The scattering of IR rays for different portions of the flow namely, single-phase water, void with liquid film and at the slug-bubble interface are shown in fig. 4(a), 4(b), 4(c).

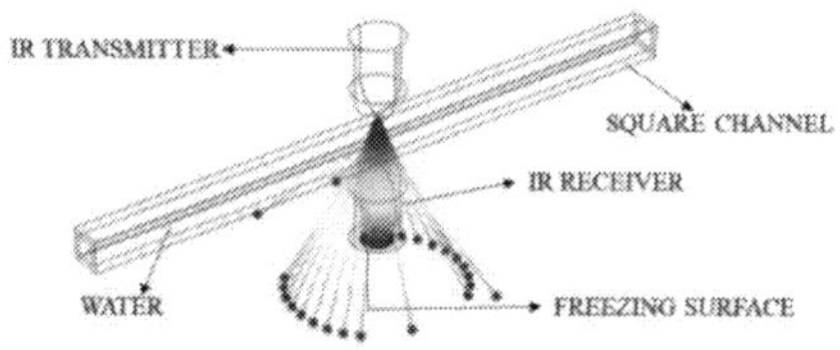

(a)

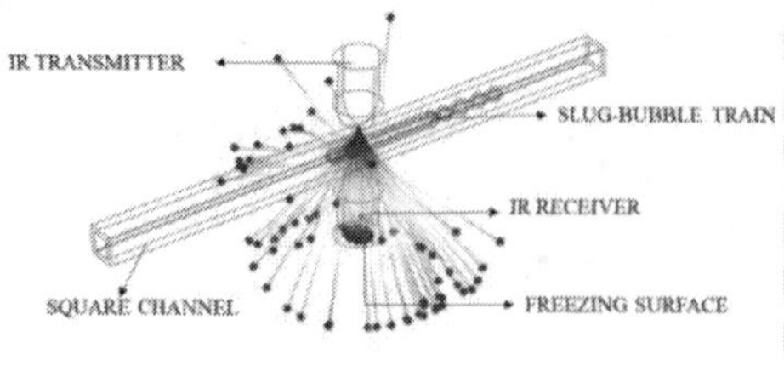

(b)

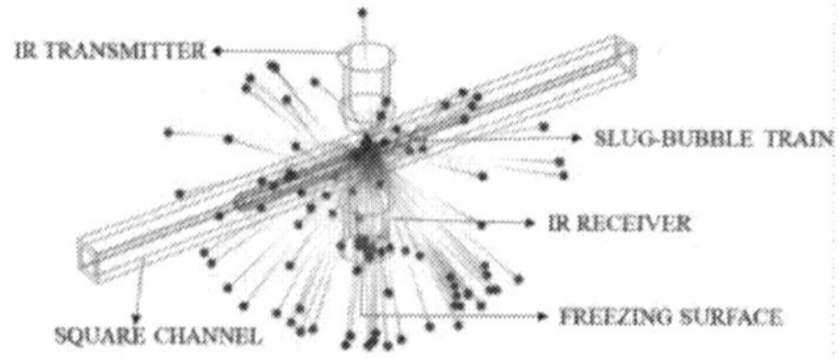

(c)

Fig. 4. 3D scattering of IR rays for test section having **(a)** only water, **(b)** slug regime and **(c)** at the point of slug-slug interference.

The number of photons received for the three cases are 73, 31, and 14 and the corresponding current values are 0.864, 0.684 and 0.471 mA respectively. For a single slug, the pattern obtained for IR sensor is explained by Mithran [9]. The length and shape of the voids also determine the variations in the number of photons received at the photo-diode. As the length of the voids decreases, like in bubbly flows, the obtained signal pattern in the plot seems to have narrow troughs. The variations in the current measured from IR sensor for the regime considered in the present study is due to the interference during refraction of the rays by the successive slugs. At the point of contact of the two successive slugs, the scattering of the rays is more when compared to other cross section of void fractions. The numerical results are in good agreement with that of the experiments. The velocity at which this two-phase flow crossed the transceiver unit may also be a reason for minor ambiguity between experimental and numerical plots. The results obtained confirm that the IR transceiver setup can be successfully employed for the characterization of slug train flow regimes. Further the numerical model developed can also be successfully employed to analyse various other flow parameters like variation in liquid film thickness, flow shape effects on IR transciever which is the scope for future work. On further extending the study, a feedback method could be developed for precise control of micro thrusters online.

5 Conclusion

In present work, the behavior of a IR sensor during air-water two-phase flow for a slug-bubble train flow regime in a square mini channel is analysed. The IR rays effects on air-water two-phase flow measured as current (mA) (experiment) is compared with that of percentage of photons (numerical). The experimental results are compared with the numerical results and validated successfully. The flow inside a mini/microchannel holds a significant role in outer space propulsion systems and many other miniaturized reactors. With precise calibration, these sensors could be used for the flow regime identifications in MEMS.

References

1. T. C. Thulasidas, M. A. Abraham, and R. L. Cerro, *Chemical Engineering Science*, vol. **52(17)**, 2947–2962, (1997).
2. B. Mehta and S. Khandekar, *11th International Conference on Quantitative InfraRed Thermography*, 11-14 June (2012), Naples Italy.
3. E. X. Barreto, J. L. G. Oliveira, and J. C. Passos, *Experimental Thermal and Fluid Science*, vol. **63**, 1–8, (2015).
4. Y. Y. Shi and M. Wang, *Flow Measurement and Instrumentation*, vol. **41**, 90–96, (2015).
5. Benjamin Greenfield, William Louisos, and Darren Hitt, *49th AIAA Aerospace Sciences Meeting including the New Horizons Forum and Aerospace Exposition, Aerospace Sciences Meetings*, https://doi.org/10.2514/6.2011-189
6. T. L. Liu and C. Pan, *Applied Thermal Engineering*, vol. **94**, 568–578, (2016).
7. S. Malhotra, A. Deshpande, and S. Ghosh, *International Journal of Multiphase Flow*, vol. **96**, 48–55, (2017).
8. S. Arunkumar, J. Adhavan, M. Venkatesan, S. K. Das, and A. R. Balakrishnan, *Flow Measurement and Instrumentation*, vol. **51**, 49–54, (2016).
9. N. Mithran and M. Venkatesan, *Flow Measurement and Instrumentation*, vol. **58**, 12–20, (2017).

Drag reduction in a class 8 truck - scaled down model

Vishwa Krishna. R[1], *Suwathy.* R[2], *Pragadeesh.* M [1], and *Venkatesan.* M[1,*]
[1]School of Mechanical Engineering, SASTRA Deemed University, Tirumalaisamudram, Thanjavur – 613401, Tamilnadu, India
[2] Department of Aerospace Engineering, Madras Institute of Technology, Chennai-600044, Tamilnadu, India

Abstract. Trucks are heavy load vehicles used mainly for commercial transport operations. There are several classes of heavy duty commercial vehicles classified based on the weight loaded. More than 50% of the engine output power in such trucks is utilized to overcome the drag. Drag force in automobiles is the resistance offered by air on vehicles at higher speeds. Class 8 trucks suffer higher drag when compared to other classes. In the present work, a numerical model is developed using a commercial code ANSYS FLUENT to predict the drag coefficient value. The effects of gap width and cab front radius with a constant fairing is analysed using the numerical model developed. A Class 8 model truck with minimal drag coefficient having constant fairing and optimized gap width between the trailer and cab is proposed.

1 Introduction

Heavy trucks which contain a cab and trailer come across more drag than conventional vehicles because of the vortex formation in the gap. The conventionally used class 8 truck is found to have bluffed and un-streamlined profile by design. This profile results in increased drag due to flow separation phenomenon. Many research works were reported on aerodynamic design of heavy commercial vehicles to reduce the resistance force and increase the engine efficiency in a feasible way. The drag force can be minimized by fairing the shape or implementing drag reducing devices. Also with proper design of geometry of the front cab of a given truck, the drag coefficient can be optimized. Such optimized geometries will result in decreased fuel consumption.

Allan [1] studied the effect of drag experimentally in a tractor trailer model. The cab front edge was rounded and the effect of it on drag was studied. Such experimental methods cannot be implemented for all possible design variations or such an implementation is not economical. With increase in computational resources, employing computational techniques is an easier solution in such cases where an optimized profile can be found out numerically. To verify the results, limited experiments can be done. Osth and Krajnovic [2] used large eddy simulation and studied the flow through the tractor trailer model numerically for four different cases. The gap width was changed and the cab front edge was rounded in and the effect was studied. Bettle et al. [3] investigated the speed of truck under the conditions of crosswind on a standard sized North American transport truck. Various aerodynamic parameters for different relative wind directions were varied and the parameter which would tend to overturn a truck was detailed. Hu et al. [4] made inferences from latter and numerically analysed the effect of aerodynamic loads with SST k-omega turbulence model in a heavy duty truck. The results showed a decrease in side force with the protection of fence and strengthening of rolling and yawing moments. Wang et al. [5] investigated the aerodynamic characteristics of three rear shapes numerically and experimentally in a wind tunnel to examine the effects

As a result of induced accidents in UK, Coleman & Baker [6] studied the aerodynamic forces experienced by a scaled down articulated lorry in a wind tunnel experimentally. The side flow oscillations were largely caused by turbulence buffeting and lift force oscillations that were caused by wake vortex shedding. Pragadeesh et al. [7] numerically analysed eight different models with constant gap width. Among the eight cases, the addition of a fairing to a truck which has its cab rounded to optimal radius was observed to have minimal drag and eddies to a convincing level.

To reduce drag, proper fairing methods like rounding the sharp edges to a certain radius, and fixing fairing plates between cab and trailer can be adopted. In the present work, co-efficient of drag is found for nine different cases in scaled down model of a class 8 truck. Among these nine cases, three cases which have got minimal drag are analysed to have minimal drag. The observed drag values are more dependent on the vortex that is formed in the gap width. The vorticity magnitude for both z-direction are calculated individually and their effect on drag is detailed. The model with minimal drag value and its design cases are presented.

* Corresponding author: mvenkat@mech.sastra.edu

2 Numerical model

2.1 Size of the domain

The scaled down model of a class 8 truck is shown in Fig. 1. It consists of a cab and a trailer placed in a domain of 8000 x 2104.5 x 1708 mm. The dimensions of the entire model is calculated based on rear box width b=0.305 m. The dimensions of the cab are 204.35 x 280.6 x 280.6 mm. The dimensions of the trailer are 762.5 x 280.6 x 305 mm. Cab and trailer are connected by a cylinder of diameter 24.4 mm. The gap width is varied in the present work. The height of cab and trailer above the ground are 64.05 mm and 152.5 mm respectively. Air with a velocity of 24.4 m/s is supplied at the inlet of the domain. Absolute pressure condition is chosen for exit of the domain.

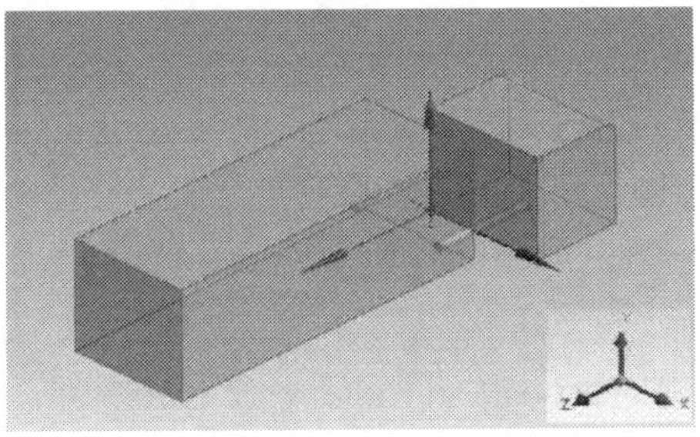

Fig. 1. Numerical model

2.2 Governing equations

The mass momentum equations along with standard k-ω turbulence model are used to model the trailer using commercial Fluent code Ansys [8] and are shown in eqn. (1), (2) (3) and (4).

$$\nabla.(v) = 0 \tag{1}$$

$$\frac{\partial(\rho v)}{\partial t} + \nabla.(\rho v v) = -\nabla p + \nabla.(\tau) + \rho g \tag{2}$$

$$\frac{\partial}{\partial t}(\rho k) + \frac{\partial}{\partial x_i}(\rho k u_i) = \frac{\partial}{\partial x_j}(\Gamma_k \frac{\partial k}{\partial x_j}) + G_k - Y_k + S_k \tag{3}$$

$$\frac{\partial}{\partial t}(\rho\omega) + \frac{\partial}{\partial x_i}(\rho\omega u_i) = \frac{\partial}{\partial x_j}(\Gamma_\omega \frac{\partial\omega}{\partial x_j}) + G_\omega - Y_\omega + S_\omega \tag{4}$$

The above equations are solved simultaneously to find velocity, pressure and turbulence variables.

The model is validated with air having its flow behaviour in turbulent regime. Hence, k- ω SST model is used in the estimation of coefficient of drag for the above model and are shown in eqns (5) & (6). The choice of this model is done based upon the validation study in which other turbulences models used didn't predict the experimental results closely. Thed model merges the Wilcox k-w model for near walls and standard k–ε model away from walls using a blending function. Eddy viscosity formulation is changed to account for the transport effects of principle turbulent shear stress. The coefficient of drag obtained is 1.025. This value obtained is in close agreement with Allan [1] which validates the present model developed. Grid independent study is done in such a way that the value of drag coefficient does not change with increase in number of grids and optimizing computational time. Based on this study 19, 48, 729 number of mesh elements is used. More details on the study are available in Pragadeesh et al. [7]. For pressure-velocity coupling, SIMPLE algorithm is used. Second order upwind momentum is used in spatial discretization.

3. Results and discussions

The study was carried out for nine models as shown in Table 1. Three models among these considered models showed minimal drag. So flow analysis of these three models is done with variations in gap width and modifications of the cab radius. Flow for rest of the models is not explained in this present study. It is found that vortex formation in the gap width resulted in increase in drag. If the vortex magnitude decreases, the drag value also decreases. Modification is done on these three models with addition of fairing and rounding the cab surface. The top surface of the plate is rounded to 600mm for all the three models considered.

Table 1. Study for nine models

S.No	Radius of cab front surface (mm)	Gap width (mm)	Coefficient of drag
1	0	110	0.945
2	0	200	0.963
3	0	230	0.968
4	24.4	110	1.032
5	24.4	200	1.008
6	24.4	230	0.998
7	48.8	110	0.995
8	48.8	200	0.952
9	48.8	230	0.925

The models are kept in such a way that air flows in Z-direction and flow separates in x-direction. Flow separation along Y-direction is minimised by rounding the cab radius and by addition of a plate (fairing) above the cab and trailer. Vorticity magnitude is measured along Z-direction. The velocity contours along Z-direction is obtained so as to understand the vortex phenomena in Z-direction. It is observed in all the three models that the velocity of flow decreases along Z-direction as air moves inwards due to flow separation. A larger vortex is formed in the gap width due to flow separation. The magnitude of vorticity is measured along Z-direction and a plot is drawn for all the three cases. Since the flow is in Z-direction, the vortex variation along Z-direction is plotted. The vorticity magnitude is defined as a pseudo vector field which is used for describing the vortex motion in a continuum in a specified area. Mathematically it can also be defined as the curl of flow field velocity vector given by (5).

$$\vec{\omega} = \nabla \times \vec{V} \quad (5)$$

3.1. Model 1 - Rounded cab front surface with medium gap width

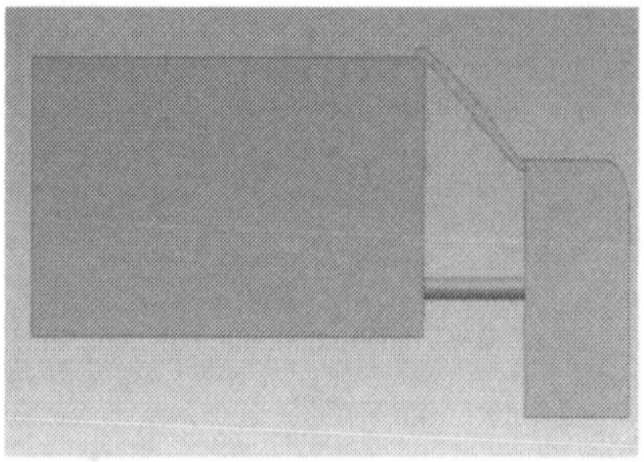

Fig. 2. Model 1

In this model, the front surface is rounded for the cab and is shown in Fig. 2. A medium gap width is maintained. The gap width is kept as 200 mm and the front surface of the cab is rounded to 48.8 mm. Since cab is rounded, flow separation is reduced. Since the top surface of the fairing is curved, the velocity along the fairing is less. It is observed from the velocity contour that velocity near the cab front surface is small when compared to sharp edge as seen in Fig. 3. The vorticity magnitude is plotted along z-direction in the gap width region and is shown in Fig. 4. The vorticity magnitude is low near the cab. Along Z-direction, the vorticity magnitude gradually increases till Z = 0.045 m due to gradual change in velocity. There are minor variations in vorticity magnitude due to lower velocity values till Z = 0.095 m. After that due to larger velocity variations, vorticity magnitude reaches its maximum at the start of the trailer. Velocity contours and their respective plots are shown below. As a result of these flow variations, the average drag coefficient obtained is 0.952

3.2 Model 2 - Sharp cab front surface with smaller gap width

In the second model, the gap width is kept as 110 mm and front surface of the cab is kept sharp as shown in Fig. 5. A larger portion of air moving over the cab has higher velocity because of this sharper edge. The velocity near the fairing is minimal due to curved geometry of fairing. There is less amount of air near the cab due to which vorticity magnitude is also low. Along Z-direction there is a steady increase in velocity due to which voriticity increases until Z = 0.025 m. Since the

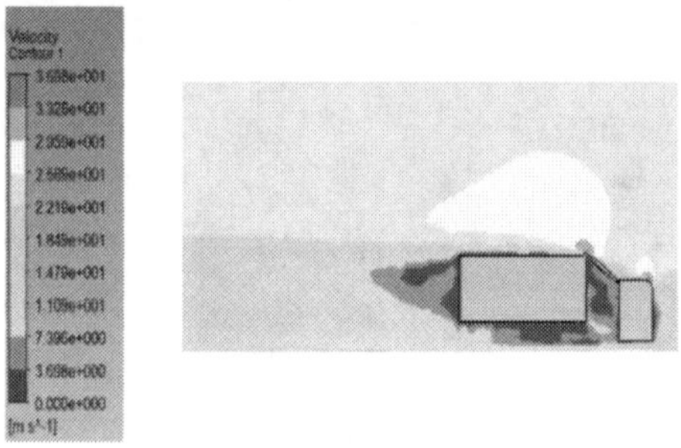

Fig. 3. Velocity contour for model 1

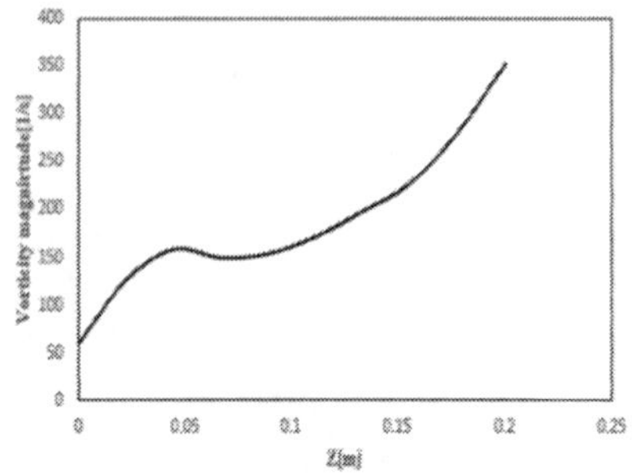

Fig. 4. Vorticity vs z-distance for Model 1

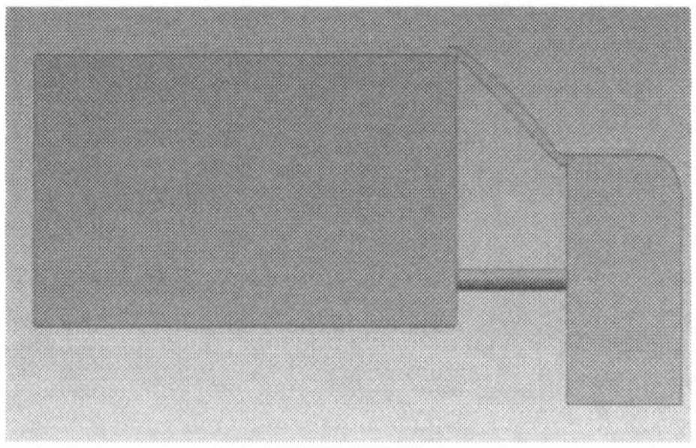

Fig.5. Model 2

amount of air after Z=0.025m decreases, the vorticity magnitude decreases till Z=0.05m. After that, magnitude of vorticity increases till Z=0.09m because of large variations in velocity. In the latter part of the gap width, voritcity magnitude starts decreasing because air will leave the gap width. The coefficient of drag obtained in model 2 is 0.945. The velocity contour and plots of vorticity magnitude vs z-distance is shown respectively in Fig 6 & Fig 7.

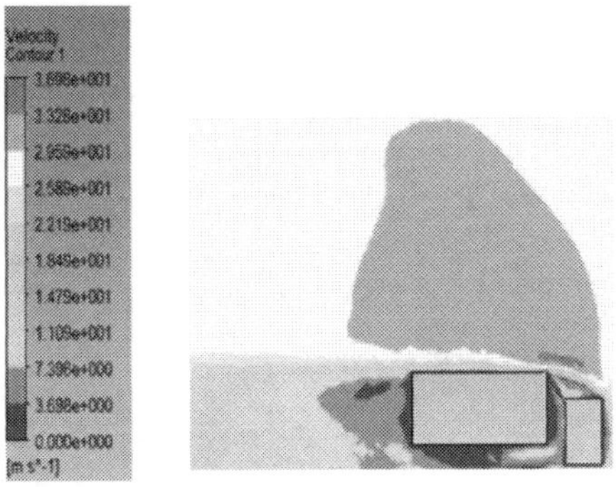

Fig.6. velocity contour for model 2

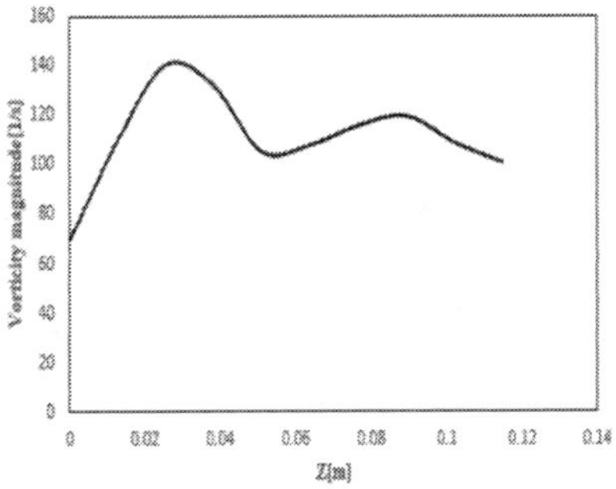

Fig. 7. vorticity vs z-distance for Model 2

3.3 Model 3 - Rounded cab front surface with larger gap

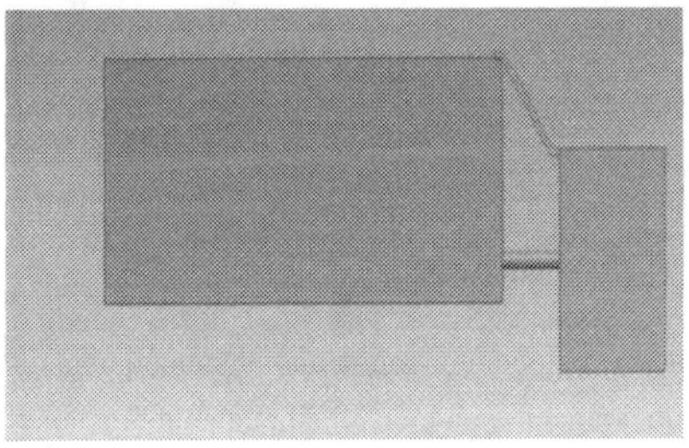

Fig. 8. Model 3

The model 3 shown in Fig. 8, has its front surface of the cab rounded to 48.8 mm and a gap width as 230 mm. The velocity value is minimal at the front surface, rounded cab and curved top fairing. Similar to the previous models, vorticity magnitude is minimal near the cab. Along the Z-direction, there are multiple regions where the vorticity magnitude increases and decreases gradually. The region where vorticity has gradual increase are from Z=0-0.025 m, Z=0.05-0.075 m and Z=0.1-0.175 m. The region where vorticity has minimal variations are from Z=0.025-0.05 m, Z=0.075-0.1 m and Z=0.175-0.2 m. The region of gradual increase is due to increase in velocity. The region of minimal variations is due to low velocity. After Z=0.2 m there is steep increase due to rapid increase in velocity. These are evident from the contours and plots given shown in Fig. 9 & 10. The coefficient of drag obtained in this model is 0.925. The velocity contours and vorticity magnitude are show in Fig. 9 & 10.

It can be seen from the plot shown in Fig. 10, that vorticity magnitude is less at start of cab and trailer near the origin. Along the Z-direction the magnitude of vorticity increases than the value which was measured at the start of trailer.

From the above three models it can be observed that, the effect of gap width, additional fairing and rounding the cab front surface has its impact on formation of vortex in the gap width. It is very well clear that the vorticity magnitude near the cab is less than the vorticity magnitude near the trailer. For a truck having smaller gap width, the vorticity magnitude drop and raise. For a truck having larger gap width, the vorticity magnitude did not show major variations. But the velocity showed minimal variations at certain locations due to which vorticity magnitude is maintained constant. The vorticity magnitude showed variations at certain locations due to gradual changes in velocity.

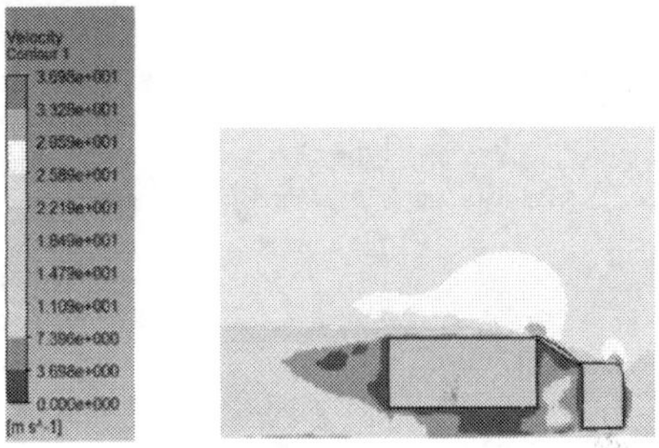

Fig.9. velocity contour for model 3

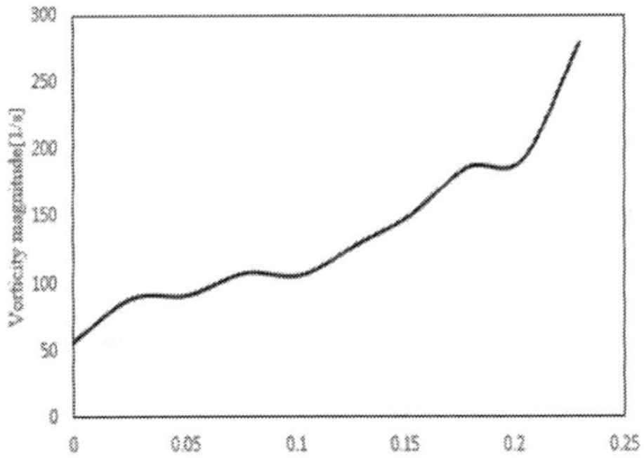

Fig.10. vorticity vs z-distance for Model 3

4 Conclusions

In the present study, a numerical model is developed using a commercial code ANSYS FLUENT to predict the drag coefficient value of class 8 trucks. The results predicted are compared with experimental results available in literature for validation. The effects of gap width and cab front radius along with a constant fairing are analysed using the validated numerical model. It was observed from the analysis of three models, that minimal drag coefficient for a given truck depended upon the gap width and the magnitude of vortex. Vortex formation in the gap played an important role in drag reduction. Third model which a rounded cab front surface with larger gap width had resulted in minimal drag coefficient.

References

1. J.W. Allan, *Journal of Wind Engineering and Industrial Aerodynamics*, no. **9** **(1-2)**: 125-136 (1981).
2. J. Östh, and S. Krajnović, *Journal of Wind Engineering and Industrial Aerodynamics* **102**: 36-47(2012).
3. J. Bettle, A. G. L. Holloway, and J. E. S. Venart, *Journal of Wind Engineering and Industrial Aerodynamics* **91, no. 5**: 573-592 (2003).
4. H. Xing-jun, Q. I. N. Peng, L. I. A. O. Lei, G. U. O. Peng, Jing-yu WANG, and Y. A. N. G. Bo, *Journal of Hydrodynamics, Ser.* ***B*** **26, no. 3**: 394-399 (2014).
5. Y. Wang, Y. Xin, Zh Gu, Sh Wang, Y. Deng, and X. Yang, *Journal of Applied Fluid Mechanics* **7, no. 4** (2014).
6. S.A. Coleman, and C. J. Baker, *Journal of Wind Engineering and Industrial Aerodynamics* **53, no. 3**: 401-429 (1994).
7. M. Pragadeesh, R. Suwathy, M. Venkatesan, *In International conference on design and manufacturing* (IIITDM Kancheepuram December 2016), (Paper No.64).
8. ANSYS FLUENT Theory Guide: ANSYS. Inc. (2013).

Investigation of efficacy of low length-to-diameter ratio and nozzle pressure ratio on base pressure in an abruptly expanded flow

Fharrukh Ahmed[1,*] and S. A. Khan[2]

[1]Research Scholar, Bearys Institute of Technology, Mangalore, Karnataka, India
[2] Department of Mechanical Engineering, Faculty of Engineering, IIUM, Kuala Lumpur, Malaysia

Abstract. This study has been carried out to assess the efficacy of the flow regulations in the form of tiny jets to regulate the pressure in the base region of an abruptly expanded duct. Four tiny jets of 1mm diameter placed at 90° intervals at 6.5 mm distance from the main jet in the wake region of the base were employed as flow management mechanism. The experiments were conducted at the inertia level of M = 2.5 & 3.0. The jets from the nozzles were expanded abruptly into a circular duct with four cross-sectional areas of 2.56, 3.24, 4.84 and 6.25. The L/D ratio of the enlarged duct considered was from 10 to 1 and experiments were conducted for Nozzle Pressure Ratio (NPR) from 3 to 11. Since the jets Mach numbers are high and the highest NPR tested was 11 which imply that the flow remains over expanded, even though, with increase in the NPR, the level of over expansion will decrease. It is well known that for over expanded nozzles an oblique shock will be formed at the nozzle lip, which in turn will result in the increase of the base pressure once it passes through the shock wave. From the results it is observed that for the NPRs 3 and 5 there is no appreciable gain in the base pressure, and hence, control employed as tiny jets are not effective, however, at NPR 7, 9, and 11 there is remarkable change in the base pressure values. This clearly indicates that NPR plays a significant role to decide on the magnitude of the base pressure and the control efficacy of the flow regulation mechanism as the tiny jets. It is found that the present method of flow regulation mechanism can be used as effective regulator of the base flows in an abruptly expanded duct. The control does not alter the nature of the flow in the enlarge duct.

Keywords. Base pressure, active control, abrupt expansion, nozzle pressure ratio.

1 Introduction

Base Flow instabilities are a grave concern to the design of high performance aerodynamic vehicles and efficient propulsive systems (rockets, aircraft bodies, re-entry vehicles, ramjet, and thrust augmenting ejectors). The base flow is characterized by large flow separations, resulting in the low pressure circulation zone at the base, which have many dispensable outcomes. The transonic/supersonic flow proceeding over the body of the vehicle collaborates with the exhaust nozzle jet in such a way that the turbulent layer and re-circulatory shock system deteriorates the performance of the vehicle, and the base flow is the cue. For the prospective aerodynamic vehicles (re-entry vehicles, scram-jet), advanced nozzles and propulsion systems, the performance turns more pertinent to the external flow, and therefore the base flow plays even greater role. In order to comprehend the parameters of base flow aerodynamics, more experiments are carried out and new applications are chosen. In case of flow separation, the pressure in the wake region is usually substantially smaller than the free stream ambient pressure, and the flow will get separated and will reattach with the solid wall of the enlarged duct. The reattachment point and reattachment length will depend upon the Mach number, the diameter ratio, Nozzle Pressure Ratio (NPR), length-to-diameter ratio and the type of the boundary layer at the nozzle exit. The flow field at the base which is very complex is one of the significant and complex problems in fluid dynamics. The flow at the base will be wave dominated consisting of diamond shock and barrel shock. When the flow at the nozzle exit is under the influence favorable pressure gradient there will be an expansion fan located at the nozzle lip, where as in the case of correctly expanded case still waves are bound to be there at the nozzle exit, however, for correctly expanded case these waves are weak. The enquiry that befalls is the influence of pressure gradient on the base drag. From the previous studies, it has been shown that at lower Mach numbers, base drag comprises only of the ten percent of the total drag of the body, whereas at Transonic/Supersonic Mach numbers, the major contribution of the total drag comes from base drag, which is sixty percent of the total drag. Hence, the slight increment in the base pressure could induce substantial decrease in the base drag; therefore enhancing the performance and reduce the fuel consumption.

Nomenclatures	
P_b	Base Pressure
P_w	Static Pressure
L/D	Length to Diameter ratio
NPR	Nozzle Pressure Ratio
A_2/A_1	Area Ratio

*Corresponding author: fharukh@gmail.com

1.1 Literature review

To study and define the given flow phenomenon, various experiments have been conducted by the researchers. In order to simplify the understanding of base flow analysis, the comprehensive study of approaching free stream Mach numbers, character of expansion phenomenon and nozzle characteristics is achieved by our experimental study. To further increase of the base pressure, which decreases the base drag, one can think of different geometrical modifications, like boat-tails, additional cavities, sting and discs, or application of base bleed and base combustion. However, the studies of base drag reduction with active control has not been studied much, therefore we will study the problem with an internal flow.
Viswanath[1] explored experimentally the estimation of drag for various models like multi-step after-bodies that considers the idea of managed separated flows in the transonic as well supersonic Mach numbers. The significant geometrical variables which are influencing the drag of such after-bodies were considered to be of interest, and their influences were tested through an experimental study by considering various variables. Their outcome indicate that multi-step after-bodies can be designed that give considerable decrease in the net drag and it may go beyond 50 % in comparison to the cases having blunt bases; nevertheless, in comparison to the bodies with the boattailed after-bodies cases of a given base area, the multi-step after-bodies have resulted into higher drag. Therefore, the certain flow behavior involving the flow separation as well as the reattachment on multi-step after-bodies was considered based on flow visualization studies.
Singh and Rathakrishnan[2] concluded from their research the efficacy of various tab parameters on the flow exiting from the converging nozzle at Mach M = 1. They concluded that the passive control in the form of tabs are greatly efficient to control the center line decay and hence the core length of the jets, and they observed that the tabs can reduce the core length by 80 per cent. Asymmetric flows in symmetric ducts with sudden expansion were induced by the disturbances engendered at the edge of the expansion and augmented in the shear layers(Vijayaraja, Senthilkumar et al.)[3]. Also, they demonstrated that the strength of fluctuating energy in the low Reynolds number can be greater than the counterpart turbulent flow. The study of closed-cavity laminar flows at adequate Reynolds number was carried out by Acrivos [4]. They observed that the flow becomes real turbulent once the critical Reynolds number is exceeded. Experimental analysis of the flow field in suddenly expanded combustion chamber was studied by Khalil.[5] For highly turbulent flow field, they verified the eminency of laser Doppler anemometer for measuring the reverse flow.
Khan and Rathakrishnan([6]-[9]) did experimental examination to assess the effectiveness of micro jets for various level of expansion to regulate the base pressure in abruptly expanded ducts at moderate and high supersonic speeds. The result thus produced showed that on the positive side the highest enhancement is152 % percent at Mach M = 2.58. The result also indicated that the control mechanism does not augment the nature of the flow field. Lovaraju et al. [10] conducted the experiments to study the efficacy of tabs and a wire which is projecting normally into the flow at the nozzle exit, as a passive control on the characteristics of an axi-symmetric sonic jets which were operated at various level of expansion, and their studies suggests that the efficacy of passive controls either in the form of cross-wire or the tabs on the under expanded sonic jet shows that, both the passive controls are effective in decreasing the core length significantly.
Rathakrishnan, Ahn et al. [11] presented the outcome of an experimental studies on the flow from a converging-diverging nozzle and its behavior for slanted entry at a 15° at the inertia level of M = 2.94 and later the flow was exposed to various supersonic Mach numbers in the range from M = 1.6, 1.8 and 2.0.

2. Experimental Setup

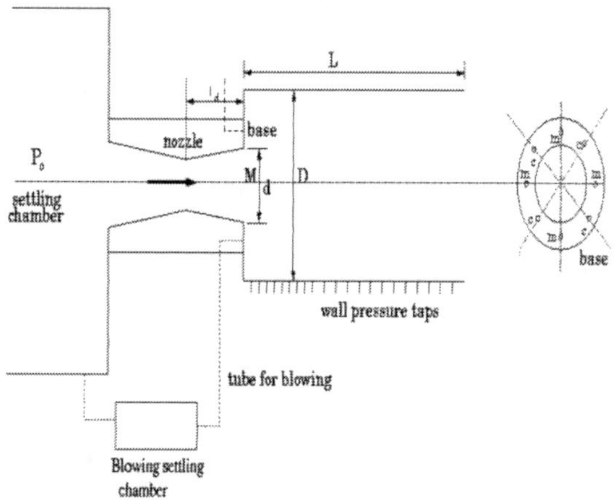

Fig. 1. Experimental setup

The investigation was performed with a full scale experimental model consisting of pipelines, pressure transducers and the settling chamber. In order to expand the gas through the experimental model, it is first allowed to go through regulating valves. The experimental model is a nozzle with an augmented duct. The flow leaving the model is subjected to ambient air. Figure 1 depicts the experimental setup.In the outlet boundary of the nozzle, 8 holes of 1mm diameter each are drafted. In the present arrangement the regulation of the pressure the separated region was accomplished by injecting the air through control chamber.

Fig. 2. Experimental Apparatus

3 Results and discussion

The measured values consisted of the base pressure (P_b), the static pressure (P_w) along the wall of the duct and the NPR in the main settling chamber as well as the control pressure. The obtained pressures were transformed to non-dimensional data by multiplying them with the inverse of the ambient pressure. The percentage change in base pressure as a function of L/D ratio was shown in the Fig.3 ((a) to (j)). The effect of area ratio with Mach number and the NPR on the percentage change in base pressure were presented in the Figure 3 ((a) to (j)). Further, while analyzing the results it should be kept in mind that the jets exiting from the nozzles were over expanded and hence an oblique shock will be present at the nozzle lip. Even though there was a reduction in the level of adverse pressure gradient with the increase in nozzle pressure ratio. It could be visualized that, when there was an oblique shock positioned at the nozzle exit the shear layer coming out of the nozzle will be deflected towards the nozzle center line rather than towards the base region. This will results in enhancement of the base/wall pressure after the shock and also this will delay the reattachment of the flow with the duct wall, which in turn will result in a longer reattachment length as compared to a case without a shock. It was well known that the reattachment length was a parameter strongly influencing the base vortex, the increase or decrease of reattachment length will modify the base pressure flow field. It was also, evident that increases in area ratio and in turn increase in relief to the flow simply indicates that relaxation space existing for the flow was increasing. This sort of relief will make the shock waves at the nozzle lip to spread relatively more freely with the increase of relief at the lip of nozzle and the area ratio as shown in (Fig. 3 (a) - (j). From the results it is seen that at the NPR 3 and 5 the micro jets were ineffective as shown in the Figs 3 (a) to (d).

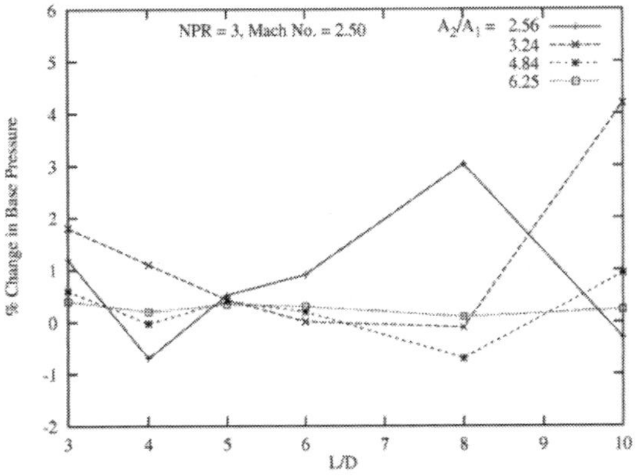

(a)

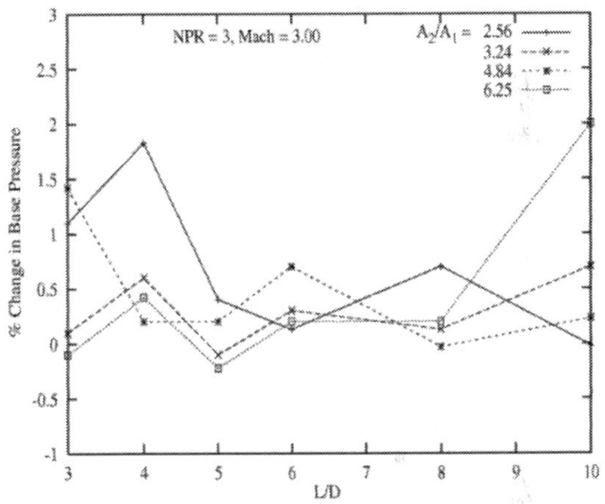

(b)

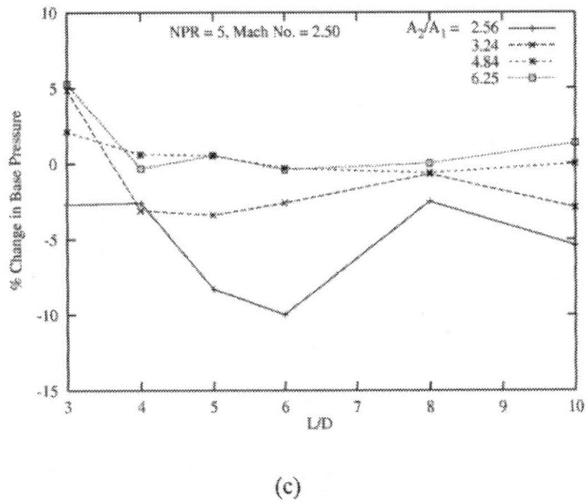

(c)

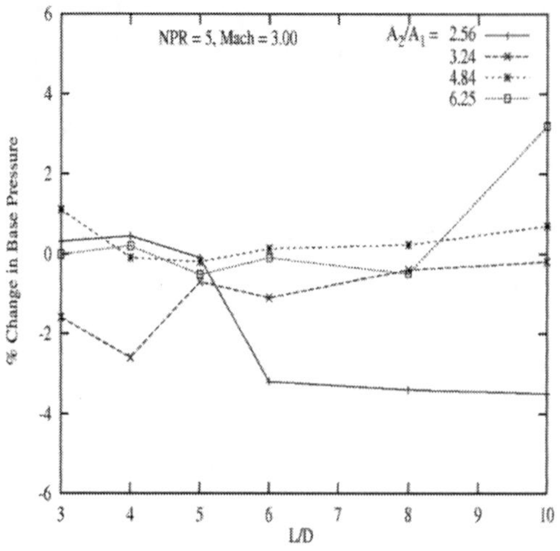
NPR = 5, Mach = 3.00
A_2/A_1 = 2.56
3.24
4.84
6.25
% Change in Base Pressure
L/D

(d)

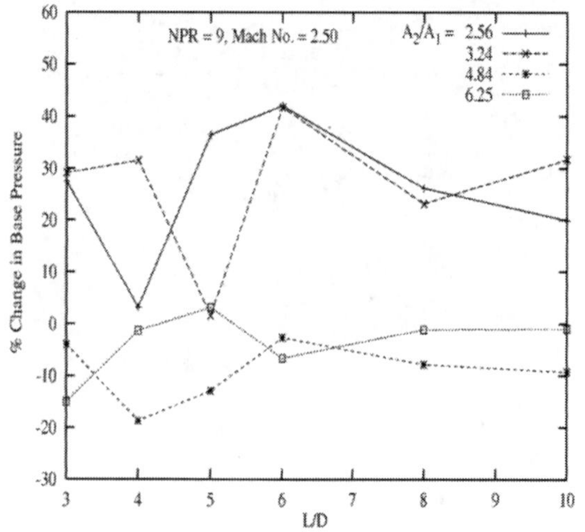
NPR = 9, Mach No. = 2.50
A_2/A_1 = 2.56
3.24
4.84
6.25
% Change in Base Pressure
L/D

(g)

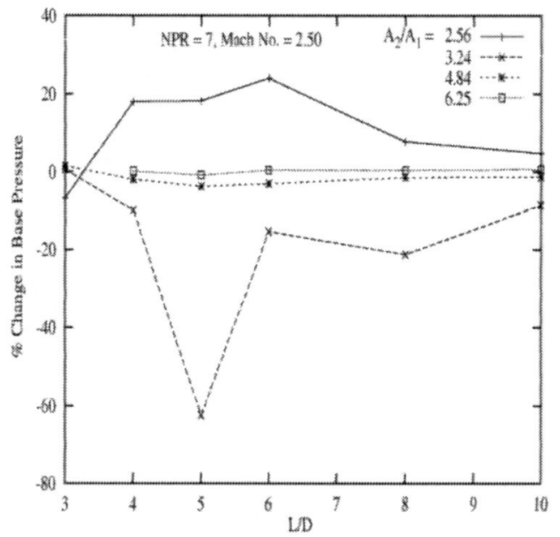
NPR = 7, Mach No. = 2.50
A_2/A_1 = 2.56
3.24
4.84
6.25
% Change in Base Pressure
L/D

(e)

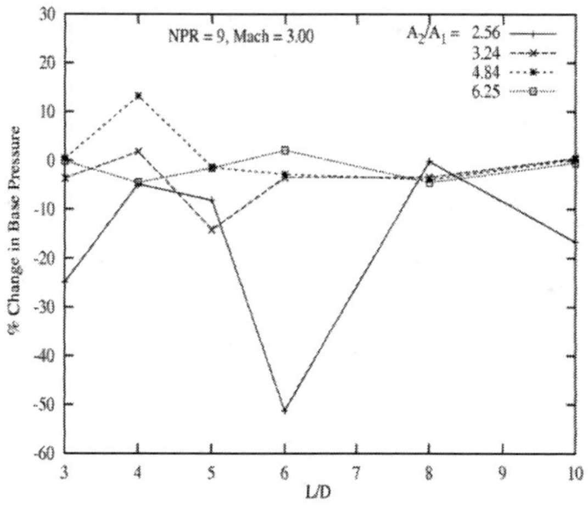
NPR = 9, Mach = 3.00
A_2/A_1 = 2.56
3.24
4.84
6.25
% Change in Base Pressure
L/D

(h)

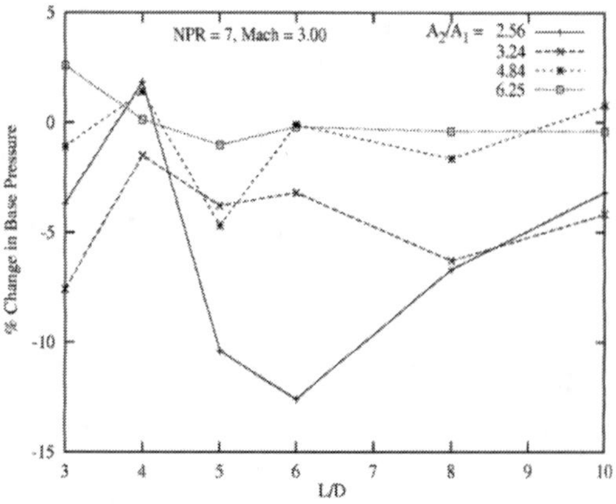
NPR = 7, Mach = 3.00
A_2/A_1 = 2.56
3.24
4.84
6.25
% Change in Base Pressure
L/D

(f)

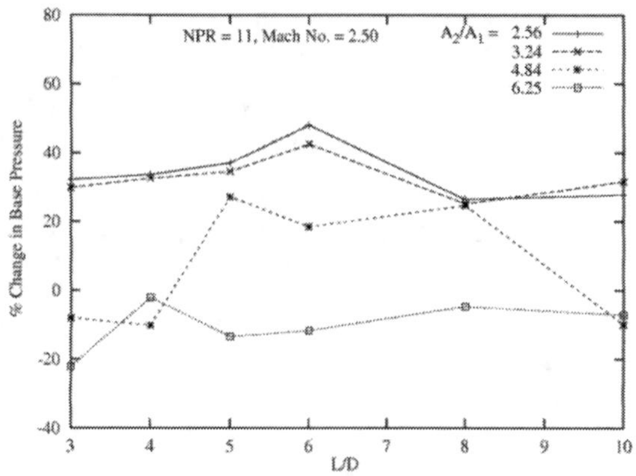
NPR = 11, Mach No. = 2.50
A_2/A_1 = 2.56
3.24
4.84
6.25
% Change in Base Pressure
L/D

(i)

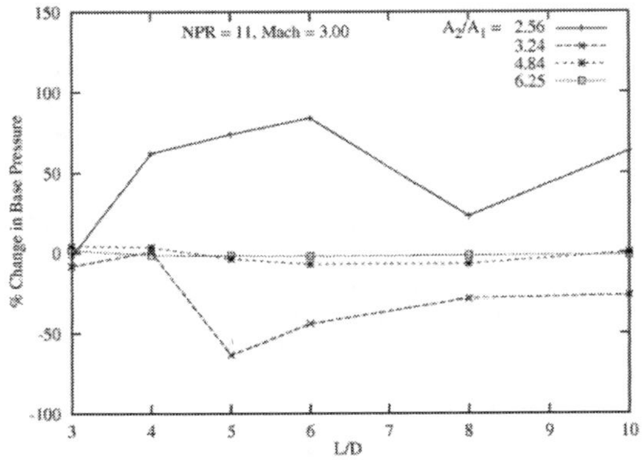

(j)

Fig. 3. Percentage change in base pressure with L/D ratio

The results for the NPR 7 were shown in the Figures 3(e) to (f)). From the results it was observed that for lower area ratios namely 2.56 and 3.24, even at NPR 7 the micro jets had become effective and more so for lowest area ratio namely 2.56. This effectiveness results in decreasing the base pressure and for the higher area ratios 4.84 and 6.25, while the control mechanism in the form of tiny jets were not at all effective. The reason for this trend may be due to the larger area avaialble to jet and vortex positioned at the base was unable to make any significant effect on the base flow field.Under these conditions when control was activated they were unable to alter the base flow field, rather they go unnoticed since for these higher area ratios the micro jets were closed to the main jet and away from the base region.

The Results for the NPR 9 were shown in the Figs. 3 (g) to (h). They exhibit the similar results as discussed above for NPR 7. In this case due to increase in the NPR, the level of over expansion had marginally come down and the same had been reflected in the magnitude of the per cen tage change in the base pressure which had marginally improved. It was also seen for the highest area ratios namely 4.84 and 6.25, the gain had increased as compared to the NPR 7 for same area ratio. These results clearly reflect the impact of NPRs on the base pressure level in the base region, reattachment point/length, and the flow establishment in the duct.

The results for the NPR 11 were shown in the Figs. 3(i) to (j). This was the highest NPR for which the tests were conducted, tests were unable to be conducted at higher NPR or for correct and under expanded case due to the limitation of the experimental set up. From the Fig. 3(i) it was seen that at Mach 2.5 for area ratio the maximum gain was forty eight percent at L/D = 6, and at other L/Ds it remains between thirty to forty percent range. For area ratio 3.24, this gain remains in the range between twenty to thirty eight percent. One interesting observation was that the micro jets which were not effective for the area ratio 4.84 and 6.25 now became effective at the NPR 11 due to the reduction in the level of over expansion, which had resulted in the reduction of strength of the oblique shock wave. For area ratio 4.84 as high as forty percent increase had been achieved. In the case of the highest area ratio namely 6.25, the gain was on the negative side and remains between ten to twenty percent. This may be due to the further increase in the area ratio leading to the weakening of base vortex, and incresae in reattachment length, which was small for lower area ratios. Since the NPR remains the same and due to increase in the relief, this trend was expected.

Another point to be noted that the with the change in the enlarged duct, the back pressure also will influence the flow field in the base area, and hence, the magnitude of the base pressure will significantly altered. As it was evident from the above discussion that the flow was dominated by the presence of shock waves of variable strength, hence, dealning with this type of flow one had to take note on a case to case basis generizing the conclusions.

The results for Mach 3.0 are presented in the Figure 3(j). Here, again the effffectiveness of the control in the form of micro jets were observed only for lower area ratios, whereas, for the larger area ratios 4.84 and 6.25 it remains ineffective. This could be due the level of expansion at the Mach number as the NPR for correct expansion was quite high as compared to the Mach 2.5 while all other parameters remain the same.

Finally, in view of the above, we can conclude that apart from the effect of shock or expansion wave at the nozzle exit, the relief available due to variation of area ratio, location of micro jets, level of expansion and L/D ratio will affect the base pressure. Furthermore, it should be kept in mind that in case of lowest area ratio 2.56, the micro jets in the base region were located at middle of the base area, whereas for area ratio 3.24 (and also for area ratios 4.84 and 6.25) the micro jets were closer to the main jet. This was due to the pitch circle diameter for the control mechanism was kept same for all the area ratios. This was also the reason for peculiar trend in the above discussed results.

4 Conclusions

From the above results it could be concluded that in supersonic regime, the inertia level had a very strong effect on the base pressure. For a fixed value inertia level, the NPR, which controls the level of expansion had a major role to play on the control efficacy of the tiny jets. Also, it was seen that with further increase of the NPR, the flow regulation mechanism becomes more efficient in enhancing the base pressure with Mach number. In view of above, it could be said, the control mechanism in the present study to control base pressure was effective for Mach M = 2.5 for NPRs 7, 9, and 11 for lower area ratios. The effectiveness of the micro jets was only marginal for higher area ratios namely 4.84 and 6.25, for all the inertia levels and nozzle pressure ratios of the present investigation. Further, it was concluded that the results were required to be analyzed on the case by case basis, hence, no definite rule could be established. One had to study on case by case basis and also on what is needed and to identify the set of

parameters which will lead to maximum increment or decrement of the base pressure.
The flow field in the suddenly expanded duct was dominated by the presence of the waves both strong as well as the weak waves, since the jets exiting the nozzles are over, under and correctly expanded, and hence, no definite trend can be drawn. The reflection of the waves from the wall, recompression and recombination will continue to take place in the base region of the duct wall. Thereby, making the flow oscillatory for some combinations of parameters, that would otherwise exhibit similar trends as expected when compared with and without control cases of the wall pressure field. The micro jets had demonstrated that they could be considered as efficient active controllers of the suddenly expanded flow field. Hence, increasing or decreasing the base suction to appreciable level for range of variables of the present study. From the results it was seen that the nozzle pressure ratio (NPR) plays a key role in deciding the magnitude of base pressure with and without control, in the supersonic regime too. All the non-dimensional measured values exhibited in the present study are within an uncertainty range of 2.6 % on the either side. All the tested values are reproducible in the range of 3 per cent on the either sides.

References

1 P. Viswanath, AIAA Journal. **39**(1), 73-78 (2001).
2 Singh, N. K. and E. Rathakrishnan, International Journal of Turbo and Jet Engines. **19**(1-2): 107-118 (2002).
3 K. Vijayaraja, C. Senthilkumar, S. Elangovan, E. Radhakrishnan, International Journal of Turbo & Jet-Engines. **31**(2), 111-118 (2014).
4 J. Acrivos, Journal of Fluid Mechanics. **112**, 127-150 (1981).
5 K. H. Khalil, *Flow, mixing and heat transfer in furnaces: the science & applications of heat and mass transfer reports, reviews & computer programs* (Elsevier, Technology & Engineering, 2014).
6 S.A. Khan, E. Rathakrishnan. International Journal of Turbo and Jet Engines. **20**(1), 63-82, (2003).
7 S. A. Khan, E. Rathakrishnan, International Journal of Turbo and Jet Engines **21**(4), 233-254 (2004).
8 S. A. Khan, E. Rathakrishnan, International Journal of Turbo and Jet Engines **21**(4), 255-278 (2004).
9 S. A. Khan, E. Rathakrishnan, Aircraft Engineering and Aerospace Technology. **78**(4), 293-309 (2006).
10 P. Lovaraju, S. Clement, E. Rathakrishnan, Shock Waves. **17**(1-2), 71-83 (2007).
11 E. Rathakrishnan, J. Ahn, International Review of Mechanical Engineering. **8**, 1-10 (2014).

MATEC Web of Conferences **172**, 01005 (2018) https://doi.org/10.1051/matecconf/201817201005
ICDAMS 2018

Sloshing response of partially filled rectangular tank under periodic horizontal ground motion.

Amiya Pandit[1*], and *Kishore Chandra* Biswal[2]

[1,2]Department of Civil Engineering, NIT Rourkela, 769008, India

Abstract : The event caused due to the movement of the unrestricted free surface liquid in container due to any external excitation is known as sloshing. The problem of liquid sloshing phenomenon in stationary or in moving container is a great matter of concern for many researchers and engineers. The containers may range from a cup of milk, ponds, and lakes to fuel tanks of launching vehicles and cargo ships carrying variety of liquids such as oil, liquefied natural gas, and chemical fluids. As a result of sloshing there is spillage of liquid from the containers and it causes structural instability and structural damage. Due to these reasons, liquid retaining structures which are special in construction and in function from an engineering point of view must be constructed well to be resistant against oscillation of the liquid due to external excitations. The amount of liquid participating in the sloshing motion depends on the shape of tank, the liquid depth, internal objects, if any, orientation, duration, amplitude, and frequency contents of external excitations. This present study focuses on the forced vibration analysis of partially filled two-dimensional rigid rectangular tank numerically. A Finite element (FE) code in two dimensions is developed to understand the behavior of sloshing. This method is competent enough of evaluating both impulsive and convective response of tank-liquid system in terms of base shear and hydrodynamic pressure distribution along the walls of the containers.

1 Introduction

Sloshing of liquid in partly filled containers impose dynamic pressure along the walls of the containers, which results structural instability. So the study of sloshing in containers like fuel tanks running on the roads, cargo ships carrying chemicals or oil, fuel propellant of rockets and the overhead tanks containing liquids is a great matter of concern for different departments of engineering. In order to secure the containers in addition with the vicinity the complex behavior of sloshing should be analyzed precisely. The amplitude of slosh depends on some of the extensive parameters such as amplitude and frequency of excitation, shape and size of tank, liquid depth and liquid properties. Housner [1] performed analytical solution for both rectangular and cylindrical shape container considering a spring mass model to define the liquid sloshing. But it was highly objectionable as this concept was not fitting to arbitrary shape containers. So different mathematical models have been developed to investigate the dynamic response of liquid in containers. Aslam [2] approached finite element method to find out the sloshing displacement and hydrodynamic pressure in partially liquid filled annular tank and validated the numerical results with the experimental results which was in a good agreement.

* amiyaranjanpandit1991@gmail.com

Veletsos et al. [3] numerically determined the response of circular cylindrical tank excited vertically, and calculated the hydrodynamic force at the base of the tank for flexible and rigid condition. Koh et al. [4] evaluated the dynamic response of three-dimensional rectangular liquid storage tanks, including free surface sloshing, subjected to horizontal ground excitation using the boundary element method. Faltinsen et al. [5] performed analytical analysis on 3-D tank at resonant condition and validated with the experimental results. Biswal et al. [6] evaluated the sloshing frequency in cylindrical tank without baffle and again applying both with flexible and rigid baffles. Finite element were used to discretize both liquid and structural domain. Arafa [7] formulated to study the sloshing behavior in partially filled rectangular tank without or with baffles placing at different position inside tank using finite element. The study exhibited, the baffles inside suppress the sloshing effect. A pressure based finite element method is employed by Mitra and Sinhamahapatra [8], to study the effect of submerged component on the sloshing effect on rectangular tank under earthquake excitation. Chen et al. [9] studied the nonlinear sloshing behavior in cylindrical and rectangular tank using the boundary element method and second order Taylor series expansion and compared with the experimental results obtained using a small scale model under horizontal harmonic and recorded earthquake

 https://doi.org/10.1051/matecconf/201817201005

excitation. Virrela et al. [10] computed and compared both linear wave theory and nonlinear wave theory using finite element package ABACUS and found that nonlinearity has no significance effect on the sloshing frequency. Nayak and Biswal [11-13] numerically conducted both linear and nonlinear analysis to evaluate the base shear, base moment and pressure distribution along the tank wall as well as the block wall, under seismic excitation.

A limited study has been carried out to distinguish between impulsive and convective component of liquid sloshing for rectangular study numerically. Present investigation gives an overall idea regarding these components under harmonic excitation with different frequency and amplitude of excitation.

2 Mathematical Formulations

Figure 1 shows the problem geometry for rectangular tank, showing the Cartesian coordinate system o-xz, where origin is at the midpoint of the free surface with z axis pointing vertically upward and x axis along the direction of length (L) of the tank. The liquid is considered to be inviscid, incompressible and flow is assumed to be ir-rotational. For the linear analysis wave breaking is not considered. The governing equation for motion of liquid in terms of *velocity potential*, $\phi(x, z, t)$ is thus expressed by Laplace equation:

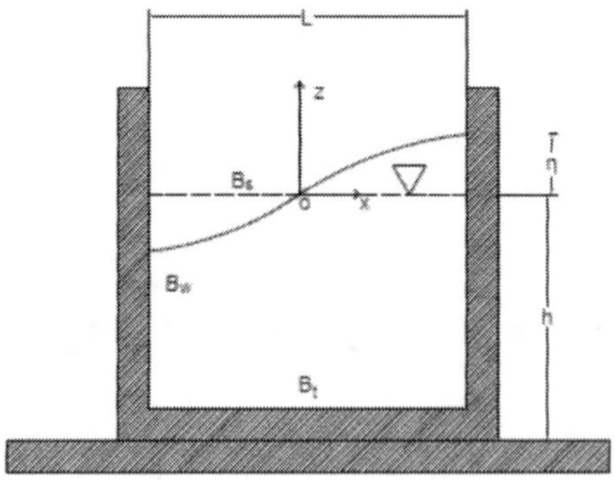

Fig. 1. Schematic diagram of rectangular tank.

$$\nabla^2 \emptyset = 0 \qquad (1)$$

At the free surface, combination of kinematic and dynamic boundary condition gives rise to

$$\frac{\partial^2 \emptyset}{\partial t^2} + g\,\frac{\partial^2 \emptyset}{\partial t^2} = 0 \quad \text{on } B_s \qquad (2)$$

$$\frac{\partial \emptyset}{\partial n} = V_n \quad \text{along the side walls of the container} \qquad (3)$$

$$\frac{\partial \emptyset}{\partial n} = 0 \quad \text{at the bottom of the tank} \qquad (4)$$

where V_n is the velocity normal to the walls of the container.

3 Finite element formulation of liquid domain

The liquid domain Ω bounded by $B = B_t \cup B_s \cup B_w$ is discretized by four nodded iso-parametric quadrilateral element as shown in Figure 2, and velocity potential is considered as the degree freedom. The velocity potential is approximated as

$$\emptyset \approx \emptyset(x, z, t) = \textstyle\sum_{j=1}^{n} \emptyset_j N_j(x, z) \qquad (5)$$

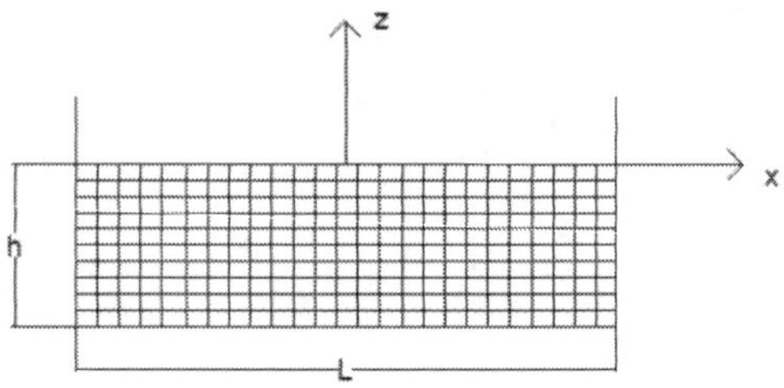

Fig. 2. Typical finite element mesh for fluid domain

Where ϕ_j are time dependent nodal velocity potentials, N_j are the shape functions and n is the number of nodes. Application of Galerkin's weighted residual method to Laplace equation gives rise to

$$\int_B N_i \left[\frac{\partial^2 \emptyset}{\partial x^2} + \frac{\partial^2 \emptyset}{\partial z^2}\right] dB = 0$$

(6)

On application of product rule and after rearrangements we

$$\int_B \left(\frac{\partial N_i}{\partial x} \textstyle\sum_{j=1}^{n} \frac{\partial N_j}{\partial x} + \frac{\partial N_j}{\partial z} \textstyle\sum_{j=1}^{n} \frac{\partial N_j}{\partial z}\right) \emptyset_j dB - \int_B \frac{\partial}{\partial x}\left(N_i \frac{\partial \emptyset}{\partial x}\right) + \frac{\partial}{\partial z}\left(N_i \frac{\partial \emptyset}{\partial z}\right) dB$$

(7)

Applying the boundary conditions in equation (7)

$$\frac{1}{g}\int_{T_{fs}} N_i \sum_{j=1}^{n} N_j \ddot{\emptyset}\, dT_{fs} + \left[\int_B \frac{\partial N_i}{\partial x} \sum_{j=1}^{n} \frac{\partial N_j}{\partial x} + \frac{\partial N_i}{\partial z} \sum_{j=1}^{n} \frac{\partial N_j}{\partial x}\right] \emptyset_j dB = \int_{T_{sw}} N_i \boldsymbol{V_n} dT_{sw} \quad (8)$$

$$[M_f]\{\ddot{\emptyset}\} + [K]\{\emptyset\} = \{F\} \quad (9)$$

where $[M_f]$ is the free surface matrix

$[K_f]$ is the fluid coefficient matrix

$\{F_i\}$ is the force vector

$$M_{ij} = \frac{1}{g}\int_{T_{fs}} N_i N_j\, dT_{fs}$$

$$K_{ij} = \int_B \left(\frac{\partial N_i}{\partial x}\frac{\partial N_j}{\partial x} + \frac{\partial N_i}{\partial z}\frac{\partial N_i}{\partial z}\right) dB$$

$$F_i = -\int_{T_{sw}} N_i V_n\, dT_{sw}$$

The hydrodynamic pressure p and slosh amplitude η at any instant are obtained from given equation

$$p = -\rho \frac{\partial \emptyset}{\partial t} \quad (10)$$

$$\eta = -\frac{1}{g}\frac{\partial \emptyset}{\partial t} \quad (11)$$

The base shear S_b can be evaluated using

$$S_b = \int_{T_{sw}} p n_x\, dT_{sw} \quad (12)$$

where, n_x is the unit normal vector.

4 Results and discussion

A finite element code in MATLAB has been developed to evaluate the sloshing frequency, free slosh displacements, hydrodynamic pressure on tank wall and base shear for partially filled rectangular tank under horizontal excitation. The density of liquid = 1000 kg/m^3

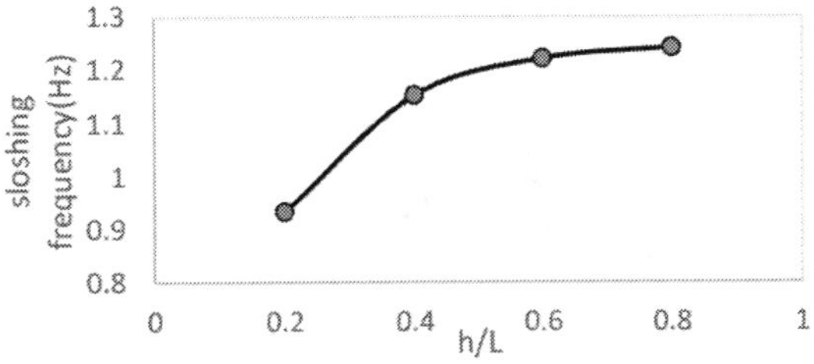

Fig. 3. Variation sloshing frequency

Figure 3 shows as the depth of liquid increases sloshing frequency increases initially and becomes less significant later on. The tank of length (L) 0.6 m, liquid depth 0.08 m and width is kept unity for this present study. The natural sloshing frequency is found to be 0.7181 Hz. The amplitude of excitation (a) is kept 5 mm and the harmonic ground motion applied to the tank is given be

$$y = a\sin\omega_1 t + a\sin\omega_2 t$$

where ω_1, ω_2 = excitation frequency in rad/sec. The excitation frequency is 0.8 times of natural sloshing frequency so that linearization assumption can be maintained appropriately.

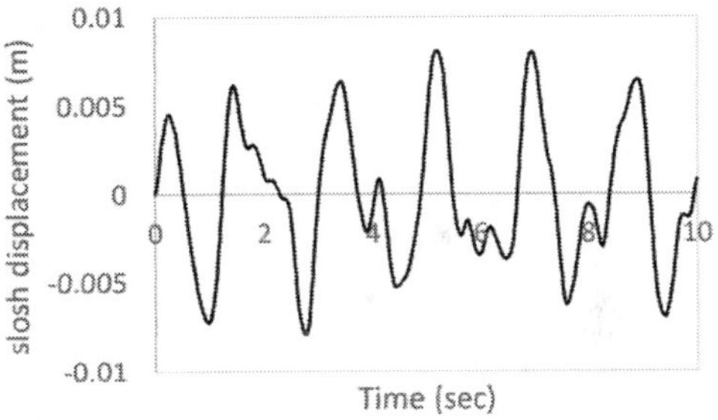

Fig.4. Time history of slosh displacement

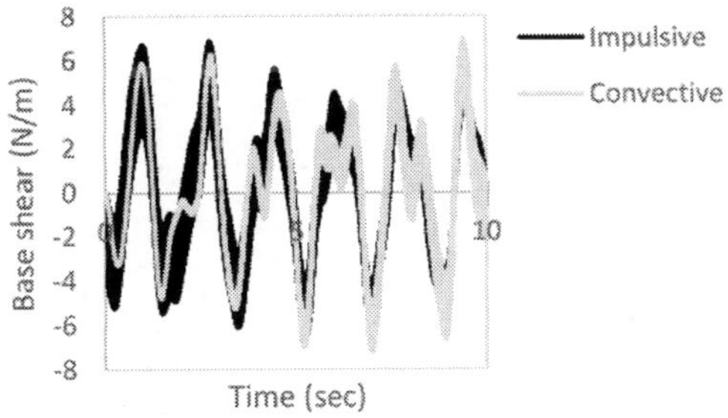

Fig. 5. Time history of base shear

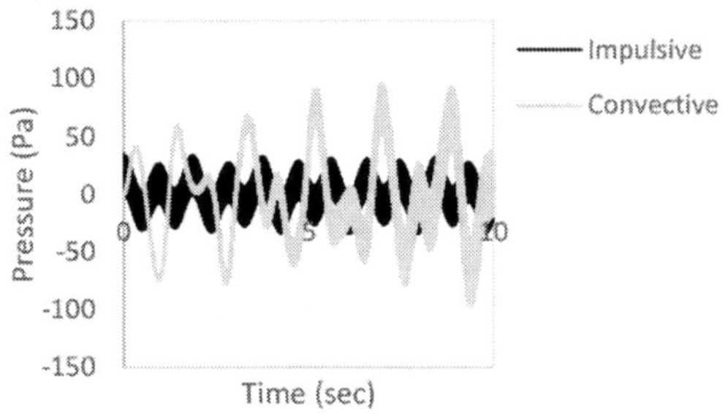

Fig. 6. Time history of pressure

The amount of liquid effectively participates in sloshing is termed as convective component of contained liquid and rest is termed as impulsive component of contained liquid.

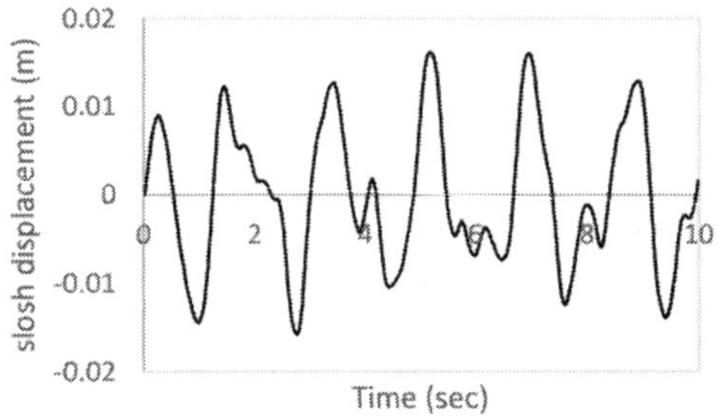

Fig. 7. Time history of slosh displacement

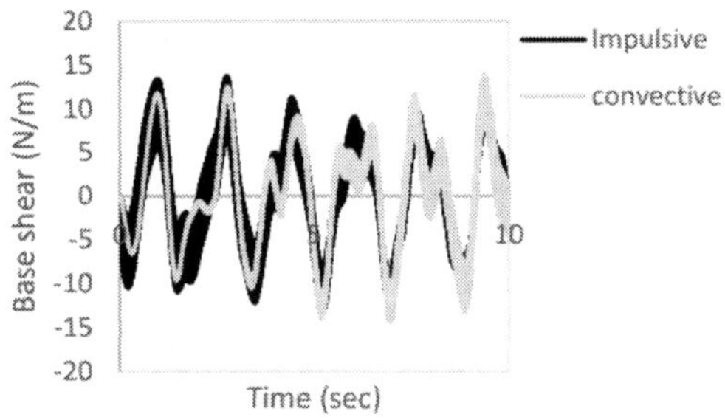

Fig. 8. Time history of base shear

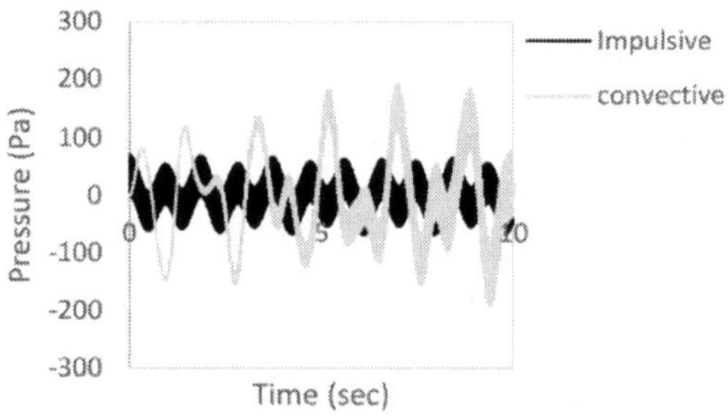

Fig. 9. Time history of pressure

Figure 4 and 6 shows the time history of slosh displacement from the free surface of liquid on left wall at different amplitude of excitation i.e. 5 mm and 10 mm. As it can be clearly visible as the amplitude of excitation increases the slosh displacement increases considerably. Same behavior can be marked out for base shear generated and pressure acting on the left walls of the containers. From figure 5 and 8 it is visible that when excitation starts the convective component is less activated but as period of excitation has increased, the convective component of liquid starts to participates in sloshing more actively. As the amplitude of excitation has increased the response of base shear and pressure have also increased.

Since the accurate calculation of base shear and pressure induced due to sloshing is necessary in order avoid failure of tank liquid system. Again to achieve more accuracy the classification between impulsive and convective component is carried out numerically in the present analysis.

5 Conclusion

The numerical model effectively predicts the slosh displacement, base shear and pressure induced due to liquid sloshing in 2D rectangular container. The result shown in the present study, compares for same mass of liquid and without changing any dimension of the rectangular tank under harmonic excitation with amplitude of excitation 5 mm and 10 mm respectively. Higher value for amplitude of excitation enhance the participation of liquid in container as a consequence of which the dynamic response also increased. The importance of the present investigation can be analyzed by virtue of the classification between the impulsive and convective component of liquid sloshing.

References

1. G.W. Housner, Bull. Seis. Soc. America, **53**, 2 (1963).
2. M. Aslam, Int. J. Num. Meth. Eng, **17**, 159-170 (1981).
3. A.S. Veletsos, J. Str. Eng., **112**, 6 (1986).
4. C.G. Koh, S. Mahatma, C.M. Wang, J. Eartquake. Eng. Str. Dyn, **23**, 17-31 (1994).
5. O. M Faltinsen, O.V. Rognebakke, A.N. Timokha, J. Fluid. Dyn., **487**, (2003).
6. K.C. Biswal, S. K. Bhattacharya, P. K. Sinha, J. Sou. Vib., **259**, 177-192 (2003).
7. M. Arafa, J. Vib. Cont., **13**, 7 (2007).
8. S. Mitra, K.P. Sinhamahaptra., J. Sou. Vib., **304**, 361-381 (2007).
9. Y. H. Chen, W. Whang, C. Koh, Eatq. Eng, Str. Dyn., **36**, 1701-1717 (2007).
10. J. C. Virella, O.V. Prato, C. A. Godoy, J. Sou. Vib., **312**, 442-460 (2008).
11. S. K. Nayak, K.C. Biswal, J. Eatq. Eng., **17**, 1023-1062 (2013).
12. S. K. Nayak, K.C. Biswal, Str. Eng. Meca., **47**, 5 (2013).
13. S.K. Nayak, K.C. Biswal, J. Sou. Vib., **368**, 148-173 (2016).

MATEC Web of Conferences **172**, 01006 (2018) https://doi.org/10.1051/matecconf/201817201006
ICDAMS 2018

Numerical investigation of hybrid blend design target bullets

V R Muruganantham [1], *T* Babin[2*]

[1,2]Dept. of Mech. Engineering, Kumaraguru College of Technology, Coimbatore, Tamil Nadu, India

Abstract. In the past few decades, researchers are involved in the studies of high-energy penetrator geometry and also the materials used for causing brutal damages on enemy targets. The highly focused area in this field is the shape of the nose and penetration angle of the penetrator. Hybrid blend design target bullets is one among the best aerodynamic design as it is having two different shapes within the nose. The design of these bullets plays an influential role in terms of aerodynamic characteristics and its dynamic performance capability on hitting the targets. Objective of this research is to perform a comparative analysis of 3 different geometrically varied shapes of these bullets by numerical simulation using CFD, based on its performance characteristics such as aerodynamic co-efficients, shock wave existence, Turbulent dissipation rate and normal drag force on the bullet surface.

1 Introduction

One of the exciting challenges for shooters is getting precision and accuracy over the targets. Utilization of technology leads to researches in guns and ballistics to develop such bullets. According to recent researches, Very Low Drag (VLD) bullets have shown significant improvement because of its low drag and efficient performance under high winds. In addition, VLD bullets have very high Ballistic Co-efficients (BCs). Bryan Litz [1], a Ballistician found that these VLD bullets are highly sensitive to seating depth. This is due to the nose design it has, and further researches on the nose ogive design leads to development of hybrid types of bullets. The advantage over the combined tangent and secant ogive design is lies in the improvement of accuracy for long-range shooters on targets. The reason behind the improvement of hybrid bullets is the combined strength of tangent and secant ogive design, in which the tangent ogive from bearing surface of bullet has minimal sensitive to seating depth, whereas the continuous secant ogive gives the high BC and also very less drag similar to VLD type bullets. In the current research series, this paper contributes the numerical investigation of hybrid blend design target bullets to obtain its aerodynamic characteristics and comparative performance analysis.

1.1 Geometry

1.1.1 Secant-tangent Ogives

Ogive is the term, which describes the point from where the curve approaches the bullet diameter. Ogive is usually characterised by its length of arc radius. For, a constant nose length, ogive can be of tangent curve or secant curve. If the ogive is said to be perfectly tangent, then the curve is defined by its own very specific radius. And, if the radius is longer than that, then the curve will obtain secant ogive. Secant ogive is usually varied from tangent radius to very long radius. With reference to Bryan Litz [1], it is to be noted that the secant ogive results minimal drag particularly when the radius is twice the tangent ogive radius. Pictorial representation of both curves are given in the Figure 1.

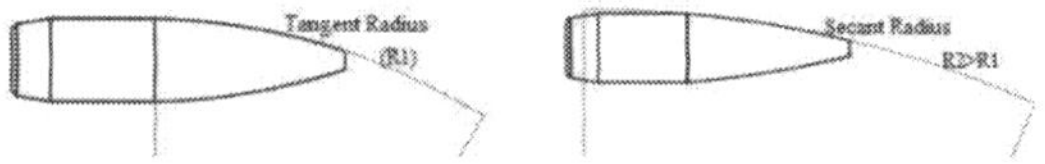

Fig. 1. Tangent ogive vs. Secant ogive.

1.1.2 Design of Hybrid Target Bullet

Hybrid blend design bullets for target hunting applications are developed first by Bryan Litz [1] after his research on substantial effect of sensitivity over seating depth of secant ogive type bullets and its target accuracy. On continuous improvement, the combination of tangent ogive into the secant ogive came into existence, which results in very less sensitivity on seating depth and low drag. From bearing surfaces, tangent ogive starts the curve and subsequently secant ogive follows as shown in the figure 2.

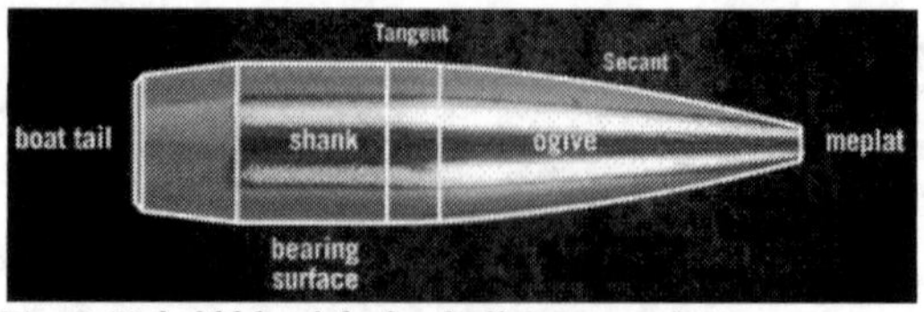

Fig. 2. Hybrid blend design bullet nomenclature.

* Corresponding author: babinn4340@gmail.com

1.2 Model Specification

Three cases of hybrid bullets are taken for CFD analysis from Berger reference sheet [2]. Dimensional specifications measured in inches for the three cases are discussed in the Table 1.

Table 1. Bullet dimensions for three cases

Type/description	6.5 mm 140 gr	7 mm 180 gr	30 Cal 230 gr
Bullet Dia.	0.264	0.284	0.308
Boat Tail	0.224	0.245	0.280
Nose Length	0.768	0.852	0.859
Base to Ogive	0.704	0.468	0.875
Bearing Surface	0.415	0.430	0.529
Sectional Density	0.287	0.319	0.346

2 Methodology

Computational Fluid Dynamics is the tool for solving the set of PDEs governing the fluid motion by numerical methods. Among various commercial packages, ANSYS FLUENT 18.2 (2D) is chosen for numerical simulation of hybrid blend design target bullets. Graphical representation of aerodynamic properties are obtained from CFD POST 18.2. Numerical simulation has been carried out for different speeds ranging from subsonic to supersonic, in which the fluid is assumed to be taken as compressible medium.

2.1. CFD Modelling and Meshing

Far-field boundary around the bullets is chosen as flow field for external compressible flow analysis. Parabolic domain of 11 m extent is developed around the bullet, which is 5 m far from the flow entry. Discretization of domain was made with the help of ICEM CFD 18.2 for O-grid meshing. To get better accuracy, O-grid mesh is chosen and elements are counted around 1, 65,000 after grid-independent study. Y+ value to capture boundary layer influence on aerodynamic performance is estimated and found to be 0.000481 mm. Figure 3. Shows the meshed domain with full-view of bullet.

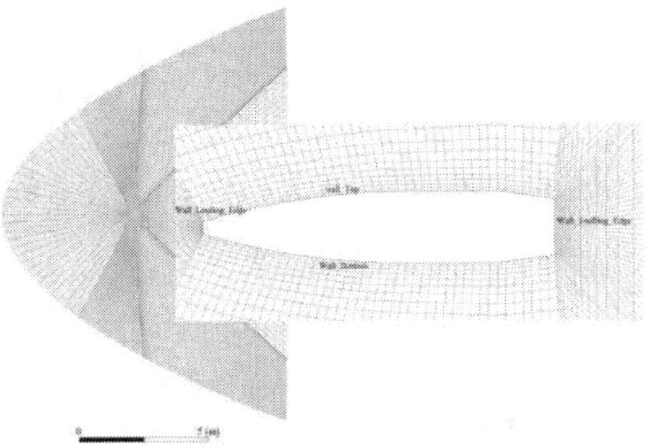

Fig. 3. Discretization of domain.

2.2 Governing Equations

Governing equations are the PDEs containing fluid variables for solving the fluid flows. One of the most recently developed model, especially for compressible external fluid flow applications by Spalart-Allmaras [3] called one-equation turbulence model is used for CFD analysis. This model was designed specifically for wall bounded aerodynamic flows and shows better accuracy for inflation layers. The one-equation turbulence model [3] is given by the following equation (1):

$$\frac{\partial}{\partial t}(\rho\tilde{\nu})+\frac{\partial}{\partial x_i}(\rho\tilde{\nu}u_i)=G_\nu+\frac{1}{\sigma_{\tilde{\nu}}}\left[\frac{\partial}{\partial x_j}\left\{(\mu+\rho\tilde{\nu})\frac{\partial\tilde{\nu}}{\partial x_j}\right\}+C_{b2}\rho\left(\frac{\partial\tilde{\nu}}{\partial x_j}\right)^2\right]-Y_\nu+S_{\tilde{\nu}}$$

Where, G_v is the Turbulent Viscosity Production term
Y_v is the Destruction term (Shear flow region)
And, S is the scalar measure of deformation Tensor.

According to Eleni. D. C [4], the Spalart-Allmaras shown greater accuracy over the experimental results comparing to other two-equation turbulence models. Hence, this model can be used 2D compressible external fluid flow applications. With one-equation model, energy equation is enabled to study the temperature gradient effects because of high Reynold's number.

2.3 Boundary Conditions and simulation Methodology

For Compressible fluid flow analysis, reference pressure is taken as zero, so that the corresponding gauge pressure (equal to total pressure) is applied on the far-field condition. Mach number is the major boundary condition, which is obtained from article by Petzal [4] and flow type is categorised. Ideal gas is chosen as material for flow domain. Three coefficient Sutherland law of viscosity model is well suited for high Reynolds number compressible flows. To get faster convergence, implicit scheme is used rather than going for explicit scheme. In addition, Coupled second order upwind scheme is applied especially for applications dealing with high-speed aerodynamics. Table 2. Shows the boundary values and solver schemes used for modelling simulation of hybrid blend design bullets.

Table 2. Boundary conditions and Solution Methodology.

	Value/ Method
Gauge Pressure	101325 Pa (Ref. Pressure = 0 Pa)
Mach Number	0.54 (Subsonic Flow)
	0.90 (Transonic Flow)
	2.27 (Supersonic Flow)
Solver	Density-Based, Steady
Formulation scheme	Implicit, coupled
Material	Ideal-gas
Viscosity model	Sutherland
Temperature	300 K
Convergence criteria	$1e^{-05}$

3 Results and Discussion

Results are taken with the convergence of 1E-05, an approximate of 15,000 iterations to achieve such convergence. Numerical simulations helps to visualise the flow distribution over the domain and to plot the technical parameters along the co-ordinate to compare the results. Numerical simulations has been conducted for three different cases at three Mach zones. For aerodynamic performance comparison, drag force is estimated and the velocity contour plots were taken for observing the flow distribution around the bullet. Similarly, the turbulent dissipation contour plots based on vorticity are taken and discussed.

3.1. Velocity Distribution

3.1.1. Subsonic Speed (0.54 Mach)

The velocity contour graph of three cases at subsonic speed of 0.54 Mach for all three cases are presented in the Figures 4, 5 and 6. The plots clearly describes the pattern of wave zones near top and bottom edges of bullet. Since, three cases exhibit similar hybrid ogive profile, the flow pattern remains same. However, as far as the size of bullet increases, area covering wave zone and wake zone length increases.

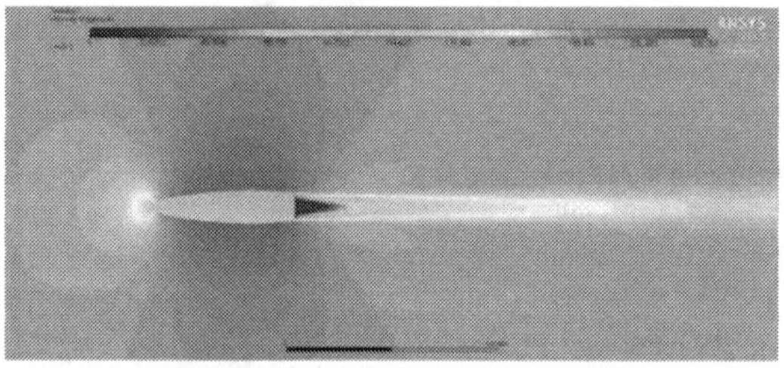

Fig. 4. Velocity contour of first case

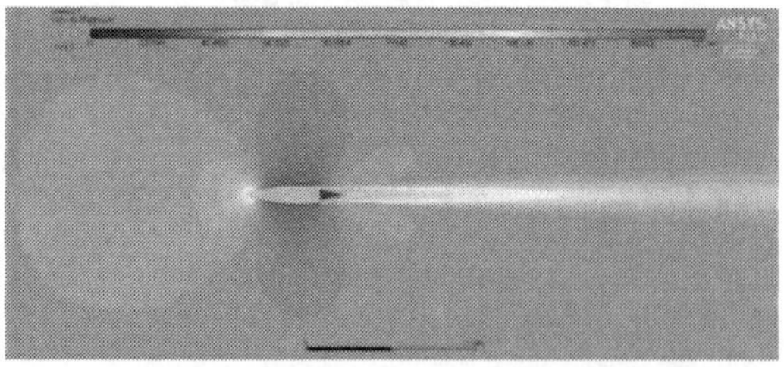

Fig. 5. Velocity contour of second case

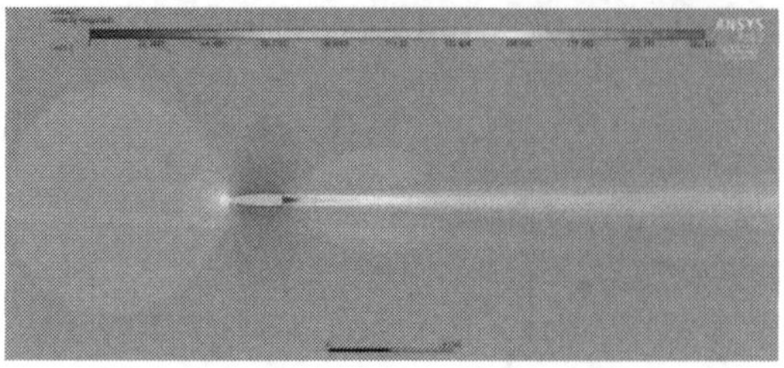

Fig. 6. Velocity contour of third case

3.1.2 Transonic Speed (0.90 Mach)

Since, they attain similar flow distribution for the three cases, the second case alone presented for transonic speed in the Figure 7. At Transonic speed, the shock wave region was formed at the top and bottom walls, it shows the transition from subsonic to supersonic speed.

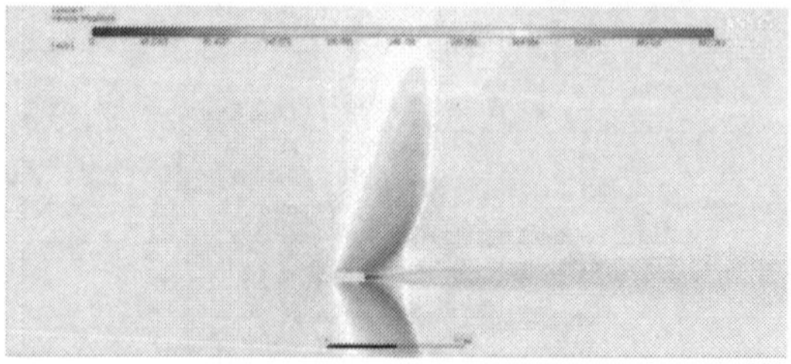

Fig. 7. Velocity Contour of second case (0.90 Mach)

3.1.3 Supersonic Speed (2.27 Mach)

On moving to high-speed case, the contour graph in the Figure 8. Shows the supersonic wake existence around the bullet in the direction of flow.

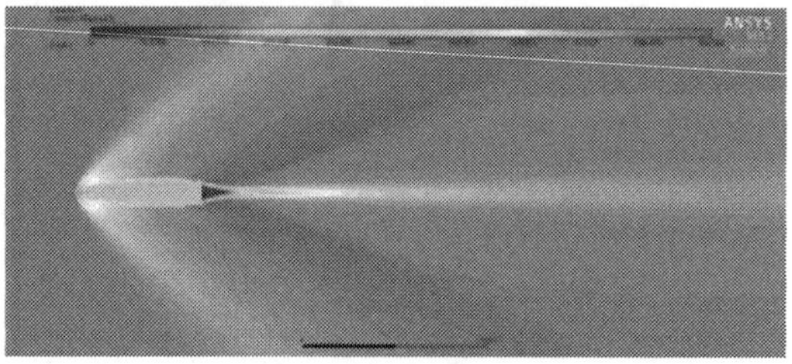

Fig. 8. Velocity contour of second case (2.27 Mach)

3.2. Aerodynamic forces and co-efficients

Drag force (F_d) is the force acting normal to the surface of bullet, which is a function of speed and the surface area of contact. Figure 9. Compares the drag force against Mach number for all three cases. Since, the drag force is directly proportional to square of velocity as expressed in following relation (2), the F_d increases.

$$F_d = \frac{1}{2}\rho v^2 C_d A \qquad (2)$$

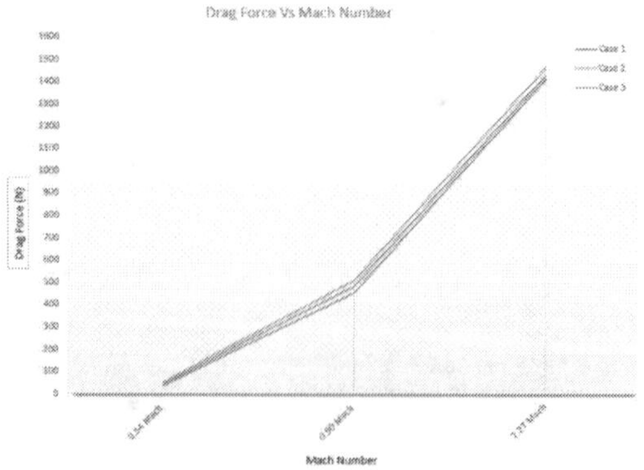

Fig. 9. Drag force Vs Mach Number

The following relation (3) gives co-efficient of skin friction for the fully turbulent and compressible fluid flow.

$$C_{D_{f}} = \frac{0.455(\log_{10} \mathrm{Re})^{-2.58}}{(1+0.21M^2)^{0.467}} \frac{S_{wet}}{S_{ref}} \quad (3)$$

C_D due to friction is completely dependent of Reynolds number and Mach number for a given profile. Figures 10, 11 and 12 presents the co-efficient of skin friction for hybrid blend design bullets at three different Mach number. It clearly shows that the value of skin friction drag increases with increase in Mach number.

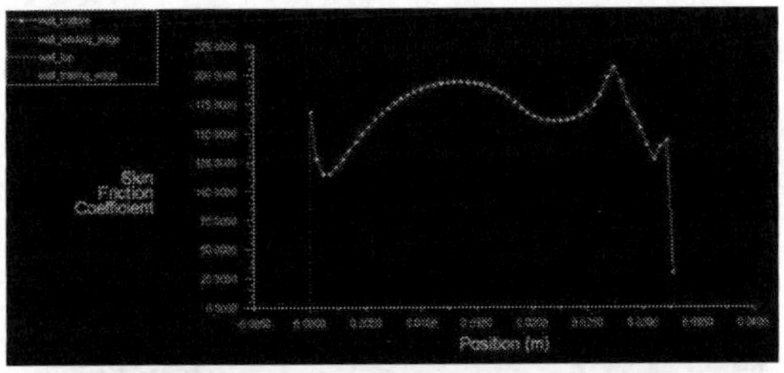

Fig. 10. Skin-friction coefficient at 0.54 Mach

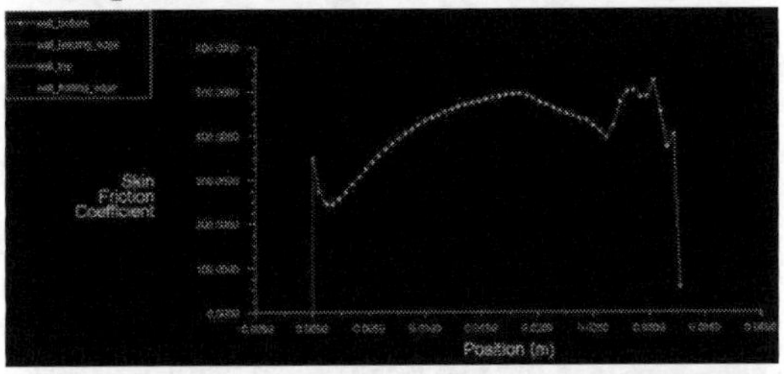

Fig. 11. Skin-friction coefficient at 0.90 Mach

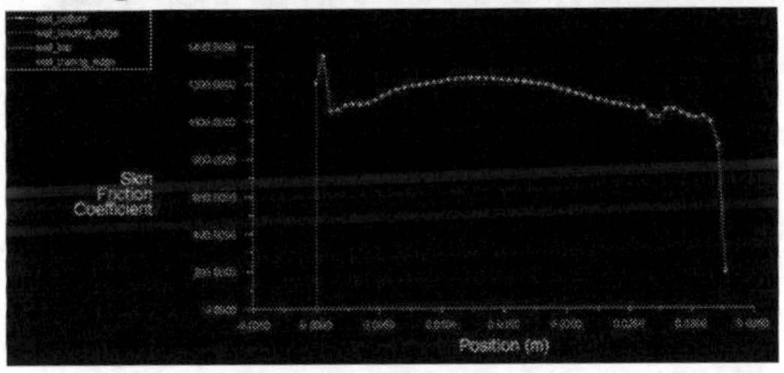

Fig. 12. Skin-friction coefficient at 2.27 Mach

3.3. Turbulence Properties

Turbulence produced behind the bullet is a major influence on the performance of bullet. Turbulence viscosity ratio determines the length of its influence behind the bullet.

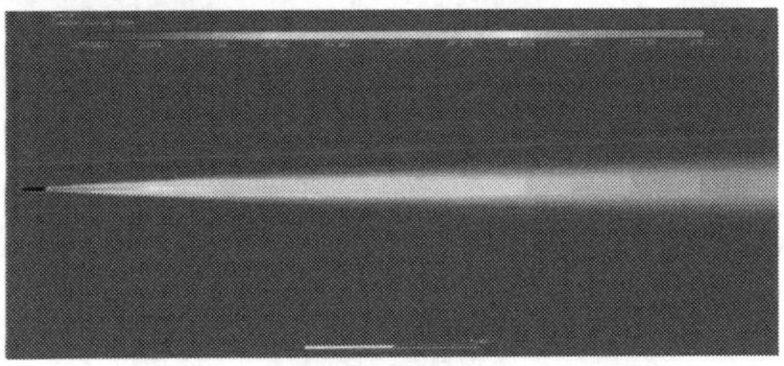

Fig. 13. Viscosity ratio at 0.54 Mach

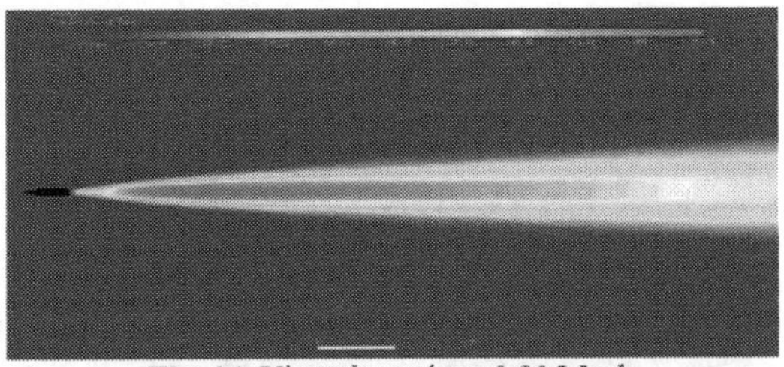

Fig. 14. Viscosity ratio at 0.90 Mach

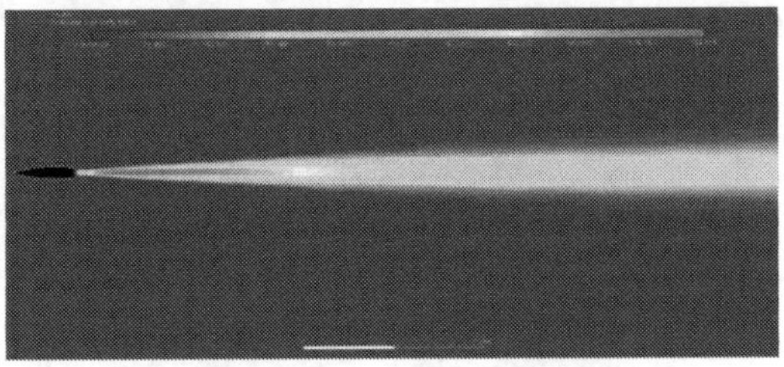

Fig. 15. Viscosity ratio at 2.27 Mach

Figures 13, 14 and 15 shows the viscosity ratio at different Mach numbers. It is to be noted that, turbulence produced is more in case of transonic speed. Due to shock wave existence, the turbulence is produced at greater influence.

3.4 Conclusion

Three cases of hybrid blend design bullets had been taken and numerical investigation has been carried out. Flow distribution, drag force, skin friction coefficient was monitored and turbulence properties are also investigated. Key findings of this research are discussed below:

a) Velocity profile remains constant for the three different cases. It shows that, for hybrid ogive profile, velocity distribution is independent of size of bullet.

b) Drag force increases with increase in Mach number and it proves the drag force relation.

c) Skin-friction drag increases correspondingly as Mach number increases. Profile of skin-friction coefficient has been validated against mathematical relation.

d) Turbulence effect is almost three times larger in case of transonic speed comparing to subsonic and supersonic conditions.

References

1. Litz, Bryan. Characteristics of Different Bullet Ogive Designs. App. Ballistics. **ABDOC-2.1** (2015)
2. Berger. Bullet Product Catalog. Mesa, AZ (2018)
3. P. R. Spalart, S. R. Allmaras. A One Equation Turbulence Model for Aerodynamic Flows. **AIAA-92-0439** (1992)
4. Petzal, David E. How Fast is the Speeding Bullet. Field and Stream **97** (1992)

Recent advancements in shape optimization of aero spiked high speed re-entry vehicle using CFD

Harish Panjagala[1], *E L N Rohit* Madhukar[2], and *I Ravi* Kiran[2]

[1]Faculty of Mechanical Engineering, Koneru Lakshmaiah Educational Foundation, Vaddeswaram, Guntur, Andhra Pradesh, India – 522502.
[2]UG Scholar, Department of Mechanical Engineering, Koneru Lakshmaiah Educational Foundation, Vaddeswaram, Guntur, Andhra Pradesh, India – 522502.

ABSTRACT: Due to increasing demand of High Speed Re-entry vehicles for Space activities within the world, a serious issue associated with the method of deceleration down a vehicle is by the intense heat generated because of development of stronger shocks at the nose. The price of thermal protection systems (TPS) to cut back the warmth generated by the return vehicles is extremely high. In this paper, the ultimate outcome is to cut back the aero heating which is achieved by introducing a spike at frontal region of the nose. Additionally, this spike avoids the deterioration and preserves the structural integrity of space vehicle over elevated temperatures. Further, four totally different geometries of tip specifically Blunt, Slender, Snap and Pan for the aerospike has been introduced and their impact on performance is evaluated and compared with the vehicle having TPS. Hence, usage of aerospike in return vehicles is the most successful and economical over different protection system.

1. INTRODUCTION

Re-entry Vehicle moves at terribly high speed in suborbital, that the wave generated is extremely robust and therefore the temperature behind that shock are greatly high. And this whole heat is transmitted to the vehicle body. It is required to scale back that heat from the body, and to dissipate the warmth in to air. These hot gases then flow subsequently within the boundary of the re-entry vehicle. Therefore, our major objective of re-entry vehicle design is to act as protection of the vehicle to avoid from the severe mechanics of heating.
The objective of re-entry vehicle design is to reduce the warmth which fits into the vehicle and maximize that heat, that get into the air outside the vehicle. Thus, properly designed re-entry vehicle can act as protective covering and reduces the mechanics of air drag. Concerning the role of wind tunnels in fluid propulsive studies, in 1946 von Neumann aforesaid "Indeed, to a great extent, experimentation in fluid dynamics is performed under conditions wherever the underlying physical principles don't seem to be unsure, wherever the quantities to be ascertained are fully determined by known equations. Therefore, wind tunnels, for instance, are used at present, a minimum, as computing devices to integrate the PDE's of fluid dynamics."
F.F.J. Schrijer, F. Scarano & B.W. van Oudheusden [1] conducted an experiment on Reattachment and Separation of Boundary layer on a Blunt shaped Cone-Flare through QIRT. Transient HT measurements are administrated on the cone flare model during a short period hypersonic facility at Mach 9 through Quantitative Infrared Thermography (QIRT). Infrared vision camera was used to compute the temperature of surface and the temperature statistics were effectively related to heat transfer co-efficient specifically by two different reduction techniques. Z. Jiang, Y. Liu & G. Han [2] conducted experiment to attain effective wave for reducing the drag underneath non-zero attack angles and additionally to avoid from severe aero heating, a new concept of the Non-ablative TPS for high speed re-entry vehicles was planned and lateral jets are developed at the spike region as a renovating system of shock at front facet of hypersonic vehicles. The spike works as alternative of the bow shock at nose of a blunt vehicle, the lateral jet works to scale back the heat generated at the front of spike and diverts the shock away from the vehicle in conical shape. For the conceptual design and CFD validation the experimental visualization of flow & pressure measurements was conducted in a hypersonic wind tunnel.
D. Dirkx & E. Mooij [3] investigated the abstract design of a high-speed re-entry vehicle whose effect on performance is calculated by variation of shape and geometry. The two distinct categories of form optimisation in vehicles. .i.e. Capsule and winged vehicle. By victimization the local-inclination strategies and flight mechanics characteristics of these vehicles were analyzed. Entry flight regimes at Mach 3 were computed by assumptive boundary conditions. It has conjointly been studied that depending on the geometry of the structure there's important drop on pressure and temperature of body move in compressed medium however at stagnant purpose, at that the structure is modified between slendered nacile and once more compared to a blunt nacile based on the work of P. Harish et al [4]. It has been studied that there's a crucial role of temperature in supersonic and hypersonic flight period due to compressible effects and the thermal load reduction strategies like counter flow ejection of fluid

panjagalaharish@kluniversity.in

jet of lower temperature collectively of the preventive estimation based on the work of Y.Y. Zeng et al [5]. It was additionally studied that there's sudden amendment of properties in fluid at the nacile portion of the re-entry vehicle because of operational speed and a spread of shocks like bow shock at nose of the vehicle. This case is analyzed by comparing with spike and without spike by observing decreasing pressure data in these high-speed vehicle based on work of C. Nataraj et al [6]. Flow analysis is performed using ANSYS FLUENT 17.2.

2. MODELING AND ANALYSIS

2.1 Modelling by using Solid Works Software

The profile of the winged hypersonic re-entry vehicle is modelled through Solid works 2016 tool. The dimensions taken for the high speed moving body is as shown in the fig2.1.1.

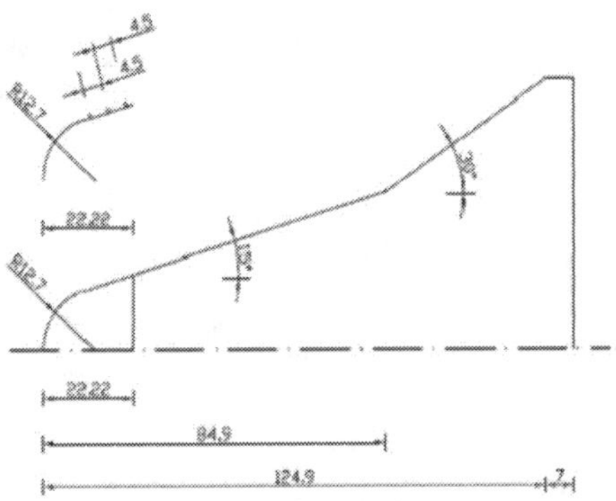

Figure 1 2D geometry of blunt body

Solid works 2016 modeling software has been used for constructing the appropriate 3D model with available dimensions. This can be done by revolving the geometry about the center line as shown in the figure 1.The Domain is created by using ANSYS FLUENT 17.2 Package. The face of the domain is extended by 1.25*l* and behind the geometry by1.5*l* times. The top and bottom of the domain are enlarged by 1.25*l* times, where *l* is the major length of the geometry model .i.e.131.9mm.

2.2 Meshing by using Ansys Fluent 17.2

Before going to the flow analysis over the Vehicle the Vehicle model was imported from Solid works 2016 to ANSYS FLUENT 17.2. Before starting of the mesh the boundary layer is to be created around the Vehicle body. And then mesh the faces of the body by using ANSYS FLUENT 17.2. To create mesh of the domain all triangles method is inserted to map elements. Then quality of the generated mesh is checked.(figure 2&3)

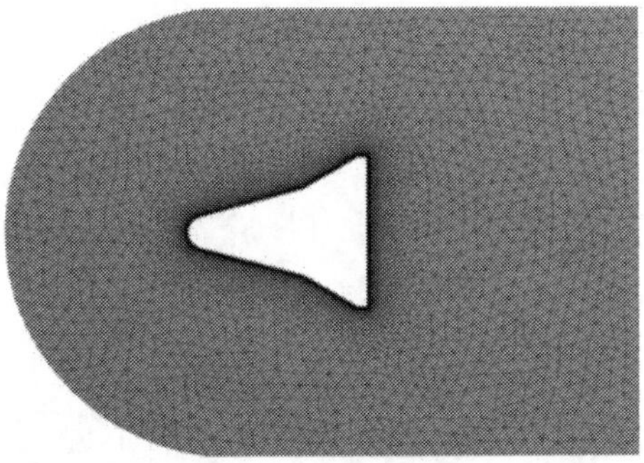

Figure 2 Mesh Model of the vehicle

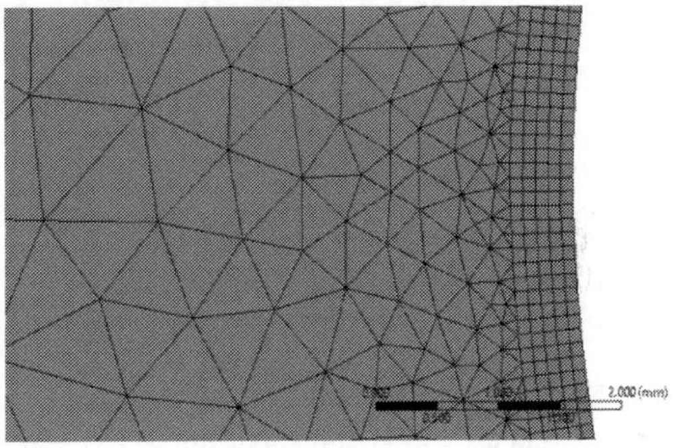

Figure 3 Boundary Mesh of the vehicle.

.2.3 Boundary Condition Calculations

After creating and meshing of the Vehicle in ANSYS then we are Exporting the model in to the ANSYS FLUENT 17.2 for the flow analysis for the flowing references.

At Mach number (M) =9.1
Gamma (γ) = 1.4
Stagnation pressure=p_0 in bar
Static temperature=T in Kelvin
Stagnation temperature=T_0 in Kelvin
From Alan-pope formula:

$$p = \frac{p_0}{\left(1+\frac{\gamma-1}{2}M^2\right)^{\frac{\gamma}{\gamma-1}}} \quad (1)$$

$$= \frac{50}{\left(1+\frac{1.4-1}{2}9.1^2\right)^{\frac{1.4}{1.4-1}}}$$

p = 0.002205 bar
(Stagnation Pressure)

From Wegener formula:

$$log_{10}\,p = \frac{-605.4}{T} + 4.114 \quad (2)$$

$$T = \frac{605.4}{4.114 - log_{10}0.0022025} \times 0.5556 \quad (3)$$

T= 49.6763 + 273 k (Static temperature, k)

$$T_0 = T(1+\frac{\gamma-1}{2}M^2) \quad (4)$$

$$= 49.6763(1+\frac{1.4-1}{2}9.1^2)$$

T_0=873.446K (Stagnation temperature, k)

2.4 Flow Analysis using Ansys Fluent 17.2

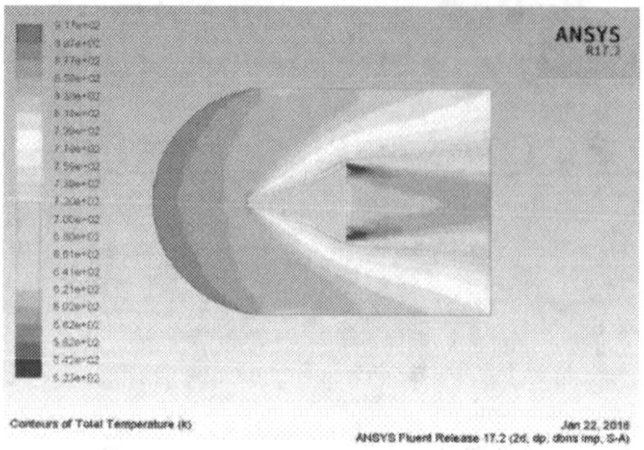

Figure 4 Temperature Contour of Blunt Body

The results for the Blunt re-entry vehicle are indicated. The temperature values along with corresponding contours and graphs are shown in figures 4 & 5.

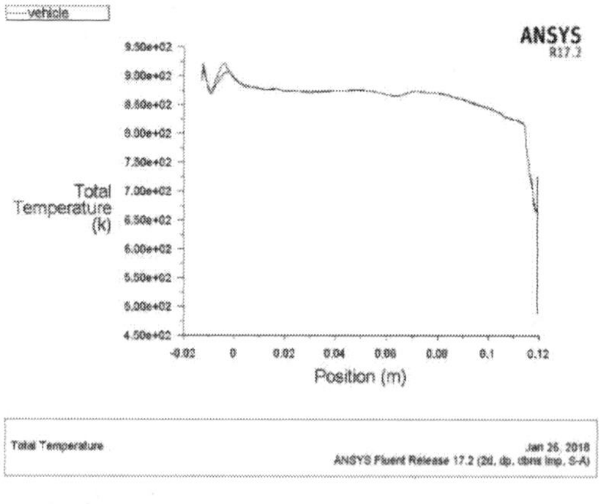

Figure 5 Temperature Plot of Blunt Body

The temperature at the nose for the re-entry hypersonic vehicle was 908 k.(figure 5)

3. VALIDATION AND VERIFICATION

The experiment was done at Delft University of Technology, Netherlands. The wind tunnel was operated at a free stream Mach no. 9.1.The total pressure P_0 maintained in wind tunnel is ranging between 25 and 75 bars. A temperature controlled electrical heating system is used to heat the fluid (dry air) to a temperature of 873 K. (figure 7)

The figure 6 shows the experimental scleren graph. This graph shows the shock wave pattern and the fluent generated shock wave pattern for the same experimental input so we are comparing the both graphs for validation.

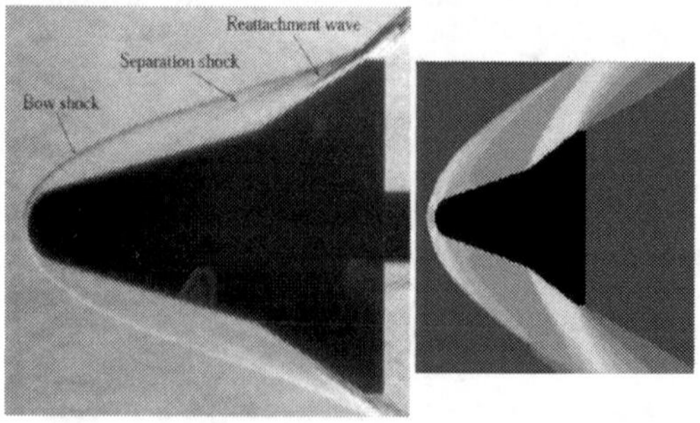

Figure 6 Contrast of Experimental Result with analytical result.

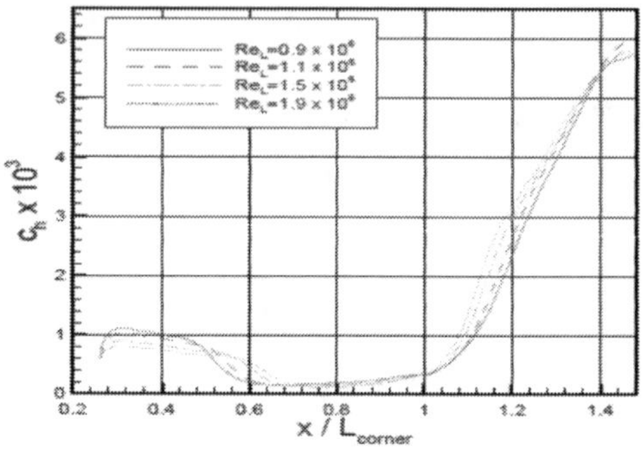

Figure 7. Heat transfer plot for the experimental model.

3.1 Calculation of Heat Transfer Co-efficient (C_H):

The Heat Transfer Rate is given by the relation

$$q = \rho c l \left(\frac{dT}{dx}\right) \quad (5)$$

Where the values of ρ (density), c (specific heat) and l (thickness) of the skin are

'ρ' Stainless Steel density = 8000 Kg/m3

'c' Stainless Steel specific heat = 502 J/Kg K

'l', thickness = 10 mm

And the co-efficient of heat transfer is calculated by using the equation

$$C_H = \frac{dQ/dT}{\rho_\infty V_\infty (h_0 - h_w)s} \quad (6)$$

Where, ρ_∞, U_∞, and Cp_∞ are called as density, velocity and specific heat at constant pressure respectively under free stream condition. Stagnation temperature is To and wall temperature is Tw. From the figure shown in figure 8 it can be computed that there is less than 5%deviation among Simulation and practical results. Hence, the taken case is validated. The simulation in continued in ANSYS 17.2 with the validated boundary conditions to decrease cost and save time.

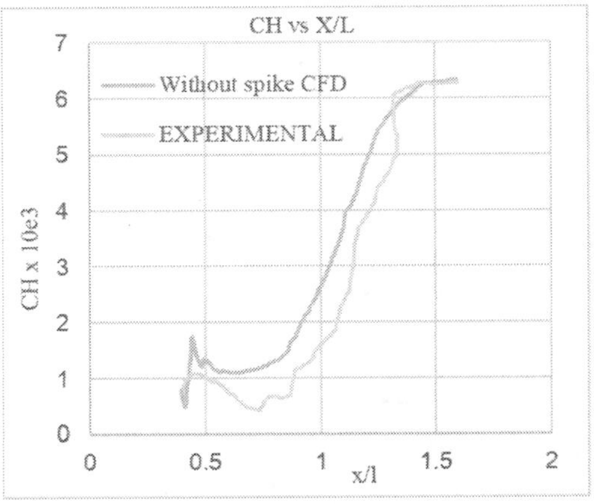

Figure 8. Experimental Result vs Simulation results

4. TOPOLOGY OPTIMIZATION

From the without spike results, it is observed that at nose region the temperature is very high, so that the simulation is concentrated on reducing the Temperature at nose. Thus, spike is introduced for the same model and for same input and successfully found reduction in nose temperature compared to without spike one. The following figures shows the mesh model four different designs of tip of the spike.

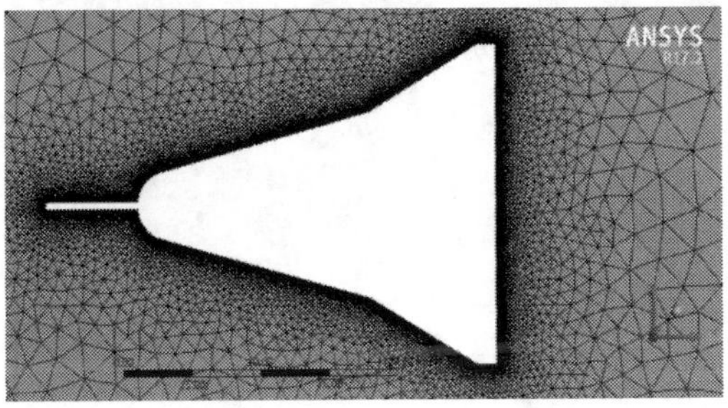

Figure 9. Mesh Model of Blunt Spike Vehicle

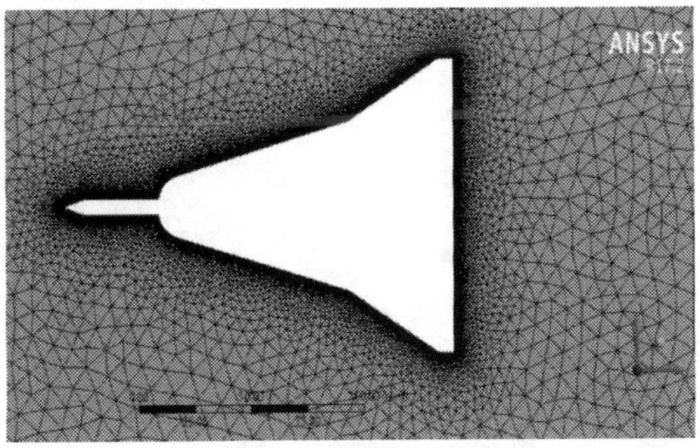

Figure 10. Mesh Model of Slender Spike Vehicle

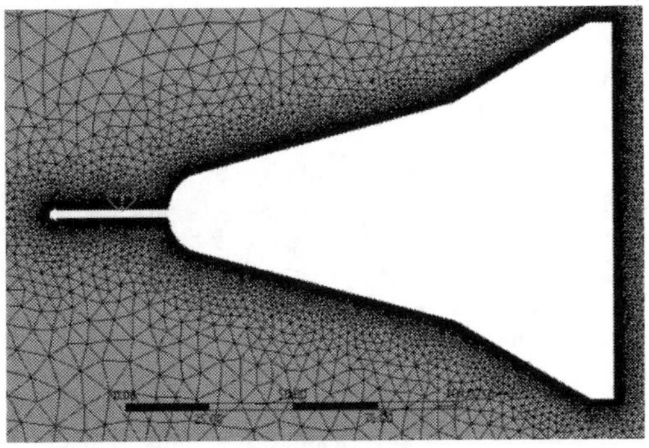

Figure 11. Mesh Model of Snap Spike Vehicle

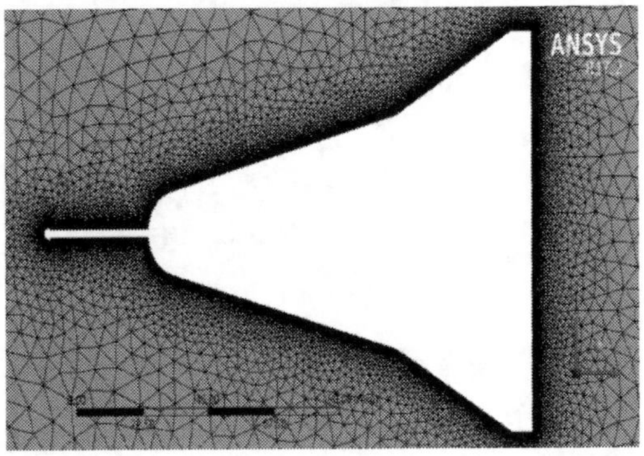

Figure 12. Mesh Model of Pan Spike Vehicle

By changing the form and geometry of spikes its effect on presentation is evaluated, during this study, the form optimisation of two categories of spikes has been studied: A spike with tip as slender & blunt (figure 9 & 10). On one facet, slender spike style reduces the drag however a lot of heating. On the opposite facet, blunt spike style produces a lot of drag. However they're choosable as aero heating reduces. To avoid intense shocks thereby minimizing the temperature, spikes are built at the nose region.

Modeling, Meshing & Fluent analysis is performed in same way as that of without spike model for the same boundary conditions. A spike of four designs was modelled at the frontal region of the high speed vehicle. Blunt spike with 40mm length and having blunt radius at tip of 3mm. slender spike of length 40 mm and having slender angle at tip of 17.06^0.The spike length is same for Snap spike (figure 11) with radius equal to 4mm.Similarly,Pan spike (Figure 12) is having a slender angle of 17.06^0 with slant length 4.47mm.

5. RESULTS AND DISCUSSION

5.1 Blunt Spiked Vehicle

For Blunt, spiked body the following pressure profile and Temperature profile is analyzed (figure 13 & 14). The temperature observed for the high speed Blunt spike vehicle was 812 K (figure 15). Due to the spike the generated shock wave scatters away and gradually lowers temperature along structure of vehicle.

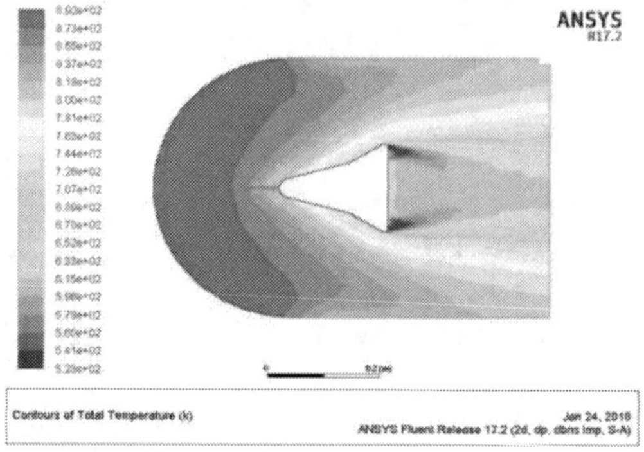

Figure 13. Temperature Contour of Blunt Spike Body

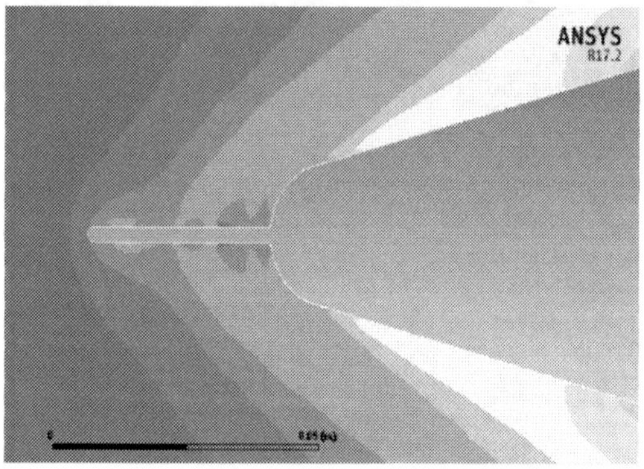

Figure 14 Enlarged View of Temperature profile

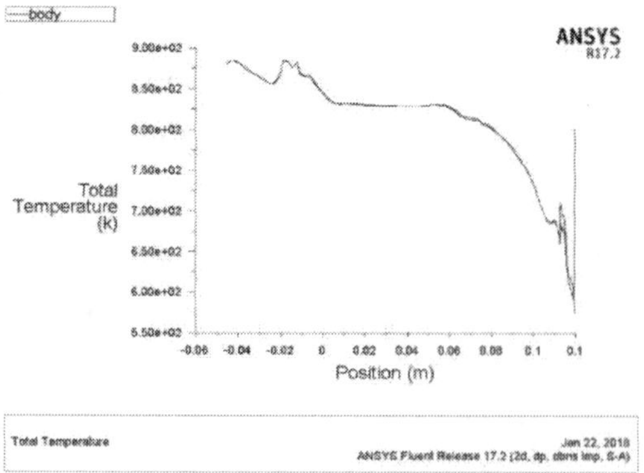

Figure 15 Temperature plot of Blunt Spiked Vehicle

5.2 Slender Spiked Vehicle

For slender spiked body Temperature profile is shown in figure 16 & 17. The temperature observed for the slender spiked vehicle was 848k.(figure 18)

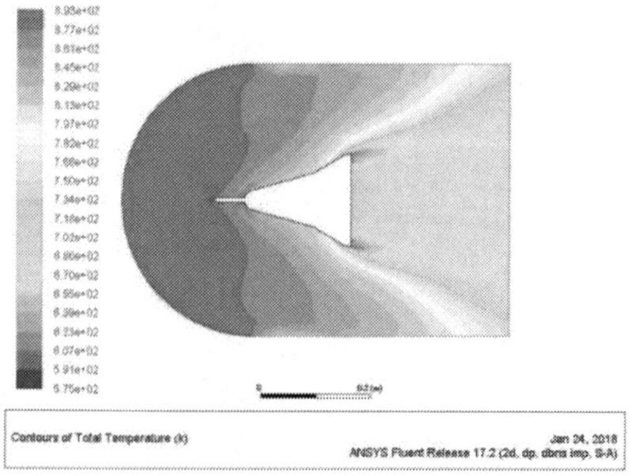

Figure 16 Temperature Contour of Slender Spike Vehicle

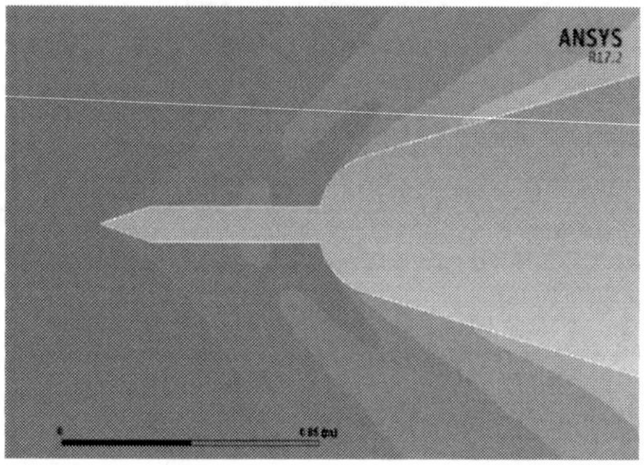

Figure 17 Enlarged View of Temperature profile

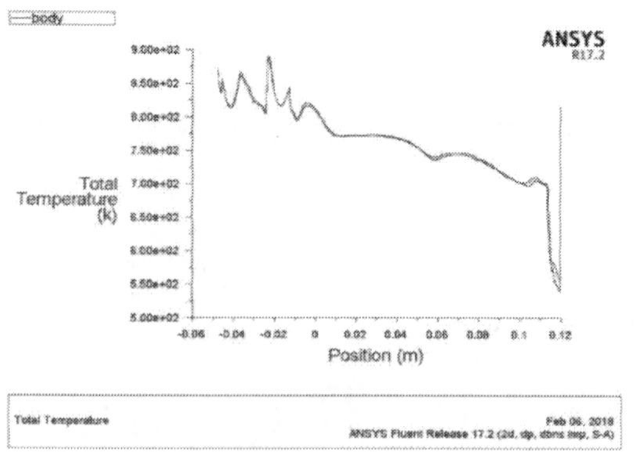

Figure 18 Temperature plot of Slender Spiked Vehicle

.5.3 Snap Spiked Vehicle

For Snap spiked body the analysed Temperature profile is shown in the figure 19 & 20. The temperature computed for the snap vehicle was 820k.(figure 21)

https://doi.org/10.1051/matecconf/201817201007

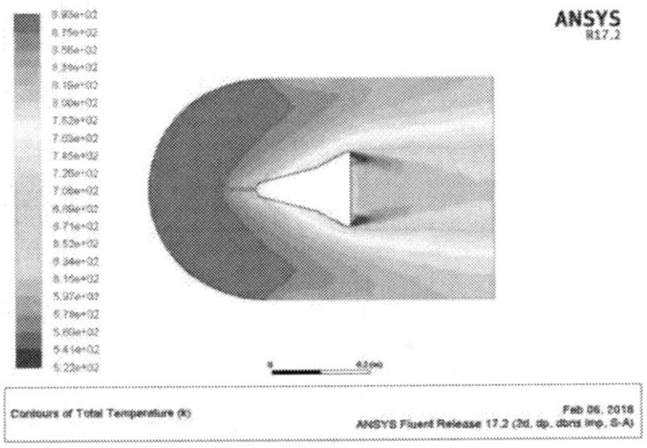

Figure 19 Temperature Contour of Snap Spiked Vehicle

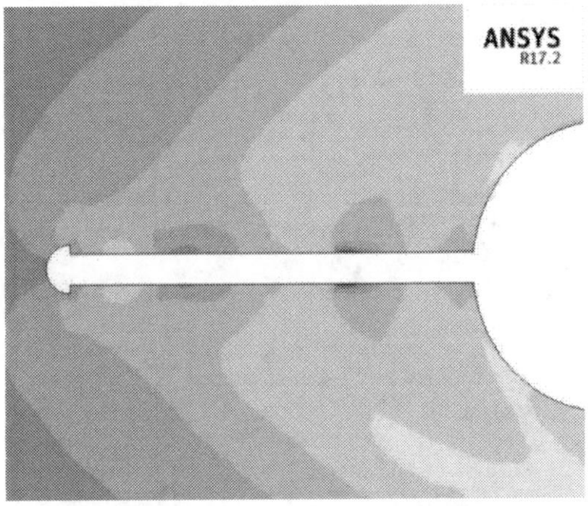

Figure 20 Enlarged View of Temperature profile

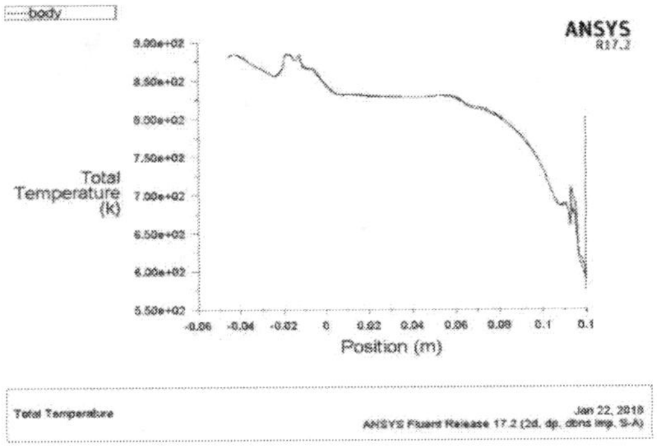

Figure 21 Temperature plot of Snap Spiked Vehicle.

5.4 Pan Spiked Vehicle

The temperature computed for the snap spiked vehicle was 820K (figure 24).It can be clearly observed in the figure 22 & 23.

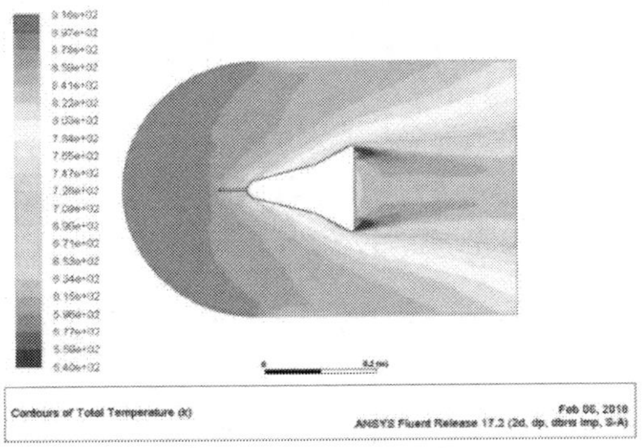

Figure 22 Temperature Contour of Pan Spiked Vehicle

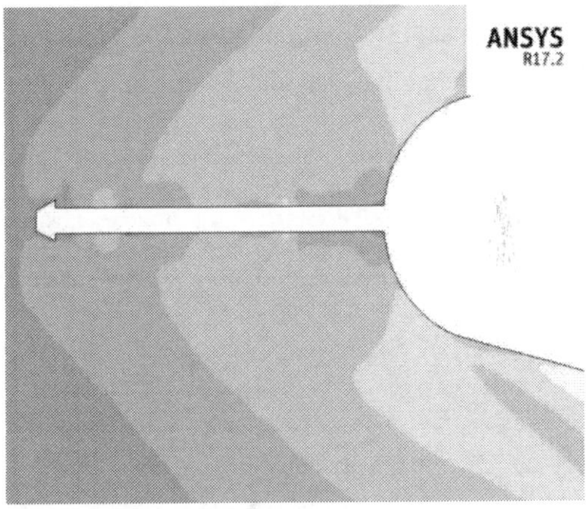

Figure 23 Enlarged View of Temperature profile

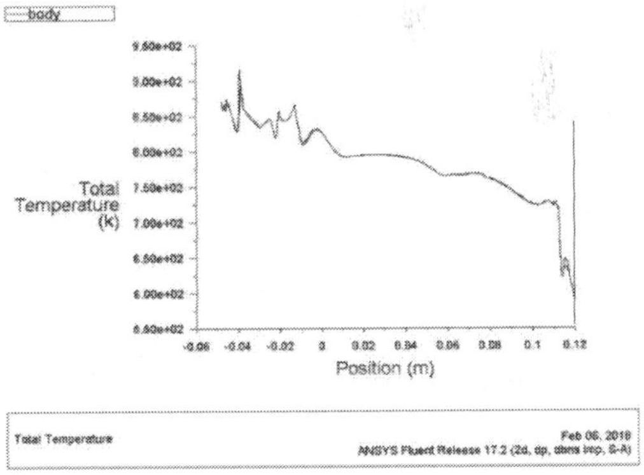

Figure 24 Temperature plot of Snap Spiked Vehicle.

The figure 25 shows temperatures of without spike and with spiked vehicle that includes Blunt, Slender, Snap, and Pan Spike respectively with respect to their corresponding positions. Clearly, the nose region of without spiked vehicle is at a higher temperature than that of its other parts. However, there isn't much reduction of temperature throughout the body .In addition, spiked vehicles has showed a better response in the reduction of temperature especially at the nose.

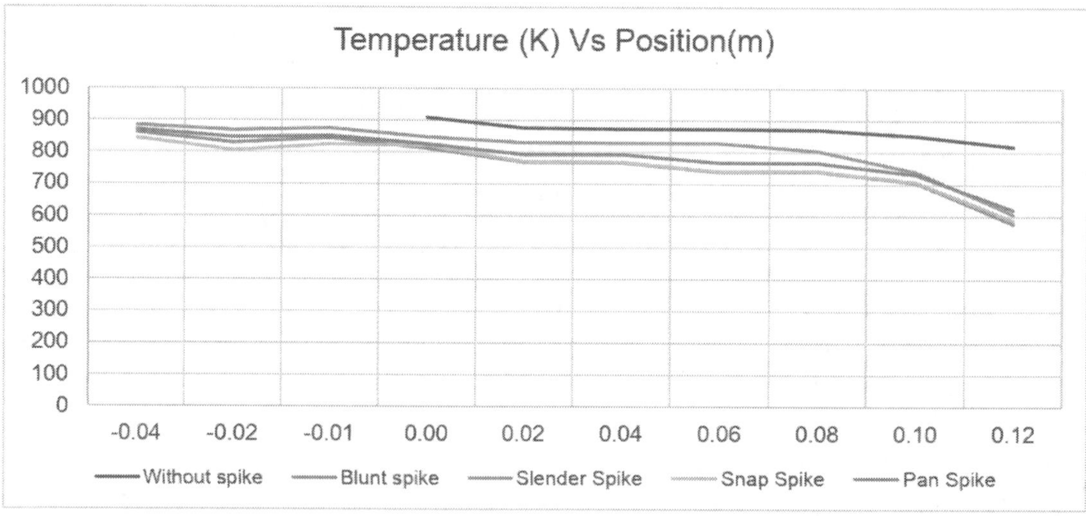

Figure 25 Temperature Comparison of Blunt, Slender, Snap and Pan and without spike

Moreover, blunt spike vehicle is much more effective followed by Snap spike then by Pan Spike. The slender one was least effective which shows similar trend as of without spike vehicle and was below it.

6. CONCLUSION

Keeping in view of future requirements for Hypersonic field of research as well as to generate data base towards extending some of existing static temperature, correlations, a detailed CFD study of hypersonic laminar surface flows on several spiked models will have relevance to reentry missions.

In this paper, design and optimization for the re-entry vehicle in hypersonic flight regime at Mach no 9.1 is carried out. While analyzing the Blunt, Slender, Snap and Pan shaped spikes, it is noticed that by introducing Blunt spike there is 13% reduction of heat at the nose, slender spike is around 7.5 %, Snap spike is around 11% and Pan spike is around 9.5% when compared with the without spike model. Finally, it can be concluded that as a result of spikes heat is reduced at nose. The Slender & Pan spiked vehicles have comparatively less heat reduction than with Snap & Blunt spiked vehicles. So, Blunt and Snap spikes are comparatively preferable over two.

7. REFERENCES

[1] F.J Schrijer Ferry and Fulvio Scarano and W. Bas. Van Oudheusden, Hypersonic Boundary Layer Separation and Reattachment on a Blunted Cone-Flare using Quantitative Infrared Thermography, *AIAA International Space Planes and Hypersonic Systems and Technology* **6967** pp.1-8 (2003)

[2] Jiang Zonglin and Liu Yunfeng and Han Guilai, A new concept of the Non-ablative Thermal Protection System (NATPS) for hypersonic vehicles, **51**, pp. 584-590 (2013).

[3] D. Dirkx, E. Mooij, Optimization of entry-vehicle shapes during conceptual design, *Acta Astronautica* **94**, pp.198–214 (2014)

[4] P. Harish, K. Rajagopal, Aero thermodynamic Design optimization of a hypersonic winged re-entry vehicle using CFD, International Journal of Engineering and Technology **3** pp. 1-8 (2014).

[5] P. Harish, H N. Naveen and Dr. K Rajagopal, Recent advancements in aerothermodynamic Design Optimization of Spiked hypersonic Winged Re-Entry Vehicle using CFD,International Journal of Engineering Research & Technology **3** , pp. 877-882. (2014).

[6] Y.Y. Zeng and N.A. Ahmed, Computational fluid dynamics investigation of counter flow ejection effects on a body moving with high speed, *Proceedings International Conference on Mechanical Engineering and Material Science* **27**, pp.410-413 (2012).

[7] C.N.Nataraj and H. Manjunatha, Aerothermodynamic Design and Optimisation of Hypersonic Research Reentry Vehicle, International Journal of Science and Research **3** , pp.1234-1238 (2014).

[8] A. Peter Gnoffo, K. James Weilmuenster, H. Harris, Computational Aerothermodynamics design issues for Hypersonic Vehicles, AIAA **36**, pp.1108-1113 (2001).

[9] P. A. Gnoffo, Planetary-entry Gas Dynamics, Annual Review of Fluid Mechanics **31**(2001).pp. 459-494.

[10] E.H. Hirschel, C. Wieland, Selected Aerothermodynamic Design Problems of Hypersonic Flight Vehicles, International Journal of Computational Fluid Dynamics, **25**, pp. 74-86 (2009).

[11] M. Grant, I. Clark, R. Braun, Rapid Simultaneous Hypersonic Aerodynamic and Trajectory Optimization Using Variational methods, *Conference AIAA Atmospheric Flight Mechanics*, **6640** (2011).

[12] D. J. Kinney, Aero-Thermodynamics for Conceptual Design, ***42nd*** *AIAA Aerospace Sciences Meeting and Exhibit*, Reno, Nevada, January (2004)

[13] Sahoo Niranjan, S. Saravanan, Jagadeesh, K P J Reddy, Simultaneous measurement of aerodynamic and heat transfer data for large angle blunt cones in hypersonic shock tunnel, *Sādhanā Academy Proceedings in Engineering Sciences* **31** , pp. 557-581 (2006).

[14] John. D. Anderson, Jr. *Computational Fluid Dynamics*, (McGraw-Hill Book Co., International Editions, 1995).

[15] Versteeg H. K. and Malalasekara W., *Introduction to Computational Fluid Dynamics*, (Long Man Group Ltd., 1995).

Numerical model for concentration measurement during decomposition of H_2O_2 over silver catalyst

Sanjay Vasanth.KB[1], *Gokul Raj*.G[1], *Venkatesan* M[1]*
[1]School of Mechanical Engineering, SASTRA Deemed University, Thanjavur 613401

Abstract. Minimal thrust of the order of few Newton is produced using micro thrusters, which are used for controlling the orientation and trajectory of the satellite or spacecraft. Controlled thrust production of the order of milli Newton is still a challenging task. Green propellants like Hydrogen peroxide are utilized in such applications. The products of this green fuel after decomposition are water in the form of steam and oxygen. The present work involves the simulation of hydrogen peroxide decomposition on silver catalyst which can be used in space propulsion. The simulations are done for flow of 30% hydrogen peroxide concentration when flowing inside a mini channel of size 3.4mm. Helical shaped silver catalyst is positioned inside the tube to enhance the decomposition. The model involves the concentration measurement of the reactants and products and after passing over the catalyst surface. The results will help in choosing the geometry and position of the catalyst to have effective decomposition of the monopropellant.

1 Introduction

In space propulsion, minimal thrust is required for propelling a spacecraft due to the absence of drag forces. As the technology keeps developing, many researchers and organisations have moved towards developing green fuels that are efficient and provide the same amount of thrust produced by conventional fuels. Hydrogen peroxide is such a type of green fuel that provides controlled amount of thrust for manoeuvring and trajectory setting in outer space. Hydrogen peroxide has found applications in different fields like medicine where 6% H_2O_2 is used as an antiseptic and in industries as a bleaching agent. Hydrogen peroxide is also eco-friendly and is used in advanced oxidation process (AOP), in textile industry to give high degree of brightness to the fibres without compromising their mechanical properties, as a monopropellant or bipropellant. High concentration of hydrogen peroxide is used for rocket propulsion along with other fuels. Hydrogen peroxide of varying concentrations is important for many applications. In propulsion where the concentration will be varying after passing through the catalyst, the formed products decide the amount of thrust. Hence the concentration measurement becomes important. Some of the works reported on hydrogen peroxide based propulsion systems and concentration measurements are detailed below.

Widdis et al. [1] developed a model based on MEMS-based micro propulsion using a monopropellant hydrogen peroxide. It was observed that the incomplete decomposition of hydrogen peroxide was because of the less wetted area on the catalytic walls. Adam et al. [2] evaluated the performance of hybrid sounding rockets in which hydrogen peroxide was used as an oxidizer. The study was done for optimal burn duration and selection of nozzle. Goszner et al. [3] studied the decomposition of hydrogen peroxide on silver and gold alloys for concentrations from 5 to 0.05 mol/lit and found that the catalyst behavior was basically determined by the oxygen affinity. It was concluded that gold and silver had similar properties as a catalyst. Bao-Jin et al. [4] measured 30% concentration of hydrogen peroxide using an optical device with the help of refractive index variation. Ryan et al. [5] developed a numerical model to study the hydrogen peroxide decomposition in catalytic micro reactors. It was found that due to the absence of non-mixing mechanism in micro scale the decomposition was slow in micro channels. A new technique was developed by introducing gas bubbles which significantly increased decomposition by 57% for the given length. Berna et al. [6] developed a COMSOL model and applied it to isothermal, steady state and 3D PEM fuel cells to observe parameters like concentration profiles, current density profiles and polarization curves. Duncan et al. [7] detailed the measurement of oxygen (O_2) concentration in its gaseous form with a paramagnetic analyzer. Baldikova et al. [8] used non-woven fabric fibers, which easily absorbed the manganese dioxide micro particles. The composite

*Corresponding author: mvenkat@mech.sastra.edu

https://doi.org/10.1051/matecconf/201817201008

material of low-cost was used as a recoverable catalyst for hydrogen peroxide decomposition. Mohith et al. [9] measured the concentration of hydrogen peroxide experimentally based on the change in optical property and refractive index. An IR transceiver circuit was used to measure the concentration. Kang et al. [10] studied high concentration hydrogen peroxide to design high-speed underwater jet engines. Hydrogen peroxide decomposition was the primary fluid used for propulsion and a huge amount of propulsive force was successfully generated and could be used for unmanned underwater vehicles. Lennon et al. [11] worked on the multiphase catalytic decomposition of H_2O_2 into H_2O and O_2 and investigated a way to reduce the risk of thermal runaway. Thirumalikumaran et al. [12] analysed the products after catalytic decomposition of hydrogen peroxide with silver catalyst inside mini channels. The above literature details the various works reported on decomposition of hydrogen peroxide using silver as a catalyst. Many types of sensors are reported in literature for concentration measurement when the fluid does not undergo catalytic decomposition. However works done for online measurement of concentration during catalytic decomposition is limited. In the present work a numerical model is developed using COMSOL Multiphysics to analyze the effects of geometry of the catalyst silver on decomposition rates. The concentration variation during the flow process is also detailed. Such a numerical model is essential for understanding the decomposition phenomenon, which in turn will control the thrust required for micro propulsion devices used in Nano satellites.

2 Numerical model

The numerical model is developed using COMSOL Multiphysics software. Initially as a validation exercise the catalytic decomposition process is carried out in a batch process. Later a 2D and a 3D model is developed with the same package to analyse the variations during the flow process.

2.1 Zero Dimensional Model

The zero dimensional model assumes a batch type reactor process with constant volume. The rate expression is of first order for the given reaction and Arrhenius expression is used as given in Goszner [3] et al. The values for activation energy are given as 52200 J/mol and frequency factor of collision as 0.18 $\times 10^9$ (m^3/s mol). The reaction is taken from the experiment done by Thirumalikumaran et al. [12] The decomposition reaction is,

$$H_2O_2 \xrightarrow{decomposition\ by\ catalyst(Ag)} H_2O + 0.5O_2 \quad (1)$$

The surface species (catalyst) is kept with locked concentration. The governing equations in reaction engineering are,

$$\frac{dc_i}{dt} = R_i + R_{ads,i}\frac{A_r}{V_r} \quad (2)$$

Here for a species i, C_i is the concentration, R_i is the rate of the species, $R_{ads,i}$ is the rate effect of surface species on i, A_r is the surface reaction area and V_r is the reactor volume. The study is carried out with time steps of 0.1 to 1 second, the results are plotted in a 1D graph.

2.2 2D Model Straight Channel

After verifying the energy balance and volume based reactant and products, the model is further developed to simulate the decomposition in a mini channel. Here the hydrogen peroxide will flow over a 2D catalytic wall for a specific length. The dimensions are same as that of Widdis et al. [1]. The flow velocity is set as 1.2 m/s and unidirectional, the concentration of hydrogen peroxide is set as 100%. Part of the channel is insulated with wall type boundary condition and the remaining with catalytic walls. The catalytic walls are set with open boundary. The outlet pressure is set as 1atm. The physics used in the geometry is similar to the one used in the 3D model explained in the next section. The mesh is given as normal mesh. The geometry is given below,

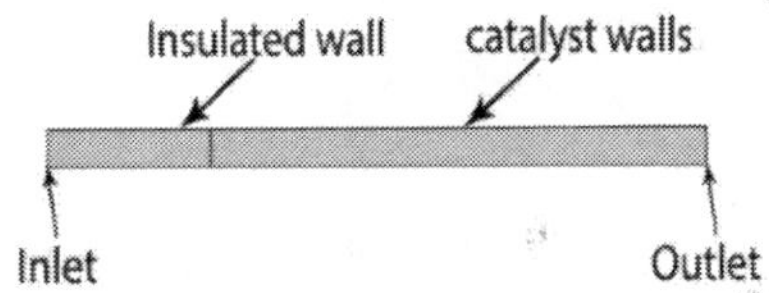

Fig. 1. Geometry of the straight channel.

2.3 3D Model with Helical Catalyst

After validation of the process involved using a 2D model, a 3D model is developed. The main geometry is developed in COMSOL Multiphysics software which consists of a cylindrical mini channel having diameter of 3.4 mm and length of 60 mm. The catalyst is of helical structure (spring) having 10 turns and has a diameter of 3 mm. The helix is positioned about 20 mm from the inlet. Silica glass is the chosen material for the mini channel walls and the catalyst is silver (Ag).

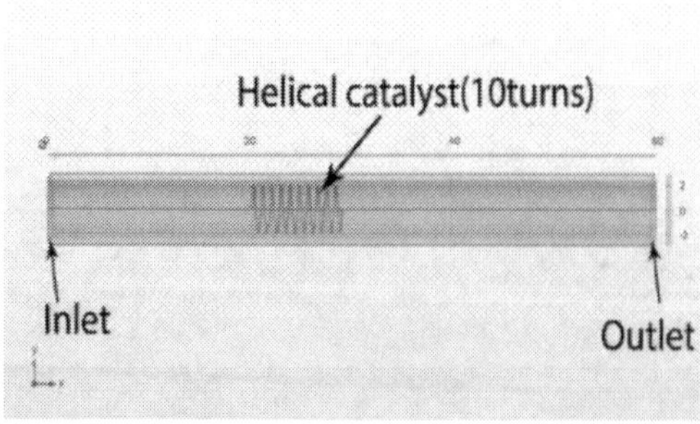

Fig. 2. Numerical model of 3D geometry with helical catalyst.

The 3D component is developed using the following physics. Transport of diluted species is used to compute the concentration field. The governing physics involves the use of Fick's law for transport by diffusion and adsorption as the adsorption happens on the catalytic surface. The products are formed when they are desorbed from the catalyst surface. The equations governing the transport of diluted species are,
Fick's law,

$$\nabla \cdot (-Di\nabla Ci) + u \cdot \nabla Ci = Ri \tag{1}$$

$$Ni = -Di\nabla Ci + uCi \tag{2}$$

Here for the species i, N_i is molar flux, D_i is the diffusion coefficient, u is the velocity of the fluid and C_i is the concentration. Transport mechanisms also include convection. Diffusion coefficients for the species are specified as 4 x 10^{-4} m/s. No flux is given for the outer walls of the channel. Initial values are given as concentration of the species. Inlet concentration of hydrogen peroxide is given as 8823.52 mole/m^3(for 30% concentration), water as 38888.89 mole/m^3, and oxygen as 0 mole/m^3. The flux discontinuity is specified over the catalyst boundary for the surface reaction of hydrogen peroxide to happen. Heterogeneous catalysis takes place in this type of catalytic decomposition. The same reaction and values used in the zero dimensional model is taken and the species from the reaction are H_2O_2, H_2O, Ag and O_2. This physics acts over all the domains. The governing equations in chemistry are,

Reaction rate,

$$R_i = \sum_j R_{ij} \tag{3}$$

Reaction rate over the surface,

$$R_{ads} = \sum_j R_{ads,j} \tag{4}$$

Arrhenius equation,

$$k = A(\frac{T}{T_{ref}})^n \cdot e^{\left(-\frac{E}{RT}\right)} \quad T_{ref} = 293K \tag{5}$$

In this R_i *and* R_{ads} denote the rates of the reactants and rate with the surface species. In Arrhenius equation k is the reaction rate, A represents the frequency factor for the reaction, T represents the temperature, T_{ref} represents the reference temperature, n represents the forward temperature exponent, E denotes the activation energy, R denotes the gas constant with value 8.314 J/mol. Navier-Stokes equation and continuity equation are employed in this physics. The inlet velocity is specified as 0.01m/s and in the outlet it is given as suppressed backflow and pressure is set as zero. The governing equations are,

Navier-stokes equation,

$$\rho\left(\frac{\partial u}{\partial t} + u \cdot \nabla u\right) = -\nabla p + \nabla \cdot \left(\mu(\nabla u + (\nabla u)^T) - \frac{2}{3}\mu(\nabla \cdot u)I\right) + F \tag{6}$$

Continuity equation,

$$\frac{\partial \rho}{\partial t} + \nabla \cdot (\rho u) = 0 \tag{7}$$

Here p is the pressure of the fluid, u is the velocity of the fluid, ρ is the fluid pressure and μ is the fluid dynamic viscosity. In the Navier-stokes equation the various terms corresponds to the (left side) pressure force, (right side) viscous forces and the external forces applied. Mesh is given as finer. The 3D model is studied as time dependant from 0 to 1s with time steps of 0.1.

3 Results and discussion

3.1 Zero dimensional Model

The concentration distribution from the reaction engineering in zero dimensional plot is shown in a 1D plot. The graph is plotted considering the inlet concentration of hydrogen peroxide and other species. The reaction has proceeded in the expected way and the products are formed as shown in the graph. The study is time dependant and starts from 0 to 1 second with time steps of 0.1. The graph with the concentration variation with respect to time from reaction engineering is shown below,

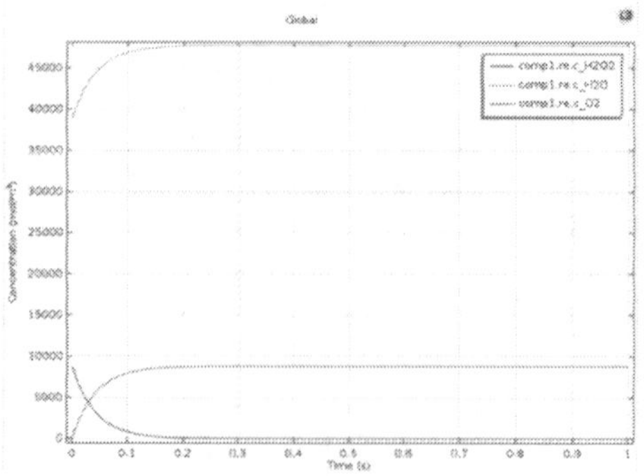

Fig. 3. Concentration plot in 1D for 30% concentration of hydrogen peroxide.

From the reaction-engineering graph, it is found that the concentration of hydrogen peroxide decreases in a parabolic trajectory to lower concentrations and reaches a zero value indicating a complete decomposition. Similarly the concentrations of water and oxygen has increased in a parabolic trajectory to higher concentrations. The concentration of hydrogen gets exhausted in 0.2 s and goes to near zero for the rest of the time, similarly for water the main change has happened in 0.2 s. From the reaction, the amount of water concentration coming out is higher.

3.2 2D Model Straight Channel

The catalyst reaction is modelled with the wall as the catalyst and the fluid that is in contact with the wall is only decomposed. In microscale, the transport is only due to diffusion. The results that are from this model are similar to the straight channel model by Widdis et al. [1] which validates the present numerical model developed.

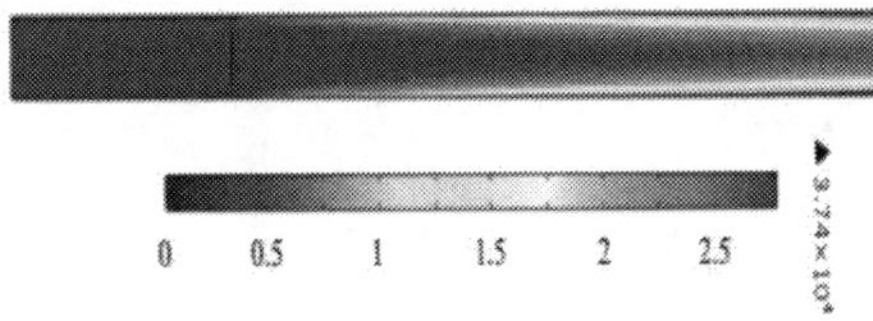

Fig. 4. Hydrogen peroxide concentration (mol/m^3) in straight channel (surface plot).

In this channel, the decomposition of hydrogen peroxide is very much less. As the decomposition in this channel is incomplete, a better geometry of catalyst can improve this discrepancy. A better computational model is developed with the help of COMSOL Multiphysics in 3D geometry.

3.3 3D Model with Helical Catalyst Reaction

The results from the 3D component are studied as time dependant and the 3D plots are plotted for the concentrations of H_2O_2, H_2O and O_2 and velocity after 1s.

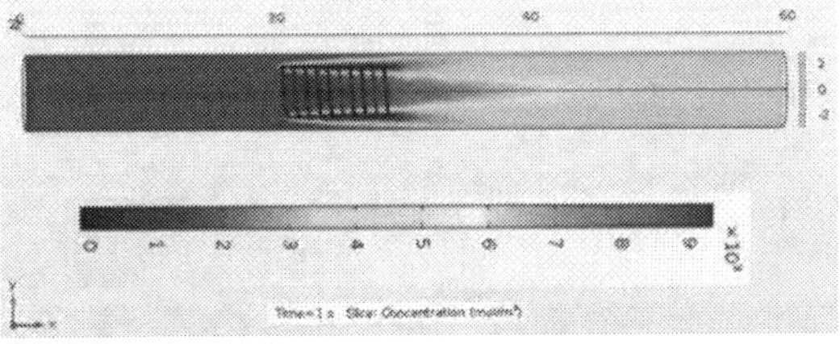

Fig. 5. Concentration of H_2O_2 (mol/m^3)

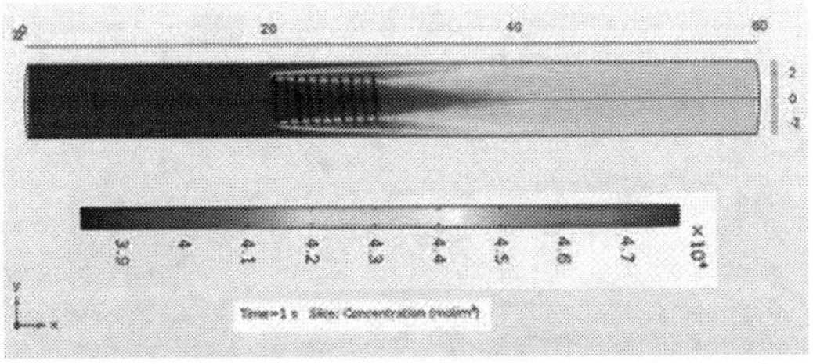

Fig. 6. Concentration of H_2O (mol/m^3)

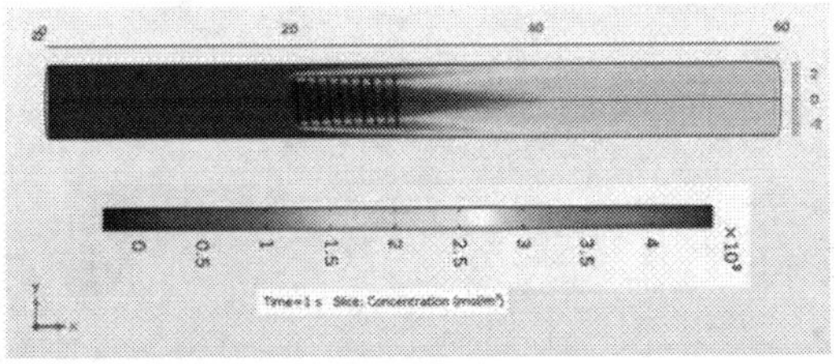

Fig. 7. Concentration of O_2 (mol/m^3)

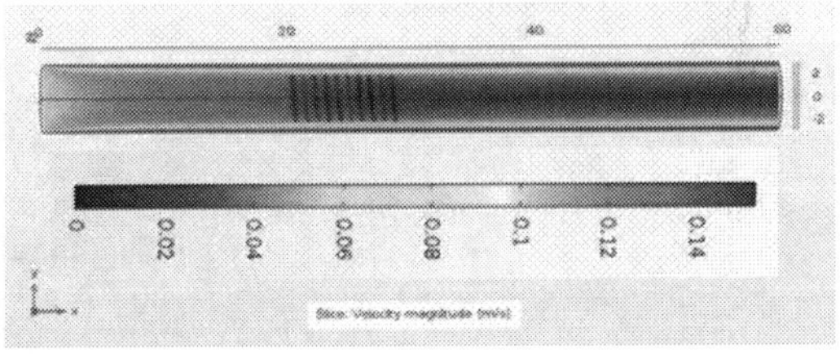

Fig. 8. Velocity plot in (m/s).

From the plots, it is seen that there is a significant decrease in the concentrations of H_2O_2, and concentration of H_2O and O_2. The decomposition is much better in this model with a helical silver catalyst and this serves as a better model for the decomposition of hydrogen peroxide. The helical catalyst improves the surface area that is exposed to the hydrogen peroxide thereby improving the reaction and solving the discrepancy. The increase in number of turns of the helix will increase the reaction further. The velocity in the channel has increased and has reached nearly 0.15 m/s. The pressure drop is also significant.

4 Conclusion

A numerical simulation model is developed using COMSOL Multiphysics software. The model that is developed is validated with the model by Widdis et al.[1] A 3D model is developed to simulate the decomposition reaction of hydrogen peroxide on a silver catalyst for the measurement of the reactants and products. The percentage of unused hydrogen peroxide concentration coming out after reaction from the catalyst is also known and this can help in placing the number of helical catalysts or increasing the number of turns in the catalyst so that maximum reaction of hydrogen peroxide can happen. This model can be used for concentration measurement. Future work is to match the numerical values with the experimental values of concentration.

Acknowledgement:

We would like to sincerely thank Mr. Ravichandran. V, PG Senapathy Centre for Computing Resource, IIT Madras for providing us the computational facilities.

References

1. Stephen J. Widdis, Kofi Asante, Darren L. Hitt, Michael W. Cross, Walter J. Varhue, and M. Ryan McDevitt, *IEEE/ASME transactions on mechatronics*, **vol. 18**, no. 4,(august 2013).
2. Okninski, Adam, *Acta Astronautica* (2018).
3. K. Goszner and H. Bischof, *Journal of catalysis* **32**, 175-182 (1974).
4. Bao-jing Peng, Chao-fu Ying, Hui-qun Ye, Yong Zhao, Yun-tao Liu, *Proceedings of SPIE* **Vol. 5642**, (SPIE, Bellingham, WA, 2005) doi: 10.1117/12.571033.
5. M. Ryan McDevitt, Darren L. Hitt. *Proceedings of the International Conference on New Trends in Transport Phenomena Ottawa*, Ontario, Canada, (May 15-16 2014) Paper No. 83
6. Sezgin, Berna, Dilara Gulcin Caglayan, Yilser Devrim, Thomas Steenberg, and Inci Eroglu. *I, International journal of hydrogen energy* **41, no. 23**: 10001-10009 (2016).
7. Alastair Duncan , Oliver W Pratt, *Royal college of Anaesthetists* CPD Mtrix: 1A03, 2A04.
8. Baldikova, Eva, Kristyna Pospiskova, Mirka Safarikova, and Ivo Safarik, *Materials Chemistry and Physics* **203**: 280-283 (2018).
9. P. Mohith, S. Vijaya Raghavan, P. M. Siva, N. Mithran, M. Venkatesan, *017 IEEE International Conference on Computational Intelligence and Computing Research, Coimbatore*, Ref. No. ICCIC1700092, (December 14-16, 2017).
10. Kang, Hongjae, Seungkwan Baek, Byeonguk Ahn, Yongtae Yun, and Sejin Kwon, *Ocean Engineering* **153**: 193-200 (2018).
11. Lennon, E., A. A. Burke, M. Ocampo, and R. S. Besser, *Journal of Power Sources* **195, no. 1**: 299-306. (2010)
12. Thirumalikumaran S.K, Suwathy R and Venkatesan M, *Journal of Aerospace Technology and Management* Accepted for publication (2017).

Methods of computational modeling of coronary heart vessels for its digital twin

Ilya Naplekov[1], *Ivan* Zheleznikov[1,*], *Dmitry* Pashchenko[1], *Polina* Kobysheva[1], *Anna* Moskvitina[1], *Ravil* Mustafin[1], *Maria* Gnutikova[1], *Alina* Mullagalieva[1], *and Pavel* Uzlov[1]

[1]Samara State Technical University, 443100 Samara, Russia, 244 Molodogvardeyskaya str.

Abstract. In this work, methods of numerical modelling of the coronary vessels system of the human heart have been studied. This investigation includes transient flow of the liquid – blood and dynamics of zones of shear stress at vessels. The main goal of the research is obtaining of hemodynamic and shear stress for creating the digital twin of coronary heart vessels. The results were obtained for low Reynolds numbers about 20 of three-dimensional laminar flow. With this Reynolds number the turbulent flow of the blood is modelled by Realizable k-ε model, and SST models to the narrowing, expansions, and blocks inside the vessels. Loads caused by the additional energy consumption because of the turbulent flow of the blood (increase in arterial blood pressure) have been analyzed. A two-dimensional model of a separated vessel with fixed blood back-flow prevention is developed. Presence of a turbulent flow core is discovered. By the means of stress-strain properties of the model, visual representation of the wearing process of the blood back-flow preventer, and heart diseases progression is obtained.

1 Introduction

The concept of the digital twin, suggested in 2002, carries the idea of fusion between the virtual and the physical worlds. Digital twin is a model of a physical product in the field of production and design. «The vision of the digital twin itself includes more or less all information which could be useful in all – the current and subsequent – lifecycle phases» [1], such vision is represented in the work of S. Boschert and R. Rosen. Moreover, E. J. Tuegel and co-authors confirmed it. [2] .In their article, a model represented the idea of how a digital twin can be used to predict the lifetime of an aircraft structure and to ensure its structural integrity. In other words, the value of a digital twin is expressed in predicting local damage through the calculations of temperatures and structural deflections in conditions met during a flight. Nowadays, offers of methods for integrating digital twins are being put forward from all over the world. For example, Ramtilak A. [3] in his work describes the Digital Twin Spark Ignition (DTS-i) model successfully launched on two products - 150 DTS-i and 180 DTS-i. The main purpose of these models is to improve the characteristics of four-stroke cycle engines in a wide area of usage and to comply with the 2005 emission standard in India.

Anyways, the implementation of ideas embedded in the concept of a digital twin, requires the extensive use of numerical methods of computer modeling and CFD-modeling (Computational Fluid Dynamics) [4]. This method provides the possibility of obtaining a visual representation of the nature of the processes in the object under study and the possibility of investigating its various designs without creating expensive experimental installations. In 2017, PTC and ANSYS introduced a solution that will quickly add the results of ANSYS engineering calculations to applications created on the PTC platform for ThingWorx's Industrial Internet of Things (IIoT) .The development of a connector between these two technical engineering platforms will allow customers to transfer and use the results of calculations when operating physical products in the real world of things. Also, it will allow them to match intelligent finite element models with physical products throughout the life cycle and manipulate the results in the real world. In practice, the software product ANSYS has been frequently used for the development of digital twins. For example, it can be used to create a model of an actually existing pump. That is what Chris MacDonald with his colleagues from ANSYS and PTC exactly did [5] .They believe: "Companies can use digital twins to detect and isolate faults, optimize asset operation, and generate insights to improve the next generation of the product. In 2012, NASA published the work "The digital twin paradigm for the future NASA and US Air Force vehicles" [6], which assessed : "Future generations of NASA and US Air Force vehicles will require lighter mass while being subjected to greater loads and more extreme days. Current approaches for certification will likely be unable to address these extreme requirements. To address the shortcomings of conventional approaches, a fundamental paradigm shift is needed. This paradigm shift, the Digital Twin ..." As mentioned

* Corresponding author: zheleznikov_is@icloud.com

https://doi.org/10.1051/matecconf/201817201009

above, the potential for application field of digital twins is quite wide. The digital twin finds its widest application in different technical systems. However, recently, digital twins appear in bioengineering and biomechanical systems [7, 8]. In this article, some methods of numerical simulation of the human coronary vessel system for creating its digital twin are considered. For effective application of CFD-modeling one can use number of commercial programs, such as ANSYS, FlowVision, ComsolMultiphysics, etc., and open source programs such as OpenFOAM, Salome, Code Saturn, and others.

How can a digital twin of human vessels be useful? Taking into account the pulsation nature of the velocity inside the vessels, non-Newtonian properties of the blood, and the presence of turbulent flow, it becomes possible to predict the behavior of the object of investigation with some accuracy. Such data as the flow in the areas of narrowing, expansion, and obstacles inside the vessel, the dynamics of the zones of shear stress, etc. is extremely valuable in the study of the development of diseases of the human cardiovascular system. Damage to endothelial cells, given as a wall material, is caused, in particular by turbulent blood flow in bifurcations (Virchow's triad - can be the cause of thrombosis, an extremely dangerous disease). The identification of the turbulent flow core allows observing a direct correlation with the wear of valve flaps that prevent the reverse flow of blood (phlebeurysm, an extremely common disease). The data required for the analysis are an array of hemodynamic and mechanical characteristics. Obtaining this data is the purpose of the work presented. Whether it is necessary to develop a realistic virtual model of biological objects to reflect the real and digital worlds, overcoming the gap between design and production (for example, production of stent for removal of thrombus) or not, will depend on comprehensive research and improvement of various methods of numerical modeling [9].

2 Methods

One of the most important tasks in the development of a numerical model of the heart-vessel system is the maximally accurate adherence to the physical nature of the vessel. The validation criterion of a numerical model in relation to a physical process or object is the higher the more perfect the specific algorithm and the methodology for performing calculations. This was demonstrated in a study by Xinrong and co-authors, "Influence of the boundary conditions on aorta blood flow simulations" [10], as well as in the study "Oscillatory flow in human arteries", by Giannoglou and his colleagues [11]. In their works, the term "Computational Cardiovascular Dynamics" is used to show the possibility of obtaining high dependability of mathematical methods. This term combines Bioengineering and Computational Fluid Dynamics, as part of Computational Mechanics, as well as phenomena of hemodynamics and the theory of mechanics for the numerical analysis of cardiovascular systems. Thus, in order to apply the digital twin to diagnose and treat cardiovascular diseases, it is required to use a set of specific techniques from the cross disciplines. In the present paper, for these purposes, the various computational domains and numerical setups were used.

The objects of research are part of the system of coronary heart vessels, a two-dimensional vessel with fixed leaves in two positions, and with the presence of obstacle on the way of the blood flow.

For each computational domain are marked following elements: the blood flow inlet and outlet, as well as vessel walls. Creating of geometry is an extremely important stage of the work, which determines the level of correspondence of a particular algorithm with the actual physical process. The geometry is created according to the peculiarities of the human vessel system. Increasing the accuracy of the solution is accomplished by adapting the mesh in the Meshing module of ANSYS. For numerical simulation, a constant increase in the number of elementary cells is often an unjustified solution. Saving the processing power while keeping the quality of the solution is achieved by adapting the mesh in the context of the simulated process. For example, for the region of the computational domain in which there is a transition of energy from the ordered state (the absence of vortices) to the disordered, and then to the heat (dissipation near the solid wall), the Inflation function is used. The Inflation function allows scaling the grid near the wall in a logarithmic order. Such a grid structure allows turbulence models to perform calculations in the boundary layer. The presence or absence of vortices, their dimensions, the degree of dissipation, and the degree of influence of the geometry of the vessel on blood distribution are controlled by the solver settings and the grid structure.

The general view of the geometry and the computational grid is shown in Fig. 1

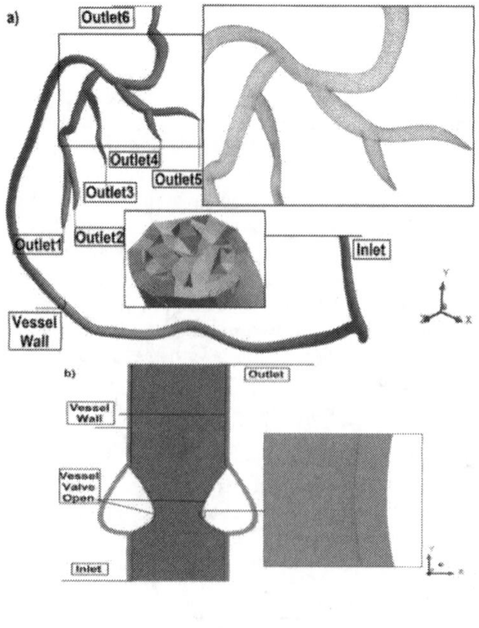

https://doi.org/10.1051/matecconf/201817201009

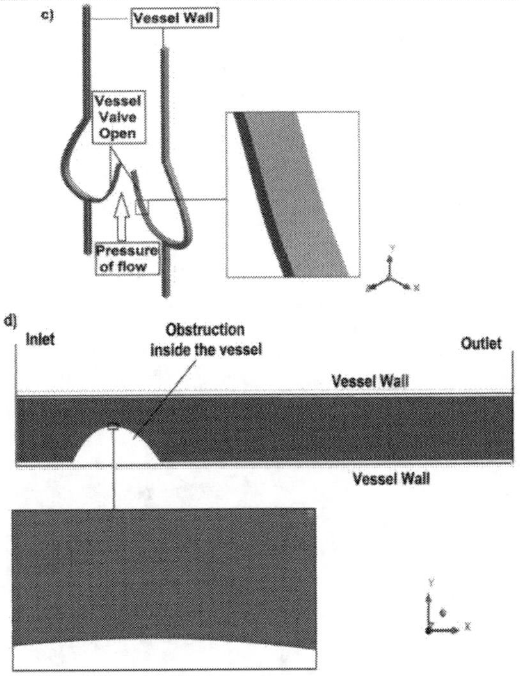

Fig. 1. a) coronary heart vessels, b) a two-dimensional model of a vessel with fixed valve position (preventing back flow of blood) position 1 1, c) a two-dimensional model of a vessel with fixed valve position 2, d) a two-dimensional model of a vessel with an obstruction in the blood flow (for example , thrombus). The total number of grid elements: a) 510399, b) 121567, c) 9242, d) 159188

In the Numerical Setup of the Fluent, the mathematical description of the flow process is expressed by a system of differential equations consisting of continuity equations, the law of conservation of momentum, and the law of conservation of energy, generally described in Ansys Fluent 14.0: Theory Guide [6].

The task is solved in a non-stationary form (Time-Transient) in two-dimensional and three-dimensional computational domains. At the Inlet area, the main calculation parameter for the blood flow (Solver Type - pressure-based) is defined. The velocity has a pulsating character [13]. Fig. 2 shows the electrocardiogram in the normal conditions (b) and the velocity pulsation diagram given in model (a). In the model there are no low amplitudes P-wave, T-wave. The main amplitude of the complex QRS-complex is transmitted with a period of 0.5 sec. Velocity Formulation: Absolute - the flow is mostly not subjected to rotation; vortices, if present, are not global, but local. Equation of energy is connected - Energy Equation On.

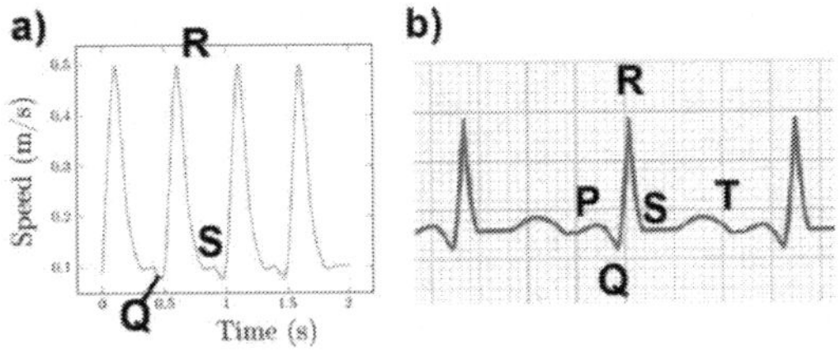

Fig. 2. a) Graph of the velocity pulsations given in the model, b) The electrocardiogram in normal conditions.

Viscosity. Calculations of blood viscosity are conducted according to Hemorheology. Hemorheology studies the properties of the blood flow and its elements (plasma and cells) [14]. Laminar flow prevails in the vessels. The Reynolds number takes on a value in the region of 50 [15]. With this type of flow, the blood moves with cylindrical layers whose axis coincides with the axis of the vessel. In this case, the bloodstream is quasi-static. Therefore the balance between the inflow of energy from the mean motion and the transfer of energy to heat is maintained, the profile is uniform. However, with a decrease in blood viscosity, constrictions, dilations, bends, obstructions in the flowline, increased blood pressure, and when the inner surface of the vessel becomes coarse and rough, the laminar flow becomes turbulent [16-17]. Turbulent flow is characterized by the presence of vortices. These vortices significantly increase the internal friction of the blood. The volume velocity of the blood flow is no longer proportional to the pressure gradient, but to the square root of the last mentioned. To increase the volume flow velocity, for example, twice, it is necessary to increase the pressure gradient by about 4 times. This means that with a turbulent type of blood flow, the load on the heart increases significantly [18].

3 Results and discussion

In the calculations, both laminar and turbulent components are included when consider turbulence model equations. One of these components is the turbulent viscosity. It is the component in the equation, which ceases to be constant. Throughout its flow, its variation does not depend on the position in space, except for the wall layer [19-20]. For numerical simulation of processes in a turbulent core, wall layer and turbulence, turbulence models are used. The main ones are related to three types: DNS, RANS, LES.

In the presented study, there is insufficient density of the computational mesh for the use of DNS and LES. In addition, there is no effect of energy exchange between vortices and their pulsations. Of all the models of the RANS type, the universal group is k. In practice, k-ε models are the best for the widest range of Reynolds number.

However, case records of k-ε models have shown that in the calculation of boundary layers with a positive (negative) pressure gradient, all models of this type tend to overestimate the generation of the kinetic energy of turbulence. This fact leads to a fundamentally incorrect description of such flows [21-22]. This disadvantage is quite serious since it makes it practically impossible to correctly calculate the flows with a detachment from a smooth surface. For this reason, the calculation was carried out for two models: k-ε Realizable, and k-ω SST model that combining the main advantages of the groups k-ε Standart and k-ω Standart.

The Realizable model is based on k and ε, where k is responsible for the pulsation of the kinetic energy, and ε is responsible for how fast this energy dissipates into heat. Compared to Standard, it contains an alternative formulation for turbulent viscosity and modified transport equation for the dissipation rate obtained from the exact equation for the transfer of the mean-square vorticity fluctuation. The term "Realizable" means that the model satisfies certain mathematical limitations of the Reynolds stress, consistent with the physics of turbulent flows, in contrast to the k-ε Standard and RNG.

Model SST (Shear Stress Transport). This is a model of the group k-ω. Unlike Standard, it is stable in pure form and, unlike k-ε, underestimates the dissipation at the walls. This model copes well with the calculation of near-wall parameters. The determination of turbulent viscosity is modeled taking into account the transfer of turbulent shear stress [23]. The SST model is considered more accurate and reliable for a wider class of flows (for example, negative pressure gradient flows, aerodynamic profiles, transonic shock waves) than the Standard. The group of models k (both ε and ω) incorrectly reflect heat and mass transfer due to the fact that they are isotropic (properties do not change in the core). Anisotropic models are DNS and LES. The reasons for their incompatibility with the chosen methodology are mentioned above.

The functional of the used models used includes the connection of a variety of improved near-wall functions, which is useful in the numerical modeling of turbulent processes [24]. These models are a two-layer model with the so-called expanded functions of the wall. Their work is possible if the wall mesh is thin enough to be able to solve a viscous sublayer (typically with the first near-wall node placed at $y+\approx 1$) [25]. In the chosen technique, this is meaningless, because the Near-Wall Treatment parameter is set to the Standard Wall Function.

As a fluid material, blood is used, the walls are epidermis. For the epidermis, the following parameters are entered: Density – 1300 kg/m^3, Specific Heat – 3650 J/(kg·K), Thermal Conductivity – 0.2 W/(m·K). For blood: Density – 1060 kg/m^3, Specific Heat – 3750 J/(kg·K), Thermal Conductivity – 0.478 W/(m·K) [27]. Blood considered as a non-Newtonian fluid. Hence, the calculation of its Molecular Viscosity cannot be performed in the classical way [28]. There are many models of non-Newtonian fluids, but as in the case of models of turbulence, the universal one has not been found. These are some of them: Carreau (1991), Power Law (1991), Non-Newtonian Power Law (1992), Casson (1993) [29]. They are based on the calculation of the viscosity through the Shear Rate, the temperature (in the study the temperature is constant), and the constants k and n. To specify non-Newtonian blood parameters for both flow modes, one need to use a special command: /define/models/viscous/turbulence-expert > turb non-newtonian [30]. Then set the constants for viscosity calculation: Consistency Index, k – 0.01669 (kg·s^{n-2})/m, Power-Law Index, n – 0.52815, Viscosity Limits – 2.8894e^{05}…100 kg/(m·s) [31, 32]. In this case, the viscosity is calculated independently of the temperature, according to the Non-Newtonian Power Law (1992). Blood pressure is 13332 Pa. Number of iterations of hybrid initialization is set to 10. The size of the time step is 0.01 s, the number of steps is 50, the number of iterations per time step is 40.

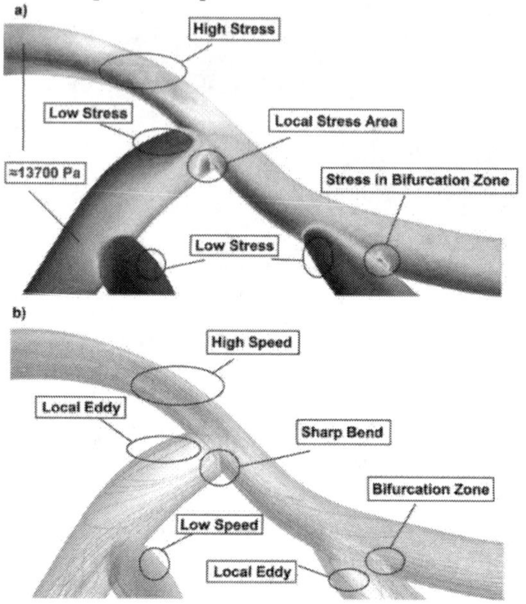

Fig. 3. Wall Shear Stress contours, b) Velocity Streamlines.

The area of increased mechanical stress corresponds to the area where the blood flow velocity is maximal. On the entire surface of the vessel at a time of 0.1 s the blood pressure assumes a value of about 13,700 Pa. In places where local vortices are formed and the minimum blood velocity, the stress is low. Strong tension is observed within a bend of the vessel, as well as in the bifurcation zone of the vessel. At a time of 0.1 s, the Reynolds number varies throughout the coronary vessel system from 14 to 79.

The Strain Rate as a function of velocity pulsation is shown (for times 0 s, 0.1 s, 0.2 s, 0.3 s, 0.4 s) in Fig. 5. The maximum strain rate coincides with the peak of the velocity-time diagram. The presence of vortices in the regions of expanding, bifurcations and after bending is demonstrated. The models of turbulence k-ε Realizable and k-ω SST are equally good at modeling vortices. Fig. 6 shows the velocity profile in the cross-section of the vessel for the models k-ε Realizable and k-ω SST. It can be seen how the SST mode, unlike the Realizable model, slightly underestimates the velocity decrease near the wall (the transition is smoother). The Realizable model increases the velocity of the flow inside the flow core in comparison with the SST model. For both models, there is a general overestimation of the blood flow velocity. This phenomenon is caused by their isotropy. Both models neglect the flow properties in the core, applying averaging, instead of calculating.

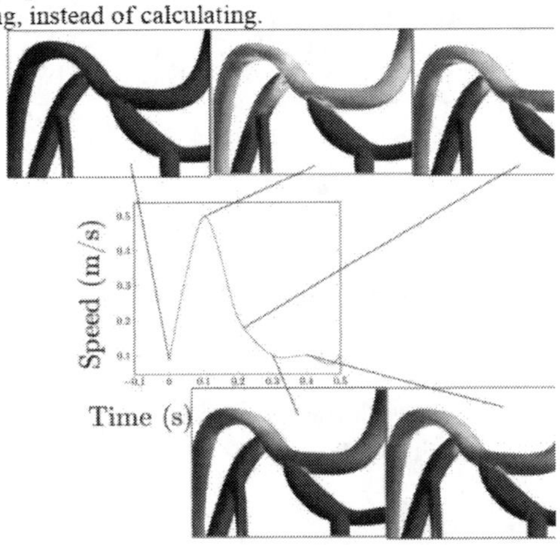

Fig. 4. Dynamics of the Strain Rate relative to the period of pulsating blood flow (for times 0 s, 0.1 s, 0.2 s, 0.3 s, 0.4 s).

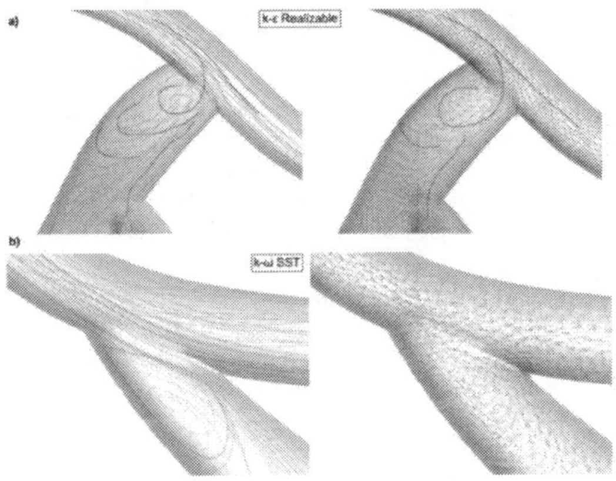

Fig. 5. a) Flow lines and vortex vectors for k-ε Realizable, b) Flow lines and vortex vectors for k-ω SST.

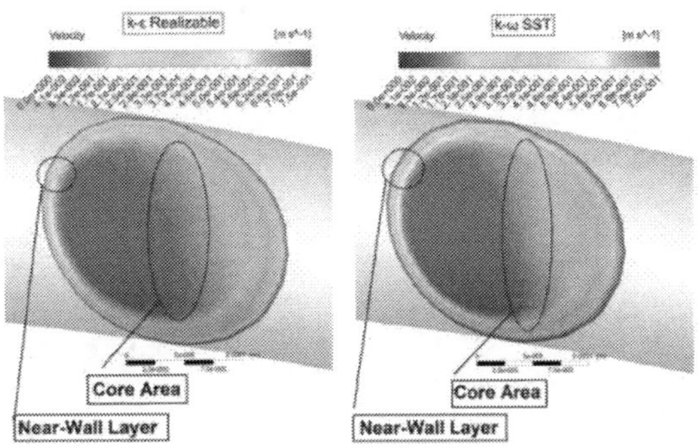

Fig. 6. The velocity profile in the cross-section of the vessel for the models k-ε Realizable and k-ω SST.

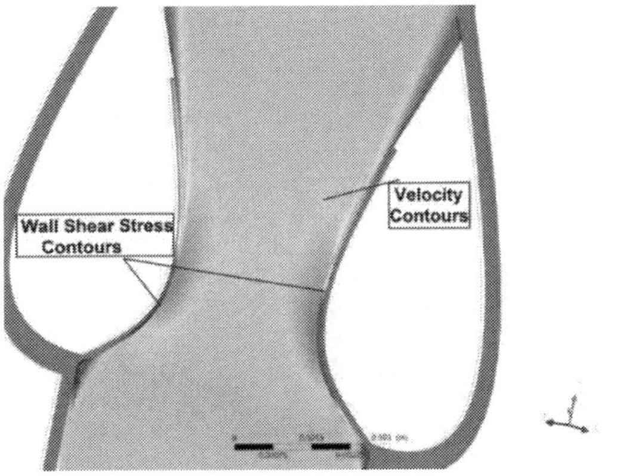

Fig. 7. Velocity Contours inside the vessel and shear stresses of the valve walls.

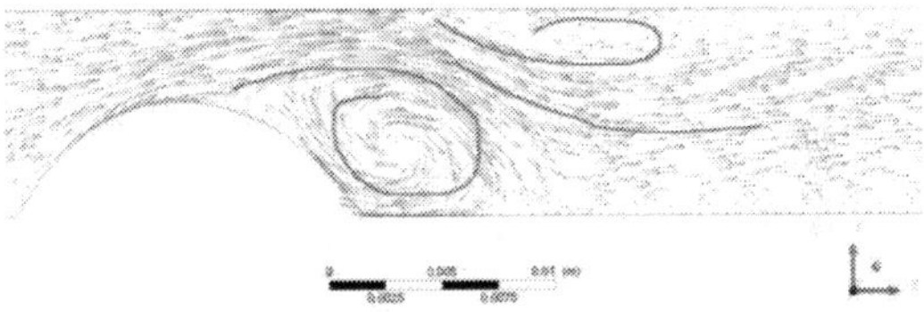

Fig. 8. Formation of the vortex flow inside the vessel.

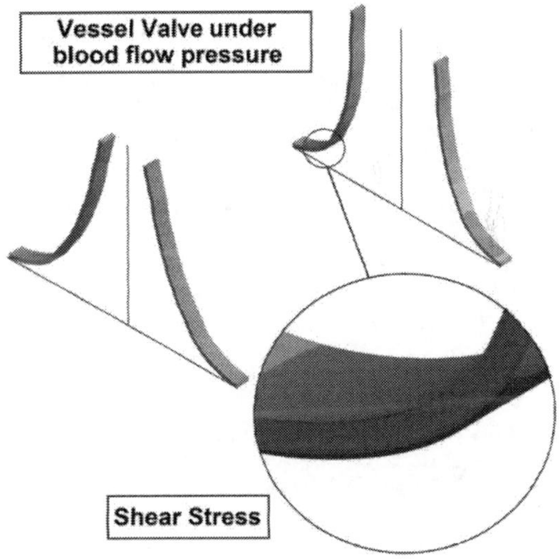

Fig. 9. An opening of the valves of the vessel under pressure of blood flow and the stress inside the valve.

Fig. 7 shows Velocity Contours and shear stress plotted on the model of a two-dimensional vessel with fixed flaps, which prevent the reverse flow of blood. With a high-velocity gradient, the flap walls are subject to strong stress. Fig. 8 shows the formation of a vortex flow inside the vessel. Fig. 9 shows the distribution of shear stress on the flap of the vessel under the blood flow pressure.

4 Conclusions

The development of a realistic virtual model of the coronary vessel system is based on numerical modeling. During computational modeling, it is important to consider the model not only as an object but also as a process. The methods of mapping the physical process in the digital model are considered above. Turbulent blood flow in bifurcations refers to the causes of damage to the walls of the vessel. As a consequence, thrombus appears. The thrombus reduces the pore of the vessel, acts as an obstacle in the path of blood flow. This phenomenon also leads to the formation of a vortex flow inside the vessel. With a vortex type of blood flow, the load on the heart increases significantly, the blood pressure increases. Under the pressure of blood flow, the walls of the valves, which do not allow blood to flow in the opposite direction, wear down and eventually weaken. If a certain amount of blood is able to flow in the opposite direction, the veins undergo pathological changes. Stresses that cause physical pathologies are shown on a digital model. According to the mechanical characteristics of the coronary vessel model, a visual representation of the progression of heart disease has been obtained. Production, prevention, or timely treatment of diseases of the human cardiovascular system depends on the reliability of a digital twin (for example, production of a stent for the removal of a blood clot).

This paper is financially supported by the Development Program of the Samara State Technical University up to 2020.

References

1. S. Boschert, R. Rosen *Mechatronic Futures: Challenges and Solutions for Mechatronic Systems and their Designers* (Springer International Publishing 2016), 59-74
2. E. J. Tuegel, A. R. Ingraffea, T. G. Eason, S. M. Spottswood, International Journal of Aerospace Engineering, **2011** (2011)
3. A. Ramtilak, A. Joseph, G. Sivakumar, and S. Bhat, SAE Technical Paper, **№. 2005-26-008,** 2005
4. D. Pashchenko, International Journal of Hydrogen Energy. **42** 14926-14935 (2017)
5. URL: https://www.ansys.com/about-ansys/advantage-magazine/volume-xi-issue-1-2017/creating-a-digital-twin-for-a-pump. (date of circulation on 01.02.18)
6. E. Glaessgen, D. Stargel, 53rd AIAA/ASME/ASCE/AHS/ASC Structures, Structural Dynamics and Materials Conference 20th AIAA/ASME/AHS Adaptive Structures Conference 14th AIAA 1818 (2012)
7. I. A. Uchida, R. D. Rowe, American journal of human genetics, **9,** 133 (1957)
8. D. Tomar, S. Agarwal, International Journal of Bio-Science and Bio-Technology, **6**, 69-82 (2014)
9. B. Schleich, CIRP Annals, **66,** 141-144 (2017
10. X. Cao, J. Wang, R. Wang, X. Zhang, J. Tang Journal of Tsinghua University (Science and Technology) **54,** 700-705 (2015)
11. G. Giannoglou, J. V. Soulis, D. K. Fytanidis, Information Technology and Applications in Biomedicine (ITAB), **10**, 1-4 (2010)
12. Ansys Fluent 14.0: Theory Guide (Date of circulation 29.08.17 г.)
13. A. S. Popel, P. C. Johnson, Annu. Rev. Fluid Mech, **37,** 43-69 (2005)
14. O. K. Baskurt, M. R. Hardeman, M. W. Rampling, *Handbook of hemorheology and hemodynamics* (IOS press, 2007) **69**
15. J. A. Moore, D. A. Steinman, C. R. Ethier, Journal of Biomechanics, **31**, 179-184 (1997)
16. W. Koenig, E. Ernst, Atherosclerosis **94**, 93-107 (1992)
17. E. Ernst, Atherosclerosis **59**, 263-269 (1986)
18. S. Moore Journal of biomechanics **39,** 1454-1463 (2006)
19. L. Formaggia, D. Lamponi, A. Quarteroni, Journal of engineering mathematics **47**, 251-276 (2003)
20. S. J. Sherwin International Journal for Numerical Methods in Fluids, **43**, 673-700 (2003)
21. S. Chien, Biorheology, **8**, 35-57 (1971)
22. P. L. Lip, Ophthalmology, **108**, 705-710 (2001)
23. L. Morris Journal of Biomechanical Engineering, **127**, 767-775 (2005)
24. C. A. Taylor, T. J. R. Hughes, C. K. Zarins, Computer methods in applied mechanics and engineering. 158 155-196 (1998)
25. I. E. Vignon-Clementel, Computer methods in applied mechanics and engineering, **195,** 3776-3796 (2006)
26. O. K. Baskurt, O. Yalcin, H. J. Meiselman, Clinical hemorheology and microcirculation, **30**, 169-178 (2004).
27. Y. I. Cho, K. R. Kensey, Biorheology, **28**, 241-262 (1991)
28. N. Fraysse, G. M. Homsy, Physics of fluids, **6**, 1491-1504 (1994)
29. A. Lindner, Journal of Fluid Mechanics, **469**, 237-256 (2002)
30. D. Pashchenko, MATEC Web of Conferences. – EDP Sciences, **145** (2018)
31. D. S. Sankar, K. A. Hemalatha, Applied mathematical modeling, **31**, 1847-186 (2007)
32. D. Paschenko, International Journal of Hydrogen Energy. **42(49)** (2017)

Study of Flow Induced Noise in Vertical Inline Pump Using Lighthill Analogy

Stephen Christopher[1,2*], *Shouqi* Yuan[2] , *Ji* Pei[2], *G Xing* Cheng[2], and *Wang* Yiyun[2]

[1] Department of Mechanical Engineering, Saveetha School of Engineering, SIMATS, Chennai 602 105, Tamil Nadu, India
[2]National Research Center of Pumps, Jiangsu University, Zhenjiang 212 013, Jiangsu Province, China

Abstract. The pump user demands for specification along with noise and vibration magnitude for extreme operating conditions apart from best efficiency flow. In order to predict the noise level in pumps, the entire unit of pump has to be considered for acoustic evaluation. Hence, the full three dimensional model of vertical inline pump which consists of inlet pipe, impeller, and volute casing is considered for detailing the flow induced noise using Lighthill analogy. The performance characteristics of pump are well matched between experimental and computational results and the overall sound power level at best efficiency flow using computation was compared with empirical relations and found to be agreed well. The streamline patterns along the inlet pipe shows the effects of recirculation and return flow at partload condition and confirm that the utilization of flow passage is not well streamlined even for nominal flow rate. Further, the detailed study of inlet pipe cross sections reveals the nature of different noise sources such as dipole and quadrupole sound levels along with Proudman source power level. The intensity of turbulence and vortices in flow passage is predicted well using quadrupole source and Proudman sound power levels.

1 Introduction

Many researchers have identified the various sources of noise inside the pump. The occurrence of noise is mainly due to the interaction between two domains such as impeller and volute, in addition, noises are due to the flow nature inside the pump. The details of flow physics and their sources of noise are discussed by Gülich [1]. A centrifugal pump produces hydraulic noise corresponds to vane passing frequency and its harmonics and at shaft frequency with related harmonics [2]. Further it was confirmed that the main source of noise generation from pumps is from vane tongue interaction [3,4]. The detection of noise radiation from the two dimensional model of centrifugal pump was analyzed by Langthjem et al. [5] using Lighthill acoustic analogy and concludes that there was a good agreement between numerical and experimental readings. Jiang et al. [6] studied the feasible method to couple fluid structure interaction with flow induced noise in turbomachines using Large Eddy Simulation (LES) based CFD and FEM. The efficiency of noise radiation from a centrifugal pump for different blades shapes was determined by Huang [7] using numerical simulation hybrid with Ffowcs Williams-Hawkings (FW-H) acoustic model.

The three dimensional unsteady flow was simulated with $k-\varepsilon$ turbulence model and coupled with LMS Virtual Lab to estimate the sound field inside the pump and the results concluded that dipole source was the main cause for flow induced noise [8]. The noise generation by the blades as well as from the volute of centrifugal pump was captured using FW-H model and found to be well agreed with experimental findings [9].

The gap in literature is that, most of the researcher concentrated on impeller and volute but not on the suction pipe of centrifugal pump or the inlet pipe of vertical inline pump. Hence in this work, commercial CFD code called ANSYS CFX 17.0 is used for determination of flow induced noise in all components of vertical inline pump using Lighthill analogy. The relation between dipole, quadrupole sources of noise as well as the Proudman sound power using Lighthill analogy are analyzed for inlet pipe for various flow conditions.

2 Numerical Approach

2.1 Lighthill Analogy

The analogy for prediction of sound from the flow field was detailed by Lighthill [10] from the equations of continuity and momentum and hence deduced to wave equation. The flow noise was predicted successfully for implementation of acoustic analogy derived by Ffowcs Williams and Hawkings to all fluid machinery applications [11]. A generalized version of the Lighthill equation that represents the noise sources is given by Williams, et al. [12] as follows:

$$\underbrace{\frac{\partial^2 \rho'}{\partial t^2} - c_s{}^2 \frac{\partial^2 \rho'}{\partial x_i{}^2}}_{A} = \underbrace{\frac{\partial}{\partial t}\{\rho v_n \delta(f)\}}_{B} - \underbrace{\frac{\partial}{\partial x_i}\{\tau_{ij}\delta(f)\}}_{C} + \underbrace{\frac{\partial^2 T_{ij}}{\partial x_i \partial x_j}}_{D} \quad (1)$$

Eq. (1) represents the wave equation (A) with three sources of noise term in flow, they are monopole (B), dipole (C), and quadrupole (D). The mass flow

* Corresponding author: christo_spc@yahoo.com

fluctuation is related to monopole source and it is called as thickness noise due to non uniform flow, dipole is the pressure fluctuation in flow field (or loading noise). Both monopole and dipole are the surface source terms, whereas quadrupole is volume source term. Quadrupole is mainly because of turbulence or vortices or in other term fluctuating shear stress in the flow domain. But, for low Mach number applications, direct noise emission from quadrupole source is of no practical importance in water [13]. Conversely, Hanson & Fink [14] proved that the substantial agreement between test data was obtained only by including quadrupole source term. Lighthill stress tensor (T_{ij}) is defined as the summation of momentum flux tensor $(\rho v_i v_j)$ and fluid stress tensor (τ_{ij}) and subtracted by stress in the uniform acoustic medium at rest $(c_s^2 \rho' \delta_{ij})$.

Proudman [15] derived a method to extract the noise radiation from isotropic turbulence at low Mach number with high Reynolds number from Lighthill theory for complete flow field. The total acoustic power (P_A) which is the function of time averaged kinetic energy and its dissipation rate per unit volume is given in Eq. (2).

$$P_A = \alpha \rho \varepsilon \left(\frac{\sqrt{2k}}{c_s} \right)^5 \tag{2}$$

Sarkar and Kussaini [16] calibrated the constant (α) to determine sound power level which is also known as broadband noise source in the flow domain. Eq. (2) is rewritten with reference to acoustic reference power in dB to determine Proudman sound power level (Psp). The fluctuation of near field pressure and velocity plays a vital role in determination of far field noise levels. Hence the present work is discussed based on near field parameters distribution throughout the flow domains of vertical inline pump.

2.2 Flow Domain and Discretization

The test pump was designed for head of 20 m and for flow rate of 0.0014 m^3/s at 2910 rpm. The resultant impeller is of double curvature radial flow with six vanes. The diameters of suction and delivery pipes are same, which is about 80 mm with the length of 10 times the pipe diameter to have uniform flow at the entry of inlet pipe and to predict the exact flow nature at the exit of diffuser. Figure 1 shows the mesh of low specific speed pump ($n_s = 36$) of inlet pipe, impeller, and volute casing.

Using ICEM CFD 17.0, the flow models was discretized with structural mesh by proper consideration of y^+ near the wall and at critical surfaces of vertical inline pump which was between 5 to 10 and maximum of 80 was adapted for other surfaces of domains. Lakshminarayana [17] stated that the solution from flow equations are strongly depends on the parameters during setup conditions such as grid size, turbulence model, time step, boundary conditions and so on. Based on the grid sensitivity studies, the following number of nodes is used for simulation they are 0.39 millions, 0.42 millions, and 0.5 millions for inlet pipe, impeller, and volute casing with radial discharge respectively and 1.56 millions is for suction and delivery pipes. The head coefficient and efficiency discrepancy between the maximum number of nodes and the optimized nodes are 0.4% and 0.3% respectively.

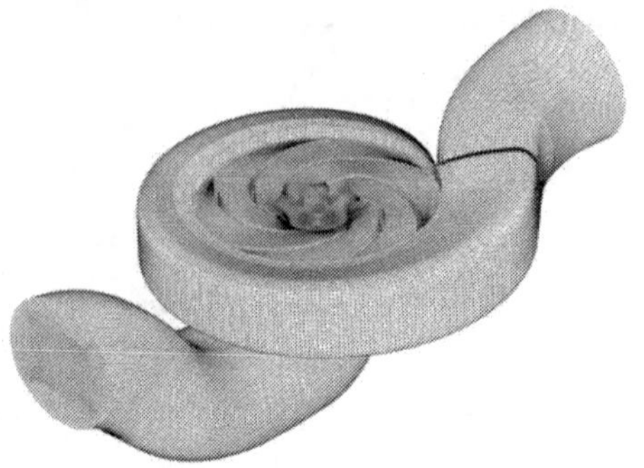

Fig. 1. Mesh of vertical inline pump

2.3 Setup Conditions for Simulations

The steady state simulations were performed from 50% to 150% of nominal flow rate (Qn) in the step of 10%, whereas for transient case, 50%, 80%, 100%, 120%, and 150% of Qn were solved. The total pressure (stable) and mass flow rate are given at the inlet of suction and outlet of delivery pipes respectively, which was taken from the real working condition of pump. The results from steady state were used as initial guess for transient analysis with time step of 1.72 × 10-4 s for the impeller rotation of 3 deg. Around 1.5 times the impeller rotation was simulated to extract the results for acoustic studies with water as working fluid at the reference pressure of 101.325 kPa and at temperature 25°C. The interface between stationary and rotating domains was set to 'transient rotor stator' for transient case while for steady case it was 'frozen rotor'. Shear stress transport (SST) model was considered for identification of turbulence in flow domain especially in near wall region and for determination of flow separation [18]. The computed results were converged, when the residual value is lower than or equal to 1e-4.

2.4 Choice of Planes and Points

Figure 2 shows the choice of planes (from sections 1-1 to 7-7 are denoted as Cs 1 and Cs7 correspondingly) across the inlet pipe along with the mean streamline starts from entry $(x/l = 0)$ to the exit $(x/l = 1)$ of inlet pipe. The choice of monitoring points and locations will be helpful to compare with the experimental findings in later stage.

https://doi.org/10.1051/matecconf/201817201010

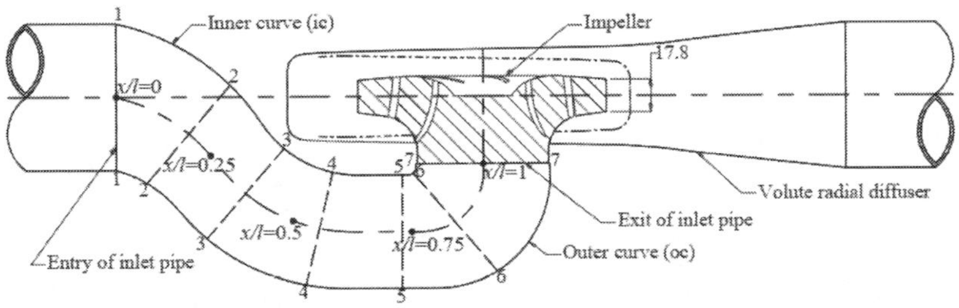

Fig.2. Sketch of vertical inline pump

3. Experimental Test Rig

The open test rig was used to conduct the performance characteristics of vertical inline pump from shut off to full open conditions of throttle valve. The pressure at the suction and delivery sides of the pump was measured with the help of pressure transmitter in the range of 0-1.6 bar (absolute) and 0-4 bar respectively. The entire tests were conducted at constant speed and it was controlled using variable frequency drive. The input power to the motor was measured through Wattmeter and the flow rate through pump was obtained with the help of electromagnetic flow meter. The uncertainty of measurement for the total head, flow rate, input power, and efficiency are within the allowable limits. The experiments were repeated to get the repetition of characteristics.

4. Results and Discussion

4.1 Performance characteristics of pump

The hydraulic characteristics using experimental (exp) readings along with the computational (cfd) results are plotted in Fig. 3 in non dimensional form. The total head, flow rate, and input power are given in terms of head (ψ), flow (ϕ), and power (ν) coefficients respectively. The best efficiency (bep) of the pump during testing is found to be 72.5%, but the hydraulic efficiency from the cfd is about 82% for the same flow with the difference in head coefficient between the simulation and the experiment is 3.8%. The agreement between head coefficient is found be well near the best efficiency point as well near low flow rate zone, whereas the deviation is found at higher flow rates. The reason for deviation is due to hydraulic head losses inside the impeller, in addition to shock losses at the entry of impeller. Based on head and power coefficients, the efficiency curves are drawn and found to have more deviation between experiment and computational values. Experimental efficiency is nothing but overall efficiency of the pump, which includes leakage losses, friction losses inside the pump, mechanical losses, and so on.

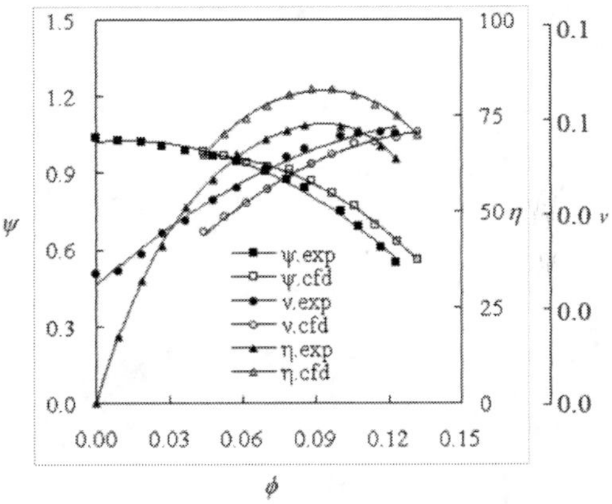

Fig.3. Performance characteristics of pump

4.2 Sound power level

Another benchmark for the computational results is with respect to sound power level (L_{WA}) obtained from the pump. Fritz et al. [19] have given a relation to found out the total A-weighted sound power level from centrifugal pump based on the input power of motor, which includes the pipelines. Similarly Schmitt and Klein [20], used to calculate the same sound power for pump (alone) and another relation to calculate the magnitude without the influence of motor and gear box [21]. The sound power level from computational results is compared with empirical formula [19-21], which are given in Table 1 and Table 2.

Table 1. Sound power level from empirical relation

Input power (kW)	Reference	L_{WA} (dB)
3.45 (at BEP)	[27]	76.6
	[28]	66.5
	[29]	77.8
3.58 (at full open)	[27]	76.9
	[28]	66.6
	[29]	78.0

* Corresponding author: christo_spc@yahoo.com

Table 2. Sound power level from computation

Harmonics	frequency (Hz)	L_{WA} (dB)
1st (fs)	48.5	71.8
6th (vpf)	291.0	53.1

The magnitudes are calculated based on best efficiency point (BEP) and maximum input power for vertical inline pump at design speed of non cavitating condition with reference to 1 kW power. From computation, the sound power level is calculated and found to be function of observer radius (1 m), acoustic pressure, acoustic impedance, and the acoustic reference pressure in logarithmic form. Table 2 shows the magnitude of sound power level at shaft frequency (fs) and vane passing frequency (vpf). The variation in magnitude of sound power with respect to empirical equations is about 6% - 8% higher with reference to [27,29] and less than 8% for [28] with respect to first harmonics (i.e., shaft frequency). This confirms that the prediction of sound power level from numerical tool is well agreeable and gives the scope for further investigation of acoustics in flow domains.

4.3 Velocity coefficient distribution

Figure 4 shows the streamline of velocity coefficient (Cv) for inlet pipe and impeller (meridional view) at 0.5Qn, Qn, and 1.5Qn. Around 30 points of equally spaced 3D streamlines are used to represent the flow stream along the inlet pipe with the gray background for clear visibility and represented in 2D form. It is inferred that the flow disturbance is more for the case of 0.5Qn than compared with Qn and 1.5Qn. It is due to the recirculation and return flow effects from the impeller and similar phenomenon were visualized by Posa et al. [22] and found to extend to the upstream of impeller. Due to this effects, noise due to turbulence increases, when flow is less than Qn. In addition, the flow passage of inlet pipe is not well designed even for the case of nominal flow rate; hence there is scope for optimizing the inlet pipe for this pump. All the streamlines and contours are given for the same time step (30.9 ms) for comparison.

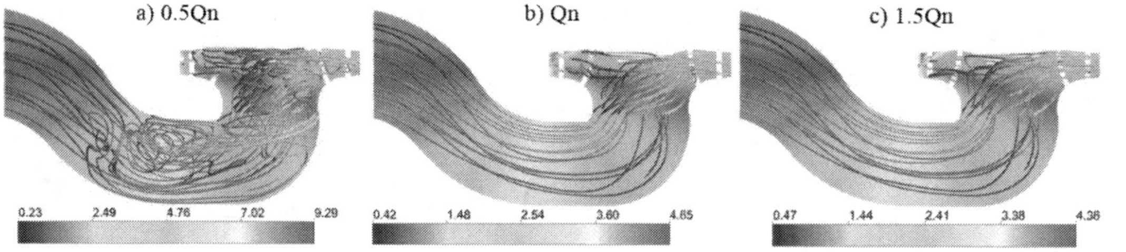

Fig. 4. Streamlines of velocity coefficient

More details about the sound source level across each cross sections of inlet pipe were discussed for 0.5Qn in the following section 4.4. In general, the flow through bends and elbow are very complex because the pressure gradients changes drastically from one to another locations, which results in increase of turbulence intensity and hence leads to flow induced noise and structural vibration. Hambric et al. [23] discovered the similar kind of phenomenon like secondary flow fields because of strong pressure gradients at the downstream of elbow. This confirms the possibility of noise generation from the turbulence in inlet pipe of vertical inline pump.

4.4 Source of noise across cross sections at 0.5Qn

The dipole (DSL) and quadrupole (QSL) sources in Eq. (1) is rewritten in normalized form as the ratio of corresponding source terms with respect to acoustic reference pressure for water (1μPa) and were calculated on the near field. As stated in section 2.1, the dipole source is directly proportional to fluctuation in pressure; hence the contours of dipole are similar to pressure contours. Figure 5 and 6, shows the contours of DSL, QSL, and Psp for the cross sections across the inlet pipe at 0.5Qn. From DSL contours, it shows that the zone of high and low pressure varies from one to other sections drastically. This creates a way for recirculation of flow inside the inlet pipe apart from the reduction of flow rate in pump. The monopole source is neglected, since its significance is low when compared with other two sources in stationary flow domain. The top of each section represents the point on inner curve (ic) for Fig. 5 & 6. The zone of low pressure is evident at the bottom of Cs2 and moves in clockwise direction until Cs5 (see Fig. 6), whereas no distinct low pressure is found in Cs6.

From Cs3 to Cs7, the uniformity of pressure pattern is not observed; it is due to reduced flow rate. A QSL contour is nothing but the indication of turbulence or vortices in the flow field. When compared with DSL, the magnitude of QSL is small, but it should not be neglected in order to have an optimized design of any flow passage. It is found that there is no direct correlation between DSL and QSL for all cross sections of inlet pipe. Each section is having its own turbulence intensity, but it is not sufficient enough to create noise. A only few proportions from fluctuating turbulence power are utilized to generate sound power [24].

 https://doi.org/10.1051/matecconf/201817201010

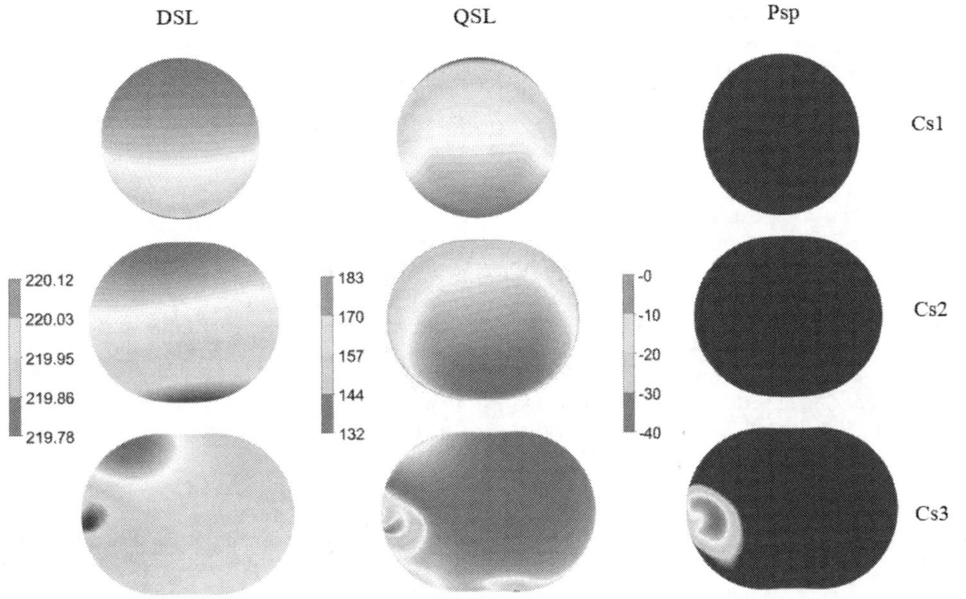

Fig. 5. Contours of DSL, QSL, and Psp at 0.5Qn for Cs1, Cs2, and Cs3

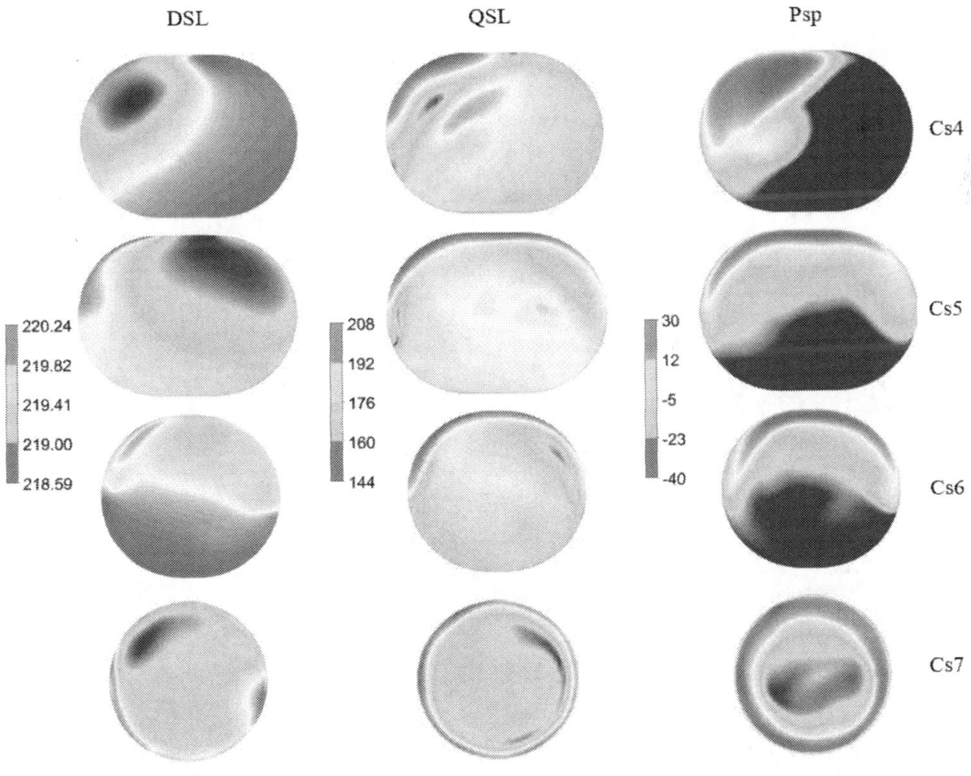

Fig. 6. Contours of DSL, QSL, and Psp at 0.5Qn from Cs4 to Cs7

Proudman sound power level (Psp) for Cs1 and Cs2 are -40 dB, which means that the turbulence is very minimal (i.e., zero). From Cs3, the intensity level of vortices increases drastically, till it reaches the exit of inlet pipe (i.e., Cs7). Psp starts from Cs3 at left hand side of section and increases in Cs4 at top position of left hand side, whereas for Cs5 and Cs6, the entire top surfaces are filled with turbulence phenomena. This may be due to sudden curvature (or short radius) of inlet pipe and also it confirms the swirl flow due to the pressure gradients and operating condition of pump. In Cs7, the intensity of vortices is observed around the periphery of cross sections and minimal at the core, this ensure the helicity (product of vorticity and velocity) at that section [25]. Even for end suction centrifugal pump, Fu et al. [26] confirmed the presence of low frequency oscillation of inlet pressure near the impeller eye by means of fast response pressure transducers. Similar kind of cross sectional studies can be extended to other flow rates for more detailing of inlet pipe flow passage.

5. CONCLUSIONS

The flow induced noise in vertical inline pump is studied using Lighthill analogy for various flow rates and examined for inlet pipe domain. Experimental characteristics of pump are well matched with computational readings and slight variation is found at off-design conditions, it may be due to the losses in pump. The comparison between empirical relations suggested by various researchers for sound power level with the computed values shows that it agrees well with the magnitude at shaft frequency. From velocity streamlines along the inlet pipe it infers that the intensity of turbulence was more for lower flow rates and less for higher flow rates. But due to existing curvature in inlet pipe, the turbulence exists even for other flow rates. The same is confirmed from cross sections contours of quadrupole source and Proudman sound power levels. This study enables the pump user to optimize the shape of flow domain with respect to flow induced noise, but proper balance as to be adapted between hydraulic, structural and acoustic aspects.

Acknowledgment

This work was supported by the state key program of National Natural Science Funds of China (Grant No. 51239005).

Nomenclature

A_2	$=\pi d_2 b_2$ outlet area of impeller, m^2
b_2	outlet width of impeller, m
c_s	speed of sound in working fluid, m s^{-1}
Cv	$=V_{cs}/V_{emp}$ velocity coefficient, non-dimensional
d_2	outlet diameter of impeller, m
DSL	$=20Log\left(dipole\ source\ term/P_{Arefpr}\right)$ dipole sound level, dB re 1μPa
f	non-zero at fluid region, and zero at the moving boundary
g	acceleration due to gravity, m s^{-2}
H	total head, m
k	turbulent kinetic energy, m^2 s^{-2}
l	total length of mean streamline in inlet pipe, m
L_{WA}	sound power level, dB re 1 kW
n	rotational speed, rpm
n_s	$=n\sqrt{Q}/H^{3/4}$ metric specific speed
P_A	Proudman sound power, W m^{-3}
Psp	$=10Log\left(P_A/P_{Arefpo}\right)$ Proudman sound power, dB re 10^{-12} W m^{-3}
P_{Arefpo}	acoustic reference power, 10^{-12} W m^{-3}
P_{Arefpr}	acoustic reference pressure, Pa
Q	flow rate, m^3 s^{-1}
Q_n	nominal flow rate, m^3 s^{-1}
QSL	$=20Log\left(quadrupole\ source\ term/P_{Arefpr}\right)$ quadrupole sound level, dB re 1μPa
t	time, s
T_{ij}	momentum flux tensor, N m^{-2}
u_2	$=\pi d_2 n/60$ peripheral velocity at outlet of impeller, m s^{-1}
v_n	normal velocity of the surface, m s^{-1}

* Corresponding author: christo_spc@yahoo.com

V_{cs} velocity at cross section of inlet pipe, m s^{-1}
V_{enip} velocity at entry of inlet pipe corresponding to Q_n , m s^{-1}
x distance along mean streamline of inlet pipe, m

Greek symbols
α calibrated constant to determine sound power level
β_2 outlet vane angle, deg
$\delta(f)$ dirac delta function describes the moving surface geometry
ε turbulent dissipation rate, m^2 s^{-3}
η efficiency, %
ν $=2P_i/\rho A_2 u_2^3$ power coefficient, non-dimensional
ρ density of water, kg m^{-3}
ρ' density fluctuation with respect to reference condition, kg m^{-3}
τ_{ij} fluid stress tensor, N m^{-2}
ϕ $=Q/\pi d_2 b_2 u_2$ flow coefficient, non-dimensional
ψ $=2gH/u_2^2$ head coefficient, non-dimensional

Reference

1. J. F. Gulich, *Centrifugal pumps* (third ed., Springer, New York, 2014)
2. H. C. Simpson, R. Macaskill, T. A. Clark, Proc. of the Institution of Mech. Engineers, Conference Proc., **181,** 84-108 (1966)
3. R. Dong, S. Chu, J. Katz, J. Turbomachinery. **119,** 506-515 (1997)
4. M. S. Howe, *Acoustics of fluid-structure interaction* (Cambridge Univ. Press, Cambridge, 1998)
5. M. A Langthjem, N. Olhoff, J. Fluids and Struc. **19,** 369–386 (2004)
6. Y. Y. Jiang, S. Yoshimura, R. Imai, H. Katsura, T. Yoshida, C. Kato, J. Fluids and Struc. **23**, 531–544 (2007)
7. J. Huang, S. Geng, R. Wu, K. Liu, C. Nie, H. Zhang, Acta Acoust. **35**, 113–118 (2010)
8. X. Yong, W. Chunlin, Y. Xiaoyong, Y. Jian, Z. Cheng, Chinese J. Drain. and Irri. Mach. Eng. **34**, 483-489 (2016)
9. M. Gao, P. Dong, S. Lei, A. Turan, Energies. **221**, 1-11 (2017)
10. M. Lighthill, *Proc. of the Royal Society A, Mathematical, Physical and Eng. Sci.* **211**, 564-587 (1952)
11. J. E. F. Williams, D.L. Hawkings, *Proc. of the Royal Society A, Mathematical, Physical and Eng. Sci.* **264**, 321-342 (1969)
12. J. E. F. Williams, D.L. Hawkings, J. Sound and Vib. **10**, 10-21 (1969)
13. K. R. Fritz, C. C. Hantschk, S. Heim, H. Nürnberger, E. Schorer, B. Stüber, D. Vortmeyer, *Flow noise*, in G. Muller, M. Moser (Eds.), (Springer-Verlag Berlin Heidelberg, 2013)
14. D. B. Hanson, M. R. Fink, J. Sound and Vibration. **62**, 19-38 (1979)
15. I. Proudman, *Proc. of the Royal Society A, Mathematical, Physical and Eng. Sci.* **214**, 119-132 (1952)
16. S. Sarkar, M.Y. Hussaini, NASA Contract Report 93-74, NASA Langley Research Center, Hampton, Virginia. 1993.
17. B. Lakshminarayana, J. Fluids Eng. **113**, 315-352 (1991)
18. ANSYS, *ANSYS CFX User's Guide*, Release 17.0, (ANSYS, Inc., Canonsburg, PA, 2017)
19. B. Stüber, K. Fritz, F. Lang, *Stand der technik bei der lärmminderung inder petrochemie*, (Umweltbundesamt, Berlin, 1981)
20. A. Schmitt, V. Klein, *Schallemission von chemie-normpumpen*, (Chemie-Anlagen und Verfahren 6, 1979)
21. *Europump leitfaden*, (Geräuschemission bei kreiselpumpen, 2002)
22. A. Posa, A. Lippolis, E. Balaras, J. Fluids Eng. **138**, 121101-1-121101-13 (2016)
23. S. A. Hambric, D. A. Boger, J. B. Fahnline, R.L. Campbell, J. Fluids and Struc. **26**, 121-147 (2010)
24. B. A. De Jong, M.S Thesis, Faculty of Mech. Eng., (Delft Univ. of Tech., 2016)
25. S. Christopher, S. Yuan, J. Pei, G.X. Cheng., J. Fluids Eng. **140**, 051201-1-051201-10 (2017)
26. Y. Fu, J. Yuan, S. Yuan, G. Pace, L. d'Agostino, P. Huang, X. Li, J. Fluids Eng. **137**, 011102-1-011102-12 (2015)

MATEC Web of Conferences **172**, 02001 (2018) https://doi.org/10.1051/matecconf/201817202001
ICDAMS 2018

Enhancement of heat transfer in phase change material using graphite-paraffin composites

Sathiyaraj. R[1], *Rakesh*. R[1], *Mithran*. N[1], and *Venkatesan*. M[1*]

[1]School of Mechanical Engineering, SASTRA Deemed University, Thanjavur - 613401

Abstract. Phase change materials (PCMs) are energy storage materials which can be used for maintaining a controlled thermal environment for various applications in earth and space. PCMs are used in advanced technologies in aerospace cooling applications like heat exchangers and heat pipes for re-entry vehicles and spacecraft. Paraffin is a phase change material (PCM) commonly used for energy storage-related applications. Paraffin wax exhibits slow thermal response due to low thermal conductivity value (~0.2 W/m K for most paraffin waxes). In the present work, an attempt is made to fabricate a composite PCM using graphite powder. Such a composite material has enhanced thermal conductivity along with reduced melting period which are desirable properties of a PCM during solid to liquid phase change process. The reduction in melting period is indicated by the difference in change in temperature measured by the thermocouples during a specified time. The temperature variation and solid-liquid interface formation during the melting process are experimentally studied. The results showed that composite graphite powder with paraffin can improve the total phase transition time.

1 Introduction

The key concern in today's world is reducing the disparity of supply and energy demand by storing it. To reduce the greenhouse emissions, there was a paradigm shift to utilize renewable sources of energy. One of the potential sources of renewable energy is the sun rays. Thermal Energy Storage (TES) is a technique used for releasing and storing of renewable energy in the form of sensible and latent heat.

Latent heat thermal energy storage is a better method as it provides a relatively high energy storage density as detailed by Agyenim et al. [1]. Latent heat storage can store the latent heat of fusion in an isothermal process which corresponds to phase transition temperature of the phase change material (PCM). A PCM initiating the melt and solidification process will release more heat during liquid to solid and absorb more heat during solid to liquid phase change because of lower thermal resistance.

The solid-liquid phase change involves relatively less volume change when compared to a solid-gas or liquid-gas phase process with less latent heat transfer. Sharma et al. [2] and Kenisarin and Mahkamov [3] detailed the desirable physical, thermal, kinetic and chemical properties of a PCM. An ideal PCM should have high specific heat and thermal conductivity in both solid and liquid state, low vapour pressure, less volume variation during solidification and should be chemically stable. The organic ones are normally preferred then inorganic type of PCMs used in Latent heat Energy Storage system.

This is because when compared to Inorganic PCMs like salt hydrates, organic PCMS have better chemical stability, biodegradability, non-toxic and are non-corrosive in nature. Thirugnanam and Marimuthu [4] detailed that Paraffin could be used in various applications as it had excellent thermal storage, high heat of fusion, inertness, compatibility with all metal containers and ability to be incorporated easily into thermal storage systems. Lin et al. [5], developed a heat exchanger based on PCM for NASA future space missions. The heat exchanger was used to store thermal energy in excess during high heat load periods by melting paraffin wax. The excess thermal energy was later rejected by the space craft's radiator, freezing the PCM which was utilized for next heating cycle. Moreover, experiments indicated that the paraffin would freeze without supercooling. Nihal and Emel [6] observed that lower thermal conductivity of paraffin reduced the heat transfer performance of charging and discharging cycle during the phase change. Many recent attempts have been made for thermal conductivity enhancement of PCM. This could be achieved by different metal foams addition resulting in increased heat discharging and charging rates in thermal energy storage as observed by [7-11]. Wang et al. [7] increased the thermal storage performance of paraffin by adding copper foam. Zhang et al. [8] investigated numerically the heat transfer characteristics during the melting process of a composite PCM. A copper foam was used to increase the thermal conductivity of the paraffin. Mills et al. [9] fabricated a composite PCM using expanded graphite (EG) and paraffin wax. The thermal

* Corresponding author: mvenkat@mech.sastra.edu

conductivity value was found to be 20–130 times larger than that of pure paraffin because of addition. Mattawee and Assassa [10] observed that the adding dispersed aluminium powder inside paraffin wax increased thermal conductivity. The charging time was decreased by 60% and with enhanced heat transfer characteristics. Jackson and Fisher [11] in their study, compared paraffin heat transfer characteristics with different metal foams like copper, aluminium and graphite having different pore sizes. It was concluded that the graphite foam combination gave better result due to its higher thermal conductivity than other metal foams.

From the above-mentioned works, it can be understood that thermal conductivity and heat transfer rate can be enhanced by using different metal foams. The manufacturing of metal foams involves a complex process and the metal foams are expensive. In the present study, graphite powder is used with paraffin wax to improve the thermal performance and heat transfer rate. The total phase transition period from solid to liquid state is improved for a Graphite/paraffin composite when compared to pure paraffin wax. This is indicated by difference in thermocouple readings for a specific period of time

2 Experimental Setup

Fig.1 shows the experimental setup. The heat transfer characteristics of pure paraffin and paraffin/graphite composite PCM is studied using a cubical glass container, constant heat flux system (mica plate heater), AC power supply and a data acquisition system. The cubical glass container has a size of 5.6 cm x 5.6 cm x 6 cm with each side having a thickness of 4 mm. Paraffin wax is filled up to a height of 5.6 cm. The top portion of the container is open to the atmosphere so that the wax can contract and expand during the phase change process. The bottom side of the glass enclosure is attached to a mica plate heater which is used to melt the paraffin.

Four K-type thermocouples placed consecutively on a small wooden piece are kept inside the glass enclosure containing wax and is evenly spaced 1cm apart to each other. The temperature difference is measured at a gap of 1 cm by these thermocouples. The thermocouples are connected to a standard NI9213 module with NI cDAQ-9174 chassis. The thermocouple temperatures are stored in a computer for further processing. The typical constant heat flux is supplied for both the cases (Paraffin and Composite PCM).

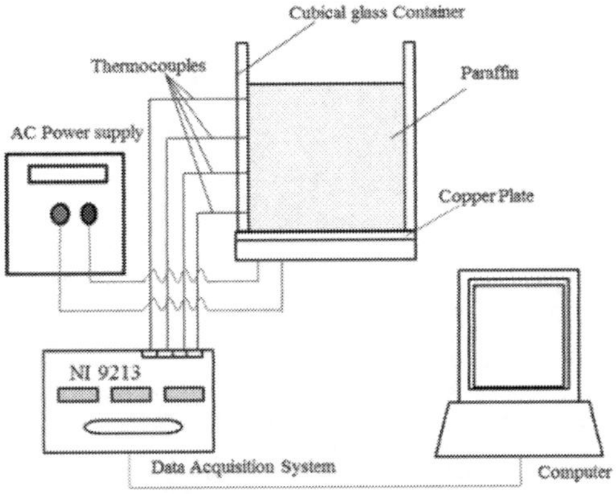

Fig. 1. Schematic illustration of the experimental setup

3 Results and discussions

The experimental data recorded by all the four thermocouples for plain wax without graphite are detailed in Fig. 2. Similarly, the temperature curve of pure paraffin with graphite powder during melting is shown in Fig. 3.

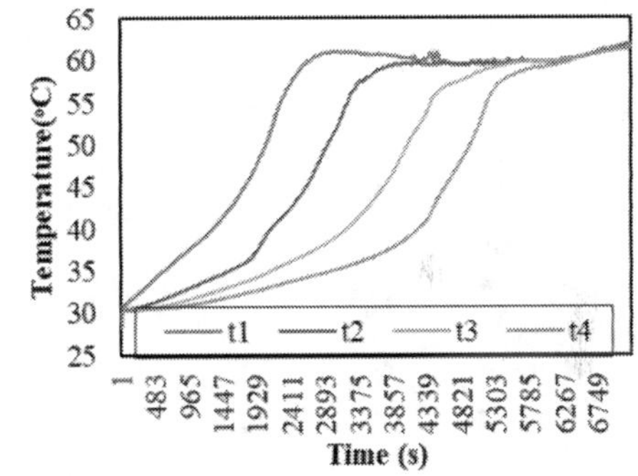

Fig. 2. Temperature curve for melting paraffin wax

During the initial study, the characterization of phase change from solid to liquid is observed for pure paraffin wax. In this case, 140g paraffin wax is filled up to a height of 5.6 cm in the glass enclosure. In the second study, the solid-liquid phase change behaviour is observed for paraffin wax with 10 % of graphite powder. In this second case of study, 14 g of graphite powder is mixed with 126 g of paraffin wax and the mixture is filled up to 5.6 cm in the glass enclosure. The phase change of solid to liquid at different stages and their characteristics are recognized from the temperature-time graph from four thermocouple temperature measurements. The temperature curve starts to rise from room temperature (28 °C) for both the cases. The melting point of paraffin wax is 58 °C.

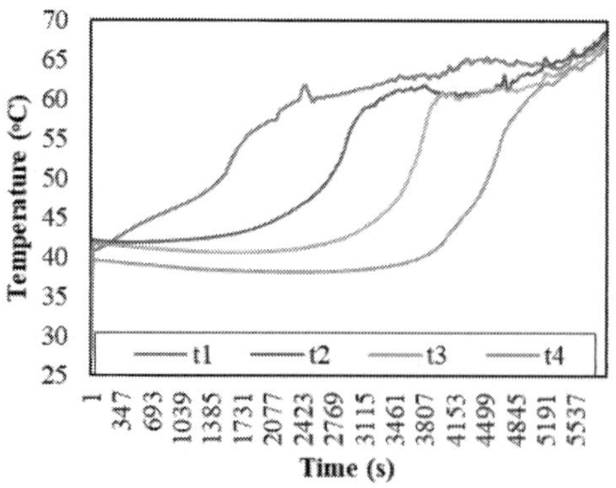

Fig. 3. Temperature curve for melting paraffin wax with graphite powder

A gradual increase in temperature is observed for all thermocouple locations but at different temperature range. Meanwhile, for the composite PCM, increase in temperature occurs at a rapid rate as the graphite powder has higher thermal conductivity (300 to 550 W/mK). For both the cases, the first thermocouple (t1) region is closer to the heating surface. Hence, the paraffin wax in both the cases at bottom side melts sooner than the top layers inside the glass container. It is observed visually that the melting process occurred layer by layer. While the melting happens at the lower layers and the paraffin at that region entirely transferred to the liquid phase, the unchanged paraffin remains solid at the top and appear to be floating over the liquid paraffin as shown in Fig. 4.

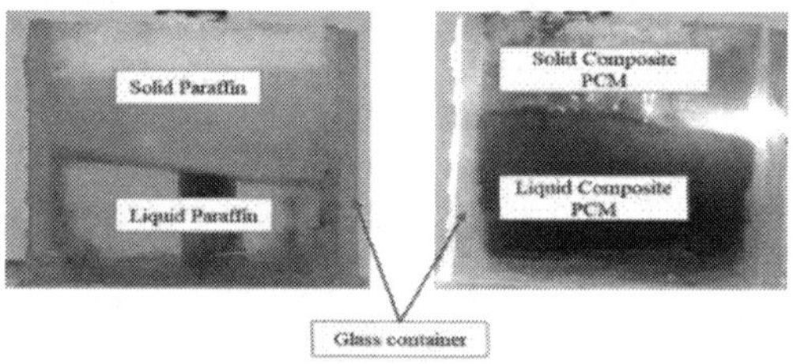

Fig. 4. Melting process of paraffin and composite paraffin

As the temperature rises above 58 °C (melting point), the paraffin wax is completely converted from solid to liquid which is observed in the recorded video. In the first case, the first thermocouple (t1) reached 58 °C at a period of 2468 s. In the second case (with graphite powder), the first thermocouple (t1) reached the melting point at 2132 s. For the other thermocouples, the melting point is reached faster in the second case when compared to the first case. From the Temperature-Time graphs, it is also observed all the thermocouples reached a common temperature of 62 °C in 7228 s, (pure paraffin) which indicates that the total volume of paraffin wax has melted completely to liquid. After this phase change process, the temperature indicated by all the thermocouples increased homogeneously. In the second case (paraffin/graphite composite), all the thermocouples reach 67 °C in 5879 s indicating that the composite PCM is in a liquid state. From the results above, it is understood that, by using graphite powder as a composite material for PCM, the heat transfer characteristics of paraffin is enhanced. The melting process is rapid and is 500 s to 1000 s slower when compared to a pure paraffin wax under similar heat load conditions. Hence, the total phase transition time is improved in composite PCM case. The composite paraffin starts to melt at lower layers initially and the recirculation of graphite powder is observed within the liquid paraffin which in turn improved the heat transfer performance. This phenomenon proves that the heat transfer is dominated by heat diffusion.

4 Conclusion

In the present work, the solid-liquid phase change process of a paraffin wax and a composite paraffin wax (graphite powder) is studied experimentally. The composite material has enhanced heat transfer characteristics with desired thermal storage. The enhancement is measured as delay in time during phase change process indicated by the temperature change of the thermocouples. The limitation of higher thermal resistance of paraffin wax is overcome by adding graphite powder. Compared to the pure paraffin, paraffin/graphite powder composite showed a better heat transfer performance because of the high thermal conductivity of Graphite Powder.

References

1. Francis Agyenim, Neil Hewitt, Philip Eames, Mervyn Smyth, *Renewable and Sustainable Energy Reviews*. **14**, pp. 615–628, (2010)
2. Atul Sharma, V.V. Tyagi, C.R. Chen, D. Buddhi. *Renewable and Sustainable Energy Reviews*. **13**, pp. 318–345, (2009)
3. Murat Kenisarin, Khamid Mahkamov, *Renewable and Sustainable Energy Reviews*. **11**, 1913–1965, (2007)
4. Thirugnanam. C, Marimuthu. P, *International Journal of Engineering and Innovative Technology*, **3(2)**, 372-376, (2013)
5. Kuan-Lin Lee, Calin Tarau, Nathan Van Velson, *47th International Conference on Environmental Systems*, pp. 16–20, (2017)
6. Nihal S, Emel O, *Thermochim Acta*. **540**, pp. 7–60, (2012)
7. Changhong Wang, Tao Lin, Na Li, Huanpei Zheng, *Renewable Energy* **96**, pp. 960–965, (2016)
8. P. Zhang a, Z.N. Meng a, H. Zhu a, Y.L. Wang b, S.P. Peng, *Applied Energy*. **185**, pp. 1971–1983, (2017)
9. F Mills A, Farid M, Selman JR, Al-Hallaj S, *Applied Thermal Eng*. **26**, pp. 1652–61, (2006)
10. Mettawee ES, Assassa GMR, *Sol Energy*. **81**, pp. 839–45, (2007)
11. Jackson GR, Fisher TS, *In: 45th AIAA Thermophysics Conférence*. p.2811. (2015)

A SIMULATION STUDY OF COOLING SYSTEM FOR HEAVY DUTY DIESEL ENGINE

Pranita Adsul[1], *Vinod* Kotebavi[1], *Sanjeev* Bedekar[2], *and Ashwini* Mishra[2]

[1]*Department of Mechanical Engineering, Amrita School of Engineering, Bengaluru, Amrita Vishwa Vidyapeetham, India*
[2] *Cummins India Office Campus, Balewadi, Pune, Maharashtra-411045, India*

Abstract. The main function of the cooling system is to control the temperature of the engine components and improve the performance of an engine. To know the flow and temperature distribution in the jacket cooling system for 6 cylinder diesel engine is analyzed using 1 dimensional method by using GT-Suite 1D simulation software package. The present work employs 1D simulation of water jacket in GT-ISE to perform a comprehensive study of mass-flow and thermal distribution over the inlet of the cooling package of a selected engine in several steady state operating points. The results show, that the suggested predictive method successfully captures the thermal effect of recirculation while reducing the necessity for calibration done by prototype testing.
Keywords. GT-Suite, 1 Dimensional, Calibration, water-jacket.

1 INTRODUCTION

The role of engine cooling system is to take away heat fast enough to keep the engine material temperature well below its melting point and increase the durability of engine [1]. Complexity of cooling system of a vehicle increases with the advancement of technology. Cooling system of an internal combustion engine involves a complex geometry of cooling jacket. Engine specification along with engine block shape and operating range determine the cooling jacket layout. A cooling jacket is not only intended to carry away heat from engine, but also helps to accomplish the optimal operating temperature at the earliest. It is very difficult to analyze the complete behavior of water jacket experimentally and theoretically, due to complex geometries and flow paths [2]. Vehicle thermal management system had fairly remained constant from last few years. In this field primary targets are to achieve quick engine warm-up, avoidance of overheating or over cooling and temperature maintenance. However recently, the need of improved efficiency had driven higher demand. Also, the engine is desired to operate with less parasitic losses [3]. In actual practice, almost all parts of cooling circuit take part in heat transfer, in other phenomenon which influences the efficiency of cooling system. For example, pressure drop influence the fluid flow rate, which determine the convective heat transfer coefficient. Therefore, there are many physical phenomenon one could model in order to capture the behavior of the system as thoroughly as possible [4].

The main objective of the present research is to prove the feasibility of a cooling system model developed in GT-Suite. GT-Suite is one-dimensional simulation software which plays major role to reduce the lead time as well as overall cost. Flow model used in GT-Suite based to solve the basic equations of thermodynamics and fluid mechanics (generally, the Navier-Stokes equations). The Navier-Stokes equation is the continuity equation and law of conservation of momentum and energy equation. These equations are solved for each element of discretised model separately, either using explicit or implicit schemes [5]. Cooling system with special kind of water jacket is simulated. Due to the complexity of the cooling system it is difficult to achieve the simulation results close to bench test [6]. Total temperature rise through water jacket has been measured and results were compared with test data.

However, one dimensional cooling system thermal fluid analysis enables the study of entire cooling system control and also optimization of piping but it does not provide detailed fluid flow analysis inside water jacket [7]. Coolant circuit consists of numerous components, and 1D thermal fluid simulation is applied to study the temperature behavior of the coolant. For both normal and severe driving conditions automotive designers should design a robust engine cooling system [8]. However, the interaction of the all vehicle components can be very complex, such that, in many cases it is very

* Corresponding author: k_vinod@blr.amrita.edu; adsul.pranita@gmail.com

difficult to predict how a new thermal management component may interact with entire vehicle power train, especially in terms of fuel economy and emission control. The challenge is to develop simulation models that can help to optimize the vehicle thermal management strategy in effort to find out a proper cost/benefit ratio for new components and control strategy [9]. Prior to the doing any engine water jacket modifications, the coolant flow characteristics such as pressure drop, water jacket wall temperature and velocity must be identified and fixed carefully [11].

Internal combustion engines are also classified according to cooling system, air cooled engines and water cooled engines. In later system water is used as a heat transfer medium which is preferred over air because of its higher specific heat capacity, density and thermal conductivity. It allows water to transfer heat to greater distance compared to less volumetric flow and reduced temperature difference [12]. Adequate temperature management strategies will be beneficial to achieve high NOx conversion level in after treatment system. Split-cooling system and other thermal management measures have potential to reduce CO_2 and increase exhaust temperature during cold start simultaneously [13].

2 METHDOLOGY

In this paper one-dimensional model of cooling system for diesel engine has been developed in GT-Suite software. The performance prediction through any tool requires so, required input parameters fed to the tool. However, care should be taken with input data to get prediction near to test data. Figure 1 illustrates the methodology followed to develop predictive model.
Inputs required

To develop 1 D model main requisites are:

1) CAD model of an engine
2) Performance details of components
3) Thermostat hysteresis curve
4) Boundary condition data

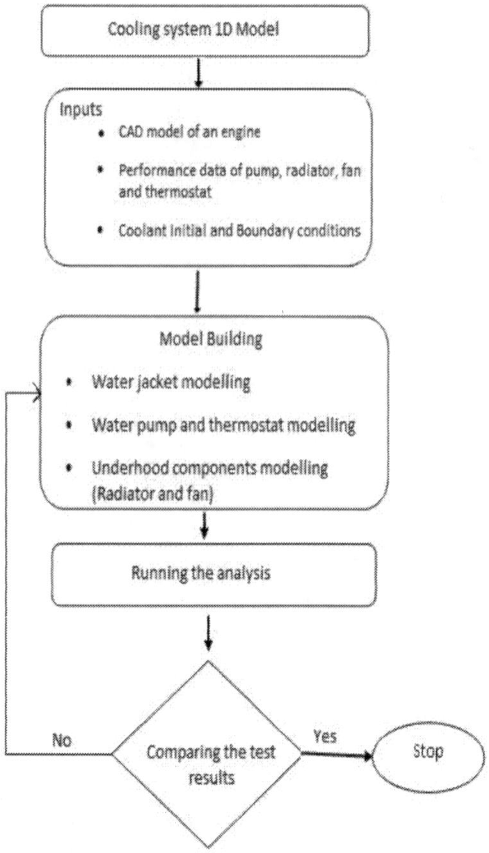

Fig. 1. Methodology adopted to develop 1 D simulation network

CAD model of an engine head and block imported in to the Spaceclaim software to extract cooling jacket. Spaceclaim is the pre-processing tool of GT-Suite. Which is used to define the flow area of coolant. Then flow path of coolant through engine head and block has been extracted [figure 2] and converted it in to the flow splits. Whole core spit divided in to six different flow spilt each one for one cylinder and one inlet to the water jacket and outlet from the water jacket to the thermostat shown in figure 2. Heat transfer coefficient (HTC) taken from the 3D CFD (Computational Fluid Dynamics) model and temperature boundary conditions are taken from the engine performance model which was developed in GT-Power. So again, the accuracy of the complete cooling system model results is depend on the accuracy of engine performance (GT-Power) model and the 3D CFD HTC results.

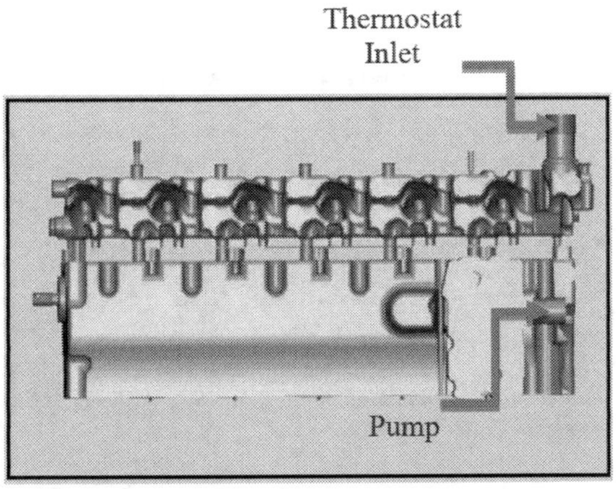

Fig. 2. Extracted core of an engine water jacket

2.1. Simulation method and Cooling Circuit theory

GT-Suite is based on one-dimensional tool which represents the flow and heat transfer in piping and other components of a cooling system. Several researchers have used GT-Suite to simulate engine cooling system, lube system etc. [14, 15]. This tool allows modeling the integration of all heat management activities arising in engine and vehicle development. Also, it provides flexibility to model for advanced concepts and has built in vehicle/engine simulation calculation of thermal loads under different conditions or under any driving cycle. Because of these advantages GT-Suite provides powerful model building facility and reduces user efforts.

Implicit solver is used to solve the Navier-Stokes equation for flow splits to describe how the temperature, pressure and density of a moving fluid are related. For the heat transfer calculation Colburn analogy is used. Always it's not possible to get satisfactory results by Reynolds analogy. Thus, Colburn empirically modified Reynold's analogy which is given by equation, introduces the j-factor.

$$St = \frac{f}{2} Pr^{-2/3} \quad (2.3.a)$$

$$j = (St)\left(Pr^{\frac{2}{3}}\right) = \frac{f}{2} \quad (2.3.b)$$

Where, f is the friction factor

The focus of the study is to develop a cooling system model with closed loop including underhood components to predict performance of the cooling system and validate the results with test data. This work will reduce the number of tests thereby saving cost, time and efforts involved. The most important parameters used to validate the analytical results are Limiting Ambient Temperature (LAT) and Coolant out temperature. LAT is defined as the maximum allowable ambient temperature to which the vehicle cooling system can work without failure. For the prediction of a standard cooling system performance of an engine at least two circuits has to be considered. Primary circuit through which heat is taken away from the engine thermal masses to the coolant through convective heat transfer and second is the cooling module in the vehicle where heat is rejected to the air passing over radiator.

Cooling circuit analysis includes, water pump, water jacket, thermostat, radiator and fan. Heat transfer from oil to coolant in oil cooler system has modelled using single flow split, where heat rate input from oil to coolant is given as single value of input which has obtained from the detailed analysis of oil cooler. This extraction of heat from oil increases coolant temperature but it does not affect much on coolant working as that of it is much effective for oil to release heat to coolant which helps to retain oil properties by maintaining oil temperature in limit. CAC has modelled in the same way as that of oil cooler. Modelling of the cooling components has an advantage of optimizing each component individually irrespective to whole complex system. A simplified diagram of the cooling circuit is represented in Figure 3.

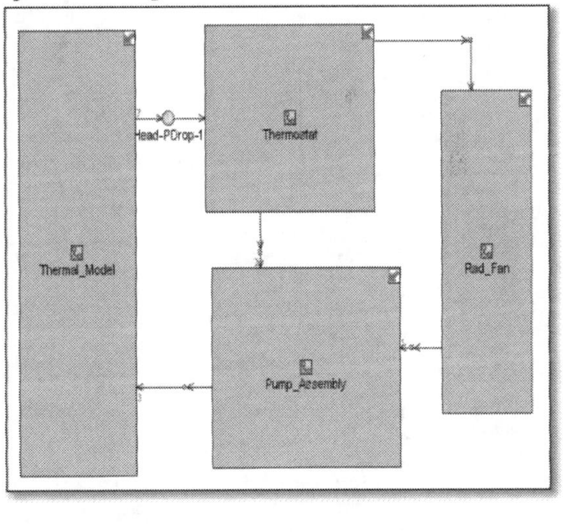

Fig. 3. Scheme of the cooling circuit

A wax thermostat has been used, in which opening of thermostat is temperature bounded. If the coolant temperature is below 82°C, coolant flows through bypass valve and again circulates to the engine block through pump. To accommodate bypass connection, there are two interdependent thermostat valves namely, 1) Master thermostat and 2) Slave thermostat. The master thermostat valve is connected to the radiator and slave thermostat valve by passes the coolant to the pump without flowing through the radiator. To compare test data arrangement has been done to keep thermostat in fully open condition irrespective of the coolant temperature. Fan has been modelled with drive ratio of 1.1 and it is engaged at any flow condition to validate the results with test data. The steady state analysis has been performed for engine cooling system and methodology validated experimentally. The deviation between the test results and analytical results found reasonable.

2.2 Engine Testing

The objective of engine testing was to validate simulation tool which has been developed for cooling system performance prediction and also to test the behavior of cooling system of an engine in worst case conditions. The combination of two different conditions such as rated speed and peak torque were used in testing to understand the effect of performance parameters on components on the end results. The details of combination of testing parameters and components are mentioned in Table 1. Vehicle level testing has been performed to test the behavior of cooling system at worst operating conditions; engine tested at rated speed and peak torque conditions. Thermostat has been kept at fully open condition and fan at fully engaged condition with engine irrespective of the coolant temperature.

Table 1. Combination used for cooling trials

		Rated Speed	**Peak Torque**
Engine		180 hp	180 hp
Make		Cummins	Cummins
Radiator	Thickness	48	48
	Fin Pitch (fpi)	16	16
	Core Size	785 X 537.2 X 48 mm	785 X 537.2 X 48 mm
Fan	Pulley ratio	1.1	1.1
	Blade Diameter (mm)	596	596

The parameters of interest during testing are engine water inlet temperature, engine water outlet temperature before radiator or top tank temperature and ambient temperature to calculate LAT.

$$LAT = T_{ref} - TTT + T_{amb} \qquad (1)$$

Where,

LAT is limiting ambient temperature

T_{ref}= Reference temperature of coolant

TTT= Top tank temperature

T_{amb}= Ambient temperature

3 RESULTS AND DISCUSSION

The model is prepared to solve the steady state condition. The simulation results were compared with test results. It is inferred that, the simulation allows quick calculation and diminutive amount of time to complete the simulation. The calculations can be accelerated with appropriate selection of the initial conditions (temperature of particular parts, system pressure). In our simulation, we focused on unfavorable cases where, underhood cooling circuit is exposed to the highest load.

Table 2 represents the comparison between simulation and test results. There is around +/-2% deviation in results. The present model is over predicting the coolant out temperature for rated speed as well as for Peak torque condition. Such results may be owed to the assumption of constant temperature of thermal masses exposed to combustion gases for one engine speed as well as it has considered that heat transfer coefficient is same over all range of temperature.

Considering the equation (2), the heat transfer coefficient is also depends on the thermos physical properties as well as on the geometrical properties, namely hydraulic diameter and velocity.

$$Nu = \frac{h \cdot D}{k} = Nu\,(Re, \Pr) \qquad (2)$$

Apart from this also, have to consider the sensitivity error of the sensors and man-made.

Table 2. Comparison of test and simulation results

Sr. No.	**Parameters**	**Limit**	**Test condition**		**Simulation**	
			Rated Speed	**Peak Torque**	**Rated Speed**	**Peak Torque**
1	Ambient Temperature	21~39	32.5	33.7	32.5	33.7
2	UCL	<100	78.4	88.7	79.4	90.4
3	LCL		73.1	81.2	73.5	80.7
4	LAT	Rate speed>46	54.1		53.1	
		Peak Torque>41		45		43.3

All the values are expressed in Degree (°C)

 https://doi.org/10.1051/matecconf/201817202002

4 CONCLUSIONS

In the present paper a steady state analysis has been performed for six cylinder diesel engine cooling system using GT-Suite 1D simulation software package. The result illustrates that, the maximum deviation between simulation and test result is nearly +/-2%. The accuracy of the proposed method is depend on the accuracy of the components characteristic in the cooling system as well as on the operating point of a components. It also inferred that, operating point for pump and fan is falling within characteristics data points. Conclusively, this kind of simulation reduces the required time compared to the 3D CFD simulation, as well as efforts and cost required for testing, the accuracy of the results are also in acceptable range. So, such studies will be helpful to reduce analysis time and overall cost. Further, this model can be modified to take in account some more components such as condenser, cabin-heater etc. which can affect the cooling system. In future, this model can be coupled with engine model and the effect of cooling system, on performance parameters like bsfc (brake specific fuel consumption), bmep (brake mean effective pressure), gmep (gross mean effective pressure) and torque of an engine can be observed with running model in duty cycle which will help in understanding the behavior of an engine cooling system as well as engine performance in transient way. Furthermore, GT-Suite is capable of building complete virtual model of a complete vehicle, from cooling system components, transmission system, accessories and tire.

Acknowledgements

Authors are immensely grateful to Cummins India Pvt. Ltd. for providing necessary facilities to carry out the present research work. We also wish to acknowledge the valuable guidance and encouragement from the Director, Head of the Department of Mechanical Engineering and teachers, Amrita school of engineering, Bangalore, during the work.

Acronyms

GT- Gamma Technology

GT-ISE – GT- Integrated Simulation Environment

1D- One Dimensional

HTC- Heat Transfer Coefficient

LAT- Limiting Ambient Temperature

FPI- Fins per inch

UCL- Upper Cooling Limit

LCL- Lower Cooling Limit

BSFC- Brake Specific Fuel Consumption

BMEP- Brake Mean Effective Pressure

GMEP - Gross Mean Effective Pressure

References

1. Umekar, Manvendra M., and D. Govindaraj., No. 2011-26-0087. SAE Technical Paper, (2011).
2. Mulemane, Aditya, and Ravindra Soman, No. 2007-01-4129. SAE Technical Paper, (2007).
3. Shutty, John, Wolfgang Wenzel, Michael Becker, Stephen Bohan, and Glenn Kowalske, No. 2013-01-0546. SAE Technical Paper, (2013).
4. Minovski, Blago B., and Lennart Lofdahl, No. 2014-01-0653. SAE Technical Paper, (2014).
5. Bolehovský, Ondřej, and Jan Novotný, *Journal of Middle European Construction and Design of Cars* 13, no. 3 (2015): 24-32.
6. Soujanya, C., and V. Sundaram. , No. 2015-26-0196. SAE Technical Paper, (2015).
7. Nessim, Waleed, F. Zhang, Zhao Changlu, and Zhu Zhenxia, In *International Conference on Mechanical Engineering and Material Science (MEMS 2012), Shanghai, Dec*, pp. 28-30. (2012).
8. Masjuki, H. H, *Engineering e-Transaction (ISSN 1823-6379)* 7, no. 2, pp 127-134. (2012)
9. Keller, Philip, Wolfgang Wenzel, Michael Becker, and Josh Roby, No. 2010-01-0806. SAE Technical Paper, (2010).
10. Kotebavi, V., Shetty, D and Sahu, D, Pollution Research Volume 35, Issue 1, Pages 159-166 (2016)
11. Qasemian, Ali, and Ali Keshavarz, *Tehnički vjesnik* 23.1: pp 257-264. (2016)
12. Umekar, Manvendra M., and D. Govindaraj, No. 2011-26-0087. SAE Technical Paper, (2011).
13. Honardar, Sharareh, et al., No. 2011-24-0176. SAE Technical Paper, (2011).
14. Mayer, A., et al. *Engine intake throttling for active regeneration of diesel particle filters*. No. 2003-01-0381. SAE Technical Paper, 2003.
15. Seider, Gerald, et al., No. 2001-01-1722. SAE Technical Paper, (2001).

Analytical Performance Analysis of Cross Flow Louvered Fin Automobile Radiator

Badgujar Pankaj R[1], *Rangarajan S*[2], *Nagaraja S. R*[3*]

[1,3] *Department of Mechanical Engineering, Amrita School of Engineering, Bengaluru, Amrita Vishwa Vidyapeetham, India*
[2] *Renault Nissan Technology and Business Centre India Pvt. Ltd. Ascendas IT Park, Mahindra World City, Chennai, Tamilnadu, India*

Abstract. The objective of the present paper is to propose an analytical model for calculating performance parameter of a radiator having rectangular tube with louvered fins. The theoretical effectiveness, heat transfer rate, outlet temperatures of both air and coolant are determined using effectiveness-NTU method. The coolant and air side pressure drop is also calculated. The proposed procedure is validated with experimetal results available in the literature and the GT model. It is found that the maximum deviation in the heat transfer rate calculated from proposed model is 10.97%, the coolant and air outlet temperatures is 2.75% and variation in pressure drop is about 3.29%.

1 Introduction

Radiator is the main component of engine cooling system. It plays an important role in keeping the engine temperatures with prescribed limits by removing heat from the hot coolant fluid. The coolant fluid takes heat from engine cylinder jacket and this hot fluid flow through the radiator tubes and gets cooled by exchanging heat with air which flows over fins and tubes. The heat transfer rate is increased by increasing the surface area for heat transfer using fins. In most of the automobile radiators louvered fins with rectangular cross section are used. Louvered fins are more efficient than rectangular fins. It is easy to attach fins to the rectangular tube rather than circular or oval shape tube and hence most of the radiators use rectangular tubes.

Matthew Carl et al [1] have proposed study on heat transfer process in automobile radiator including the detailed explanation of correlations for radiator performance. A. Khot et al [2] have proposed study on various methods for radiator performance evaluation. In this paper LMTD, Effectiveness- NTU method and experimental method for evaluation of radiator performance are studied. They have concluded that the heat dissipation increase with increase in pressure drop and flow rate. P. S. Amrutkar and aaa [3] have validated the effectiveness-NTU method using simulations. They have used ε-NTU method for thermal analysis of a radiator. P. Mounika et al [4] have analysed the performance of an automobile radiator for air speeds in the range of 15 kmph to 75 kmph. They have used ethylene glycol as coolant and concluded that even at higher speeds the given radiator with given number of fins attached to it works properly with slight compromise in decrease in efficiency of the fins used in the radiator. E Y Ng et al [5] have proposed a method to predict the heat transfer performance in car radiators. They have developed generalized analytical model accounting for airflow maldistribution. They have concluded that loss in engine cooling performance due to airflow maldistribution is relatively small when compared with the condition where airflow distribution is uniform with same total flow rate. R. Prakash et al [6] have proposed the radiator having two flat plates inside the tube which act as the nozzle. They have found that the nozzle effect provides additional cooling to the engine and hence the efficiency of the proposed radiator increased by 5.37% when comparing with existing radiators. Hiteshgiri et al [7] have found effectiveness, pressure drop and heat transfer rate of micro cross flow heat exchanger. They have found that for the same effectiveness, the heat transfer rate increases and the pressure drop decreases with rising working fluid temperature in the hot and the cold flow side. A. R. Esmaeili Sany et al [8] have studied radiator performance of passenger cars experimentally for wide range of operating conditions. They have found that heat transfer results from wind tunnel tests are 50% better than the actual conditions due to better air distribution. Vaisi. A et al [9] have studied the air-side heat transfer and pressure drop characteristics of flow over louvered fins in compact heat exchangers experimentally and concluded that the heat transfer rate tends to increase with an increasing inlet air velocity and also the performance of heat exchanger with louvered fins is affected by its geometrical parameters. Govindaraj. D [10] has used simulation tools for predicting cooling system performance. They have generated component performance characteristics through combined empirical and analytical methods in a consistent way thereby eliminating the discrepancies which would otherwise have occurred because of different testing methods. Solomon et al [11] have explained different geometric correlations for louvered fin radiator geometry.

* Corresponding author: sr_nagaraja@blr.amrita.edu

1.1 Nomenclature

Parameters	Notation	Unit
core height	B_H	m
core width	B_W	m
core thickness	B_t	m
Number of rows of tube in core depth	N_r	
Number of coolant tube in one row	N_{ct}	
Number of profiles	N_p	
Number of fins	N_f	
Louver pitch	L_p	m
Louver length	L_l	m
Fin pitch	F_p	m
Fin end radius	R_f	m
Fin thickness	F_t	m
Fin height	F_h	m
Angle of fin	α	degree
coolant tube length	Y_l	m
coolant tube cross section length	Y_{cl}	m
coolant tube cross section width	Y_{cw}	m
coolant tube thickness	Y_t	m
coolant tube pitch	Y_p	m
Coolant mass flow rate	M_c	Kg/s
Air mass flow rate	M_a	kg/s
Specific heat capacity for coolant	C_{pc}	J/Kg.k
Specific heat capacity for air	C_{pa}	J/Kg.k
Density of coolant	$\rho_{coolant}$	kg/m^3
Density of air	ρ_{air}	kg/m^3
Thermal conductivity of coolant	K_c	W/m.k
Thermal conductivity of air	K_a	W/m.k
Inlet temperature of coolant	T_{hi}	k
Outlet temperature of coolant	T_{ho}	k
Inlet temperature of air	T_{ci}	k
Outlet temperature of air	T_{co}	k
Kinematic viscosity of coolant	$\vartheta_{coolant}$	m/s
Kinematic viscosity of air	ϑ_{air}	m/s
Prandtl number for coolant	P_{rc}	
Prandtl number for Air	P_{ra}	
Pressure drop	ΔP	
Hydraulic diameter coolant side	D_{hc}	
Hydraulic diameter air side	D_{ha}	

2 Model Development

Modern automobiles use radiators with multi-louvered fins with rectangular tubes. This increases the air-side heat transfer performance by more than 60% when compared with the traditional plate fin heat exchangers. The louvered fins provide more surface area for heat transfer. The proposed model uses rectangular tube with louvered fins. The performance of the model is compared with experimental [9] and GT model results. The geometric, Reynolds and Nusselt number correlations from reference [11], effectiveness-NTU correlations [12] and pressure drop correlations [13] are used. These correlations are given below. Figure 1 and 2 show the geometry of a typical louvered fin radiator.

A. Geometry correlations

1. Fin length - $F_l = \pi R_f + \frac{F_h - 2R_f}{\cos \alpha_f}$
2. Radiator core frontal area - $Af_{r,r} = B_H B_W$
3. Coolant tube frontal area - $Af_{r,t} = Y_{cw} Y_l N_{ct}$
4. Fin frontal heat transfer area- $Af_{r,f} = F_t\, F_l N_f Y_l N_p$
5. Fin heat transfer area-

$$A_f = 2B_t\, F_l N_f Y_l N_p$$

6. Total heat transfer area on air side:

$$A_a = A_f + 2N_{ct} Y_l N_r [(Y_{cl} - 2R_t) + (2\pi R_t)]$$

7. Total heat transfer area on the coolant side:

$$A_c = [2\pi(R_t - Y_t) + 2(Y_{cl} - 2R_t)] Y_l N_{ct} N_r$$

8. Effective fin length: $L = \frac{F_h}{2}$

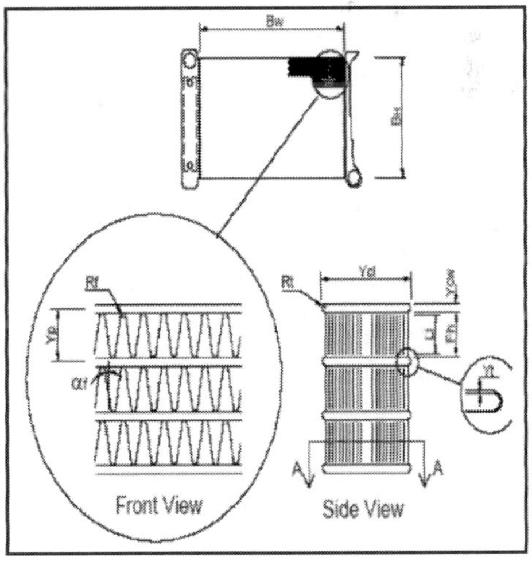

Fig.1. Geometry of Louvered Fin Radiator

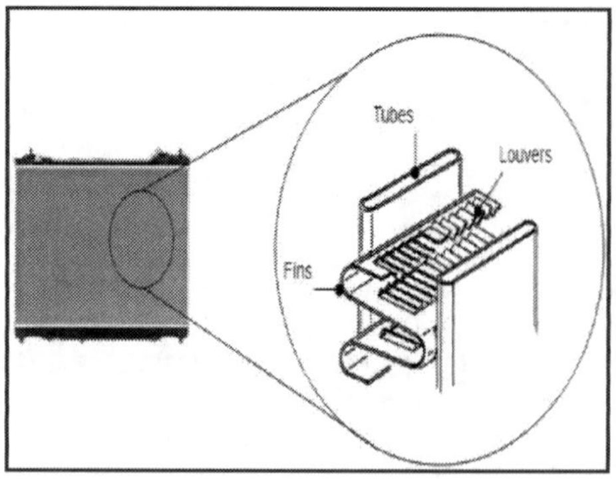

Fig.2. Tube and Louvered fin geometry description

B. Reynolds and Nusselt number correlations

1. Reynolds number for coolant:

$$Rc = \frac{V_{coolant} D_{hy}}{\vartheta_{coolant}}$$

2. Reynolds number for Air

$$Ra = \frac{V_{air} L_p}{\vartheta_{air}}$$

3. Nusselt number for air -

$$N_{ua} = 0.664\, R_a^{0.5}\, Pr_a^{0.3333}$$

4. Nusselt number for coolant-

For Different flow regime, the following equations were used for Nusselt number

i. Laminar flow (2300<Rc)

Hansen equation:

$$N_{uc} = 3.66 + \frac{0.0668\left(\frac{D_{hc}}{Y_l}\right) R_c P_{rc}}{1+0.04\left[\left(\frac{D_{hc}}{Y_l}\right) P_{rc}\right]^{0.666}}$$

ii. Nusselt number for coolant Turbulent flow (Rc>10000)

Petukhov equation:

$$N_{uc} = \frac{\frac{f}{8} R_c P_{rc}}{1.07+12.7x\left(\frac{f}{8}(P_{rc}^{0.666}-1)\right)^{0.5}}$$

iii. Nusselt number for coolant Transition Flow

$$(10000 \geq Rc \geq 2300)$$

Gnielinski equation:

$$N_{uc} = \frac{\frac{f}{8}(R_c - 1000) P_{rc}}{1+12.7\left(\frac{f}{8}\right)^{0.5}(P_{rc}^{0.666}-1)}$$

Where,

Friction factor: $f = [0.79(\ln(R_c)) - 1.64]^{-2}$

C. Effectiveness-NTU Correlations

1. Heat transfer coefficient:

$$h = \frac{Nu\, k}{D_{hy}}$$

2. Overall heat transfer coefficient:

In a radiator the overall heat transfer coefficient depends on the convective heat transfer coefficient on air and coolant side, wall conduction and the fouling factors on air and coolant side.

$$UA_{fr,r} = \frac{1}{\frac{1}{\eta_o h_{air} A_a} + \frac{1}{h_{coolant} A_c}}$$

3. Number of transfer units:

$$NTU = \frac{U A_{fr,r}}{C_{min}}$$

4. Effectiveness:

$$\epsilon = 1 - exp\{\left(\frac{1}{Cr} NTU^{0.22}\right)\left(exp^{-CrNTU^{0.78}} - 1\right)\}$$

5. Maximum heat transfer rate:

$$Q_{max} = C_{min}(T_{hi} - T_{ci})$$

6. Actual heat transfer rate:

$$Q_{act} = \epsilon Q_{max}$$

7. Coolant outlet temperature:

$$T_{ho} = T_{hi} - \frac{Q_{act}}{C_{coolant}}$$

8. Air outlet temperature:

$$T_{co} = T_{ci} + \frac{Q_{act}}{C_{air}}$$

D. Fin Efficiency corrélations

1. Fin Efficiency :

$$\eta_{fin} = \frac{\tanh(mL_c)}{mL_c}$$

Where,

$$m = \left(\frac{h_a F_p}{K_f A_c}\right)^{0.5}$$

2. Overall Efficiency:

In radiator both the finned and un-finned surfaces contribute to the overall heat transfer. The overall surface efficiency is defined for the composite structure consisting of both the fins and base structure and it can be calculated as

$$\eta_{overall} = 1 - \frac{N_{fin} A_f}{A_{fin\ base}} (1 - \eta_{fin})$$

E. Pressure drop corrélations

1. Coolant side pressure drop

$$\Delta P_{coolant} = \frac{L_{tube} N_{pass} G^2}{2 D_{hy} \rho_{coolant}} 4f$$

2. Air side pressure drop

$$\Delta P_{air} = \frac{G^2}{2\rho_i}\left((1 - \sigma_i^2 + k_c) + \frac{4fL}{D_{hy}}\left(\frac{\rho_i}{\rho_m}\right) + 2\left(\frac{\rho_i}{\rho_e} 1\right) - (1 - \sigma_e^2 - k_e)\left(\frac{\rho_i}{\rho_e}\right)\right)$$

Where,

Mass velocity $G = \rho v$

$$k_e = (1-\sigma)^2$$

$$k_c = 0.42(1-\sigma^2)^2$$

k_c and k_e are the contraction and expansion coefficients respectively.

ρ_i, ρ_m and ρ_e are inlet density, mean density and outlet density respectively.

$$\sigma = \frac{Minimum\ flow\ area}{Frontal\ area}$$

passage contraction ratio.

2.1 Proposed Analytical Model

An analytical model is proposed to find the effectiveness of louvered fin radiator. The flow chart shown in figure 3 describe the procedure for the calculation of effectiveness, heat transfer rate, outlet temperature for both coolant and air for the proposed analytical model. The flow chart describes detailed step by step calculation procedure. The inputs to the calculation are radiator geometry, mass flow rate and velocity of coolant and air, inlet coolant and air temperature as well as properties of coolant and air. From this input Reynolds number is calculated and this helps in deciding the type of flow (laminar, transition or turbulent). Based on the type of flow, equation for calculating the Nusselt number is selected. After calculating the heat transfer coefficient and fin efficiency, the overall heat transfer coefficient and the number of transfer units (NTU) are calculated. Effectiveness correlation is used for the calculation of effectiveness by using NTU. Maximum heat transfer rate is calculated from the input temperatures and minimum heat capacity. By using this maximum heat transfer rate and effectiveness, actual heat transfer rate, outlet temperature of the coolant and air are calculated. After calculating friction factor, mass velocity and hydraulic diameter, pressure drop on coolant side and air side are calculated.

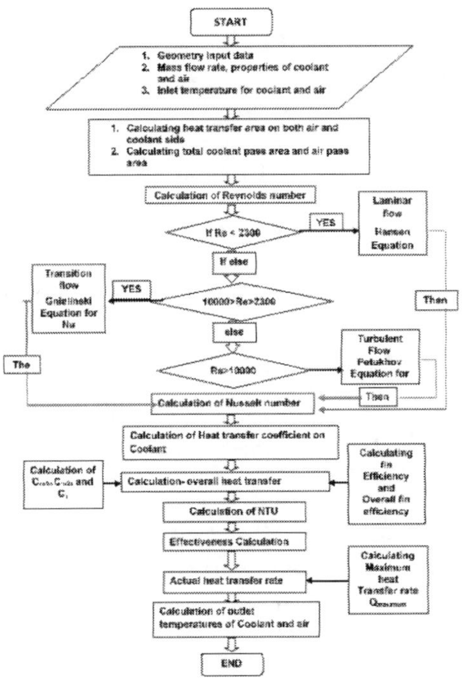

Fig.3. Process Flow Chart

3 Results and Analysis

The analytical model proposed is validated by comparing the results obtained for the model with those available in literature. Two models are used for this purpose, one experimental results of rectangular tube louvered fin radiator [9] and GT model radiator.

The input conditions for the analytical model are same as given in the reference [9]. Table 1 shows input conditions for different trials. The geometric parameters used are same as that given in the reference [9]. The comparison between the experimental [9] and the analytical results using the proposed procedure are shown in Table 2. It is observed that the calculated results obtained from proposed model closely agree with experimental results. The maximum % error in the heat transfer rate is 3.78. Figures 4 and 5 show the variation of coolant and air outlet temperatures for different trails. It can be observed that the deviation in outlet temperature of coolant is ±1 % and that of outlet temperature of air is ±2 %. The pressure drop calculations are not shown this model since in the experiments carried by Vaisi. A et al [9] have not measured the pressure drop.

Table 1. Input conditions for Coolant and air [9]

Parameters	Unit	Testing trials				
		I	II	III	IV	V
Coolant mass flow rate	kg/s	2	2	2	2	2
Air mass flow rate	kg/s	1.124	1.372	1.627	1.956	2.143
Coolant inlet temperature	°c	80	80	80	80	80
Air inlet temperature	°c	20.3	20.3	20.3	20.3	20.3
Air velocity	m/s	5.58	6.81	8.09	9.71	10.64

Table 2. Comparison between analytical and experimental results

Trials	Heat Transfer rate		Error in Heat transfer rate	Outlet temperature of water		Outlet temperature of air	
	Analytical	Exp		Exp	Analytical	Exp	Analytical
Unit	kw	kw	%	°C	°C	°C	°C
I	43.28	42.76	-1.19	73.99	73.75	54.28	55.76
II	47.40	47.49	0.19	73.29	73.15	51.29	52.13
III	50.87	50.96	0.17	72.87	72.65	48.61	49.05
IV	54.42	56.28	3.43	72.24	72.14	45.70	45.92
V	56.13	58.25	3.78	71.53	71.89	44.23	44.42

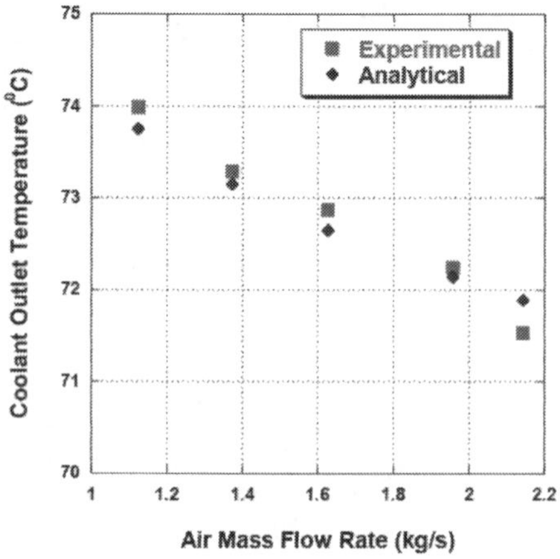

Fig. 4. Comparison between analytical and experimental coolant outlet temperature

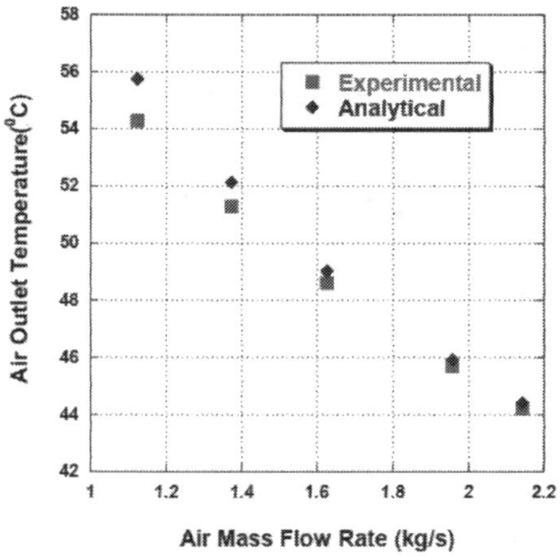

Fig. 5. Comparison between analytical and experimental Air outlet temperature

The proposed analytical model is also validated using the results for GT model. GT model is a cross flow rectangular tube louvered fin radiator used by Renault Nissan. This model also has correlations for pressure drop calculations. Table 3 shows the input parameters used for these calculations. The geometric parameters used are the same as those used in GT model. The comparison between the GT model and proposed model are as shown in table 4. It can be observed that the maximum error in heat transfer is 10.97% and in pressure drop is 3.29%.

Table 3. Input conditions for Coolant and air from GT Model

Parameters	Unit	TRIALS				
		I	II	III	IV	V
Coolant mass flow rate	kg/s	0.44	0.88	1.32	1.76	2.2
Air mass flow rate	kg/s	0.29	0.29	0.29	0.29	0.29
Inlet temperature of Coolant	°c	95	95	95	95	95
Inlet temperature of Air	°c	25	25	25	25	25

Table 4. Comparison between GT model and analytical results

Trial	Pressure drop coolant side		Error in	Heat transfer rate (Q)		Error in
	Exp	Analytical	ΔP	Analytical	Exp	Q
	bar	bar	%	Kw	Kw	%
I	0.0095	0.0093	3.07	12.27	11.06	10.97
II	0.0282	0.0286	-1.32	15.02	13.95	7.68
III	0.0565	0.0560	0.47	16.27	15.42	5.51
IV	0.0938	0.0908	3.29	16.97	16.32	3.99
V	0.1480	0.1434	3.18	17.43	16.94	2.87

4 Conclusions

Analytical model for radiator having rectangular tube with louvered fin is developed. Analytical correlations are validated with the two different models of radiator and results agree closely. Heat transfer rate and pressure drop results obtained from the analytical model compared with the GT model results. The maximum deviation in heat transfer is 10.97% and that in pressure drop is 3.29%. The analytical model developed also predicts the outlet temperatures of both coolant and air. The analytical model results are also compared with the experimental results. The maximum difference between the experimental and analytical values for heat transfer rate is 3.78%. The maximum deviation between experimental and analytical values of coolant outlet temperature is 1% and that of air outlet temperature is 2.75%. The analytical model developed is capable to predicting the radiator performance i.e. heat transfer rate, coolant and air outlet temperatures, pressure drop of both coolant side as well as air side.

We would like to acknowledge the support of Renault Nissan Technology and Business Centre India Pvt. Ltd. for permitting to carry out this work.

References

1. Matthew Carl, Dana Guy, Brett Leyendecker, Austin Miller, and Xuejun Fan, ASEE Gulf Southwest Annual Conference (2012)
2. A. R. Khot, D.G. Thombar, S. P. Gaikwad, A. S. Adadande, IOSR-JMCE, **07**, (2014)
3. P. S. Amrutkar, S. R. Patil, IOSR-JMCE, (2005)
4. P. Mounika, Rajesh K Sharma, P. S. Kishore, IJRMEE, **3**, (2016)
5. E Y Ng, P W Johnson, S Watkins, Proc. IMechE, Part D: J. Automobile Engineering, **219** (2005)
6. R. Prakash, M. Selvam, A. Alagu Pandian, S. Palani and K. A. Harish, Indian Journal of Science and Technology, **9**, (2016)
7. Hiteshgiri Goswami and Jiten Makadia, International Journal of Engineering Research & Technology, **2**, (2013)
8. A. R. Esmaeili Sany, M. H. Saidi, J. Neyestani, The Journal of Engine Research, **18** , (2010)
9. Vaisi, A, M. Esmaeilpour, and H. Taherian, Applied Thermal Engineering, **31**, (2011)
10. Govindaraj. D., No. 2011-01-2185, *SAE Technical Paper*, (2011)
11. Solomon, Samson, No. 2013-01-2810, *SAE Technical Paper*, (2013)
12. Kothandaraman, C. P., and S. Subramanyam, *Heat and Mass Transfer Data Book* (2014)
13. Kakac, Sadik, HongtanLiu, and Anchasa Pramuanjaroenkij *Heat exchangers: selection, rating, and thermal design*. CRC press, (2012)

MATEC Web of Conferences **172**, 02004 (2018) https://doi.org/10.1051/matecconf/201817202004
ICDAMS 2018

Estimation of Effective Thermal Conductivity of Two-Phase Material for Hollow Circular Model

Prateek Kumar Sahu[1], *Nisha* Netam[2*], *Lal* Chandra Shah[3]
[1,2]Department of Mechanical Engineering, National Institute of Technology Raipur, Chhattisgarh, 492010, India
[3] Department of Mechanical Engineering, Shri Shankaracharya Institute of Technology and Management, Bhilai, 490020, India

Abstract. Two-phase materials are commonly used in engineering application because of its various properties like strength, thermal conductivity, durability and toughness etc. Effective thermal conductivity (ETC) of two-phase material is the fundamental property to predict its thermal performance. Various geometry (spheres, cylinders, irregular particles) have been considered by researchers for calculating ETC of two-phase materials. Due to complex structure, hollow circular cylinder geometry is not reported yet. In this paper, two-dimensional periodic two-phase system, with hollow circular cylinder shape is considered for calculating ETC. In present work unit cell approach method is used to derive collocated parameters model for estimation of ETC. Hollow circular cylinder model with $\Psi = 0.2$ gives good result for estimating ETC with average percentage error of 6.46%.

Keywords: Effective Thermal Conductivity, Two-Phase Material, Unit cell approach, Hollow circular model.

1 Introduction

A two-phase material consists of two phases with distinct chemical and physical properties. Two phases can be solid-solid, solid-fluid and fluid-fluid. Phases are named as continuous and dispersed, and categorized into various systems based on its characteristics. Two-phase materials are commonly used in application of engineering field because of its various properties like strength, thermal conductivity, durability and toughness etc. Effective thermal conductivity (ETC) of two-phase material (TPM) is the fundamental property to predict its thermal performance. TPM are categorized into the homogeneous or heterogeneous structure. In homogeneous systems, chemical composition will be same throughout the matrix whereas, in the heterogeneous system, chemical composition will be different throughout the matrix.

Two reference model proposed by [1], for calculating ETC are series model and parallel model. Where in series model phases are in series with the heat direction and in parallel model phases are in parallel with heat direction.

$$k_e = \frac{k_s k_f}{\varepsilon . k_s + (1-\varepsilon) . k_f} \quad \text{Minimum (series)} \quad (1)$$

$$k_e = \varepsilon . k_f + (1-\varepsilon) . k_s \quad \text{Maximum (parallel)} \quad (2)$$

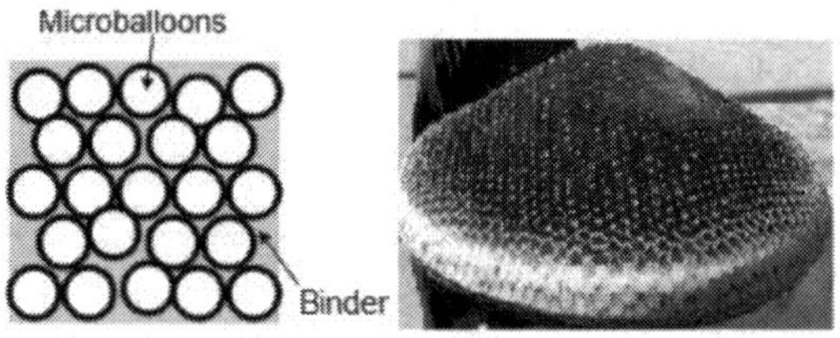

Fig 1 Schematic of two-phase material [7].

A model for calculating ETC TPM has been proposed by [2]. In this model suspension of solid sphere are considered which are not in contact.

$$K = \frac{1+2\beta v}{1-\beta v} \quad (3)$$

where β is given as:

$$\beta = \frac{\alpha - 1}{\alpha + 2} \quad (4)$$

[3] Developed the model, with the effect of (i) conduction through solid phase, contact surfaces and in voids through the fluid medium (ii) heat transfer through radiation, for estimating ETC of two-phase material. A mathematical expression for the ETC of a packed bed with the inclusion of the effect of thermal radiation, dependency on pressure, particle flattening, shaping and size distribution, along with particle contact for a cylindrical unit cell that contains a spherical particle is presented [4]. [5] Reported an analytical expression based on unit cell approach for calculating ETC of two-

* Corresponding author: nishane.me@nitrr.ac.in

phase material for the square cylinder, circular cylinder, solid cube and hollow cube. [6] Presented an analytical expression based on unit cell approach, for hexagonal and octagonal cinder model for estimating ETC. Results show average deviation of ±15.58 % with experiments results.

Various geometry (spheres, cylinders, irregular particles) have been considered by researchers for calculating ETC of two-phase materials. Due to the complex structure, hollow circular cylinder geometry is not reported yet. In this paper, two-dimensional periodic two-phase system, with hollow circular cylinder shape has been considered for deriving an algebraic equation for calculating ETC. A unit cell approach is used to calculate the ETC.

2. Methodology

Parameters on which ETC of two-phase materials depends are conductivity ratio (α) and concentration (v), contact resistance, convective heat transfer, radiative heat transfer and Knudsen effect. It also depends on particle size, shape, location, size distribution and orientation. Unit cell approach based on collocated parameter model is used to derive the expression to calculate the ETC for hollow circular cylinder model.

The ETC of the two-dimensional, hollow circular cylinder is determined for Fig. 2, with the inner radius of " b ", outer radius of " a " and connecting bar width of " c ". Due to structure symmetry, quartern of the structure is taken as an unit cell and is shown in Fig. 3.

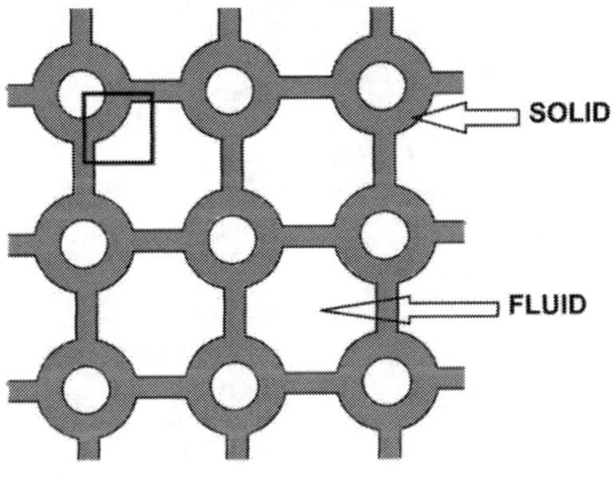

Fig. 2 Collocated Hollow Cicular Cylinder Model

The unit cell is separated into four rectangular layers as shown in Fig. 3, perpendicular to the direction of heat. Based on series approach TC of each layer is acquired. Every layer has both fluid and also solid phase in the unit cell.

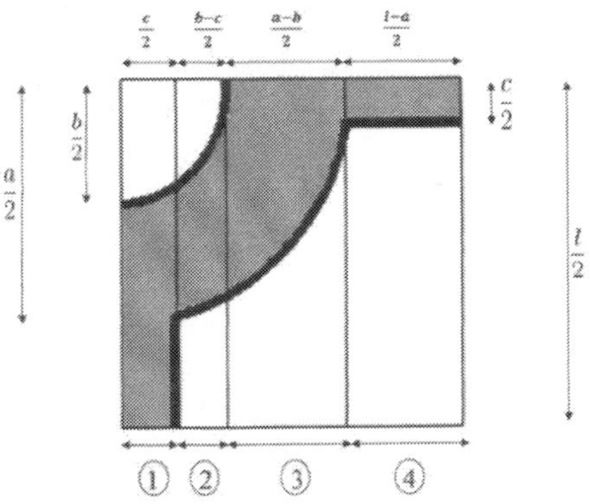

Fig. 3 Unit Cell of Hollow Circular Cylinder Model

Total resistance in unit cell is:

$$R_{total} = R_1 + R_2 + R_3 + R_4 \quad (5)$$

Here R_1, R_2, R_3 and R_4 are the thermal resistance of layer 1, 2, 3 and 4 respectively.

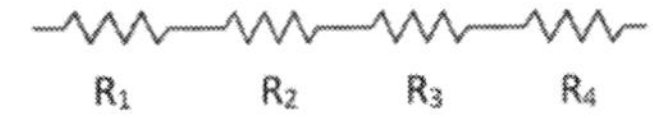

Fig.4 Thermal Resistance

Now thermal resistance for layer 1 is calculated as:

$$\text{Thermal Resistance} = \frac{\text{Thickness of rectangual strip}}{\text{Thermal conductivity X Cross secion area of strip}}$$

$$\frac{1}{R_1} = \left(\frac{K_{sf1} X\ Area}{Thickness}\right) + \left(\frac{K_s X\ Area}{Thickness}\right)$$

$$\frac{1}{R_1} = \left(\frac{K_{sf1} l\ (b/2)}{c/2}\right) + \left(\frac{K_s l\ (l-b)/2}{c/2}\right) \quad (6)$$

$$R_1 = \frac{c}{K_{sf1} lb + K_s l(l-b)} \quad (7)$$

Similarly, thermal resistance for layer 1, 2, 3 and 4 are:

$$R_2 + R_3 = \left[\left\{\frac{b-c}{K_{sf2} lc} + \frac{a-b}{K_s lc}\right\}^{-1} + \frac{2K_s l}{\pi} + \frac{2K_{sf2} l}{\pi - 2} + \frac{K_{sf2} l(l-a)}{a-c}\right]^{-1} \quad (8)$$

$$R_4 = \frac{l-a}{K_s cl + K_{sf4} l(l-c)} \quad (9)$$

Total resistance in unit cell is:

$$R_{total} = \frac{c}{K_{sf1} lb + K_s l(l-b)} + \left[\left\{\frac{b-c}{K_{sf2} lc} + \frac{a-b}{K_s lc}\right\}^{-1} + \frac{2K_s l}{\pi} + \frac{2K_{sf2} l}{\pi - 2} + \frac{K_{sf2} l(l-a)}{a-c}\right]^{-1} + \frac{l-a}{K_s cl + K_{sf4} l(l-c)} \quad (10)$$

The value of thermal conductivity at solid fluid interface for layer 1, 2, 3 and 4 is determined based on a series layer model:

$$\phi_1 = \frac{K_{sf1}}{K_s} = \frac{\alpha}{1+(\alpha-1)\delta\varepsilon} \quad (11)$$

$$\phi_2 = \frac{K_{sf2}}{K_s} = \frac{2\alpha}{(\alpha-1)\delta\varepsilon+(1-\varepsilon)(1-2\alpha)} \quad (12)$$

$$\phi_3 = \frac{K_{sf3}}{K_s} = \frac{2\alpha}{(\psi+1)\varepsilon+\alpha(2-\varepsilon-\delta\varepsilon)} \quad (13)$$

$$\phi_4 = \frac{K_{sf4}}{K_s} = \frac{\alpha}{(1-\varepsilon\psi)\alpha+\varepsilon\alpha} \quad (14)$$

Where,

Conductivity ratio $(\alpha) = \frac{K_s}{K_f}$

Length Ratio $(\varepsilon) = \frac{a}{l}$

Contact Ratio $(\psi) = \frac{c}{a}$

Hollow Shape Factor $(\delta) = \frac{b}{a}$

$$R_{total} = \frac{1}{K_f l}\left\{ \frac{\psi\varepsilon}{\delta\varepsilon\phi_1+\alpha(1-\delta\varepsilon)} + \left[\left(\frac{\delta-\psi}{\phi_2\psi} + \frac{1-\alpha\delta}{\alpha\psi} \right)^{-1} + \frac{2\alpha}{\pi} + \frac{2\phi_2}{\pi+2} + \frac{\phi_2(1-\varepsilon)}{\varepsilon(1-\psi)} \right]^{-1} + \frac{1-\varepsilon}{\alpha\psi\varepsilon+\phi_4(1-\psi\varepsilon)} \right\} \quad (15)$$

$$K_{eff} = \frac{1}{R_{total}.l}$$

$$K = \frac{K_{eff}}{K_f} = \left\{ \frac{\psi\varepsilon}{\delta\varepsilon\phi_1+\alpha(1-\delta\varepsilon)} + \left[\left(\frac{\delta-\psi}{\phi_2\psi} + \frac{1-\alpha\delta}{\alpha\psi} \right)^{-1} + \frac{2\alpha}{\pi} + \frac{2\phi_2}{\pi+2} + \frac{\phi_2(1-\varepsilon)}{\varepsilon(1-\psi)} \right]^{-1} + \frac{1-\varepsilon}{\alpha\psi\varepsilon+\phi_4(1-\psi\varepsilon)} \right\}^{-1} \quad (16)$$

$$K = f(\alpha, \delta, \psi, \varepsilon) \quad (17)$$

Here K is the non-dimensional thermal conductivity (TC) for collocated parameter of hollow circular cylinder it is a function of conductivity ratio, hollow shape factor, length ratio and contact ratio.

The solid phase portion of the unit cell is indicated in terms of concentration (v), and is given by:

$$v = \frac{\text{Area of the solid phase}}{\text{Total area of the unit cell}}$$

$$v = \frac{\pi}{4}\varepsilon^2(1-\delta^2) + 2\varepsilon\psi(1-\varepsilon) \quad (18)$$

3. Results and discussion

Non-dimensional TC is calculated for glass wool (α=10) using Eq. (16), and results are plotted between non-dimensional TC and concentration (v) for different contact ratio (c/a). Fig 5 shows the variation of non-dimensional TC with concentration (v) for hollow circular model with hollow shape factor (δ = 0.25) and different contact ratio (c/a=0, 0.1, 0.2). Fig. 6 shows the variation of non-dimensional TC with concentration (v) for hollow circular cylinder model with hollow shape factor of **δ** = 0.75 and different contact ratio (c/a=0, 0.1, 0.2). Variation of temperature on non-dimensional TC is presented in Fig. 7, with temperature non-dimensional TC is increases.

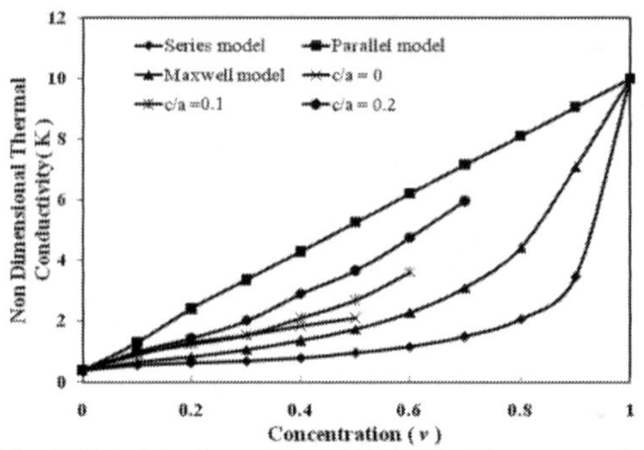

Fig. 5 Variation of non-dimensional TC with concentration (v) for hollow circular cylinder (α = 10 & δ = 0.25).

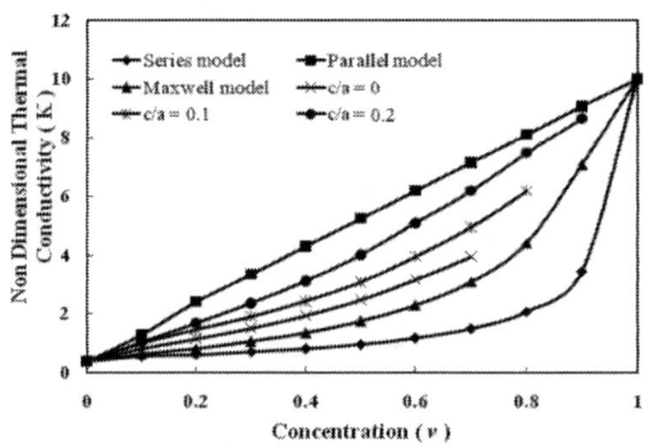

Fig. 6 Variation of non-dimensional TC with concentration (v) for hollow circular cylinder (α = 10 & δ = 0.75).

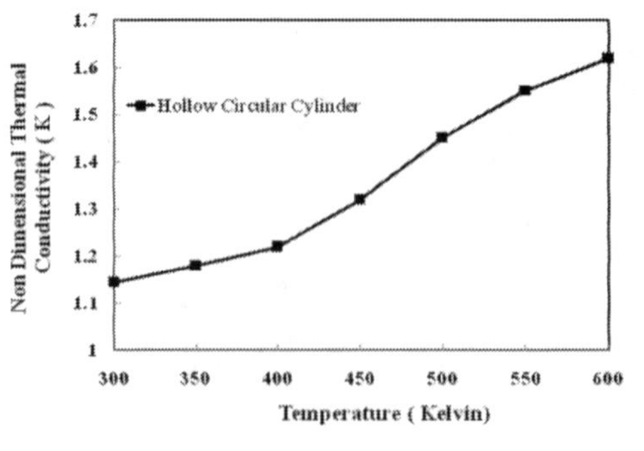

Fig.7 Effect of temperature on the non-dimensional TC of two-dimensional hollow circular cylinder (δ = 0.5, Ψ = 0.2 and ε =0.3).

Table 1 Result comparison with experimental value

Two-Phase Material	α	K_{exp}	K_{eff}	% Deviation
Glass Wool	10	1.2	1.1437	4.69
Alumina Air	20	1.6	1.7008	6.3
Air and Sand	40	2.1	2.2768	8.4
			Average	6.46

4. Conclusion

The non-dimensional TC of the two-phase material (α=10) has been calculated for concentration range (0.2 ≤ v ≤ 0.9). Results of the present model lie, between the standard model (parallel and series) and above the Maxwell model. Hollow circular cylinder model with Ψ = 0.2 gives the good result for estimating thermal conductivity with average percentage error 6.46% as shown in Table 1. The present model can be used to calculate the ETC of two-phase material for concentrations (0 ≤ v ≤ 0.9) for hollow circular geometry.

Nomenclature

K_{sf} Thermal conductivity at solid fluid interface

K_s Thermal conductivity of solid

K_f Thermal conductivity of fluid

ϕ Solid fraction

v Concentration

References

1. I. H. Tavman, Int. Comm. Heat Mass Transfer, **23**, No. 2, pp. 169-176 (1996).
2. J. C. Maxwell, *A Treatise on Electricity and Magnetism*, Clarendon Press, Oxford, p. 365 (1873).
3. S. Imura, E. Takegoshi *Heat Transfer-Japan*, 13-26(1974).
4. P. Zehner, E. U. Schlunder Chemical Engineering Technology, **42**, (1970).
5. K. S. Reddy, P. Karthikeyan, Heat Transfer Engineering, **30**, Iss. 12, (2009).
6. A. P Senthil Kumar, V. Prabhu Raja, P. Karthikeyan, Journal of scientific and industrial research, **69**, pp 356-364 (2010).
7. G. Gladysz, K. K. Chawla, *VOIDS IN MATERIALS From Unavoidable Defects to Designed Cellular Materials*, Elsevier (2015).

Performance investigation of a rotary heat exchanger installed with four ducts using computational fluid dynamics

Saneesh Nair[1], *Vinod* Kotebavi[2*] and *Alokekumar* Das[3]

[1,2] Department of Mechanical Engineering, Amrita School of Engineering, Bengaluru, Amrita Vishwa Vidyapeetham, India
[3]Sr. Manager, Volvo Group India Pvt. Ltd., Bangalore, India

Abstract. The performance characteristics of a commercially available rotary heat exchanger or were determined using CFD. A rotary air pre-heater of a thermal power plant (fuelled by natural gas), situated in Ramin, Iran, was numerically modelled with full scale dimensions using 3D approach. This model was then validated against the experimentally measured values. After a satisfactory validation of the CFD model, a design modification was applied to the model by modelling the heat exchanger with four ducts, with hot and cold fluid ducts arranged alternatively. This was done to improve the heat the heat transfer from hot gases to cold air. The medium of heat exchange (core) is modelled as porous media, since the core in the actual equipment has many number of fine passages arranged in complex manner, which would make the modelling difficult. The model was subjected to same boundary conditions as the validation model. The temperature of the outlets and effectiveness of both the models were compared and was found out that the 4-ducts model is around 12% more effective than the base model.

Keywords: Waste heat, Rotary heat exchanger, computational fluid dynamics, effectiveness

1 Introduction

As the world is seeing rising prices in the fuels, the need to save energy and utilise the lost energy has reached a high importance. In power plants mainly, there is a lot of scope of extracting waste heat from the exhaust gases. There are a lot of systems and equipments which are used for this purpose.

In industries, there are many heat recovery units such as recuperators, regenerators, rotary heat exchanger [1], economizer, heat pumps, etc. In small scale, systems like thermoelectric generators (TEG) [2], PV cells, piezoelectric devices, etc. are used to harvest energy.

The present study focuses on rotary heat exchanger. A rotary heat exchanger is a heat exchanging device used within the supply and exhaust streams of a gas handling system. In power plants, it is used for the heat recovery purpose from the exhaust gas coming out of the boiler.

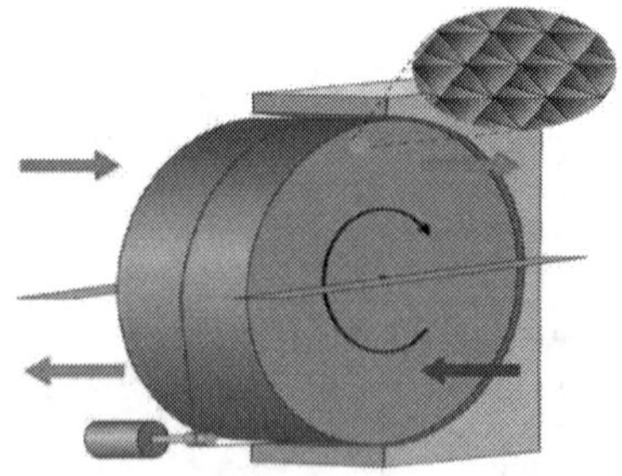

Figure 1. Rotary Heat Exchanger

As shown in the Figure 1, it consists of a rotating thermal wheel (also called core) which transfers the heat taken from hot fluid to cold fluid. The core of the heat exchanger consists of a honeycomb matrix and is made up of heat-absorbing material, which is slowly rotated. As the core rotates, exhaust gas stream transfers its heat to the fresh air stream through the core. The performance of the core depends on the thermal gradient between the two streams.

Rotary regenerators are being used very commonly used in recent decades due to its compact design, lower

* Vinod Kotebavi: k_vinod@blr.amrita.edu

operating cost in terms of heat transfer per unit volume and easy control of the heat exchange by controlling the heat wheel speed [3]. It is the one of most effective heat exchanger to extract waste heat in power plants and to cool down the electronic equipments in industries and data centres.

In one research, authors have conducted a study on the use of the rotary heat exchanger in a data centre and concluded that these regenerators are highly dependent on climate [4]. In another research, authors have used it in C.I engine in order to extract waste heat from 653K exhaust gas and concluded that a significant amount of waste heat can be extracted from the engine exhaust [5].

Studies have also indicated the rotary heat exchanger is more effective than plate type heat exchanger [6]. Some researchers have also attempted to subject the heat wheel to a varying temperature data and to study the performance of heat wheel under such conditions [7]. The inlet condition was mathematically modelled. It was found that outlet conditions also followed the same trend as the inlet conditions.

According to the studies conducted, the performance of the thermal wheel is limited to only half of its travel through the cold fluid duct, i.e. if we study the outlet temperature profile of the cold fluid, we would observe that there is a significant change in outlet temperature of the cold fluid only in the first half portion of the cold fluid duct. The second half of the cold fluid duct has the temperature close to its inlet [8].

In an attempt to solve this limitation, the current study is focused to determine how the preheater would perform if it is installed with four ducts, instead of two, with hot and cold fluid duct arranged alternatively. Clearer understanding can be achieved from the figure 8. At the end of the simulation, the performance of both the base model and '4-Ducts' model were compared on the basis of respective effectiveness.

The mathematical model of the effectiveness was done with following assumptions:

- Mass flow rate of both the streams are same, $(M_h = M_c)$.
- Knowing the properties of exhaust gas respective to Ramin power plant, Iran, is a difficult task. Thus to simplify the problem, air was chosen as the standard gas for both the streams and was different physical conditions.
- Since, there is negligible difference between the specific heat of the host exhaust gases and the cold air, thus they were assumed as equal, $(C_{pa} = C_{pg})$.

$$\epsilon = \frac{Q_{act}}{Q_{max}} = \frac{M_h\, C_{pg}\,(T_1 - T_2)}{(MC_p)_{min}(T_1 - t_1)} = \frac{M_c\, C_{pa}\,(t_2 - t_1)}{(MC_p)_{min}(T_1 - t_1)} \quad (1)$$

Where,

Q_{act} = actual heat transfer

Q_{max} = maximum heat transfer

M_h , M_c = hot & cold mass flow, respectively

C_{pa} , C_{pg} = specific heat of cold air and hot exhaust gas, respectively

$(MC_p)_{min}$ = minimum heat capacity

T_1, T_2 = inlet & outlet temperature of hot exhaust gas, respectively

t_1, t_2 = inlet & outlet temperature of cold air, respectively

Thus, considering the above assumption, above equation can be modified as

$$\epsilon = \frac{(T_1 - T_2)}{(T_1 - t_1)} = \frac{(t_2 - t_1)}{(T_1 - t_1)} \quad (2)$$

2 Modelling & meshing

The CAED modelling and meshing of the CFD model is done in Star CCM+, using dimensions mentioned in [1]. The core diameter is 9864 mm and the ducts are arranged as shown in Figure 2(a) – 2(e) below.

2.1 Base Model

The Figure 2(a)-(e) shows the CAED model of the current rotary heat exchanger. The model consists of five important parts namely; cold air inlet & outlet, hot exhaust gases inlet & outlet and the core. Core is the part where the heat transfer phenomenon takes place as the part rotates and stores the heat from the hot exhaust gases, locally and then passes it on to the cold air stream.

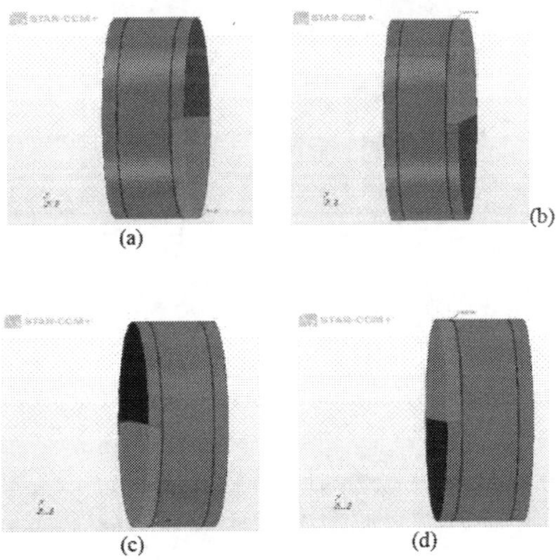

(a) (b) (c) (d)

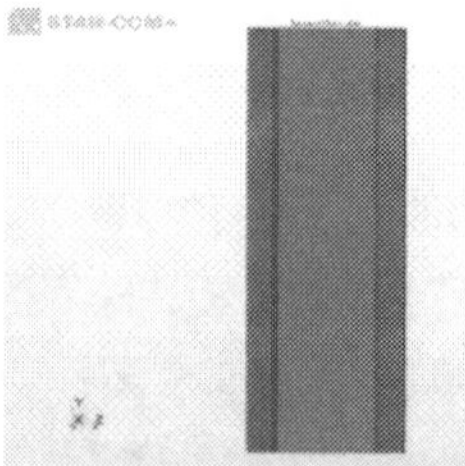

(e)

Figure 2. (a)Hot Fluid Inlet (b)Cold Fluid Outlet (c)Hot Fluid Outlet (d)Cold Fluid Inlet (e) Core

The meshing of the model is done using Surface Remesher and Polygonal Mesher. The outlets of the ducts have been extruded to some distance using Extruder Mesher. This is done to create a stabilised flow through the ducts. The Table 1 represents the mesh parameters and sizes that were set to generate a suitably fine mesh for efficient computation. Figure 3 shows the meshed base model.

Table 1. Mesh Parameters

Parameters	Values
Base Size (mm)	100
Minimum Surface Size (mm)	40
Target Surface Size (mm)	40
Extrusion length (mm)	±2500
No. of layers	10

Figure 3. Meshed Base Model

2.2 4-Ducts model

In an attempt to use most of the heat energy from the waste flue gas, the rotary heat exchanger was modified as shown in the Figure 4 below. Figure 4 shows the hot end of the 4-ducts model. The orange sections represent cold fluid outlet and the purple sections represent the hot fluid inlet. The same meshing phenomenon and the meshing parameter (as mentioned in Table 1) were used for the 4-Ducts model. The meshed model is as shown in the Figure 5.

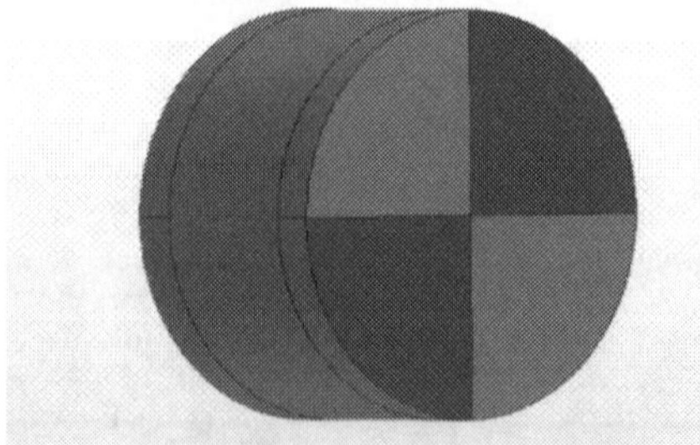

Figure 4. 4-Ducts Rotary Heat Exchanger CFD Model

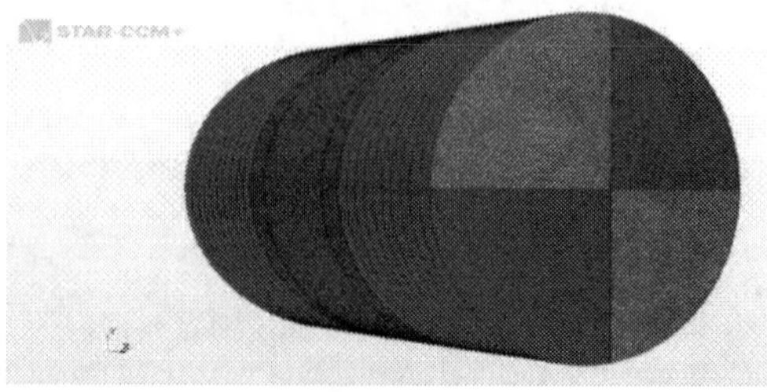

Figure 5. 4-Ducts meshed model

3 CFD Analysis

The validation and the 4-Ducts model were subjected to implicit unsteady analysis, with inlet and outlet boundary conditions set as stagnation inlet and pressure outlet respectively. The boundary parameters were set as shown in the Table 2.

Table 2. Boundary parameters

Revolution Speed	2 rpm
Inlet Air Temperature	348.9 K
Inlet Air Pressure	2970 Pa
Outlet Air Pressure	1650 Pa
Inlet Gas Temperature	634.2 K
Inlet Gas Pressure	-1890 Pa
Outlet Gas Pressure	-3110 Pa

Since, the modelling of honeycomb structures or the fine passages through the heat transfer medium is difficult, the core was treated as a porous region with a porosity of 0.8[7].

Also, the rotation of the model was induced using Rigid Body Motion (RBM) technique, in which whole of the mesh is moved by the solver in order to determine the rotational effects. Rotation rate was set at 2 rpm [1].

3.1 Validation Model

The validation of the base model was done against the experimental results got in [1] using the above mentioned models and boundary conditions. The results were in close validation to the reference experimental values as shown in the Table 3.

Table 3. Reference validation results

Results	Outlet Gas Temperature	Outlet Air Temperature
Experimental Results	393.4	534.3
Numerical Results	430.5	520.6
Error (%)	8.6%	2.6%

Figure 6 shows the temperature profile at the hot end, i.e. at the end where cold fluid exits and hot fluid enters. The absolute red region of the temperature profile represents the entry of the hot fluid. The varying temperature profile represents the cold fluid exit. As can be seen clearly from the Figure 6, only first half of the cold fluid is heated and other half is at almost the same temperature as at entry (refer the colour map).

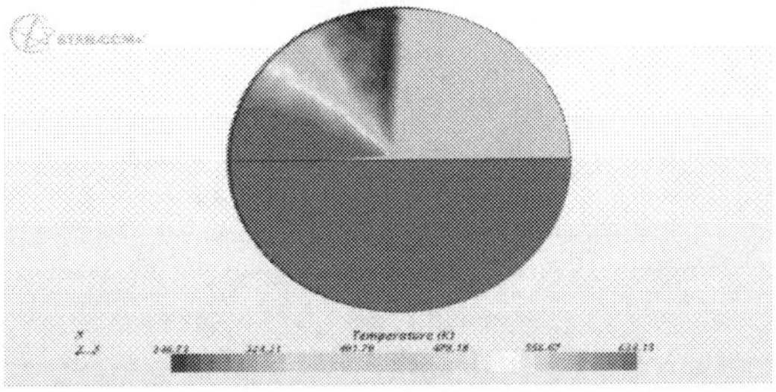

Figure 6. Hot end Temperature Profile

The Figure 6 also indicates some regions show unreasonable temperature profile (blue regions). This is believed to be an error which has occurred due to the reverse flow effects in some cells. This issue can be rectified using finer mesh sizes, but it would require more amount of computational time. Thus, temperature value has been taken from the yellow region.

3.2 4-Ducts Model

Simulation of the 4-Ducts model was done with the same physics models and boundary conditions as mentioned in Table 2.

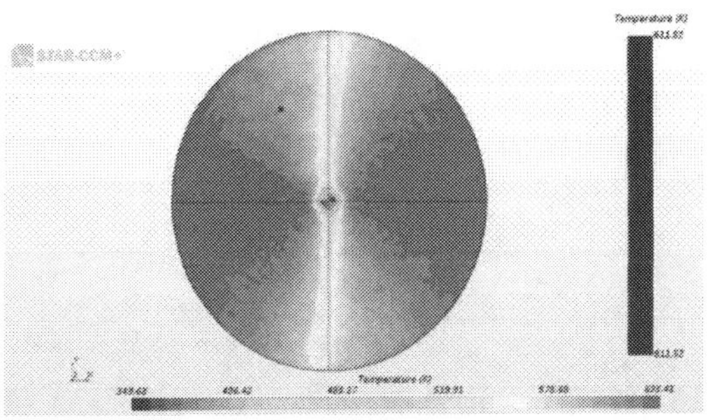

Figure 7. Hot end temperature profile

The Figure 7 shows the temperature profile of a plane placed near to the hot end but inside the heat exchanging medium. It should be noted that in RBM approach, whole of the mesh is moved with the rotational speed specified. When the simulation was stopped, the moving mesh of the heat exchange medium was stopped at position as shown in the above figure. To correct this condition, would again be requiring a lot of computational time.

However, it was deduced form the Figure 7 that the absolute red regions belong to the hot fluid inlet and the regions with colour variation (shades of blue and orange) belong to the cold fluid outlet. In order to observe the result more clearly, few different methods were adopted which are discussed below.

A point probe was kept inside the cold fluid outlet (blue dot in top left quarter of the Figure 7), which returned the temperature value of the heated air (refer the right vertical temperature bar). Thus, temperature of the cold fluid out was recorded as 611 K.

Another method involved placing a plane through the ducts as shown in the Figure 8 below.

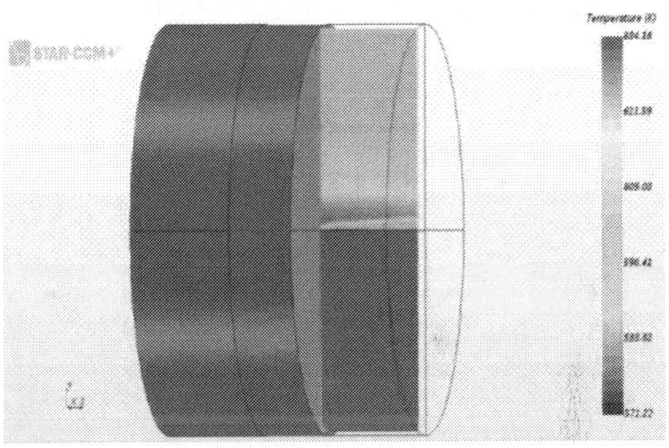

Figure 8. 4-Ducts temperature profile viewed from side

Figure 8 shows the temperature profile on a plane which is placed in the hot fluid inlet and cold fluid outlet. Absolute red region belongs to the hot fluid inlet and region with colour variation (shades of green and yellow) belongs to cold fluid outlet. The Figure 8 indicates that the cold fluid outlet temperature is around 610 K, which is same as observed in the Figure 7.

4 Results & discussions

The effectiveness of both the model was compared using the mathematical model (2). The effectiveness values were found as tabulated in the Table 4.

Table 4. Comparison of the performance of the base model and 4-Ducts model

Results	Outlet Gas Temperature (K)	Outlet Air Temperature (K)	Effectiveness (%)
Base Model Results	430.5	520.6	71.39
4-Ducts Results	395	611.52	83.81

- From Table 4, we can see that there is an appreciable increase in the effectiveness of the rotary heat exchanger by 12.42%.
- If we compare the Figure 6 and Figure 7, we can clearly see that the latter temperature profile shows very less colour variation. This indicates that the outlet temperature of the cold fluid has almost reached the inlet temperature of the hot fluid.
- From Figure 7, the point probe in the top left part of the figure indicates the temperature at that point. The point was placed inside the cold out duct. This indicates that the temperature of cold fluid has reached to around 611 K.
- In Figure 6, indicates that base model has not recovered the heat from the hot fluid as effectively as the 4-Ducts model, since the temperature of cold fluid at the outlet is nearer to its inlet condition.

The Figure 7 shows that the temperature distribution is uniform. Thus, there would be lesser thermal stress on the core material as compared to the base model.

5 Conclusions

The results obtained reflect a significant improvement over the base model. The outlet temperature of the cold fluid was greatly increased to 600 K. The outlet temperature of the hot fluid was also significantly reduces to around 395 K, which showed that more of the waste heat has been recovered as compared to the base model. Though there were reverse flows during the simulation, the end result was satisfactory.

- Such a great increase in the outlet temperature of the cold fluid establishes that it would require less fuel in the boiler. Thus the boiler efficiency would be greatly increased, reducing the load on the boiler.
- There would also be a significant reduction in the fuel consumption rate, for the same amount of the turbine output.
- If the fuel consumption is kept constant, then the gas can be superheated to a greater extent. This would increase the turbine power output significantly.
- Also, it can be deduced that there is a great increase in energy recovery which usually goes wasted with the exhaust gas.
- Above conclusions clearly represents that, if 4-Ducts rotary heat exchangers are used in a thermal power plant, there would be a great increase in the power plant efficiency.

References

1 Heidari-Kaydan, Armin, and Ebrahim Hajidavalloo, *Applied Thermal Engineering* 73.1 (2014): 399-407.
2 Rohit, G., D. Manaswini, and Vinod Kotebavi, *AIP Conference Proceedings*. Vol. 1859. No. 1. AIP Publishing, 2017.
3 Worsøe-Schmidt, P, *International Journal of Refrigeration* 14.4 (1991): 233-239.
4 Mok, Seungho, et al., *2016 15th IEEE Intersociety Conference on*. IEEE, 2016.
5 Bhojane, Monali S., and M. S. Deshmukh, *Emerging Trends in Engineering and Technology (ICETET), 2010 3rd International Conference on*. IEEE, 2010.
6 Swapnil Nilange, Bhavesh Kakulte, Bhushan Maheshwari and Jagdish Patil , *International Journal of Research in Engineering and Technology (IJRET) April 2013*, Volume: **02** Issue: 04 : 2319-1163
7 Fathieh, F., et al, *International Journal of Heat and Mass Transfer* 87 (2015): 312-326.
8 Vishal Hegana and P.R. Kulkarni, *International Journal of Advance Research In Science And Engineering (IJARSE) May 2015*, Vol. No.4: 2319-8354(E)

Evaluation of critical blockage ratio and pulse length in a pulse detonation engine using CFD and MATLAB

S. Srikrishnan[1,*], P. K. Dash[2], and V. Jayakumar[3]

[1]Mechanical Department, IARE, Hyderabad, India
[2]Aeronautical Department, MLR Institute of Technology, Hyderabad, India
[3]Mechanical Department, Saveetha School of Engineering, Saveetha Institute of Medical and Technical Sciences, Chennai, India

Abstract. A Pulse Detonation Engine (PDE) is a new invented propulsion device that takes advantage of the pressure rise inherent to the efficient burning of fuel-air mixtures via detonations. Detonation initiation is a critical process that occurs in the cycle of a PDE. A practical method of detonation initiation is Deflagration-to-Detonation Transition (DDT), which describes the acceleration of a subsonic deflagration created using low initiation energies to a supersonic detonation. The DDT process is not well understood due to a wide range of time and length scales involving complex chemistry, turbulence and unsteady pressure waves. This paper discuss about the effects of blocking ratio in the augmentation of detonation pressure and velocity inside a cylindrical tube of diameter 0.0254m and a length of 1 m. The blockages are rectangular in shape placed at 2/3rd distances of the length of the tube and the heights of the blockages are varied in terms of the diameter of the tube as 1/4th, 1/3rd, ½, 2/3rd and 3/4th the diameter of the tube. The setup is then analyzed in MATLAB using the physics of Friedlander's equation, which formulate the decay time duration of pressure across the tube length, with and without the blockage. Further, a 2D CFD analysis through ANSYS Workbench is conducted which gave the effective blocking ratio in a rectangular type of blockage placed at the 2/3rd position of the length of the tube and the results are compared. For variable pressures ranging from 1 MPa to 100 MPa input, the effective pulse length is around 0.25 seconds after which the decay of pressure and temperature attain the critical limit. Also it is found that the maximum feasible velocity occurs for an inlet pressure of 10 MPa and 2/3rd height of the blockage where the corresponding outlet velocity is 4692m/s and outlet total pressure being 10.542 MPa.

1 Introduction

The new era of detonation of air-breathing propulsion has conceived from the studies on fuel energy transformation in stabilized oblique detonation waves. The technology gradually immerged for investigations and practical development of propulsion engines operating on propagating detonations in a pulse mode [1]. Contrary to the oblique-detonation concept that is applicable to hypersonic flight at velocities comparable or higher than the Chapman-Jouguet (CJ) detonation velocity of the fuel–air mixture, the concept of pulse detonation engine (PDE) is attractive for both subsonic and supersonic flight with PDE as a main propulsion unit or as an afterburner in turbojet or turbofan propulsion system. In particular, PDE [2] based propulsion is attractive for flight Mach number up to about 3 to 4.

In a PDE, detonation is initiated in a tube that serves as the combustor [3]. The detonation wave rapidly traverses the chamber resulting in a nearly constant-volume heat addition process that produces a high pressure in the combustor and provides the thrust [4]. The operation of PDE configurations at high detonation-initiation frequency (about 100 Hz and over) [5, 7] can produce a near-constant thrust. In general, the near-constant-volume operational cycle of PDE provides a higher thermodynamic efficiency as compared to the conventional constant-pressure (Brayton) cycle used in gas turbines and ramjets [6]. The advantages of PDE for air-breathing propulsion are simplicity and easy scaling, reduced fuel consumption, and intrinsic capability of operation from zero approach stream velocity to high supersonic flight speeds [8].

The global interest in the development of PDE [9] for propulsion has led to numerous studies on detonations, particularly pertaining to its control and confinement. This is evident from the formation of collaborative teams by universities and industry worldwide [10].

Several reviews have been already presented at various meetings [1–10] and published in archival journals [11–26]. Wedge- and cone-induced oblique shock/detonation waves have received widespread attention due to their potential application in hypersonic propulsion devices.

* Corresponding author: krishnan.1987@gmail.com

https://doi.org/10.1051/matecconf/201817202006

These include ram accelerators and oblique detonation wave engines (ODWEs). Experimental observations of oblique shock-induced detonation, such as those in have shown that at least two different types of flow can arise. The first is the direct initiation of a detonation wave within a stagnation region. The second is an oblique shock wave that transitions to an oblique detonation wave a certain distance from the wedge's leading edge. Numerical simulation of wedge- and cone-induced detonation has been done recently. The above-cited studies were concerned with the body in a semi-infinite domain. Moreover, the interaction is induced by a long wedge, that is, one where the shoulder does not affect the structure of the interaction zone, an exception being. Generally, the numerical studies showed two possible ways that a detonation is achieved. In the first, an attached oblique shock in the incoming premixed combustible mixture transitions to a detonation, arising from the compression. In the second, a direct initiation is obtained.

From cited research inactivation the PDE advantages mentioned above, a number of challenging fundamental and engineering problems has yet to be solved in order to use propagating detonations for propulsion and realize. These problems deal basically with low cost achievement and control of successive detonations in a propulsion device. To ensure rapid development of a detonation wave within a short cycle time, one needs to apply (1) efficient liquid fuel injection and air supply systems to provide fast and nearly homogeneous mixing of the components in the detonation chamber (DC); (2) low energy source for detonation initiation to provide fast and reliable detonation onset; (3) cooling technique for rapid, preferably recuperative, heat removal from the walls of DC to ensure stable operation and avoid premature ignition of FAM leading to detonation failure; (4) geometry of the combustion chamber to promote detonation initiation and propagation at lowest possible pressure loss and to ensure high operation frequency; and (5) control methodology that allows for adaptive, active control of the operation process to ensure optimal performance at variable flight conditions, while maintaining margin of stability. In addition to the fundamental issues dealing with the processes in the DC, there are other issues such as (6) efficient integration of PDE with inlets and nozzles to provide high performance; and (7) efficient coupling of DCs in a multi tube PDE configuration. Among the most challenging engineering issues, is the problem of durability of the propulsion system. As the structural components of PDE are subject to repeated high-frequency shock loading and thermal deformations, a considerable wear and tear can be expected within a relatively short period of operation. The other problem is noise and vibration.

The previous paper analysed the effects of different types of blockages like triangular, semi-circular and rectangular in $1/3^{rd}$, ½ and $2/3^{rd}$ positions of the length of the tube of diameter 0.0124m and 0.25m length thus finding that the rectangular blockage is the most feasible when placed at the $2/3^{rd}$ distance of the tube [27]. This paper presents the effects of blocking ratio in the augmentation of detonation pressure and velocity inside a cylindrical tube of diameter 0.0254m and a length of 1 m. The blockages are rectangular in shape placed at $2/3^{rd}$ distance of the length of the tube and the heights of the blockages are also varied in terms of the diameter of the tube as $1/4^{th}$, $1/3^{rd}$, ½, $2/3^{rd}$ and $3/4^{th}$ of the diameter of the tube. A 2D CFD analysis through ANSYS Workbench is conducted which gave the effective blocking ratio in a rectangular type of blockage placed at the 2/3rd position of the length of the tube. Only the height of the blockages is varied and the results are compared. The setup is then analysed in MATLAB using the physics of Friedlander's equation, which formulate the decay time duration of pressure across the tube length, with and without the blockage.

For variable pressures ranging from 1 MPa to 100 MPa input, the effective pulse length is around 0.25 seconds after which the decay of pressure and temperature attain the critical limit. This result gives a predictive analysis of detonation wherein the frequency of spark for inducing controlled combustion inside the tube can be studied.

2 Methodology

2.1. Computational Method

In this study, a commercial code, 'FLUENT' which is based on the Navier–Stokes equation is utilized for this analysis. 'GAMBIT' which is the pre-processor of this study has a 'Journaling Function' for auto-grid generation. This function is useful to mesh the similar types of models repeatedly [6]. 'Realizable k–ε' is selected as a turbulence model, which is a variant of the 'Standard k–ε' model. Among the two equation Reynolds Average Navier–Stokes (RANS) turbulence models, 'Realizable k–ε' is known as the most suitable for a turbulent flow.

2.2. Modelling

It has been studied in previous paper that the rectangular type of blockage introduced in the 2/3rd position of the tube length would have the maximum detonation effect. Using the results, the current work is done to establish the critical blockage ratio to find out the most feasible height of the blockage to induce detonation in a closed pipe of diameter 0.0253m and a length of 1m. The modelling has been done in GAMBIT, where a 1m length of pipe is taken with blockages being constructed at 2/3rd distance along the length of the tube. The blockage heights have been varied as 1/4th, 1/3rd, ½, 2/3rd, 3/4th of the diameter of the tube.

2.3. Meshing

The scaled model was meshed using GAMBIT. The length of the tube was taken as the scale of the element size. The element size was taken to be 0.049 and a quad-tri type of mesh was selected to mesh the 2D faces after the extrusion of the blockages placed at the 2/3rd the distance from the origin. A total of 5 different meshes were created for the effective study of the critical blockage ratio. A sample of the mesh has been shown in Fig. 1

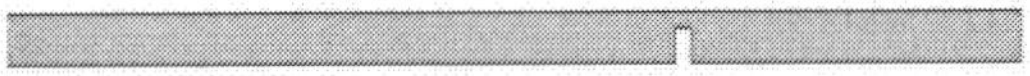

Fig. 1. Mesh created for 1m length and 2.53cm diameter tube with blockage placed at 2/3rd distance along the length of the tube.

2.4. Boundary Conditions

The listed meshes generated using GAMBIT was then imported into ANSYS 16.0 as case files. Multiphase analysis of a kerosene air vapour mixture has been done on selected mesh files to arrive at the final conclusion. Uniform pressure growth varying from 1 MPa to 1000 MPa was adapted and a coupled condition of pressure and velocity has been created at inlet. Velocity at inlet was set a uniform 0.15m/s derived from combustion velocity of kerosene air vapour combustion data (CHEMKIN). All the variation of velocity during analysis is counted in the top and bottom wall boundaries inclusive of the blockage as shown in the Fig. 1. Data computing is carried out at outflow boundary of tube and considered as a diaphragm.

2.5. Pre-processing

The model for solving the 2D analysis was considered as a k-e model without viscous heating of the surface entities to minimize the data loss of temperature rise at the surface of the tube. A multiphase analysis with primary kerosene vapour and secondary air was considered to be flowing through the inlet with an ambient velocity of 0.15m/s and ignition temperature of 1500K was done on the 2D model. Initial operating conditions were for a nominal pressure of 1MPa and linear increment of pressure was considered for evening out the acquired data to imply optimum pressure considered for achieving detonation transiting from deflagration of the mixture. The analysis was done in transient condition with a time step size of 0.01 for the duration of 10s.

2.3 MATLAB

In the calculation and presentation of the pressure blast profiles, numerous methods have been proposed. Baker reports that modified Friedlander's equation given by:

$$P(t) = P_o + P_s (1 - (t/T^+))^{-bt/T+} \quad (1)$$

Where again t is measured from the time of arrival, Ps+ is the peak side-on overpressure P_0 is the ambient atmospheric pressure, and b is the decay coefficient. Thus, a curve fitting of experimental data may be done by matching any three of four blast characteristics: peak side-on overpressure, P_s+, positive phase impulse, I^+, positive phase duration, T^+. The peak overpressure is the highest pressure the initial blast wave creates over the ambient atmospheric pressure. The positive phase impulse is the area under the pressure curve for the duration of the time period of positive overpressure, that same time period is the positive Phase duration, and the initial decay rate is a measure of how quickly the Pressure returns from the spike of peak over pressure to the ambient pressure.

As the pressure profile is of a decaying nature, pressure decay w.r.t range is presented using the following formula

$$P(R) = P_o + P_s e^{\gamma R} \quad (2)$$

P_s is the peak side-on overpressure, γ is the damping ratio, P_0 is the ambient pressure, R is distance in the space, and P(R) is the pressure with respect space again where t is measured from the time of arrival, ta. Thus, this simplified form can be altered to fit the desired profile curve. Negative pressure phase is ignored in this study. Find only one proposed functional form for this phase from the works of Brode (1955):

$$P(t) = P_o + P_s (t/T) (1 - (t/T))^{-bt/T+} \quad (3)$$

Where Po is the ambient pressure, Ps is the peak negative phase overpressure, T^- is the duration of the negative phase, and t is measured from the end of the positive phase (t_0+T). As detonations can vary in both composition and manner of destruction forces, they are usually comported by the shattering effect of sudden release of energy of each detonations or their blast pressure.

A MATLAB coding is made to study the effects of blast waves inside a tube 1m long with a diameter of 2.53cm with placing blockages at the various locations through the length of the tube. The study is done to calculate the peak pressure of detonation and as well as to study the decay time of pressure during the flow through the tube to determine the effective pulse length in a pulse detonating engine. A sample graph of the peak pressure built up and decay of blast wave is shown in Fig 2.

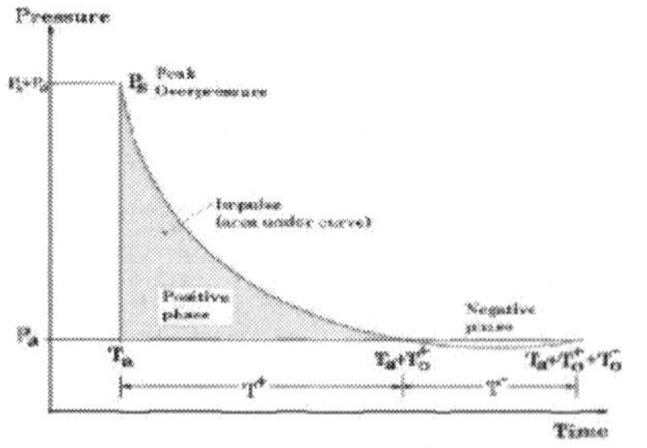

Fig. 2. Canonical blast profile using Friedlander's equation

3 Result and discussions

3.1 Formation of DDT

Deflagration-to-detonation transition (DDT) is the process by which an initially laminar flame undergoes a sequence of changes in propagation mechanism, ultimately resulting in a self-sustained supersonic detonation. While DDT is typically thought of as an acceleration process (since the combustion wave velocity can span four orders of magnitude), it is more properly considered as a process of altering the mechanism of wave propagation, with the propagation velocity reflecting the rate at which different mechanisms occur. Appreciating when these different mechanisms are at work lays a foundation for understanding and ultimately controlling DDT. The initial stages of flame kernel formation and growth as a laminar flame are dominated by diffusion of heat and radicals to the unburned layer ahead of the flame. A whole host of natural instability mechanisms act to wrinkle the flame and increase the burning rate per unit volume and, consequently, increase the propagation velocity.

3.2 Turbulent flow

A more significant role is played by the flow created ahead of the flame due to volumetric dilatation of the combustion products. In a pipe this flow becomes turbulent and a flame propagating into a turbulent flow field undergoes a dramatic increase in the burning rate. This transition to turbulence also sets the stage for feedback between the flow ahead of the flame and the flame itself. The feedback can occur via a fluid dynamic mechanism, whereby an increase in burning rate increases the flow velocity of the gas ahead of the flame, thus amplifying the intensity of turbulent flow into which the flame propagates. The unsteady flow ahead of the turbulent flame also allows for a gas dynamic feedback mechanism, provided the unsteady compression waves sent ahead of the flame increase the temperature sufficiently to produce an accelerating effect on the reaction rates. Regardless of the particular mechanism, when the combustion wave reaches very high turbulent flame speeds (600-1000 m/s), it clearly becomes a mixing limited, rather than a diffusion controlled, process. The fact that high-velocity turbulent flames propagate with mixing rates apparently greater than those required to quench the flame. When the turbulent flame reaches velocities (~ 1000 m/s) where the compression waves pushed ahead of the flame coalesce into a shock generating sufficient adiabatic compression to initiate chemical reactions, the conditions for the final onset of detonation may be present. The final initiation of detonation is usually associated with the abrupt appearance of explosion centers or "hot spots" in the shock - turbulent flame complex (the so-called "explosion in the explosion") which rapidly accelerate in the shocked gas to merge with the initial SHOCK.

Another mode of DDT due to the progressive amplification of pressure waves traversing the reaction zone at the final stages of initiation forms the detonation wave. It appears that the onset of detonation invariably involves a rapid wave amplification process. The result of this amplification process is an initially overdriven detonation, often with wave speeds exceeding the steady Chapman-Jouguet (CJ) wave speed by 50-100%, which quickly decays to a steadily propagating detonation. The observations made due to the position of the blockage placed at the 2/3rd position of the length of the tube with the variation of height of the blockage with reference to the inlet pressure the corresponding pressure and velocity at the outlet of the tube have been shown in Table 1 and Table 2.

Table 1. Variation of outlet velocity with reference to the position of the blockage and inlet pressure.

Pressure MPA	INLET VELOCITY m/s	VELOCITY AT OUTLET m/s				
		HEIGHT OF BLOCKAGE				
		0.632 cm	0.843 cm	1.265 cm	1.686 cm	1.897 cm
1	0.15	1425	1400	1433	1477	1223
10	0.15	4572	4674	4548	4692	3867
100	0.15	15244	14751	14143	14650	-

Table 2. Variation of outlet pressure with reference to the position of the blockage and inlet pressure.

Pressure MPA	PRESSURE AT OUTLET MPA				
	HEIGHT OF BLOCKAGE				
	0.632 cm	0.843 cm	1.265 cm	1.686 cm	1.897 cm
1	1.017	1.013	1.043	1.05	1.07
10	10.311	10.171	10.479	10.542	10.736
100	110.5	101.86	105.273	105.55	-

It is observed in the Table 1, 2 that, for the incremental schema of inlet pressure the variation of velocity and pressure at the exit of the tube is almost linear in nature. With regards to the exit velocity the maximum can be seen using a rectangular type of blockage placed at the 2/3rd position of the tube with 1/3rd the diameter of the tube as the blockage height. But this occurs in a high inlet pressure of 100 MPa and further there is a decrement curve for all other blockage heights at the same pressure input due to the condition of choking and reverse flow scenario can be predicted. So the most feasible height can be considered to be 2/3rd height of the diameter of the tube placed in the 2/3rd distance of the length of the tube, where at the nominal input pressure of 10 MPa, it is seen a sharp increment of output velocity and pressure. Previous experimental study shows that around 40% block of the tube diameter gives the onset of DDT and gives better performance and ignition characteristics. The current comparative analysis has shown a blockage ration of 66% of the total diameter of the tube may be the most optimum of all. The following graphs (Fig. 3, 4) shows the variation of pressure and velocity at the outlet of the tube with reference to Table 1 and 2.

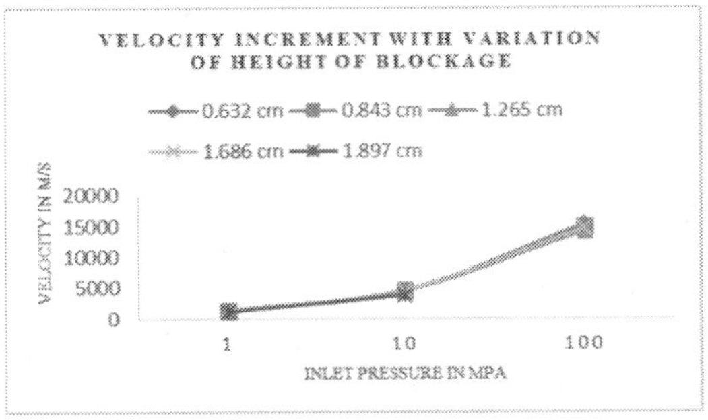

Fig. 3. Graphical representation of velocity at the outlet with variation of pressure and height

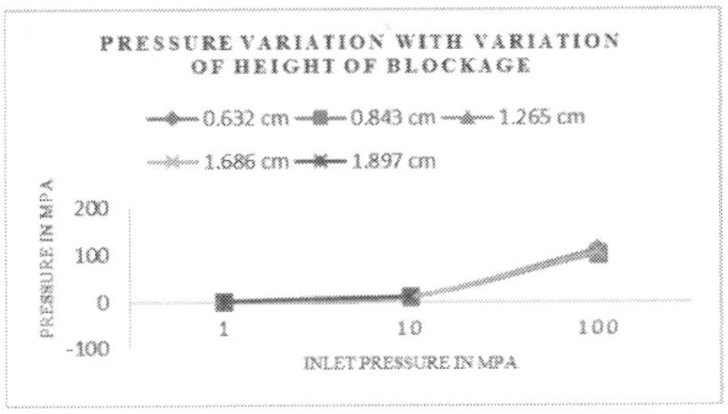

Fig. 4. Graphical representation of pressure at the outlet with variation of pressure and height

The MatLab code is generated as a combination of Friedlander's equation and the effect of peak pressure and the time of decay are studied. The variation of pressure, temperature and velocity of the flow in a tube of length 1m and diameter of 0.0254m with blockage kept at the 2/3rd distance is analysed and the results are plotted for the same with reference to time in Fig. 5 – 7.

The graphs shown above shows that the effective decay time is around 0.25 seconds from the initiation of the detonation, i.e., the pulse length for a PDE can be assumed to be around 0.25 seconds wherein the next cycle can be started. The study also shows that a maximum of 8000 m/s of velocity may decay up to 2000 m/s in the duration of 0.28 seconds. It can be further said that the effective pulse length may be optimized by the above result by designing a system to supply the air fuel mixture to the PDE at a differential pulse setting of 0.25 – 0.28 seconds for continuous thrust.

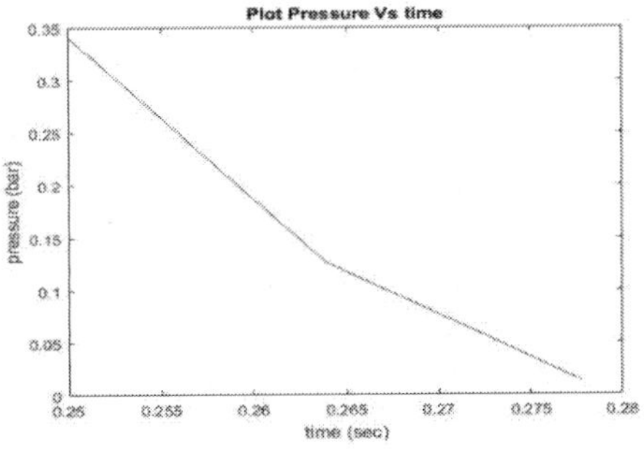

Fig. 5. Variation of pressure with reference to time

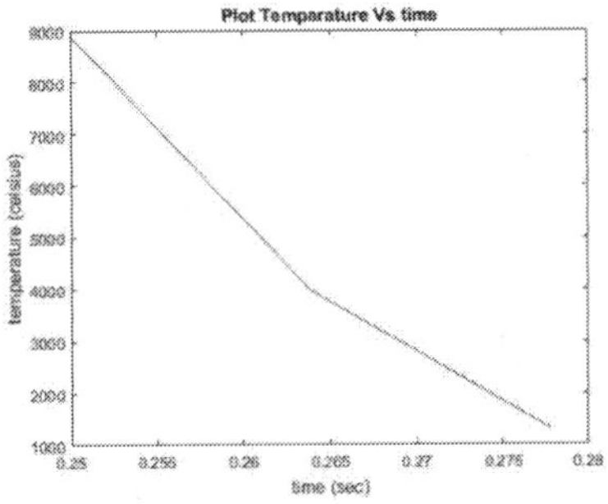

Fig. 6. Variation of temperature with reference to time

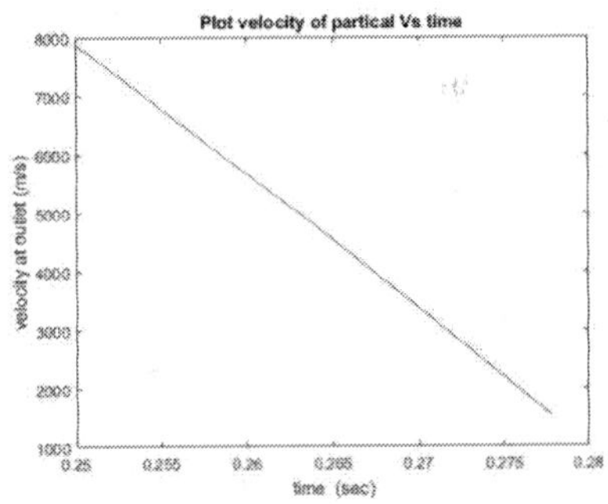

Fig. 7. Variation of velocity of particle with reference to time

3.3 Presence of obstacle

Without obstacle the run-up distance to detonation can be an order of magnitude larger and significantly more stochastic, due to increased sensitivity to the tube wall roughness, position of the igniter, etc. An obstacle blockage ratio of ~ 60% and an obstacle spacing of roughly one tube diameter as being near the point of diminishing return in obstacle spacing and blockage. Thus, the results obtained with these obstacles are believed to be the near minimum length scales required for DDT, and further optimization of the obstacles

https://doi.org/10.1051/matecconf/201817202006

wo9uld result in only minor changes in the length required for detonation formation. Due to the abrupt appearance of detonation at the final stage of the flame acceleration process, initiation is traditionally considered as a unique and well defined
event. Recent research on quasi-detonation, however, suggests that in tubes with obstacles, the same mechanisms responsible for DDT (local explosion centers resulting from shock interactions with obstacles) may also permit steady propagation velocities as low as one half of the CJ velocity. Thus, a CJ detonation is merely the limiting case of a continuous spectrum of combustion wave velocities, and the division between detonation and highly turbulent flame becomes indistinct. If a combustion wave is "denied" the mechanisms of detonative combustion, either by using a tube smaller than the characteristic cell size of the mixture or by using a tube with acoustic absorbing walls to dampen out cellular structure, then the wave will propagate at speeds of the order of one
half the CJ detonation velocity.
Direct initiation (i.e., near-instantaneous deposition of energy into the gaseous mixture, generating a blast wave of sufficient strength and intensity to decay to a self-sustained detonation) would require charges on the order of ~ 100 g of high explosive for each cycle of initiation. Other initiation sources (sparks, lasers, etc.) cannot easily provide this amount of energy (~ 100-1000 kJ) on a cyclic basis. This leaves "self-initiation," or initiation using the chemical energy of the explosive medium itself, to create the conditions necessary for the onset of detonation. The most recent estimates of the critical kernel size for planar initiation [9,10] suggest that the chemical energy available within one to two cell lengths of the explosive mixture is sufficient to initiate detonation in that mixture. The usual path of self-initiation, however, is DDT, and this typically requires an order of magnitude longer length scale before the onset of detonation occurs. For a higher hydrocarbon fuel in air (where the detonation cell size A « 5 cm), the tube diameter must be greater than A for self-sustained propagation, and typically 10 tube diameters or more are required for DDT. Thus, for hydrocarbon-air mixtures, the length required for DDT approaches or even exceeds the length scales of interest for PDE applications. The fact that, in principle, there exists sufficient chemical energy in the mixture to undergo self-initiation within one or two tube diameters suggests that a significant reduction in initiation length scales could be realized if the appropriate route to detonative combustion can be identified. The matlab code run for the program has shown the distance achievement of detonation wave and sustainability of the detonation wave. Knowledge of the mechanisms responsible for DDT and the transition between those different mechanisms is not sufficient at present to identify a "soft" initiation scheme that uses the weak igniter of DDT yet initiates on the short length scales of direct initiation. Experiments with turbulent jet initiation suggest a direction that may prove fruitful.

3.4 Laminar to turbulent transition

The role of an ignition in DDT is to create an initial flame kernel that will grow as a laminar flame. In confined tubes with ignition occurring at a closed end, the laminar flame will quickly become turbulent due to the various instability mechanisms discussed above. Since, for a fuel-air mixture, the laminar flame propagates at speeds of 30-50 cm/s, the time required for the flame to reach the tube wall and make the transition to a turbulent flame can comprise the majority of the time required for the entire DDT process. If a sufficiently powerful igniter is used, it may be possible to bypass the laminar to turbulent flame transition process entirely.
With ignition the flame is replaced with a small distance coverage in which the combustion products jet outward, the time required for DDT is reduced by a factor of more than two as shown in detonation length vs. *t* diagram (Fig 5 – 7).
Viewed as a velocity-distance *(V-x)* diagram, however, the results obtained with the different detonation transition and sustainability for detonation wave are indistinguishable. This is because, while the ignition chamber succeeds in creating a turbulent flame, the turbulent flame must still propagate the same distance before reaching velocities sufficient for the onset of detonative combustion. While the laminar to turbulent flame transition comprises the majority of the time required for DDT, it is negligible in the length scale required for DDT. While it may be argued that a very powerful igniter, or array of igniters, may be capable of near-direct initiation of detonation, studies of detonation initiation by turbulent flames emerging from obstacle-laden tubes and suggest that turbulent flame velocities in the 700-1000 m/s range are necessary for detonation initiation, with turbulent fluctuating velocities on the order of the sonic speed. This would necessitate an initiator of sonic or supersonic jets of hot gas over the entire tube area. Such an initiator would likely consist of a volume of combustible gas that must itself undergo transition to turbulent flame propagation anyway.
The length scale required for further acceleration and transition to detonation is still very large. Thus, a powerful igniter can play an important role in rapid initiation techniques, but only as a means of initially creating a highly turbulent flame. The acceleration of that flame to detonation still consumes a considerable distance. This may be capable of direct initiation upon emerging from a tube diameter smaller than the critical diameter for detonation transmission.

3.5 Sensitivity of mixture

While chemical kinetics plays a role in laminar and turbulent flame propagation, it is only with the shock-initiated reactions of the final onset of detonation that the exponential nature of kinetic rates dominates the propagation mechanism as shown through matlab codes. Compared to laminar-turbulent flame transition and the acceleration of turbulent flames, chemical sensitization or desensitization of combustible gases for detonability has received comparatively little attention to increase in the length required for transition to detonation.

As DDT is an extremely complex phenomenon, incorporating numerous different mechanisms and orders of magnitude variations in propagation velocity, it is essential that studies of DDT carefully isolate the effect being examined. The results reviewed in this paper showed that the effect of a block size and detonation sustainability for fuel-air mixtures is confined to promoting a more rapid transition from laminar to turbulent flame and does not significantly influence the run-up distance required for detonation. Only with a turbulent jet igniter of very large diameter *(d* > 10A) can the intensities and scales of turbulence necessary for the direct onset of detonation be realized. Obstacles on the tube wall can increase the intensity of turbulence in the flame and introduce transverse pressure fluctuations (shocks), and thus dramatically increase the burning rate and strength of compression waves sent ahead of the flame. This sets the stage for shock-induced combustion and the onset of detonation by the formation of explosion centers, but detonation will not occur if the tube is smaller than the characteristic detonation length scale (the cell size). It is in this final onset of detonation that kinetic rates dominate the transition mechanism and the presence of free radicals or other chemical sensitizers can have a pronounced effect. An efficacious sensitizer for hydrocarbon fuel-air mixtures has yet to be identified, in which a heavy hydrocarbon fuel is re-formed via a partial oxidation process to more sensitive hydrocarbons and free radicals, appears to be a promising way to dramatically reduce the transition length to detonation. In light of the fact that DDT is an ensemble of various mechanisms, the results reported in this paper represent the same shortcomings as prior work in DDT: examining the run-up distance to detonation alone does not isolate the various mechanisms involved. Rather, future work on DDT should be focused on defining the basic scaling laws for DDT, and then relating these laws to more fundamental parameters, such as the ratio of mixture cell size to tube diameter.

4 Conclusions

Based on the 2D CFD analysis and MatLab coding done for the tube of length 1m and diameter 0.0254m with blockages setup at the 2/3rd distance of the length of tube and by varying the height of the blockages, the following conclusions have been drawn:

1. It has been observed that for a maximum pressure of 100 MPa and 1/3rd height of the blockage the velocity at the exit is maximum at 15244 m/s, which may not be feasible for the actual working process.

2. For an inlet pressure of 10 MPa and 2/3rd height of the blockage, the velocity at the exit of the tube is found to be around 4692 m/s, which is the maximum feasible velocity. The corresponding pressure at the exit is found to be 10.542 MPa.

3. The critical blockage ratio can be found around 2/3rd height of diameter of the tube which has a steady incremental velocity schema for various pressure inputs.

4. The MatLab coding data concludes that the time taken for the inlet pressure to decay is around 0.275 seconds, so it can be incepted that 0.25 seconds may be the effective pulse length for a PDE.

The comparative study of both computational and MatLab coding evolves that the critical blockage impacts the exit velocity and pressure of the PDE, but the pulse length used for the initiation of the PDE can be an effective parameter to control the optimized performance and creation of detonation waves and propagation of the same for effective thrust production.

References

1. K. Kailasanath, AIAA paper **0470**, (2002)
2. F. Falempin, Meeting Proceedings RTO-MP-AVT-135, Paper **11**. Neuilly-sur-Seine, France: RTO, (2006)
3. S.M. Frolov, J. Propul. Power, **22,** 1162–1169 (2006)
4. W. Fan, C. Yan, X. Huang, et al., Combust. Flame **133 – 04,** 441–450, (2003)
5. Z. Wang, C. Yan, W. Fan, L. Zheng Proc. Inst. Mech. Eng. G. J. Aerosp. Eng., **223**, 721–728, (2009) doi:10.1243/09544100JAERO486.
6. Z. Wang, X. Chen, J. Huang, L. Zheng, C. Peng, Appl. Therm. Eng., **62**, 407–414, (2014)
7. Z. Wang, J. Lu, J. Huang, C. Peng, L. Zheng, Appl. Therm. Eng., **73**, 21–29, (2014)
8. C. Cathey, F. Wang, T. Tang, et al., AIAA paper, **443**, (2007).
9. Z. Wang, X. Chen, L. Zheng, J. Lu, C. Peng, J. Propul. Technol., **35**, 1434–1440, (in Chinese), (2014)
10. R. Knystautas, J.H.S. Lee, I.O. Moen, et al., The Combustion Institute, Pittsburgh, **1235–1244**, (1979)
11. F. Carnasciali, J.H.S. Lee, R. Knystautas, Combust. Flame, **84**, 170–180, (1991)
12. C.M. Brophy, L.T.S. Werner, J.O. Sinibaldi, AIAA paper, **84**, 2003-1344, (2003)
13. C. Li, K. Kailasanath, J. Propul. Power, **21**, 183–186, (2005)
14. T. Yatsufusa, M. Ohira, S. Yamamoto, et al. AIAA paper **2004-1213**, (2004)
15. S.P. Medvedev, S.V. Khomik, H. Olivier, et al., Shock Waves, **14**, 193–203, (2005)

https://doi.org/10.1051/matecconf/201817202006

16. H. Shimada, Y. Kenmoku, H. Sato, A.K. Hayashi, AIAA paper **2004-0308**, (2004)
17. D. Bradley, C.G.W. Sheppard, I.M. Suardjaja, R. Woolley, Combust. Flame, **138**, 55–77, (2004)
18. M. Li, C. Yan, J. Exp. Fluid Mech, **23**, 92–97 (in Chinese), (2009)
19. J. Yu, Y. Qin, G. Gao, J. Aerosp. Power, **26**, 1043–1047 (in Chinese), (2011)
20. H. Zeng, L. He, X. Zhang, et al., J. Appl. Mech, **27**, 543–548 (in Chinese), (2010)
21. Z. Wang, X. Chen, L. Zheng, et al., J. Propul. Technol., **34**, 422–427 (in Chinese), (2013)
22. X. Cai, J. Liang, Z. Lin, R. Deiterding, F. Zhuang, Combust. Sci. Technol., **187**, 525–536, (2015)
23. L. Zheng, C. Yan, W. Fan, et al., Journal of Combust. Sci. Technol., **9**, 344–347 (in Chinese), (2003)
24. W. Chen, Q. Zhang, W. Fan, et al., J. Propul. Technol., **32**, 296–300 (in Chinese), (2011)
25. G. Ciccarelli, S. Dorofeev, Prog. Energy Combust. Sci., **34**, 499–550, (2008)
26. Z. Wang, C. Yan, L. Zheng, W. Fan, Combust. Sci. Technol., **181**, 1310–1325, (2009)
27. S. Srikrishnan, Dr. P. K. Dash, IJMET, **8-6**, 447-454, (2017)

Impact of nano-silicon fuel additive on combustion, performance and emission of a twin cylinder CI engine

Anand Kumar S A[1,*], *Raja* V[1], *Dhivakar Raviram* R P[2], and *Sakthinathan* G[1]

[1]Department of Production Technology, Madras Institute of Technology, Chennai, India
[2]Saveetha School of Engineering, Saveetha Institute of Medical and Technical Sciences, Chennai, India

Abstract. Combustion characteristics of a fuel defines its performance and emission characteristics. Enhancement of combustion characteristics is feasible by improvisation of fuel properties. Fuel additives were used for varying fuel properties. The evolution of 'nano-concept' develops countless applications in the existing technologies. In this experiment silicon nanoparticles were synthesized by ball milling micron sized silicon particles for 45 hours. The elemental and structural characterizations for the additive material were carried out by EDS, SEM and TEM analysis. Silicon nano additive was mixed in three different weight proportions with diesel to prepare the test fuels. The fuel properties variation with the addition of nano additive were studied. Engine testing was carried out at constant 1200 rpm speed and varying load conditions. Diesel fuel added with 0.5 wt% Si nanoadditive (Si 0.5) showed maximum load carrying ability among the different test fuels. In comparison with diesel at 1200 rpm and 100% load condition, an increase in torque of 5.91% was observed and BTE was increased by 8.93% with a decrease in NO_x emission by 27.3%. Variation in the performance and emission characteristics of the fuels were the results of change in heat release rate and combustion timing with the addition of nano additives which could be studied from the combustion characteristic curves.

Keywords: Combustion Analysis, Fuel Additive, Nano-fuel, Silicon Nanomaterial

1 Introduction

The rapid development in technology increases the energy requirement which leads to the development of new energy sources as well to find an efficient way of using them. Enhancing the fuel properties results in efficient utilization of energy stored in a fuel.

The volumetric or gravimetric energy content of metals are considerably higher than the energy densities of traditional liquid fuels [1]. The combustion of metals has long been of interest to the combustion community because of their high energy density [2, 3]. Combustion of liquid fuels mixed with metals release high combustion energies. Many researchers investigated the combustion effect of such fuels called 'slurry fuels'. Slurry fuels are liquid hydrocarbon fuels loaded with micron sized metallic particles (40-80% by weight) [1].

The practical application of slurry fuels are limited, because of incomplete combustion, lower combustion efficiency, deposits in the combustion chamber and increased particulate emissions [1]. The slurry fuels doesn't form a stable solution, since the micron sized metal particles settle down by gravity. This causes incomplete combustion and increased particulate emission.

The advancement in nanotechnology enable production, control and characterization of nanized metallic particles. Nanosized metallic powders possess high specific area, which lead to high reactivity [4]. Using nanoscale energetic materials as fuel additive to enhance combustion of liquid fuels is an interesting concept.

Nano-fuels are liquid fuels in which low concentrations of metal and metal oxide (Al, B, C, Mg, Mn, Fe, CeO_2 and Al_2O_3) nanopowders (size ranges between 10-210nm) are dispersed to form a stable solution [5-7]. Nanoparticles form a more stable suspension than microparticles. While the thermophysical properties are not greatly affected since concentration and size of the particles in nano-fuels are small.

Previous studies have demonstrated significant improvements using such nano-fuels including higher energy release, increased ignition probability, improved burning rates, enhanced catalytic effects and shortened ignition delay [5, 7 & 8]. Based on these developments, research in the relatively new area of nano-fuels has

* Corresponding author: anandkumar0412@gmail.com

become a topic of significant interest. It is clear that soon many areas of combustion will be influenced by nanotechnology as a result of future fuels, propellants, pyrotechnics, explosives and reactive materials having nanoscale features.

Diesel fuel refined from crude oil is a well-known commercial and industrial fuel. In this research work, nano-fuel is prepared with silicon nanoadditive in diesel base fuel. Three test fuels were prepared with three different weight ratios of silicon nanoadditive. The prepared nano-fuels were tested in a twin cylinder diesel engine to study the effect of nanoparticle in the combustion, performance and emission of the base fuel.

2 Synthesis and Characterization of Nanoadditive

The stoichiometric combustion reaction of diesel fuel is generally represented as

$$C_{12}H_{23}+17.75O_2+66.97N_2 \rightarrow 12CO_2+11.5H_2O+66.97N_2$$

The bonds between C and H atoms in the fuel were broken and new bonds were formed with oxygen. The formation of the products CO_2 and H_2O by the oxidation of C and H atoms releases excessive heat energy. This oxidation reaction is called combustion reaction. Based on this, a material which releases more energy during oxidation reaction was chosen as additive.

Several materials like B, Al, Mg, Mn, and Fe were used as additives by several researchers [3, 5, 9-12]. In this work silicon is used, which is easily available at low cost. Silicon releases more energy than carbon and hydrogen during oxidation. The bond energy of the combustion reactants are given in the Table 1.

Table 1. Bond Energy of Combustion Reactants

Material	Bond	Bond Energy kJ/mol
Carbon	C-O	360
Hydrogen	H-O	428
Silicon	Si-O	798

Microsized silicon particles were milled in a planetary ball mill for 45 hours at 300rpm speed. Tungsten carbide vial and balls were used for milling. 20:1 ball to material weight ratio was followed.

The unmilled silicon particles were analysed in Scanning Electron Microscope to check the size of the particles. Energy Dispersive Spectrum analysis was done to confirm the chemical composition of the additive material. Transmission Electron Microscope images of the milled silicon particles were taken to confirm the particle size after milling.

3 Nano-fuel Preparation

The synthesized silicon nanoparticles were mixed with diesel in three different weight ratios (0.25 wt%, 0.5 wt% and 0.75 wt%). So that, the effect of concentration of the nanoadditives in the base fuel could be analysed. The nanosized particles formed a stable solution with the diesel neglecting gravitational effect. Figure 1 shows the base fuel and nano-fuel.

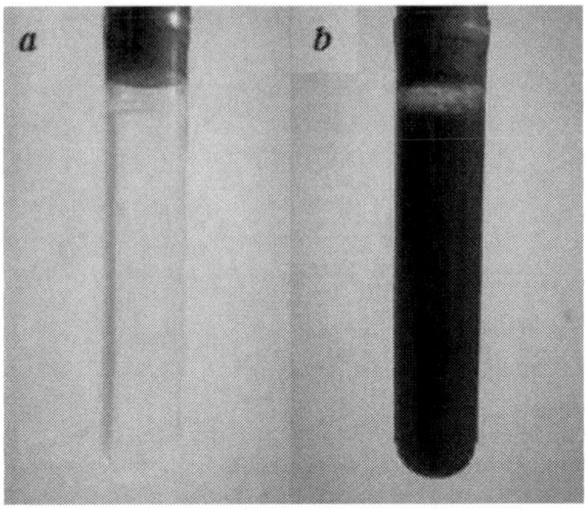

Fig. 1. Base fuel and nano-fuel

The thermal properties of nano-fuels were characterized by ASTM standard D 445 test method for kinematic viscosity, ASTM standard D 4052 test method for specific gravity, ASTM standard D 93 test method for flash point, ASTM standard D 240 test method for calorific value.

4 Engine Testing

For effective analysis of the prepared nano-fuel, it is essential to test the fuel in an engine for sufficient number of times at different load conditions. It is necessary to check the combustion, performance and emission characteristics of the fuel in an engine under actual working conditions. The experimental setup used is shown in the Figure 2.

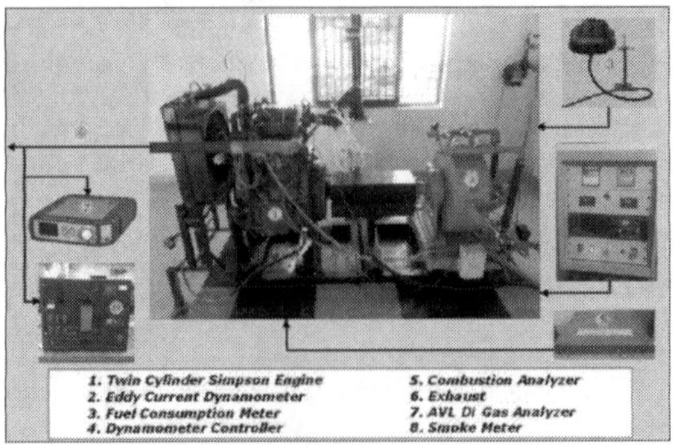

Fig. 2. Engine testing setup

A twin cylinder direct injection diesel engine of Simpsons coupled with AVL eddy current dynamometer was used for this study. The engine was operated at a constant 1200 rpm speed with varying load conditions namely 0%, 25%, 50%, 75% and 100%. The technical specifications of the engine are listed in the Table 2. The exhaust emissions were analysed by an AVL Di Gas Analyzer and a smoke meter.

Table 2. Specification of the test Engine

Engine	Simpsons S 217
Number of cylinders	2
Configuration	Vertical in-line diesel engine
Capacity	21 kW (28 bhp @ 2000 rpm)
Compression ratio	18.5 : 1
Cooling system	Water
Displacement	1670 cc (127mm*91.44mm)
Injection timing	30°BTDC @ 200 bar
Nozzle hole number	3
Nozzle hole diameter	0.3 mm

Experiments were performed in a randomized design. A total of 20 tests (4 fuel x 5 loads) were performed in three replications to get an accurate reading. Initially the engine was operated for 30 minutes to achieve steady state condition. It was confirmed by the evaluation of coolant and lubricant temperature. The accuracy of the measured parameters are given in Table 3.

Table 3. Uncertainty in the measured data

Parameter	**Uncertainty**
Engine Speed	± 4 rpm
Temperatures	± 6.5° C
Pressure	± 1 bar
Crank angle encoder	± 1°
CO	± 0.2 %
HC	± 10 ppm
NO_X	± 20 ppm

5 Results and Discussion

The structural and chemical characterization of the synthesized nanoparticles, the physicochemical properties of the prepared nano-fuel are discussed. The results of the experimental studies on the effect of the silicon nanoparticle in engine analysis is also discussed.

5.1 Nanoadditive Characterization

The particle size analysis of the silicon used in this research work was done before and after milling by SEM and TEM respectively. To confirm the chemical composition EDS analysis was done.

Fig. 3. SEM images of unmilled silicon microparticles

The SEM analysis was done before milling. The SEM images of particles before milling at two different magnifications are provided in the Figure 3. From the figure it is clarified that the particles doesn't provide uniform size and structure. The size of the particles vary in between 2 μm and 30 μm.

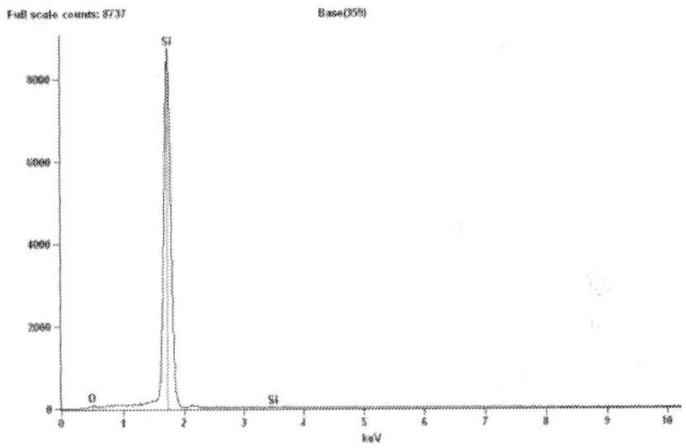

Fig. 4. EDS analysis of Silicon

The EDS spectrum of silicon is presented in the Figure 4. The elemental composition of sample material obtained from the EDS analysis is given in Table 4. The silicon used in this work is 97.89 % pure. Slight oxidation of upto 2.11 % is seen in the particles. The effect of the oxidation of silicon particles is negligible during combustion.

Table 4. EDS report for Si chemical composition

Element	**Net Counts**	**Weight %**	**Atom %**
O	481	2.11	3.65
Si	111535	97.89	96.35
Total		100.00	100.00

Silicon microparticles were milled for 45 hours in the ball mill and analyzed in TEM for testing the particle size. It could be observed from Figure 5 that, the indefinite shaped microsized silicon particles was transformed into spherical and rod shaped nanoparticles after milling. The size of the particles vary between 60 nm and 100 nm. These nanoparticles were used to prepare a stabilized nano-fuel solution.

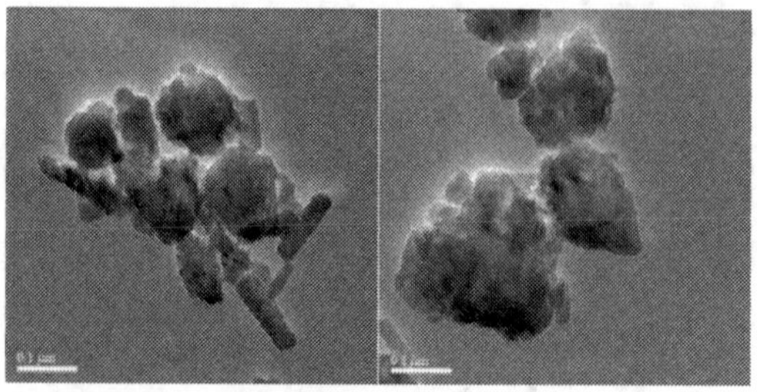

Fig. 5. TEM images of silicon nanoparticles

5.2 Nano-fuel Characterization

The test fuels prepared with three different composition of silicon nanoparticles were further analyzed to find the variation in its physicochemical properties with respect to the addition of nanoadditives. The variation in the physicochemical properties are provided in Table 5.

Table 5. Physicochemical properties of the test fuels

Fuel	Calorific Value MJ/kg	Specific Gravity	Kinematic Viscosity @40°C cSt	Flash Point °C
ASTM	**D240**	**D4052**	**D445**	**D93**
Diesel	42.891	0.831	2.73	79
Si 0.25	43.478	0.8331	2.79	76
Si 0.5	43.731	0.8352	2.82	71
Si 0.75	43.789	0.8372	2.87	73

The quantity of silicon nanoadditives added were very small (0.25 wt%, 0.5 wt% and 0.75 wt%). With the addition of nanoadditives a slight increase in the calorific value, specific gravity and viscosity were observed [13]. While the flash point is found to show a decreasing trend for the first two samples and it is increasing on further addition of nanoadditive.

The high energy content and reactivity of the silicon nanoparticles lead to the increase in calorific value of the fuel. As the wt% of the particle was increased the specific gravity and the viscosity of the fuels were also increased. Addition of silicon nanoparticles decreases the flash point of the fuel. As reported by several researchers the decrease in flash point attributes the shortened ignition delay of metals. At 0.75 wt% an increase in flash point is observed. Increase in concentration of particles leads to increase in agglomeration probability. Increase in agglomeration reduces the specific surface area of the reactive material. This is the probable cause for the reduction in flash point.

5.3 Combustion Analysis

The combustion characteristics of a fuel can be studied in detail from ignition delay of the fuels, heat release rate curve and P-Θ curve. The ignition delay of the tested fuels with respect to the different load conditions are presented in the Figure 6.

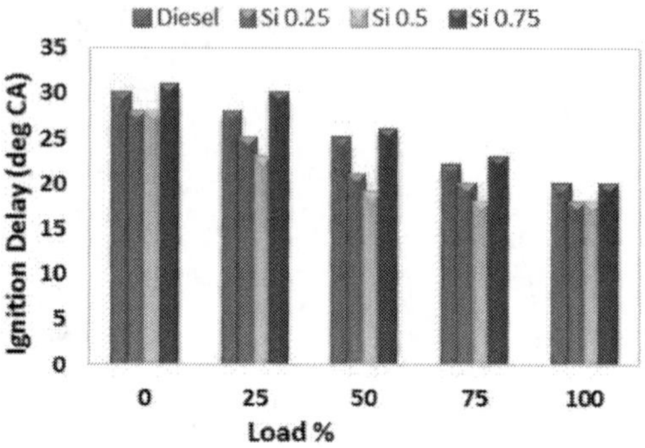

Fig. 6. Ignition delay of the fuels

For diesel without nanoadditive the ignition delay was 30° CA at no load condition and it get reduced to 20° CA at full load condition. The ignition delay for Si 0.5 fuel was 28° CA at no load condition and 18° CA at full load condition. The ignition delay of the fuels decreases with increase in load. Decrease in ignition delay was observed with increase in the quantity of Si additive upto 0.5 wt%. Further increase in the quantity of Si additive to 0.75 wt%, increases the ignition delay period. Si 0.75 fuel shows 31° CA at no load condition and 20° CA at full load condition. With increase in Si additive more than 0.5 wt% increases the agglomeration of the particles, which affects the stability of the particles in the fuel. The agglomerated particles tend to settle down in the fuel. This causes an adverse effect in the combustion of the fuel.

The heat release rate curve for the four test fuels were compared in Figure 7. The maximum heat release rate is observed from Si 0.25 fuel. But the area under heat release rate curve is equivalent to that of Si 0.5 fuel. Both the fuels releases their maximum heat energy in between 0° and 5° ATDC, which results in effective utilization of the fuel energy. The heat release rate curve for diesel shows that the heat release rate is lower than the former fuels and the maximum heat release rate occurs at 5° BTDC. The maximum heat release rate by the fuels with Si nanoadditive is the result of increase in calorific value of the fuel by the addition of nanoadditive. Si 0.75 fuel shows very low heat release rate than the other test fuels. Increase in ignition delay causes, release of maximum heat release rate at premixed combustion phase 10° BTDC. Particle agglomeration leads to incomplete combustion of fuel.

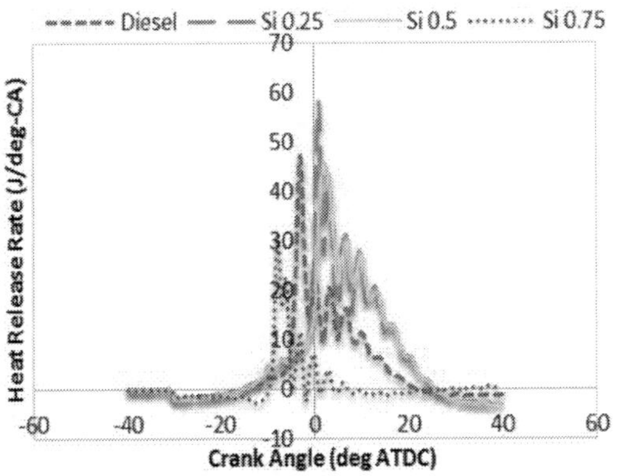

Fig. 7. Heat release rate curves comparison

The pressure rises inside the combustion chamber with respect to the crank angle for the fuels were compared in Figure 8. The pressure rise is a direct effect heat released by the fuel.

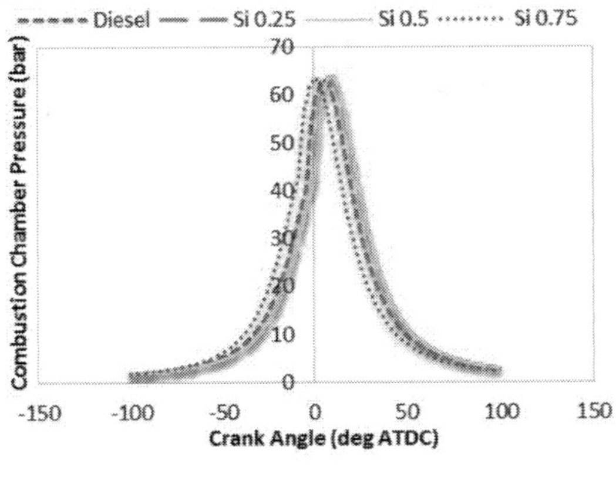

Fig. 8. P-Θ curves comparison

The maximum pressure created by the fuels during combustion are almost the same, but the time at which maximum pressure generation varies. Si 0.25 fuel and Si 0.5 fuel shows similar pressure rise curve with maximum pressure at around 10° ATDC. The maximum pressure rise for diesel fuel occurs at 5° ATDC. Si 0.75 fuel generates maximum pressure close to TDC. From the heat release rate curve and pressure rise curve it is clear that Si 0.25 fuel and Si 0.5 fuel shows better combustion characteristics than the other two fuels.

5.4 Performance Analysis

The performance of a fuel is described by thermal efficiency of the fuel. In this experiment initially the load carrying capacity of the fuels were analyzed. The fuel which produces maximum torque at 100% loading condition and 1200 rpm constant speed provides the maximum load carrying capacity. Figure 9 shows the torque generated by the test fuels. The maximum load carrying capacity is associated with Si 0.5 fuel. It produces 98.5 Nm at 100% loading condition and 1200 rpm speed. The fuel with Si 0.25 fuel produces 96 Nm and diesel produces 93 Nm torque in the same working conditions. The maximum heat release rate of the fuels loaded with nanoadditives results in high load carrying capacity.

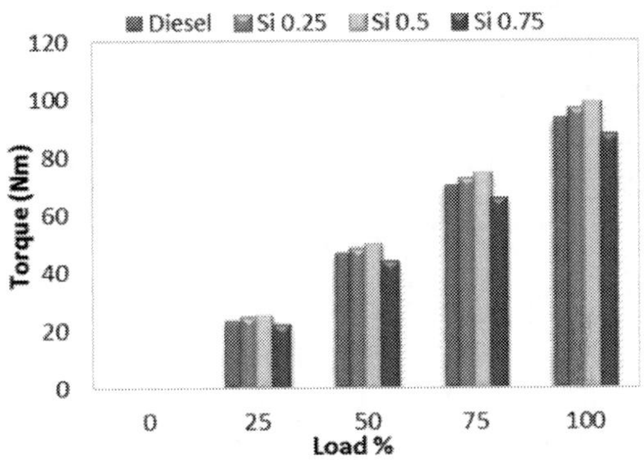

Fig. 9. Load carrying capacity of the fuels

In the following graphs, the nodal points in the curves represents the loading conditions in increasing nature through the x-axis. The graph was plotted between brake power corresponding to the loading conditions in x-axis and the performance values corresponding to the brake power in y-axis. Each fuel produces different brake powers at a loading condition, since constant speed was maintained throughout the testing. This gives more understanding of the fuel characteristics.

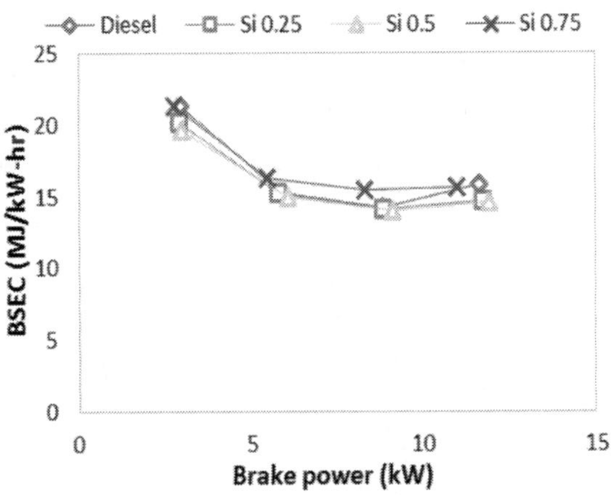

Fig. 10. Brake Specific Energy Consumption curve

Figure 10 presents the Brake Specific Energy Consumption curve for the four fuels. In general, BSEC value decreases when load was increased from 25% to 50%, with further increase in load the BSEC value follows a steady curve. This shows that frictional power is more during low load conditions, so that more energy from the fuel is supplied to overcome the frictional power. The BSEC values of diesel, Si 0.25, Si 0.5 and Si 0.75 fuels are 15.8, 14.67, 14.56, 15.65 MJ/KW-hr respectively. This clears that Si 0.25 fuel and Si 0.5 fuel provides more energy content than the other fuels. It is found that fuel with Si 0.75 fuel shows less BSEC value than diesel, which doesn't mean that it provides better performance than diesel. Because the corresponding brake power for diesel and Si 0.75 fuel at 100% loading conditions are 11.64 kW and 11.22 kW respectively. Comparatively at same brake power diesel performs better than Si 0.75 fuel.

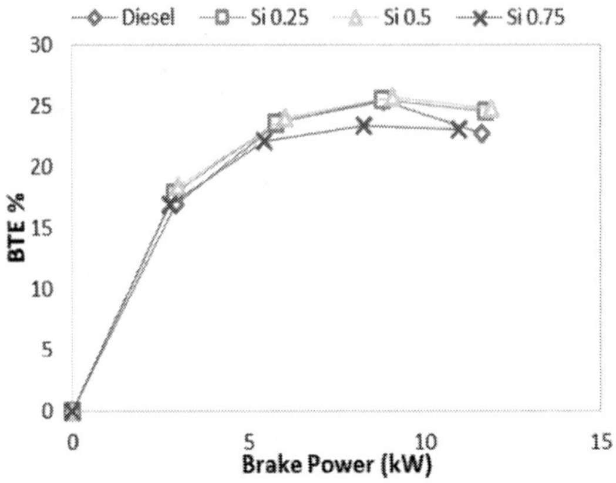

Fig. 11. Brake Thermal Efficiency curves

Brake Thermal Efficiency is inversely related with BSEC. This describes that the thermal efficiency is high when maximum energy is utilized from the fuel. From Figure 11 it is observed that Si 0.5 fuel shows maximum BTE in all the loading conditions. Si 0.25 fuel shows equivalent BTE with Si 0.5 fuel. Diesel shows similar efficiencies at midload conditions as that of Si 0.25 fuel and Si 0.5 fuel but the efficiency is lower at full load condition. This proves the better efficiency of Si 0.25 fuel and Si 0.5 fuel in full load conditions. BTE of Si 0.75 fuel is lower than the other test fuels in all the loading conditions.

5.5 Emission Analysis

Several research attempts were made by researchers to enhance the thermal efficiency of the fuel by means of fuel additives, thermal barrier coating [14] etc., it is important to enhance the thermal efficiency as well as simultaneous reduction of emission level is also important. Stringent emission norms are followed by all the countries to control the atmospheric pollution caused by vehicle emission.

The combustion nature of a fuel describe its emission characteristics. The major reason for emission during combustion are described below.

- Incomplete combustion causes CO emission.
- Poor atomization results in unburned fuel particles which are the sources for HC emission.
- High heat release rate and maximum combustion timing causes NOx emission.

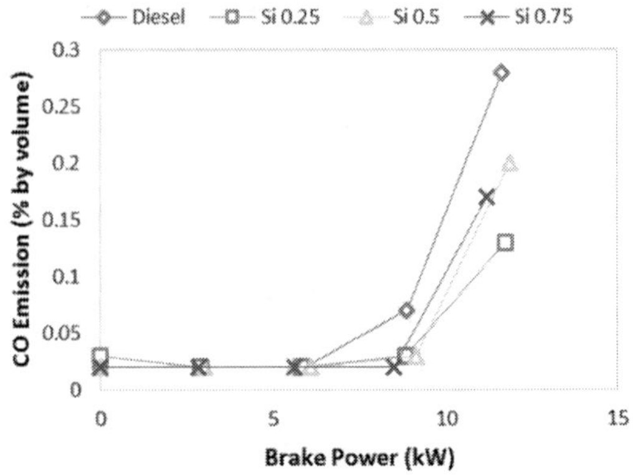

Fig. 12. Carbon Monoxide emission curves

The CO emission rate of the fuels are shown in Figure 12. The emission level is very minimum upto 50% loading condition. When the loading condition was increased further, increase in CO emission was observed. At high load conditions more fuel is injected into the combustion chamber. Increase in fuel quantity results in improper fuel mixing which causes incomplete combustion of fuel.

Among the four tested fuels, more CO emission occurs during diesel combustion. The additive particles mixed in the fuel will absorb the heat energy in the combustion chamber more rapidly than the diesel fuel, which causes evaporation of the fuel droplets. This increases the mixing of fuel with air molecules, thus reduction in CO emission is observed for additive mixed fuel.

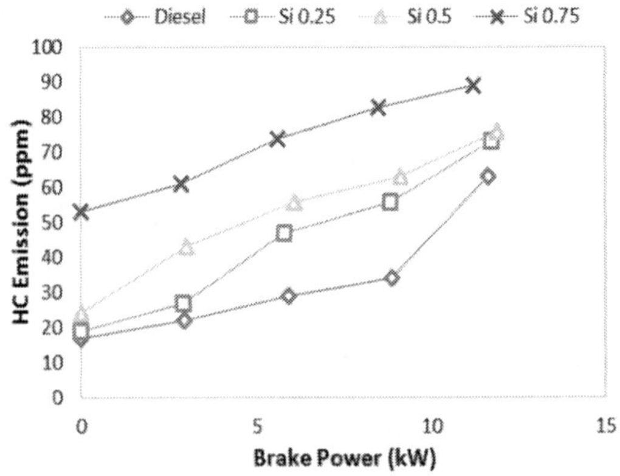

Fig. 13. Hydrocarbon emission curves

Poor atomization results in larger fuel droplets. The inner core of these larger fuel droplets often failed to evaporate and form a fuel-air mixture for combustion.

These unburned fuel particles are emitted as HC in the exhaust. Figure 13 represents the HC emission of the test fuels. Diesel combustion shows minimum HC emission in comparison with the fuels with additive. As the concentration of the additive is increased, HC emission is also increased. With the addition of nanoadditives the viscosity of the fuel increases, results in increased droplet size. This causes increased HC emission in fuels with nanoadditive.

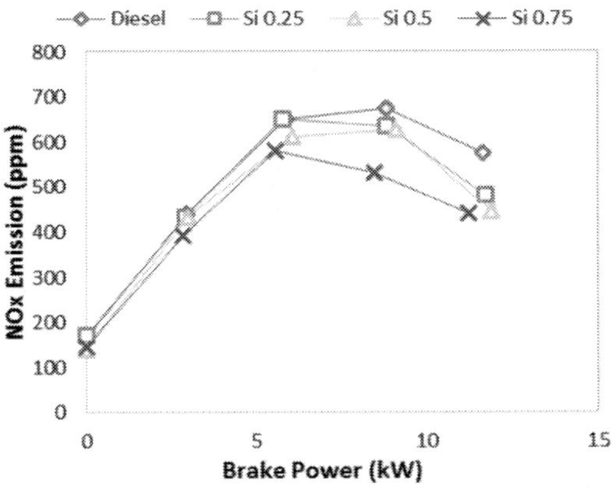

Fig. 14. NOx emission curves

NOx emission generated by the four fuels at various load conditions were presented in Figure 14. Upto 50% loading condition the test fuels follow similar characteristic curve which tends to increase with increase in load. For 75% and 100% load condition Si 0.75 fuel shows reduced NOx emission. From the heat release rate curve it is clear that the maximum heat released by Si 0.75 fuel is very minimum compared to other fuels, results in reduced NOx emission. The other three fuels shows a similar increasing trend in 75% load condition.

At full load condition fuel with Si 0.25 fuel and Si 0.5 fuel emits lower amount of NOx in comparison with diesel. This could also be explained from the heat release curve. The maximum heat release rate for diesel fuel occurs at 5° BTDC at 100% load condition. After the maximum heat release rate is observed, further increase in temperature inside the combustion chamber is possible since the piston is approaching TDC. The high temperature availability for prolonged time inside the combustion chamber results in increased NOx emission. While the Si 0.25 fuel and Si 0.5 fuel releases maximum heat energy at around 5° ATDC. This period corresponds to expansion stroke in which the temperature inside the combustion chamber tends to decrease. This results in reduced NOx emission in comparison with diesel fuel at 100% loading condition.

Smoke emitted by the test fuels at different load conditions are compared in the Figure 15. Increase in load increases the fuel intake which simultaneously increase the smoke emission. Diesel shows minimum smoke emission compared to the fuels mixed with additives. Increase in the quantity of additive increases the smoke emission. Diesel is a liquid fuel, while additives are solid particles mixed with diesel. After combustion the solid additive particles will be emitted as solid particles from the exhaust. This shows the reason for increase in smoke emission while using additives in the fuel.

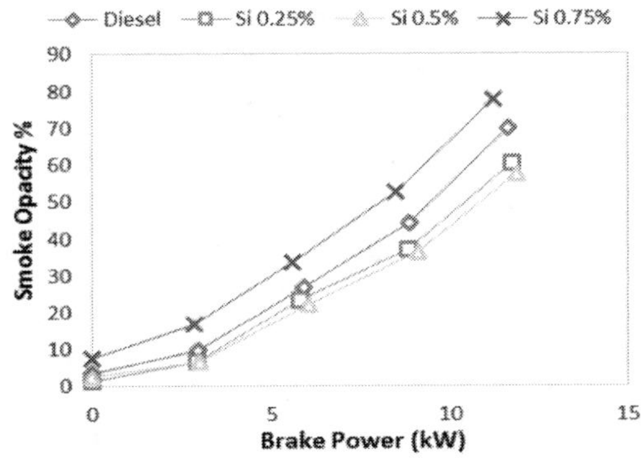

Fig. 15. Smoke opacity curves

6 Conclusion

The combustion, performance and emission of a fuel loaded with Si nanoadditive was tested in a twin cylinder diesel engine. Si microparticles were milled in a planetary ball mill for 45 hours. The particle size was reduced from 30μm to 100nm and was confirmed by SEM and TEM analysis. The size of the milled particles were in the range between 60 nm and 100 nm. The elemental composition was analyzed by EDS and the additive material was found to be 97.89% pure. Nano-fuel was prepared by adding three different concentrations (0.25 wt%, 0.5 wt% and 0.75 wt%) of silicon nanoadditives in diesel fuel. The physicochemical properties of the test fuels were analyzed and found that with increase in additive concentration a slight increase in calorific value, specific gravity and viscosity were found with decrease in flash point.

The engine analysis proved that the Si 0.5 fuel is better than the other test fuels. Si 0.25 fuel shows similar characteristics as of Si 0.5 fuel but the load carrying capacity of Si 0.5 fuel is more than Si 0.25 fuel. The combustion analysis of the fuel shows that Si 0.5 fuel provides better heat release rate and combustion chamber pressure during combustion. Si 0.5 fuel shows 5.91% increase in torque generation and 8.95% increase in BTE than diesel. In case of emission Si 0.5 fuel shows 28.57% decrease in CO emission and 27.3% decrease in NOx emission, with 20.63% increase in HC emission in comparison with diesel. From the results it is clear that Si 0.5 fuel provides better combustion, performance and emission characteristics than diesel.

References

1. I. Javed, S. W. Baek, and K. Waheed, Combust. Flame, **162**, pp. 774–787, (2015).

2. Y. Gan and L. Qiao, Combust. Flame, **158**, pp. 354–368, (2011).
3. R. A. Yetter, G. A. Risha, and S. F. Son, Proc. Combust. Inst., **32 II**, pp. 1819–1838, (2009).
4. Sadhik Basha and R. B. Anand, Alexandria Eng. J., **53**, pp. 259–273, (2014).
5. Yoon and S. W. Baek, www.ijm-me.org Int. J. Mater. Mech. Eng., **4**, pp. 44–49, (2015).
6. M. A. Lenin, M. R. Swaminathan, and G. Kumaresan, Fuel, **109**, pp. 362–365, (2013).
7. Y. Gan and L. Qiao, pp. 1–12, (2010).
8. M. Jones, C. H. Li, A. Afjeh, and G. Peterson, Nanoscale Res. Lett., **6**, p. 246, (2011).
9. E. L. Dreizin, Combust. Flame, **117**, pp. 841–850, (1999).
10. Javed, S. Wook, K. Waheed, G. Ali, and S. Oh, "*Evaporation characteristics of kerosene droplets with dilute concentrations of ligand-protected aluminum nanoparticles at elevated temperatures*."
11. T. Miller and J. Herr, AIAA, pp. 1–10, (2004).
12. Javed, S. W. Baek, and K. Waheed, Exp. Therm. Fluid Sci., **56**, pp. 33–44, (2014).
13. V. Sajith, C. B. Sobhan, and G. P. Peterson, Adv. Mech. Eng., (2010).
14. Ganapathy, S., Appancheal, A., and Velusamy, R., " SAE Technical Paper **2017-28-1981**, (2017)

Numerical Estimation of Notch Stress Intensity Factors of Sharp V-Notches

Mirzaul Karim Hussain[1,*], and *K.S.R.K.* Murthy[2]

[1]Research Scholar, Department of Mechanical Engineering, Indian Institute of Technology Guwahati, Guwahati, Assam, India-781039
[2]Professor, Department of Mechanical Engineering, Indian Institute of Technology Guwahati, Guwahati, Assam, India-781039

Abstract. In the present work a simple and efficient least squares method is implemented for accurate estimation of notch stress intensity factors (NSIFs) of sharp V-notches. Finite element (FE) stress components near a notch tip is used in the present method for determining the NSIFs. Pure mode I and mixed mode (I/II) examples are considered for numerical investigations. The mixed mode stress components are disintegrated into opening mode and shear mode stress components to separate out the mode I and mode II singularities. Thereafter, least squares method is implemented to calculate mixed mode NSIFs. The present method is easy to incorporate in existing standard finite element codes. The results obtained by the present method are found to be in good agreement with the published data.

Keywords: Notch stress intensity factor, Sharp V-notch, Finite element method, Least squares method

1 Introduction

Many engineering components and structures contain sharp or rounded V-notches and reentrant corners. Some notches are knowingly made in engineering elements to serve some specific purposes. Moreover, notches may develop due to manufacturing defects also. Cracks are also particular case of sharp notches. When in comparison to the other dimensions of notch, the notch tip radius is very small, it can be consider as sharp V-notch. The presence of notches may substantially increase local stresses at the notch tip. Due to severe local stresses, the components may get damaged well below its expected yield strength. The scenario becomes even worse in case of variable loads. Therefore, a reliable stress analysis is desired in the domain near to the notch tip to avoid catastrophic failures of engineering components and structures. Williams [1] showed that stress singularity exist for 2D elastic configuration containing sharp V-notches. In linear elastic fracture mechanics, notch stress intensity factors (NSIFs) can be used to describe the singular stress field at the notch tip. The singular stress field near at any point r near the notch tip can be written as $Kr^{1-\lambda}f_{ij}(\theta)$, $\lambda-1$ is the order of stress singularity and $f_{ij}(\theta)$ are angular components of stress field. K is defined as notch stress intensity factor of the stress field. NSIFs are used by researches as brittle fracture criteria [2-4] and fatigue strength assessment [5, 6]. Therefore, many authors studied the V-notch problems and tried to establish different methods such as semi-analytical methods [7, 8], numerical methods [9-17] and experimental methods [18, 19] for calculating NSIFs. For complicated problems numerical methods are mostly used. The various numerical methods available are boundary collocation method (BCM) [9], boundary element method (BEM) [10, 11] and finite element method (FEM) [12–17]. Chan [7] determined NSIFs for single and double edge notch problems using body force method. Zhao and Hahn [8] proposed a methodology for determining NSIFs of sharp V-notches from the SIFs of a crack problem. Ayatollahi and Nejati [12] determined the NSIFs and higher order Williams coefficients using overdeterministic method. Lazzarin [13] proposed a new methodology for computation of NSIFs from averaged strain energy density. Ju and Chung [14] and Liu et al [15] used least squares method for evaluation of NSIFs of sharp V-notched bodies. In numerical simulation shown by Ju and Chung [14] showed that the error in NSIFs are minimum if the more than two number of Williams coefficients are used. Liu et al [15] proposed a new least squares method for finding NSIFs. In their approach only one Williams coefficient is used, however, the procedure for determining the mixed mode (I/II) NSIFs is very complicated. Other than conventional FE method, Treifi et al [16] implemented fractal-like finite element method and Yi [17] used extended finite element method for evaluation of NSIFs. The least square method proposed by Liu et all [15] is the simplest, robust and very efficient for problems having one stress singularity. In the present

* Corresponding author: mirzaul@iitg.ernet.in

investigation, this method has been further explored with different components of stresses for accurate evaluation of NSIFs of sharp V-notched configurations under pure mode I conditions (only mode I stress singularity is present). However, the procedure is complicated for cases with two singularities (mixed mode cases have both modes I and II singularities). Therefore, the mixed mode stress components are disintegrated into opening mode and shear mode stress components as suggested by Ishikawa et al [20] so that the simple method proposed by Liu et al [15] could be used as well for mixed mode problems easily.

The outline of the rest of the paper is organized as follows: the formulations for the NSIFs calculation using least squares method is described in Section 2. Section 3 presents some numerical examples. Finally, the conclusions from this work are given in Section 4.

2 Theoretical background

2.1 Stress field due to sharp V-notch

Elastic stress field at any nearby point of a sharp V-notch having notch angle γ (Fig. 1) can be expressed by Eq. (1) using Williams approach [1]

$$\begin{Bmatrix}\sigma_{xx}\\ \sigma_{yy}\\ \sigma_{xy}\end{Bmatrix} = K_I A' r^{\lambda_1^I-1}\begin{Bmatrix}f_{xx}\\ f_{yy}\\ f_{xy}\end{Bmatrix} + K_{II} B' r^{\lambda_1^{II}-1}\begin{Bmatrix}g_{xx}\\ g_{yy}\\ g_{xy}\end{Bmatrix} \tag{1}$$

where K_I and K_{II} denote modes I and II NSIFs respectively, (r,θ) are the polar coordinates, A_1 and B_1 are Williams coefficients corresponding to the singularity terms and λ_1^I, λ_1^{II} are eigenvalues for modes I and II obtained from the characteristic Eq. (2). The angular functions f_{ij} and g_{ij} $(i,j=x,y)$ are given by Eqs. (3 and 4). A' and B' can be obtained using Eqs. (5 and 6).

$$\begin{aligned}\lambda_1^I \sin 2\alpha + \sin 2\lambda_1^I\alpha = 0\\ \lambda_1^{II}\sin 2\alpha - \sin 2\lambda_1^{II}\alpha = 0\end{aligned} \tag{2}$$

$$\begin{Bmatrix}f_{xx}\\ f_{yy}\\ f_{xy}\end{Bmatrix} = \begin{Bmatrix}\begin{Bmatrix}(2+\lambda_1^I\cos 2\alpha + \cos 2\alpha\lambda_1^I)\cos(\lambda_1^I-1)\theta\\ -(\lambda_1^I-1)\cos(\lambda_1^I-3)\theta\end{Bmatrix}\\ \begin{Bmatrix}(2-\lambda_1^I\cos 2\alpha - \cos 2\alpha\lambda_1^I)\cos(\lambda_1^I-1)\theta\\ +(\lambda_1^I-1)\cos(\lambda_1^I-3)\theta\end{Bmatrix}\\ \begin{Bmatrix}-(\lambda_1^I\cos 2\alpha + \cos 2\alpha\lambda_1^I)\sin(\lambda_1^I-1)\theta\\ +(\lambda_1^I-1)\sin(\lambda_1^I-3)\theta\end{Bmatrix}\end{Bmatrix} \tag{3}$$

$$\begin{Bmatrix}g_{xx}\\ g_{yy}\\ g_{xy}\end{Bmatrix} = \begin{Bmatrix}\begin{Bmatrix}-(2+\lambda_1^{II}\cos 2\alpha - \cos 2\alpha\lambda_1^{II})\sin(\lambda_1^{II}-1)\theta\\ +(\lambda_1^{II}-1)\sin(\lambda_1^{II}-3)\theta\end{Bmatrix}\\ \begin{Bmatrix}(-2+\lambda_1^{II}\cos 2\alpha - \cos 2\alpha\lambda_1^{II})\sin(\lambda_1^{II}-1)\theta\\ -(\lambda_1^{II}-1)\sin(\lambda_1^{II}-3)\theta\end{Bmatrix}\\ \begin{Bmatrix}-(\lambda_1^{II}\cos 2\alpha - \cos 2\alpha\lambda_1^{II})\cos(\lambda_1^{II}-1)\theta\\ +(\lambda_1^{II}-1)\cos(\lambda_1^{II}-3)\theta\end{Bmatrix}\end{Bmatrix} \tag{4}$$

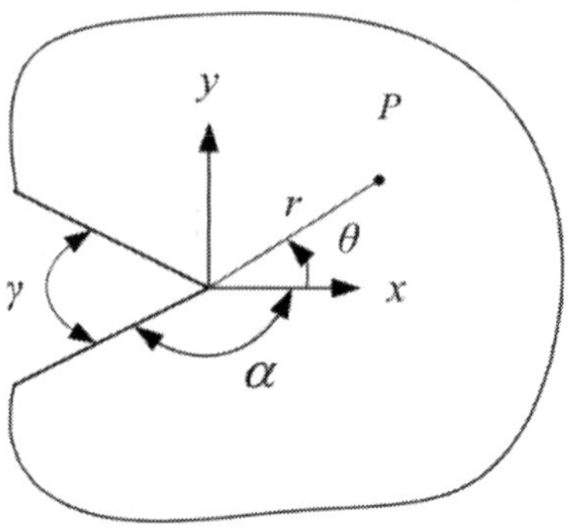

Fig. 1. Notch geometry.

$$A' = \frac{1}{\sqrt{2\pi}\left(1+\lambda_1^I - \lambda_1^I\cos 2\alpha - \cos 2\alpha\lambda_1^I\right)} \tag{5}$$

$$B' = \frac{1}{\sqrt{2\pi}\left(-1+\lambda_1^{II} - \lambda_1^{II}\cos 2\alpha + \cos 2\alpha\lambda_1^{II}\right)} \tag{6}$$

The variations of the dominating eigenvalues for both mode I and mode II are plotted in Fig. 2. From Fig. 3 it is clear that the geometry is symmetric about x axis, therefore the stress components of the FE solution at a point $P(x,y)$ can be decomposed into mode I (opening mode) and mode II (shear mode) stress components as [20]

$$\begin{Bmatrix}\sigma_{xx}^I\\ \sigma_{yy}^I\\ \sigma_{xy}^I\end{Bmatrix} = \frac{1}{2}\begin{Bmatrix}\sigma_{xx}^u+\sigma_{xx}^l\\ \sigma_{yy}^u+\sigma_{yy}^l\\ \sigma_{xy}^u-\sigma_{xy}^l\end{Bmatrix} \text{ and } \begin{Bmatrix}\sigma_{xx}^{II}\\ \sigma_{yy}^{II}\\ \sigma_{xy}^{II}\end{Bmatrix} = \frac{1}{2}\begin{Bmatrix}\sigma_{xx}^u-\sigma_{xx}^l\\ \sigma_{yy}^u-\sigma_{yy}^l\\ \sigma_{xy}^u+\sigma_{xy}^l\end{Bmatrix} \tag{7}$$

where σ_{xx}^I, σ_{yy}^I, σ_{xy}^I and σ_{xx}^{II}, σ_{yy}^{II}, σ_{xy}^{II} are finite element stress components, where suffix I and II indicate opening mode and shear mode. Suffix u and l represents stress components at points $P(x,y)$ and $P'(x,-y)$.

Therefore, the mode I and mode II stress components can be written as

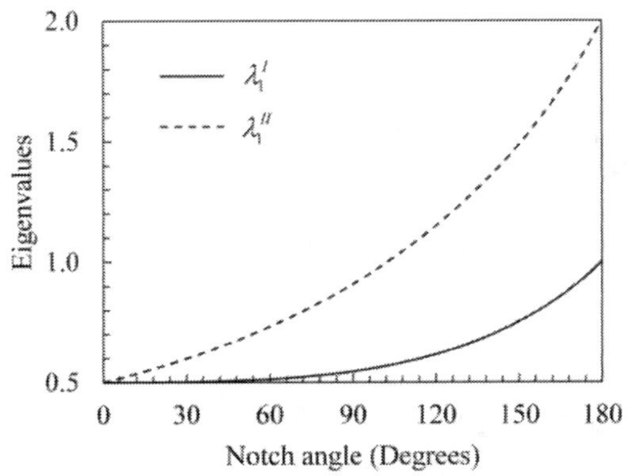

Fig. 2. Dominant eigenvalues for Modes I and II.

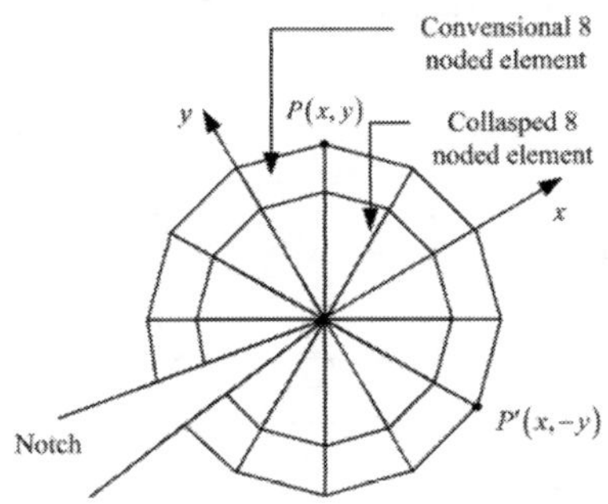

Fig. 3. A typical mesh around the notch tip.

$$\begin{Bmatrix} \sigma_{xx}^I \\ \sigma_{yy}^I \\ \sigma_{xy}^I \end{Bmatrix} = K_I A' r^{\lambda_1^I - 1} \begin{Bmatrix} f_{xx} \\ f_{yy} \\ f_{xy} \end{Bmatrix} \quad (8)$$

$$\begin{Bmatrix} \sigma_{xx}^{II} \\ \sigma_{yy}^{II} \\ \sigma_{xy}^{II} \end{Bmatrix} = K_{II} B' r^{\lambda_1^{II} - 1} \begin{Bmatrix} g_{xx} \\ g_{yy} \\ g_{xy} \end{Bmatrix} \quad (9)$$

For the present analysis, mode I NSIFs is determined from the stress component σ_{yy}^I and mode II NSIFs from σ_{xy}^{II}.

2.1 Determination of NSIFs by least squares method

In the elastic plane problem, the stress component at the vicinity of the notch tip σ_{ij} ($ij = yy$ for symmetric deformation mode and xy for anti-symmetric deformation mode) can be expressed as

$$\sigma_{ij} = K_k A_{ij} r^{\lambda_1^k - 1} \; (k = I, II) \quad (10)$$

where $A_{ij} = A' f_{yy}$ for mode I and $A_{ij} = B' g_{xy}$ for mode II conditions. Taking logarithm at both sides of Eq. (6)

$$\ln|\sigma_{ij}| = \ln|K_k| + \ln|A_{ij}| + (\lambda_1^K - 1)\ln r \quad (11)$$

In order to determine the NSIFs accurately, n number of nodes are considered in the direction $\theta = 60°$ for mode I and $\theta = 0°$ for mode II. The error in FE stresses and the analytical stresses can be given by

$$R = \sum_{n=1}^{N} \left[\ln|\sigma_{ijn}| - \ln|K_k| - \ln|A_{ij}| - (\lambda_1^K - 1)\ln r_n \right]^2 \quad (12)$$

Using least squares method, partial differentiations of the function R, with respect to $\ln|K_k|$ must be equal to zero, i.e.

$$\frac{\partial R}{\partial \ln|K_k|} = 0 \quad (13)$$

From Eqs. (12) and (13), one can get

$$-2\sum_{n=1}^{N} \left[\ln|\sigma_{ijn}| - \ln|K_k| - \ln|A_{ij}| - (\lambda_1^K - 1)\ln r_n \right] = 0 \quad (14)$$

From Eq. (14), finally the NSIFs can be obtained as

$$K_k = \frac{\exp\left[\frac{1}{N} \sum_{n=1}^{N} \ln|\sigma_{ijn}| r_n^{1-\lambda_1^k} \right]}{|A_{ij}|} \quad (15)$$

The normalized notch stress intensity factors can be described as

$$F_I = \frac{K_I}{\sigma\sqrt{\pi} a^{(1-\lambda_1^I)}} \text{ and } F_{II} = \frac{K_{II}}{\sigma\sqrt{\pi} a^{(1-\lambda_1^{II})}} \quad (16)$$

3 Results and discussions

In this section, numerical examination for determination of mode I and mixed mode (I/II) NSIFs for bench mark problems is carried out. Eight node quadrilateral elements are modeled in ANSYS for finite element analysis. Three problems viz. (a) Double Edge Notched plates under uniaxial Tension (DENT), (b) Center Notched plates under uniaxial Tension (CNT) and (c) Angled Double Edge Notched plates under uniaxial Tension (ADENT) are considered for present analysis. First two Problems are examples of pure mode I condition and third problem is example of mixed mode condition. Plane stress conditions are assumed for all the problems.

Percentage relative error is calculated as

$$\%\text{rel. error} = \left| \frac{\text{comp. value-ref. value}}{\text{ref. value}} \right| \times 100 \quad (17)$$

where the 'comp. value' refers to NSIFs obtained using present method and 'ref. value' is the available analytical or numerical value.

3.1 Mode I examples: Double Edge and Center Notched plates under uniaxial Tension

The geometry and loading considered for both DENT and CNT are as follows: $w = 10$, $h = 10$ and $\sigma = 1$ as shown in Fig. 4. The notch length $a = 4$, notch angles $\gamma = 0°$, $30°$, $60°$ and $90°$ are considered. Material properties considered are Young's modulus $E = 1$ and Poisson's ratio $\nu = 0.25$. The dimensions and loading are considered to be self-consistent. Due to symmetry in geometry as well as in loading conditions one quarter of the geometry is modelled for both DENT and CNT specimens with appropriate boundary conditions show in Fig. 4. Fig. 5 shows a typical FE mesh used for mode I problems containing 529 elements (NE) and 1700 nodes (NN). 10 number of the consecutive corner nodes along the line $\theta = 60°$ are considered and NSIFs are calculated using the stress component σ_{yy}. The first four nodes are not selected because the near tip elements may have large errors. Mode I NSIFs can be calculated using Eq. (15).

For convergence study, DENT and CNT geometries with notch length $a = 4$ and notch angle $\gamma = 60°$ are considered. Three meshes are created as shown in Fig. 6 with increasing number of elements and the results are listed in Tables 1 and 2. It is found that with each mesh refinement the error in NSIFs are decreasing. Therefore, for all the problems the fine mesh (notch tip element size = 0.0033) is employed. The NSIFs for DENT and CNT for various notch angles are listed in Tables 3 and 4. The NSIFs obtained are found to be very accurate.

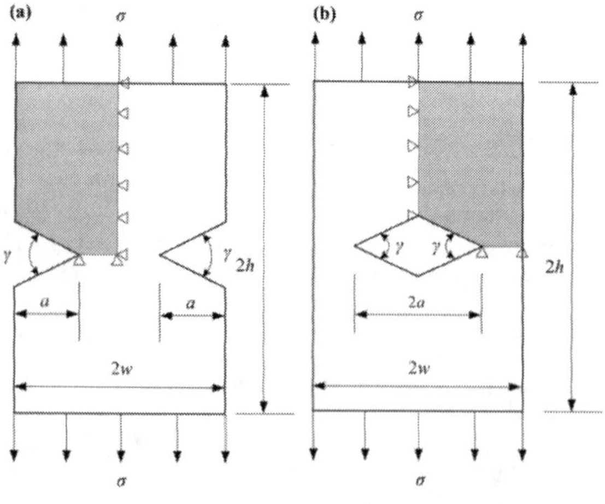

Fig. 4. (a) Double Edge Notched plate under uniaxial Tension (DENT) and (b) Center Notched plate under uniaxial Tension (CNT) specimens.

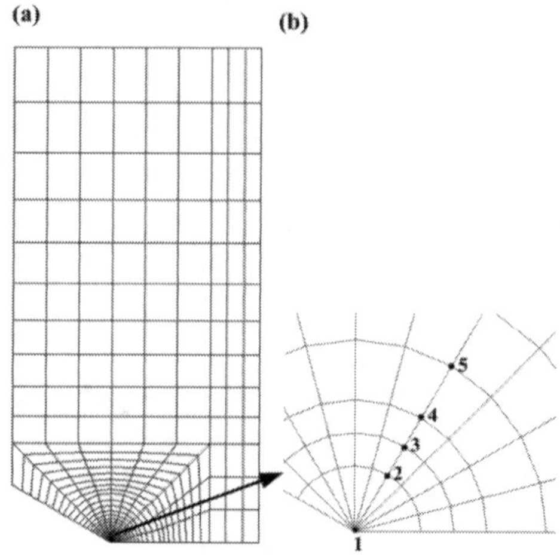

Fig. 5. (a) Typical FE mesh used for mode I specimens ($\gamma = 60°$) and (b) enlarge view of the notch tip portion.

Table 1. Convergence for DENT specimen ($a/w = 0.4$, $\gamma = 60°$).

Notch tip element size	NE	NN	F_I		
			Present	Ref [12]	%rel. err
0.0400	286	957	1.2159		1.49
0.0133	340	1141	1.2036	1.198	0.47
0.0033	529	1700	1.2004		0.20

Table 2. Convergence for CNT specimen ($a/w = 0.4$, $\gamma = 60°$).

Notch tip element size	NE	NN	F_I		
			Present	Ref [16]	%rel. err
0.0400	286	957	1.2852		1.92
0.0133	340	1141	1.2722	1.261	0.89
0.0033	529	1700	1.2709		0.79

Table 3. NSIFs for DENT specimens for different γ.

$\gamma(°)$	F_I		
	Present	Ref [16]	Ref. [12]
0	1.1368	1.132	1.133
30	1.1481	1.143	1.144
60	1.2004	1.191	1.198
90	1.3234	1.323	1.322

Table 4. NSIFs for CNT specimens for different γ.

$\gamma(°)$	F_I	
	Present	Ref. [16]
0	1.1150	1.109
30	1.1575	1.151
60	1.2709	1.261
90	1.4939	1.499

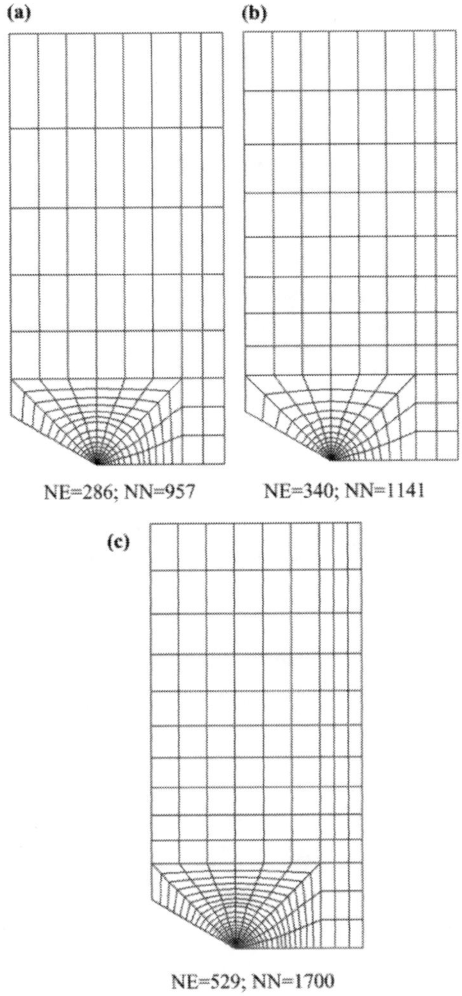

Fig. 6. Sequence of FE meshes used for DENT and CNT specimens for convergence study ($\gamma = 60°$).

3.2 Mixed Mode (I/II) examples: Double Angled Edge Notched plates under uniaxial Tension

The geometry and loading for Double Angled Edge Notched plates under uniaxial Tension (ADENT) is as follows: $w = 10$, $h = 25$ and $\sigma = 1$ as shown in Fig. 7. The notch length $a = 4$, notch angles $\gamma =$ 0°, 30°, 60° and 90° and notch inclination angle $\beta =$ 0°, 15°, 30° and 45° are considered for the present analysis. For $\beta = 0°$ the specimen will be under pure mode I condition while for other angles of β, the specimen will be under mixed mode conditions. Young's modulus and Poisson's ratio are considered as $E = 1$ and $\nu = 0.25$. The dimensions and loading are considered to be self-consistent. Because of the symmetry in geometry and loading conditions one half of the geometry is modelled for FE analysis. For calculating F_I, σ_{yy} at 10 consecutive nodes along each line $\theta = 60° / -60°$ are considered. In the similar way, σ_{xy} at 10 number of nodes along $\theta = 0°$ are considered for finding F_{II}. In mixed mode the starting node number is taken as 4. A typical FE mesh for $\beta = 15°$ and $\gamma = 30°$ is shown in Fig. 7 containing 957 elements (NE) and 3113 nodes (NN). The normalized NSIFs F_I and F_{II} for various ADENT specimens are listed in Tables 5-7. The effect of notch angle, notch inclination angle and notch length can be observed in Tables 5-7. The results obtained in the present method are compared with available results and good agreement with the literature is found.

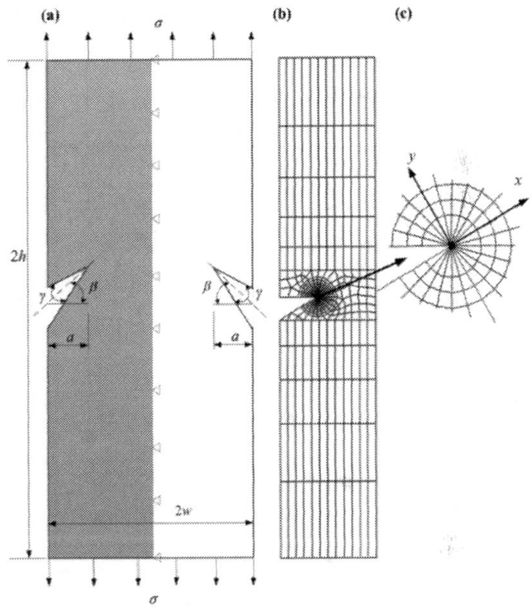

Fig. 7. (a) Angled Single Edge Notched plate under uniaxial Tension (ADENT), (b) typical FE mesh used for mixed mode specimens ($\beta = 15°$ and $\gamma = 30°$) and (c) enlarge view of the notch tip portion.

Table 5. Values of F_I and F_{II} for $a/w = 0.4$ and $\beta = 15°$.

$\gamma(°)$		F_I	F_{II}
0	Present	1.0980	0.1806
	Ref [7]	-	-
	Ref [12]	1.0979	0.1789
30	Present	1.1077	0.2486
	Ref [7]	1.107	0.248
	Ref [12]	1.1076	0.2474
60	Present	1.1526	0.3387
	Ref [7]	1.153	0.338
	Ref [12]	1.1531	0.3373
90	Present	1.2583	0.4269
	Ref [7]	1.259	0.426
	Ref [12]	1.2591	0.4241

Table 6. Values of F_I and F_{II} for $a/w = 0.4$ and $\gamma = 30°$.

$\beta(°)$		F_I	F_{II}
0	Present	1.1481	0.000
	Ref [7]	-	-
	Ref [12]	1.144	0.000
15	Present	1.1077	0.2486
	Ref [7]	1.107	0.248
	Ref [12]	1.1076	0.2474
30	Present	1.0000	0.4522
	Ref [7]	1.000	0.4522
	Ref [12]	1.0005	0.4491
45	Present	0.8380	0.5740
	Ref [7]	0.839	0.574
	Ref [12]	-	-

Table 7. Values of F_I and F_{II} for $\beta = 30°$ and $\gamma = 60°$.

a/w		F_I	F_{II}
0.2	Present	0.9994	0.5912
	Ref [7]	1.000	0.589
	Ref [12]	-	-
0.4	Present	1.0238	0.5968
	Ref [7]	1.024	0.596
	Ref [12]	1.0263	0.5944
0.6	Present	1.1396	0.6307
	Ref [7]	1.139	0.631
	Ref [12]	-	-

4 Conclusions

In this paper, a least squares method proposed by Liu et al. [15] is employed for determination of the mode I with different stress components. The mixed mode stress components are disintegrated into opening mode and shear mode stress components to separate out the mode I and mode II singularities. As a result, the stress terms will have only one singularity (either mode I or mode II) even in case of mixed mode problems. Therefore, the mixed problem can be considered as single stress singularity and can be proceed accordingly. Two mode I examples and one mixed mode (I/II) example are considered for explaining the working of the present method. The results obtained are in good agreement with the reference values. The method is simple, efficient and easy to implement in the existing finite element codes for simultaneous estimation of mode I and mixed mode (I/II) NSIFs. The present method can be extended for blunt V-notch problems also.

References

1. M.L. Williams, J. Appl. Mech. **19**, 526–528 (1952).
2. M.L. Dunn, W. Suwito, Int J Solids Struct **45**, 6468-**34**, 29, 3873-3883 (1997).
3. A. Seweryn, Eng. Fract. Mech. **47**, 673–81 (1994).
4. Carpinteri A. Eng. Fract. Mech. **26**, 143–55 (1987).
5. P. Lazzarin, P. Livieri, Int. J. Fatigue **23** 225–32 (2001).
6. C. Fischer,W. Fricke,C.M. Rizzo, Int. J. Fatigue **84**, 59-66 (2016).
7. D.H. Chen, Int. J. Fract. **70**, 81-97 (1995).
8. Z. Zhao, H.G. Hahn, **43**, 511-8 (1992).
9. Gross B, Mendelson A, Int. J. Fract. Mech. **8**, 267–76 (1972).
10. C Cheng, Z Niu, H Zhou, N Recho, Eng. Anal. Bound. Elem. **33,** 1145–51 (2009).
11. Z Niu, C Cheng, J Ye, N Recho, Int. J. Solids Struct. **46**, 2999–3008 (2009).
12. M.R. Ayatollahi, M. Nejati, Int. J. Mech. Sci. **53**, 164-77 (2011).
13. P. Lazzarin, F. Berto, M Zappalorto, Int. J. Fatigue **32**, 1559-67 (2010).
14. S.H. Ju, H.Y. Chung, Int. J. Fract. **148**, 169-83 (2007).
15. Y. Liu, Z. Wu, Y. Liang, X. Liu, Eng. Fract. Mech. **75**, 4793-803 (2008).
16. M. Treifi, S.O. Oyadiji, D.K.L. Tsang, Eng. Fract. Mech. **76**, 2091-2108 (2009).
17. G. Yi, T. Yu, T. Quoc, C. Ma, S. Hirose, Theor. Appl. Fract. Mech. **89**, 35-44 (2017).
18. M.R. Ayatollahi, M. Nejati, Mater Des **32**, 561-9 (2011).
19. T. Kondo, Y. Kurabe, T. Sasaki, T. Kurahashi, Y. Miyashita, Eng. Fract. Mech. **124–125**, 248 -61 (2014).
20. H. Ishikawa, H. Kitagawa, H. Okamura, *Mechanical Behaviour of Materials,* **3**, 447–55 ICM (1979).

Effect of material properties on optimal radial strain gage locations in sharp V-notched configurations

Pranjol Paul [1,*], *K.S.R.Krishna* Murthy [2], and *Debabrata* Chakraborty [3]
[1]Research scholar, Department of Mechanical Engineering, Indian Institute of Technology Guwahati, Guwahati - 781039, India
[2, 3] Department of Mechanical Engineering, Indian Institute of Technology Guwahati, Guwahati - 781039, India

Abstract. A simple yet reliable and powerful methodology using only one strain gage has been recommended for appropriate determination of notch stress intensity factor (NSIF) for sharp V-notched configurations subjected to mode I condition. The methodology is supported by strong theoretical postulates, and it permits the gage to be pasted prominently apart from tip of the notch thus avoiding various problems associated with singularities. Unlike the conventional methodologies, the recommended strain gage methodology also proposes optimal radial strain gage locations which are beneficial in appropriate determination of NSIFs. A FEM based numerical approach is adopted for obtaining optimal radial gage locations *a priori* for the aforesaid configuration. The optimal radial gage locations are observed to be influenced by parameters viz. the notch angle, the ratio of notch length to width of the plate and also material properties. Results were already published by the authors to establish that the optimal radial gage locations are influenced by the notch angle and the ratio of notch length to width of the plate. In this conference paper, a case is studied with a completely different material to check whether material properties influence the graphical trends of results or not.

1 Introduction

In engineering components, we may often come across stress raisers such as sharp V-notches, rounded tip V-notches and re-entrant corners. In some cases, notches are intentionally introduced in the mechanical parts such as gears, screws, bolts and nuts in order to achieve defined engineering objectives. In addition, notches also exist as a result of manufacturing processes such as anisotropic chemical etching while fabrication of microelectromechanical components [1] and injection moldings for manufacturing polystyrene products [2].

A sharp V-notch is a V-notch which has a tip radius significantly smaller than the dimension of its sides. Supported by the eigen function expansion method, Williams [3] proved that singularity of stress occurs at the tip of a sharp V-notch as $\sigma_{ij} = K^V r^{\lambda'-1} f_{ij}(\theta)$ where the order of the singularity $\lambda'-1$ and the angular functions $f_{ij}(\theta)$ are functions of the notch angle γ, and K^V is the notch stress intensity factor (NSIF). For a given set of geometrical and loading conditions, K^V illustrates the state of the stress in the vicinity of the notch tip identical to the stress intensity factor (SIF) of cracked configurations in linear elastic fracture mechanics (LEFM). It is dependent on the geometry of the body and boundary conditions.

In case of engineering components made of brittle and quasi-brittle materials, the static and fatigue load carrying capacities reduce due to the occurrence of sharp V-notches. Several researches had been attempted to recommend the brittle fracture criterion at the sharp V-notches [4-12]. Because of widespread use of K^V, significant efforts had been put in the development of several numerical [13-16] and experimental [17-22] methods for determination of NSIFs for different notched configurations.

Several experimental techniques such as strain gage [17, 18], photoelastic [19], caustics [20], coherent gradient sensing [21] and digital image correlation [22] had been developed to determine NSIFs under mode I and mixed mode conditions.

In a recently published research article , the authors (Paul *et al.* [23]) have postulated a novel and robust methodology in the domain of LEFM to determine the NSIF of sharp V-notched configurations by employing a technique using only one strain gage. The work [23] aimed to propose a novel strain gage technique based on the multiparameter strain series for appropriate determination of mode I NSIF (K_I^V) of sharp V-notched configurations made of isotropic materials. The proposed technique requires only one strain gage for the determination of K_I^V which ought to be pasted apart from the notch-tip. Another aim of the study is to develop a methodology (supported by theoretical foundations) that reflects the occurence of a valid or optimal radial gage location and allows prior estimation of such optimal location for appropriate determination of K_I^V. The analysis [23] also aimed at relating the optimal radial gage locations with the geometry of the sharp V-notched configuration.

Paul *et al.* also targets at relating the valid or optimal radial gage locations with the material properties and to confirm this observation a study similar to [23] is accomplished with a different material and the results are suitably shown in the successive section.

2 Theoretical background

The mathematical formulation of the proposed single strain gage technique and the theoretical foundation for obtaining optimal radial strain gage locations for

* Corresponding author: pranjol@iitg.ernet.in

appropriate determination of NSIF are briefed in this section. Illustration of these are available in the research article published by Paul *et al.* [23].

2.1. Strain analysis of sharp V-notches

The entire work is accomplished with an assumption of an isotropic and homogenous arbitrary two dimensional body with a sharp V-notch under mode I loading and plane stress conditions as shown in Fig. 1.

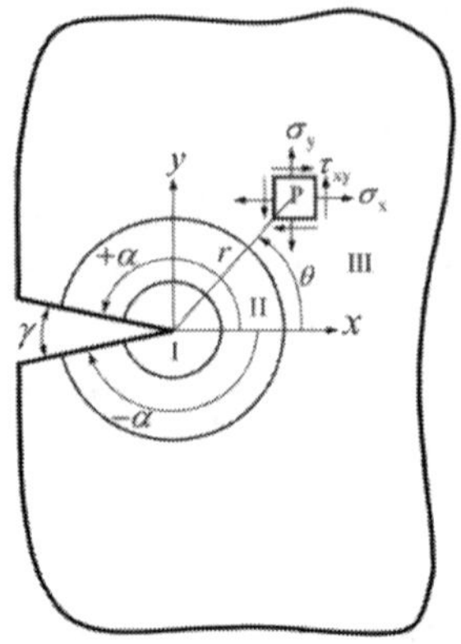

Fig.1. Typical zones (I, II and III) around the notch tip of a single edge notched tensile (SENT) configuration

Assuming stress-free notch surface boundary conditions ($\sigma_\theta = \tau_{r\theta} = 0$ at $\theta = \pm\alpha$, as shown in Fig. 1), it can be deduced that the stress components in polar coordinates under mode I loading condition is given by

$$\begin{Bmatrix} \sigma_r \\ \sigma_\theta \\ \tau_{r\theta} \end{Bmatrix} = \sum_{n=1}^{\infty} \mathrm{Re}\left\{ \frac{\lambda_n' A_n'}{r^{1-\lambda_n'}} \begin{bmatrix} (3-\lambda_n')\cos(\lambda_n'-1)\theta + (\lambda_n'\cos 2\alpha + \cos 2\lambda_n'\alpha)\cos(\lambda_n'+1)\theta \\ (\lambda_n'+1)\cos(\lambda_n'-1)\theta - (\lambda_n'\cos 2\alpha + \cos 2\lambda_n'\alpha)\cos(\lambda_n'+1)\theta \\ (\lambda_n'-1)\sin(\lambda_n'-1)\theta - (\lambda_n'\cos 2\alpha + \cos 2\lambda_n'\alpha)\sin(\lambda_n'+1)\theta \end{bmatrix} \right\} \tag{1}$$

where n is the order of the term in the infinite series, λ_n' is eigenvalue under mode I condition and A_n' is coefficient of the Williams eigen function expansion. The satisfaction of the aforementioned boundary conditions also results into a characteristic equation for the solution of the eigenvalue λ_n' as

$$\lambda_n' \sin 2\alpha + \sin 2\lambda_n'\alpha = 0 \tag{2}$$

The positive root of the Eq. (2), λ_n', represents the order of singularity ahead of the notch tip under mode I loading condition which is influenced by the value of α and hence it is also influenced by the notch angle γ (Fig. 1). Usually numerical methods are applied to solve the Eq. (2). λ_n' can be a real or a complex number which is influenced by the magnitude of the notch angle and can be expressed as

$$\lambda_n' = \lambda_n + i\lambda_n^* \tag{3}$$

The first eigenvalue λ_1' is always real (i.e., $\lambda_1' = \lambda_1$ and $\lambda_1^* = 0$) and less than unity till $\gamma < 180°$ that represents singular stress behavior, and $\gamma = 180°$ represents a notchless solid i.e., $1-\lambda_1 = 0$ in Eq. (2). The second and higher order values of λ_n' will be real or complex numbers according to the notch angle. As the stresses should be real always, the Williams expansion coefficients A_n' can be either real or complex numbers according to the notch eigenvalues. Description about the nature of λ_n' and A_n' are available in Ref. [23]. Hence, the Williams coefficients A_n' can also be expressed as

$$A_n' = A_n + iA_n^* \tag{4}$$

For $n=1$, the coefficient $A_1' = A_1$ ($A_1^* = 0$ because $\lambda_1^* = 0$ for $n=1$) represents singular term and is related to the NSIF K_I^V as follows

$$K_I^V = \lim_{r\to 0}(\sqrt{2\pi} r^{1-\lambda_1}\sigma_y(\theta=0)) = \sqrt{2\pi}\lambda_1(1+\lambda_1-\lambda_1\cos 2\alpha - \cos 2\alpha\lambda_1)A_1 \tag{5}$$

For $n>1$, A_n' are non-singular higher order coefficients. Using the stress transformation law, the stress components in Cartesian coordinate system (Fig.1) can be obtained under mode I loading as [23]

$$\begin{Bmatrix} \sigma_x \\ \sigma_y \\ \tau_{xy} \end{Bmatrix} = \sum_{n=1}^{\infty} \mathrm{Re}\left\{ \frac{\lambda_n' A_n'}{r^{1-\lambda_n'}} \begin{bmatrix} (2+\lambda_n'\cos 2\alpha + \cos 2\alpha\lambda_n')\cos(\lambda_n'-1)\theta - (\lambda_n'-1)\cos(\lambda_n'-3)\theta \\ (2-\lambda_n'\cos 2\alpha - \cos 2\alpha\lambda_n')\cos(\lambda_n'-1)\theta + (\lambda_n'-1)\cos(\lambda_n'-3)\theta \\ -(\lambda_n'\cos 2\alpha + \cos 2\alpha\lambda_n')\sin(\lambda_n'-1)\theta + (\lambda_n'-1)\sin(\lambda_n'-3)\theta \end{bmatrix} \right\} \tag{6}$$

Considering plane stress condition and utilizing the constitutive relationship, the strain components at a point $P(r,\theta)$ are given by

$$\begin{aligned} 2G\varepsilon_x &= \sum_{n=1}^{\infty} \mathrm{Re}\left\{ \frac{\lambda_n' A_n'}{r^{1-\lambda_n'}} \begin{bmatrix} 2\kappa\cos(\lambda_n'-1)\theta + \lambda_n'.\cos 2\alpha.\cos(\lambda_n'-1)\theta + \\ \cos 2\lambda_n'\alpha.\cos(\lambda_n'-1)\theta - (\lambda_n'-1).\cos(\lambda_n'-3)\theta \end{bmatrix} \right\} \\ 2G\varepsilon_y &= \sum_{n=1}^{\infty} \mathrm{Re}\left\{ \frac{\lambda_n' A_n'}{r^{1-\lambda_n'}} \begin{bmatrix} 2\kappa\cos(\lambda_n'-1)\theta - \lambda_n'.\cos 2\alpha.\cos(\lambda_n'-1)\theta - \\ \cos 2\lambda_n'\alpha.\cos(\lambda_n'-1)\theta + (\lambda_n'-1).\cos(\lambda_n'-3)\theta \end{bmatrix} \right\} \\ 2G\gamma_{xy} &= \sum_{n=1}^{\infty} \mathrm{Re}\left\{ \frac{\lambda_n' A_n'}{r^{1-\lambda_n'}} \begin{bmatrix} 2(\lambda_n'-1).\sin(\lambda_n'-3)\theta - \\ 2(\lambda_n'\cos 2\alpha + \cos 2\lambda_n'\alpha).\sin(\lambda_n'-1)\theta \end{bmatrix} \right\} \end{aligned} \tag{7}$$

where $\kappa = (1-\nu)/(1+\nu)$ for plane stress condition and G is the rigidity modulus. The strain series in Eq. (7) is an infinite series.

2.2. Development of single strain gage technique

For measuring surface strains with the help of strain gages, it is essential to locate a suitable zone around the notch tip. Zone I which is basically a small zone around the notch tip (Fig.1) is not appropriate for measurement of strains owing to high strain gradients and 3D state of stress. As the solutions of notch problems do exist in the form of infinite series, zone III which is the farthest zone is also not appropriate for the strain measurements. This is due to the fact that a large number of unknown coefficients needs to be included in the strain series for representing the strains at those remote locations and it also needs large number of strain gages. In addition, singular field may not exist at such distances. Therefore, the intermediate zone or zone II is an appropriate region for strain measurements. Here, it is presumed that strains in zone II contain the first term (singular term) of the Williams infinite series [23] and a few non-singular higher order terms. In the present investigation, zone II is assumed to be the region around the notch tip in which strains can be appropriately represented by the multiparameter strain series as shown in Eq. (7). The strains are comprised of a singular term and two non-singular higher order terms when the infinite strain series is considered up to $n = 2$ as per Ref. [23].

From Fig. 2, the normal strain component along the direction a that makes an angle β with the notch axis at a point P can be obtained using strain transformation as

$$2G\varepsilon_a = r^{\lambda_1 - 1} A_1 f_1(\theta, \beta, \lambda_1, \nu, \gamma) + r^{\lambda_2 - 1} A_2 f_2(\theta, \beta, \lambda_2, \lambda_2^*, \gamma, \nu, r) + r^{\lambda_2 - 1} A_2^* f_2^*(\theta, \beta, \lambda_2, \lambda_2^*, \gamma, \nu, r) \quad (8)$$

where ε_a is the normal strain along the longitudinal direction of strain gage, the functions f_1, f_2 and f_2^* are the functions corresponding to the first order and the second order terms of the infinite strain series. The first order term of the infinite strain series always comprises real eigen value and thus real Williams coefficient of expansion. On the contrary, the second order term may comprise complex eigen value and thus complex Williams coefficient of expansion according to the magnitude of the notch angle as mentioned in preceding sections. Thus, in case of notched configurations leading to complex eigen value and Williams coefficient, the functions f_2 and f_2^* will co-exist. However, in case of the notched configurations leading to real eigen value and Williams coefficient, the function f_2 will exist alone and the function f_2^* will not exist. More in details would be discussed in successive paragraphs.

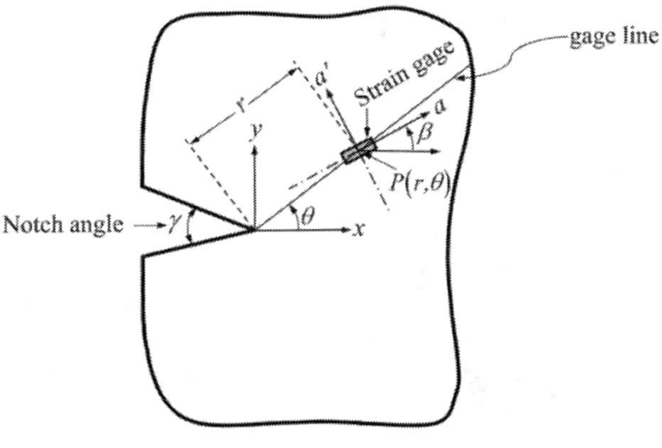

Fig.2. Schematic diagram depicting the orientation of strain gage in a sharp V-notched configuration

If A_1 is computed using Eq.(8), then the mode I SIF K_I^V can subsequently be determined using Eq. (5). Straightaway one can apply three strain gages against three unknown coefficients A_1, A_2 and A_2^* to yield three simultaneous equations using Eq.(8) to solve for A_1, A_2 and A_2^* and resultantly the mode I SIF K_I^V. However, a meticulous manipulation has been accomplished to reduce the number of strain gages to only one which is the novelty of the published work in Ref.[23]. This methodology has significantly minimised the effort and also it is understood to be cost-effective in case of experimental work in similar domain. Finally, Eq. (8) could be reduced to the following equation

$$2G\varepsilon_a = r^{\lambda_1 - 1} A_1 f_1(\theta, \beta, \lambda_1, \nu, \gamma) \quad (9)$$

where

$$f_1(\theta,\beta,\lambda_1,\nu,\gamma) = \lambda_1 \left\{ \begin{array}{l} \cos^2\beta \begin{bmatrix} \cos\theta(\lambda_1 - 3) + \cos\theta(\lambda_1 - 1) \\ [2\kappa + \cos\lambda_1(2\pi - \gamma)] \end{bmatrix} - \\ \cos\theta(\lambda_1 - 3)\sin^2\beta + \\ 2\kappa\cos\theta(\lambda_1 - 1)\sin^2\beta - \\ \cos\theta(\lambda_1 - 1)\cos\lambda_1(2\pi - \gamma)\sin^2\beta - \\ \sin\theta(\lambda_1 - 3)\sin 2\beta - \\ \cos\lambda_1(2\pi - \gamma)\sin\theta(\lambda_1 - 1)\sin 2\beta - \\ \lambda_1\cos^2\beta\cos\theta(\lambda_1 - 3) + \\ \lambda_1\cos^2\beta\cos\gamma\cos\theta(\lambda_1 - 1) + \\ \lambda_1\cos\theta(\lambda_1 - 3)\sin^2\beta - \\ \lambda_1\cos\gamma\cos\theta(\lambda_1 - 1)\sin^2\beta + \\ \lambda_1\sin\theta(\lambda_1 - 3)\sin 2\beta - \\ \lambda_1\cos\gamma\sin\theta(\lambda_1 - 1)\sin 2\beta \end{array} \right\} \quad (10)$$

Eq. (9) now contains only one unknown coefficient A_1 which can be measured using single strain gage. It can be conveyed here that gage line is the line, making an angle θ with the notch axis, on which strain gage is pasted in the direction of β at a radial distance r from the tip of the notch (Fig. 2). Therefore, by placing one strain gage at a point defined by $P(r,\theta)$ along the gage line and orienting it at an angle of β with respect to the notch axis, the mode I NSIF K_I^V can be conveniently determined using Eq. (9) and Eq. (5). It is to be noted that although Eq. (9) comprises only one coefficient, yet the equation on the right hand side is actually a three parameter representation of ε_a as expressed in the Eq. (8). Consequently, strain gage can be placed at longer radial distances with respect to the notch tip and this would reduce various effects arising in the strain gage technique. Determination of θ and β for a given Poisson's ratio ν is done through a systematic approach by eliminating the functions f_2 and f_2^* (Eq. (8)) i.e. by equating these functions to zero. In case of $\gamma \le 40°$, only the function f_2 exists due to the existence of real eigen value and real Williams coefficient. Hence, in such case, there will be only one equation if the function f_2 is equated to zero, thus leading to infinitely many solutions for θ and β. In case of $\gamma > 40°$, the functions f_2 and f_2^* will co-exist due to the complex nature of eigen value and corresponding Williams coefficient. In such case, there will be two equations if the functions f_2 and f_2^* are simultaneously equated to zero., thus leading to a unique set of θ and β. All these as discussed can be followed from the published work of Paul *et al.* [23].

2.3. Assessment of the valid or optimal radial strain gage locations

The Eq. (9) clearly shows that prior estimation of valid or optimal radial location of the strain gage is essential for the appropriate determination of NSIF K_I^V. Interestingly, the published work of Paul *et al.* [23] exhibits the theoretical basis for obtaining valid or optimal radial strain gage locations. These locations are optimal in the sense that they can be selected in a range as follows

$$r_{\min} \le r \le r_{\max} \tag{11}$$

where

$$r_{\min} = 1.25 \times \text{ thickness of the plate} \tag{12}$$

Thus, any radial strain gage location at distance r from the notch tip along the gage line if satisfies the criterion of Eq. (11) should be understood as a valid or optimal radial strain gage location. Eq. (11) helps to ensure that the gages are not pasted beyond $r_{\max}$, however the gage reading is expected to obey Eq. (9). The procedure to determine the maximum permissible radial strain gage distance $r_{\max}$ can be obtained in details from the published work as given in Ref.[23]. However, it can be briefly mentioned here that the value of $r_{\max}$ is obtained from the plot of $\ln(\varepsilon_a)$ versus $\ln(r)$ and the plot of $\ln(\varepsilon_a)$ versus $\ln(r)$ is achieved through finite element simulation.

3 Results and discussions

It is studied from the preceding sections that determining valid or optimal radial strain gage locations *a priori* is very much essential for obtaining NSIF of a specific two-dimensional sharp V-notched body. A hint has been given above with appropriate mathematical equations for showing the procedure to determine the valid or optimal radial strain gage locations. In the published work of Paul *et al.* [23], it can be observed that the maximum permissible radial strain gage distance $r_{\max}$ varies with notch length to width ratio a/b. For a given notch angle γ, the $r_{\max}$ value increases as a/b increases from 0.2 to 0.4, however it starts decreasing as a/b increases further from 0.4 to 0.8. Furthermore, it is also observed that the $r_{\max}$ value increases when notch angle γ increases for a given notch length to width ratio a/b. All these observations [23] were found for a particular material with $E = 2917$ MPa and $\nu = 0.382$. However, it is crucial to know whether material properties play any role in such graphical trends of $r_{\max}$ or not. For this study, a similar problem is accomplished with a different material having $E = 200$ GPa and Determination the results are subsequently presented in this conference paper. Sharp V-notched SENT configurations are again considered for the present analysis as same as Ref.[23]. Finite element simulations have been carried out to determine the maximum permissible radial strain gage distance $r_{\max}$. A typical FEA mesh (Fig.3) which is used for the finite element simulation can be seen below

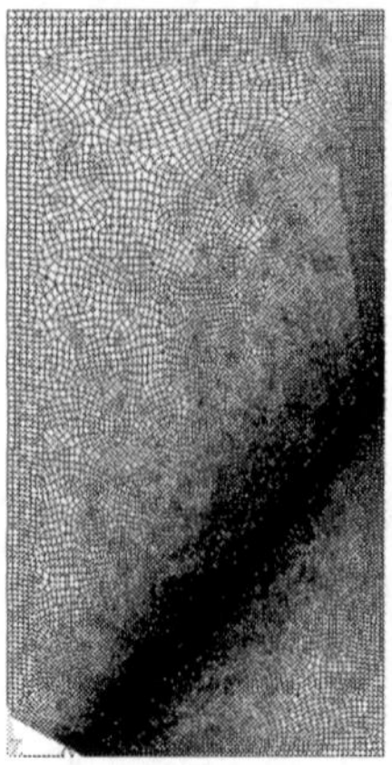

Fig.3. A typical FEA mesh in its finest state with the inclined darkened portion representing the strain gage line along which the strain gage is placed at certain optimal or valid location

The geometric, material and loading parameters associated with the problem of sharp V-notched (SENT) configurations subjected to mode I loading under plane stress condition are presented in Table 1 and Table 2 respectively.

Table 1 Geometric, loading and material parameters for notch angles 30° and 60° respectively

Notch angle (γ)	First eigen value (λ_1')	b (mm)	h/b	E (GPa)	ν	σ (MPa)
30°	0.5015	150	2	200	1/3	100
60°	0.5122	150	2	200	1/3	100

Table 2 Gage line orientation θ and strain gage orientation β for the notch angles 30° and 60° respectively

Notch angle (γ)	Gage line orinetation (θ)	Stran gage orinetation (β)
30°	50.49°	52.54°
60°	50.93°	44.50°

It was already discussed in preceding sections that the maximum permissible radial strain gage distance r_{max} value is usually obtained from the plot of $\ln(\varepsilon_a)$ versus $\ln(r)$ and the plot of $\ln(\varepsilon_a)$ versus $\ln(r)$ is obtained through finite element simulation. A straight line having a negative slope has to be superimposed on the plot of $\ln(\varepsilon_a)$ versus $\ln(r)$. The point at which the straight line representing the ideal solution deviates from the plot of $\ln(\varepsilon_a)$ versus $\ln(r)$ representing the actual solution obtained from finite element simulation would give us the r_{max} value as shown in Fig.4. The graph in Fig.4 needs to be observed from right hand side to left hand side to obtain the point of deviation. While observing from right hand side to left hand side of the graph (Fig.4), if the percentage relative error between the ideal solution and the finite element solution is found to be falling just below 0.5% then the point of deviation is said to be reached. The schematic as shown in Fig.4 can be regarded as the usual way of obtaining the r_{max} value. The plots (Fig.4) are obtained for the problem presented in this section for notch angle γ = 60°. It can be concluded from Fig.4 that for a specific notch angle the r_{max} value rises when a/b increases from 0.2 to 0.4 but the same dips when a/b increases further from 0.4 to 0.8. It is one of the major observations. Exactly similar observation was also made in Ref.[23] with a different material .

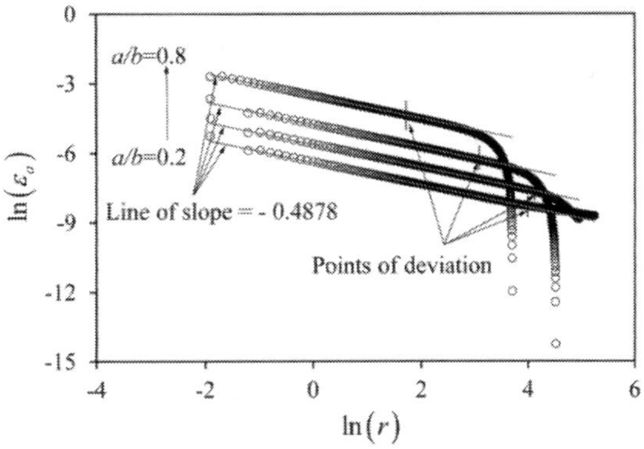

Fig.4. Variations of $\ln(\varepsilon_a)$ with $\ln(r)$ along the gage line for different a/b ratio for the SENT configuration with the notch angle γ = 60°.

Another important observation is made from Fig.5. Fig.5 shows that the trend of obtaining r_{max} value with a/b varying from 0.2 to 0.8 remains consistent for a completely different notch angle also. In addition, it is also observed from Fig.5 that as the notch angle increases, the r_{max} value also increases for a specific a/b ratio. Exactly similar observation (Fig.5) was also made in Ref.[23] with a different material . Table 3 presents the numerical results (the r_{max} values) in support of the above three important observations thus made.

Table 3 Values of r_{max} for different a/b ratio from 0.2 to 0.8 and for notch angles 30° and 60° respectively

Notch angle (γ)	a/b	r_{max} (mm)	r_{max}/b
30°	0.2	17.9347	0.119564667
	0.4	50.6119	0.337412667
	0.6	9.0448	0.060298667
	0.8	3.5155	0.023436667
60°	0.2	54.0228	0.360152
	0.4	61.7558	0.411705333
	0.6	21.9829	0.146552667
	0.8	5.5998	0.037332

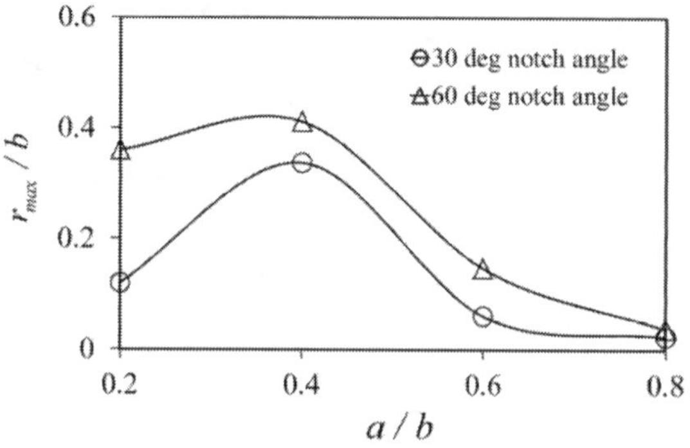

Fig.5. Variation of r_{max}/b with a/b for the SENT configurations for notch angles 30° and 60° respectively

4 Conclusions

As seen from the results section of this paper, even after changing the material, the graphical trends of the r_{max} values remain consistent as same as reported in Ref. [23]. This is a crucial observation presented in this paper. Material independence of the graphical trends is a completely novel finding and it is no where published earlier. However, the r_{max} values of course changed with different material. The graphical trends which we have discussed all about in the results section can be summarised as : for a specific notch angle the r_{max} value rises when a/b increases from 0.2 to 0.4 but the same dips when a/b increases further from 0.4 to 0.8, such trend remains consistent for different notch angles and the r_{max} value increases for a specific a/b ratio as the notch angle increases.

References

1. M.L. Dunn, W. Suwito, S. Cunningham, Eng. Fract. Mech., **57,** 417-430 (1997).
2. K. Tomari, S. Tonogai, T. Harada, H. Hamada, K. Lee, T. Morii, Z. Maekawa, Polym. Eng. Sci., **30,** 931-936 (1990).
3. M.L. Williams, J. Appl. Mech.,**19,** 526-528(1952).
4. M.L. Dunn, W. Suwito, S. Cunningham, Int. J. Solids. Struct., **34,** 3873-3883 (1997).
5. M.L. Dunn, W. Suwito, S. Cunningham, C.W. May, Int. J. Fracture, **84,** 367-381 (1997).
6. A. Seweryn, Eng. Fract. Mech., **47,** 673-681 (1994).
7. A. Carpinteri, Eng. Fract. Mech., **26,**143-155 (1987).
8. Z. Yosibash, E. Priel, D. Leguillon, Int. J. Fracture, **141,** 291-312 (2006).
9. S.Q. Hou, J.Q. Xu, Int. J. Mech. Sci., **88,** 202-207 (2014).
10. Z. Yosibash, A.Bussiba, I. Gilad, Int. J. Fracture, **125,** 307-333 (2004).
11. P. Lazzarin, R. Zambardi, Int. J. Fracture, **112,** 275-298 (2001).
12. M.P. Savruk, A. Kazberuk, Int. J. Fracture, **161,** 79-95 (2010).
13. M. Stern, E.B. Becker, R.S. Dunham, Int. J. Fracture, **12,** 359-368 (1976).
14. M.R. Ayatollahi, M. Nejati, Int.J. Mech. Sci., **53,** 164-177 (2011).
15. B. Gross, A. Mendelson, Int. J. Fract. Mech., **8,** 267-276 (1972).
16. T. Yu, L. Shi, J. Strain. Anal. Eng., **47,** 95-103 (2012).
17. T. Kondo, M. Kobayashi, H. Sekine, Exp. Mech., **41,** 1-7 (2001).
18. T. Kondo, Y. Kurabe, T. Sasaki, T. Kurahashi, Y. Miyashita, Eng. Fract. Mech., **124-125,** 248-261 (2014).
19. M.R. Ayatollahi, M. Nejati, Mater. Design, **32,** 561-569 (2011).
20. J.N.Prassianakis, P.S. Theocaris, J. Phys. D: Appl. Phys., **13,** 1043-1053 (1980).
21. X.F. Yao, H.Y. Yeh, W. Xu, Int. J. Solids Struct., **43,** 1189-1200 (2006).
22. M.R.Y. Dehnavi, I. Eshraghi, N. Soltani, Polym. Test., **32,** 778-784 (2013).
23. P.Paul, K.S.R.K.Murthy, D.Chakraborty, Theor. Appl. Fract. Mec., **94,** 57-70 (2018).

A Pioneering Tactic for Sustainable Waste Management: Design of a Cost-Effective Recycling Machine

A. Agarwal[1]*

[1]Department of Mechanical Engineering, University of Botswana, Gaborone. Botswana

Abstract. In this paper we discussed the new design, fabrication and testing of a mechanical can crusher taking into deliberation the force analysis and ergonomic factors that will enable people to use this mechanical can crusher especially in Botswana rural areas. The process of crushing cans is of vital importance as it saves a percentage of space for can disposal and also makes it easier for these cans to be transported to recycling centers or landfill areas. The methods and processes used to bring this project to life include welding, cutting, bending, drilling and assembly operations. This project study is mainly about generating a new concept of a mechanical can crusher that is compact to transport from one location to another and makes can crushing easy and fascinating to the general public so as to promote people buy into crushing in Sub-Sahara market. Even though there are numerous sorts of the mechanical would crusher be able to as of now in presence, this model provides a more practical usage than previous one with the desired mechanical advantage of 10, machinability to crush the entire cans to 25% of their original size and 100% efficiency for the entire rounds.

Keywords- Fabrication, cans, crusher, size reduction, mechanical advantage.

1 Introduction

Botswana is arguably the fastest growing economy in Africa. The Republic of Botswana is facing a lot of problems due to this rapid growth. Botswana is hard hit by an ever increasing problem of pollution. This problem is perpetuated by a number of factors including lack of garbage collection facilities (such as garbage collection trucks), poor town planning resulting in landfills being unable to keep up with the amount of garbage being piled up in them, delayed garbage collection by the city councils resulting in bins overflowing and lastly by peoples ignorance towards keeping the environment clean [1]. The majority of these cause factors can be minimized by employing the mechanical can crusher as it provides a quick fix solution to a persistent problem. Jesse M. wright developed the crusher be able to in 1937 to diminish the volume of the jars and licensed it in 1938. Capacity is regularly an issue and jars devour part of room, in this way expanding all out volume of waste [2]. Primarily the mechanical can crusher`s purpose is to reduce the size of the can so as to save space and allow for a storage of more cans and for recycling purposes. In dealing with problems of pollution it is essential to first understand the various types of pollution. There is a wide variety of pollution including water pollution, sound pollution and air pollution but to name a few. In this paper however, we shall deal with land pollution, solid waste pollution to be more specific. The problem of solid waste pollution in Botswana has reached a point where it cannot be solved within a short time frame. As a way of trying to reverse the negative environmental implications of solid waste pollution there is urgent need for all stakeholders to come up with innovative measures which will deal with the problem at grass root level [3]. Solid waste has negative effects on sectors of the economy specifically the tourism and agricultural sectors. In the tourism industry/sector solid waste robs the earth of its natural beauty making the land unattractive to tourists and as a result of this the number of tourists will decline along with the revenue they bring in. In the agricultural sector solid waste is problematic mostly to cattle farmers. Cattle can swallow solid waste such as plastics and cans and this result in beef being condemned at abettors such as the Botswana Meat Commission. A failure to meet market demands results in customers looking for new suppliers and these results in the closure of the abattoirs [3]). In Botswana at this current point in time the only can crushing platform is provided by a company called Collect-a-can which crushes cans only at an industrial (large) scale. The researchers feel that this does not solve the persistent problem of pollution as people continue to litter and garbage bins continue to pile up due to lack of solid waste management facilities such as garbage collection trucks [4]. This model however will solve the problem from a grass root level as it will take more time for bins to fill up. Also we believe that new technologies will provoke peoples buy in into waste separation and recycling as people will be fascinated by this new technology.

2 Literature Review

The can crusher is a basic idea that has already been established and tested. In this project study the focus is to improve the can crushers that are already in existence in the market. By definition Figuring out, additionally got back to building, is the procedures of removing learning or plan data from anything man-made and re-delivering it or re-creating anything in light of the extricated data. Transfer is the minimum economical and the slightest favored choice in the waste administration chain of importance. The waste administration was initially created to remove undesirable material from individuals' settlements. It was only during the recent times that the waste is moving in the upward direction towards resource recovery in many developed countries, but disposal is still the preferred final destination for most developing countries. The commonly practiced disposal methods are Landfilling and Incineration (United Nations Environment Programme, 2005).
The Collect-A-Can is a company which is involved in buying and collection of metal cans from the city as well as in the landfill. This company also has its mother plant in South Africa. The company welcomes cans from all parts of the society in Gaborone. The company has agents throughout Botswana, who is involved in buying and collection of cans. The agents buy the cans at 0.42 Pula per kilogram and sell it to the company for 1 Pula per kilogram. The collectors in the landfill are also paid based upon the amount of cans they collect. The recovered metal cans are then subjected to processing where it is crushed and pressurized in the form of a bundle which are later sent to South Africa for further processing and material recycling. The company has future plans to enhance their method and performance of collection. They plan to introduce drop off machines for the can in the city's main areas such as shopping malls and commercial areas, buy back centers in the rural part of the country, and the company has also planned to set up a plant.

2.1 Existing can crushers

The main aim of a can crusher is to crush used cans to reduce their size ultimately saving storage space for waste. While they are different types of can crushers with varying components to fit their design and purpose, there are certain features that are similar across all of them. These features include a crushing mechanism which allows for a can to be fed into the crusher and crushed. Following are the existing types of can crushers-

2.1.1 Pneumatic can crusher

Here the crushing of the cans is done by using compressed air channeled into a pneumatic cylinder with a piston coupled to a shaft and compression plate at the other end. When the air is pressurized into the cylinder the piston is pushed out and in turn pushes the compression plate that compresses the can against a stationary wall. To relief the can pressurized air is released into the atmosphere and the piston retracts [5]. It is cannot be used for heavy duty crushing.

2.1.2 Hydraulic can crusher

A similar working principle is followed as in pneumatic can crusher with the difference being the medium used to transfer power. In a hydraulic can crusher a cylinder is needed to but instead of air, hydraulic fluid is pressurized into the cylinder to push the compression plate that crushes the can but this time to relief the can. The hydraulic fluid is released back into the system (closed system) once the crushing action has been completed. Hydraulic fluid is not readily available, expensive and some of the transmission fluids are not environmentally friendly.

2.1.3 Manual can crusher

The last type of can crusher is the manual can crusher which is the most popular type of can crusher. This popularity is due to it the fact that it is cheap and simple to design. Compared to its cousins the pneumatic and hydraulic crushers the manual can crusher uses human effort as a power source instead of compressed air or compressed hydraulic fluid. To smash a can it is loaded into the crusher and human effort is needed for the crushing process. Manual can crushers can be designed to be operated by foot or hands depending on the users' preference [6]. Manual can crusher is an inexpensive type of can crusher which requires minimal maintenance during its life.

2.1.3.1 Hand operated

This kind of crusher is able to be worked by hand physically. This type of can crusher is made of either wood and metal or even plastic. Metal made can crushers are mostly used, but are relatively expensive when compared to those made from wood or plastic.

2.1.3.2 Foot operated

The crusher be able to is anything but difficult to utilize and accompanies a help bar which is intended to keep jars from being constrained out of the crusher be able to by the weight of the foot plate being pushed down. The crusher likewise includes a puncturing screw which can be acclimated to help pounding harder jars.

3 Problem statement & Scenario in Botswana

Solid waste in general refers to all the wastes that are generated from household, commercial, industrial and agricultural sectors. The system relating to environmental protection in Botswana is still considered to be at infancy stage. Until 2002 , there was no proper record of the composition of solid waste that was generated annually in Gaborone landfill [4].The major instrument used for solid waste management in Botswana is the Waste Management Act (Laws of Botswana,2011) and Botswana's strategy for waste management (Botswana Environmental Information system,2011). The waste management act provides some regulatory measures for the waste collection disposal and recycling. The local authorities are responsible in making these regulations effective, in reality [3]. The efforts of various stakeholders such as The Botswana Government and NGO's such as Collect-A-Can and Tshomarelo Tikologo to try and sensitize people about proper solid waste disposal seem to be in vain as pollution is still a persistent concern. People continue to litter and to make matters even worse there are poor waste disposal facilities such as the inadequacy and unreliability of garbage pickup trucks resulting in waste build up. This waste unfortunately ends up over spilling and polluting the surrounding even more. At the end of the design and fabrication process the end model is expected to be able to reduce the scrape volume of the aluminum cans which are a main constituent of the solid waste. By reducing the scrape volume of the aluminum cans this will provide more storage space for solid waste in garbage bins resulting in less over spillage and less pollution into the environment. The can crusher will reduce the cans volume to increase storage space and reduce pollution. The manual can crusher will be more mechanically advantageous to help human efforts and has a hopper to temporarily store cans and act as a semi-automated feeder mechanism when the crushing operation is being carried out. Following are the base line objectives-

i. To design and fabricate a mechanical can crusher at a low cost.
ii. To plan and create a mechanical crusher be able to that has a programmed nourish container.
iii. To design and fabricate the can crusher that has storage to locate the can after crush.
iv. To improve people's buy-in into recycling, waste separation and proper solid waste management.

4 Methodology-

Can crusher works on the principle of slider crank mechanism. In this turning movement of hand wheel is changed over into responding movement of slider i.e. A wrench is an arm connected at right points to a turning shaft by which responding movement is granted to or gotten from the pole. It is utilized to change over round movement into responding movement, or some of the time responding movement into roundabout. The arm might be a bowed segment of the pole, or a different arm connected to it. Appended to the finish of the wrench by a rotate is a pole, normally called an interfacing pole. The finish of the bar appended to the wrench moves in a roundabout movement, while the opposite end is generally obliged to move in a straight sliding movement, in and out. The following are the proposed models from which the selection will be made. The criteria used is choose a model looks at the model that satisfy most of the project requirement these include a compact crusher, attractive to the consumer while maintaining a low cost to fabricate.

4.1 Concept 1

In order to achieve the best possible mechanical advantage a gear system is the best power transmission means available. The proposed model was selected for the above stated reason and also because it has an automatic feed hopper which increases the storage capacity of the entire device.

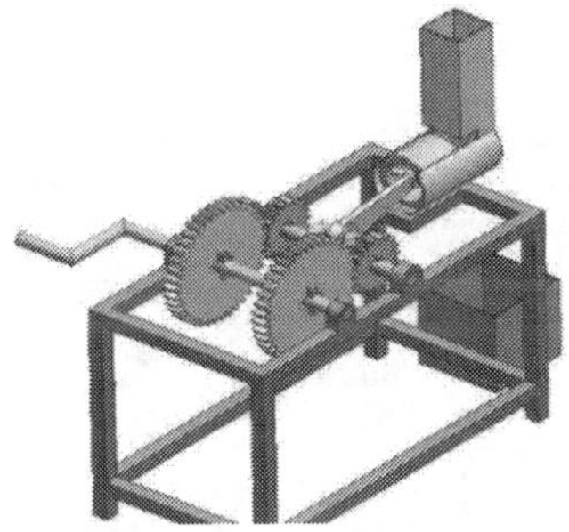

Fig. 1: The first proposed model.

The proposed model as shown in figure 1 was seen to be wasteful in the use of material. Firstly for the frame instead of using square bars for it would be wiser to use angle spacers as they are made of less material and can also offer the structural rigidness that is required from the frame. It is also clear that there are two sets of gear trains on the can crusher.

The use of the one set of a gear train is as effective as using two so it would be wiser to use only one set as the same results will be achieved will a much lower budget as also guards against material wastage .The crusher lever has the same diameter as that of the shaft. This is also unnecessary as to achieve mechanical advantage the crusher leaver has to have a smaller diameter as compared to the shaft diameter. In addition to this this excess material on the diameter is wasted.

4.2 Concept 2

The chain driven system is cheaper than a gear system and it is for that reason that we proposed this model. The chain maintenance is also cheaper and chain failure is also less as compared to the gear driven system. The second concept as shown in figure 2 was an improvement from the first concept.

Fig. 2: The second proposed model

The input from the first sprocket (driving sprocket) to the driven sprocket insured that mechanical advantage was implemented correctly. The only problem with this concept was that the power transmission mechanism does not offer the best alternative in terms of input power by the user. A gear train mechanism is more advantageous as compared to a chain driven device so for these reasons this concept was side-lined.

4.3 Concept 3

The gear driven system offers the best mechanical advantage. We chose this model because it offers the best mechanical advantage at a cheaper cost as it only utilizers one set of a gear train. And also the storage capacity is maximized.

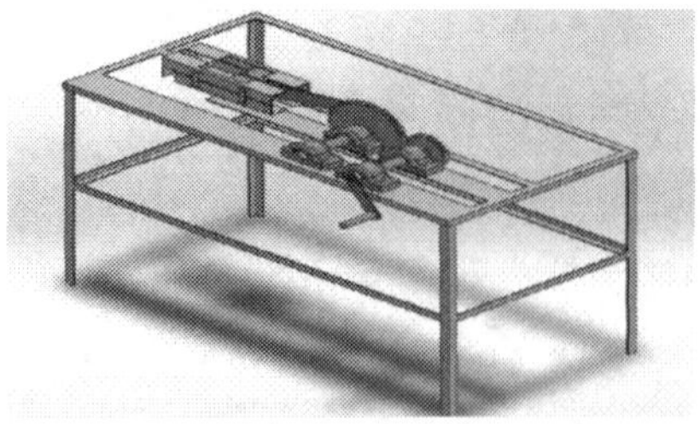

Fig. 3: Approved model of the design.

The third concept was seen to be the best alternative as it satisfied all the objectives tasked to the project. The concept as shown in figure 3 offers maximum exploitation of the concept of mechanical advantage as the gears are arranged in the correct manner. Also, the design is user friendly, as it is in no way complex to the user to understand. It also offers other advantages which will be discussed in depth in subsequent chapters.

5 Analysis and design consideration

After a model concept design has been chosen a though mathematical analysis of the model is done. This will help in making the right decisions in terms of the material selection to meet the objective of the project. The mathematical analysis will touch in detail on the force that the crusher will need to crusher a standard aluminum can. After the force has been established, analysis of the components dimensions to ensure they manage the force exerted. These components include the drive shafts and gears driving the crank mechanism as shown in figure 4.

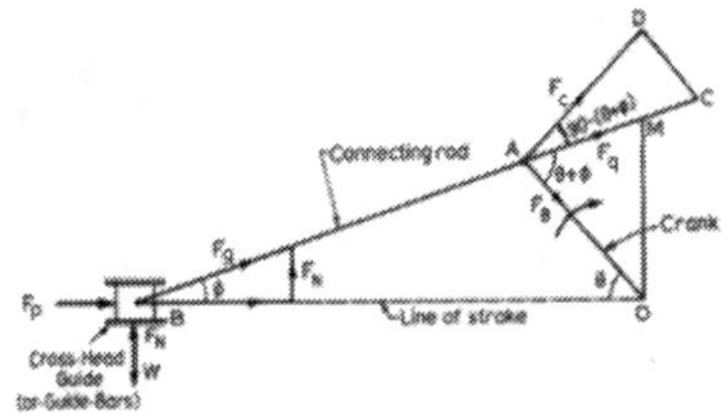

Fig. 4: Four bar linkage mechanism

Mechanical Advantage = Output Force / Input Force

Specifications of Can, Can volume = 330ml

Height = 115mm

Diameter = 66mm

Thickness of aluminum can = 0.1mm

Yield compressive strength of Al = 90Mpa

Therefore force applied by the slider to crush the can should be ≥ 170N

The crushing force the product exerts on the drink-can shall be at least 170N. Since we want our model in safe zone so we are considering a factor of safety and taking output crushing force 200N

Input Force applied on hand wheel = 20N

n = l / r = length of connecting rod / radius of crank = 290mm / 75mm = 3.86 ≈ 4

Crank angle at particular instant θ = 60°

$l \sin \Phi = r \sin \theta$

$\cos \Phi = 1/n \times (n^2 - \sin^2\theta)^{1/2}$

$= 1/4 \times (4^2 - \sin^2 60)^{1/2}$

$= 0.9764$

$\Phi = \underline{12.5°}$

Force acting along connecting rod F_q

$F_q = F_p / \cos \Phi \quad = 200 / \cos 12.5$

$= \underline{204.9\ N}$

Crank Effort $T = F_c \times r$

F_p = Axial force along sliding rod
$F_c = F_p \times \sin(\theta + \Phi)/\cos\Phi = 200 \times \sin(60 + 12.5)/\cos 12.5 = 195.4$ N
T = 195.4 × 0.075 = 14.65 Nm
ω = 2πN / 60 = 0.628 rad. /sec
Speed ratio = rpm of hand wheel/rpm of crank = N_1/N_2
Gear ratio = 4 $4 = N_1/6$
N_1 = 24rpm
Power Transmission from hand wheel to crank is equal $T_1 = 14.65 \times 6 / 24$
T_1 = 3.66 Nm
Calculation for hand wheel
T = F × r If we are taking radius of hand wheel = 235mm
Then, 3.66 = F × 0.235 F = 15.57 N
Mechanical Advantage = Output Force /Input Force = 200/15.57 = 12.8 ≈ 10
Therefore this satisfies the desired mechanical advantage of 10.
As we want to crush 6 cans in one minute, so from power transmissions equation and from balancing point of view
Gear ratio = 4 No. of teeth in gear = 120 and, no. of teeth in pinion = 30
Pitch circle diameter of gear = 190mm
Module = pitch circle dia. / no. of teeth in gear = 190 / 120 = 1.58 ≈ 2
For use with highly reliable materials where loading and environmental conditions are not severe and where weight is an important consideration therefore the factor of safety is taken as 1.5
Torque = Crank effort * Radius of crank But T = F * R   Then, 4.96 = F × 0.235
F = 21.1N,
Torque = force x length of lever
= 21.1 x 272

CANS	LENGTH (mm)
1	28.8
2	27.7
3	29.9
4	30.1
5	29.0
6	27.0
7	29.1
8	28.8
9	29.2
10	30.3
11	30.1
12	27.4

T = 5.74Nm
For shafts Using Gorge Sign approach-
Permissible stress equation

$$\sigma = \frac{S_{yt}}{n} \quad (1)$$

$$\sigma = \frac{215}{1.5} = 143.33 \text{ MPa}$$

To find the shaft diameter

$$\sigma \geq \frac{M}{z} \;;\quad Z = \frac{\pi d^3}{32} \quad (2)$$

The equation to find the diameter of the shaft

$$d \geq \sqrt[3]{\frac{32M}{\pi\sigma}} \quad (3)$$

$$d \geq \sqrt[3]{\frac{32(14.65)}{\pi 143.33\times10^6}} = \mathbf{0.0134 \text{ m or } 13\text{mm}}$$

6 Testing and Result discussion

Once the desired design had been achieved and the specification of the material and dimensions had been done, fabrication processes were employed to bring the product to reality. These processes include wielding, cutting measuring and marking, drilling and grinding. These processes were chosen as they were more cost effective coinciding with the objectives of the work. The model as shown in figure 6 is operated by only one person and can crush up to 360 cans an hour at full load.
A can is fed into the crushing compartment and the lever is rotated to crush the can. Once the can has been crushed in falls into the collection chamber were it will be stored whilst awaiting transportation to the recycling facilities

Figure 5: Finally fabricated can crusher.

The can crusher is relatively easy to use as it requires only the turning of the lever for the crushing operation to take place. The cans were started off with the one and then it was fed into the crusher by hand to be crushed. The can was measure to see if the met the desired length of 25% the original size.
Height of the can = 115mm.

The results of the length of the can after crushing is shown in table 1-.

Table 1: The results of the length of the can after crushing.

Number cans crushed to desired size = 12
Total number of cans crushed = 12

$$efficiency = \frac{no.\,cans\ crushed\ to\ desired\ size}{total\ no.\,of\ cans\ crushed}$$

$$efficiency = \frac{12}{12} = 100\%$$

$$cycle\ time = \frac{processing\ time}{number\ to\ be\ processed}$$

$$cycle\ time = \frac{60\ seconds}{3\ cans} = \mathbf{20\ sec.\ per\ can}$$

The actual product was measured against the desired design to test if it met the required engineering standard and objective of the project. After the can crusher satisfied the design tests were ran to find if it met the required performance objectives. After testing the performance of the can crusher it was determined that the throughput of the can crusher was able to crush 3 cans per min. The Can- before and after crushing are shown in figure 6.

We managed to crush 12 cans successfully with no glitch in the operation of the crank mechanism. And we achieved to crush the entire cans to 25% of their original size, as seen in figure with 100% efficiency for the entire rounds. The model was designed to crush cans of a known capacity only (330ml).

Fig. 6 can- before and after crushing.

Only beverage cans are to be crushed by the model. In terms of storage capacity the design does not offer limitless storage and requires regular emptying. The can crusher can only crush one can at a time since its intended application was for small scale usage.

Conclusion

The fabricated model was successful in achieving its intended application. This model was however a foundation stone of can crushing design and fabrication in the University of Botswana and invites future scholars to improve the design were possible. The can crusher is able to crush 6 cans a minute and it can be placed in any location without diminishing its crushing capacity. The can crusher should be designed to be able to crush cans of all beverage capacities (330 ml, 440 ml etc.). The crusher should not only be restricted to crushing beverage cans but should also crush other food stuff cans such as bean and fish cans. The crusher should also be designed in a manner that would allow it to crush multiple cans at a time so as to improve its efficiency. A motor powered by solar panels can be used to crush the cans to completely eliminate human effort.

References

1. R. P. F. T. J. Government, "scoping report for the state of waste," p. 5, March (2013).
2. kartikey desai courtney jones, sarah kasi and jon ko, "can crusher project," 4th period-tech.
3. Kgathi d.l. & bolaane b., waste management & research, vol. 19, pp. 342-353, (2001).
4. Ketlogetswe c. & mothudi t.h., resources conservation and recycling, vol. 44, pp. 333-342, (2005).
5. Surve qais, sayed saif, sayed jafar, shaikh nadeem, international journal of mechanical and industrial technology, pp. 155-159, 2015.
6. R. Kumar, "product innovation," (2005)
7. N. Delson, 6 october 2016. [online]. Available: http://www.maelabs.ucsd.edu/mae_guides/machine_design/machine_design_basics/mech_ad/mech_ad.htm.
8. Properties-of-aluminium; 2016. [Online]. Available: http://www.aluminiumdesign.net/why-aluminium/properties-of-aluminium/.
9. George larsen nelson, in analysis of the four-bar linkage, wiley, new york, technology press of the massachusetts institute of technology, 1951, p. 730.
10. Ning gu, in computational design methods and technologies, igi global, (2012).
11. a. Elfasakhany, j. Marquez, e.y. rezola, j. Benitez, "international journal of mechanical engineering and technology," p. 112, 2012.
12. J. J. Ward, applied mechanics, uk, macmillan education uk, 1972, pp. 48-60.
13. R. K. Bansal, "theory of machines," in theory of machines 4th ed, bangalore, india, laxmi publications, 2006, pp. 553-560.
14. "Three useful techniques in four-bar linkage design: a review," asme, 1968, p. 4.
15. P. M. F. Ashby, "materials selection ¬ what is materials," 4 march 2010. [online].

Available:
http://www.azom.com/article.aspx?articleid=5134
16. P. M. F. Ashby, "materials selection ¬ what is materials," 4 march 2010. [online]. Available: http://www.azom.com/article.aspx?articleid=5134.
17. I. G. A. P. Francis, "office for national statistics uk material flow review," uk material flow review, january 2005.
18. R. P. F. T. J. Government, "scoping report for the state of waste-2," p. 5, march 2013.
19. C. Houska, "metal for corrosion resistance: part 2," the construction specifications institute, 99 canal center plaza, suite 300, alexandria, 2000.
20. P. Michael pfeifer, "5 materials selection steps for superior products," in materials enabled designs, 1st edition the materials engineering perspective to product design and manufacturing, 2009.
21. Handbook of mechanical engineering 2nd, s. Chand publishing empowering minds, pp. 351-352; (2008).
22. A. Slocum, 2008. [online]. Available: http://web.mit.edu/2.75/fundamentals/fundamentals%20book%20pdf/fundamentals%20topic%204.pdf
23. A. -. F. A. N.-f. M. Stockist, 17 may 2005. [Online]. Available: http://www.azom.com/article.aspx?articleid=2863.
24. M. H. B. M. Zulkifli, "design of a recycle bin tin can crusher," pp. 12-15, (2008).
25. Kgosisele e. & zhaohui l., "an evaluation of waste management in botswana: achievements and challenges," journal of america science, 2010.
26. "waste management act from laws of botswana," 1998. [online]. Available: http://www.laws.gov.bw/volume%2014/chapter%20%2065%2006%20waste%20management.pdf. [accessed 01 05 201

Study on Fatigue Life of Heat-treated Aluminium alloy coated with Castor oil.

A.Sivasubramanian[1,*], *T.S.Kirubasankar*[2], and *S. Vinoth kumar*[3]
[1]Department of Mechanical Engineering, VSB Engineering College. Karur
[2] Department of Mechanical Engineering, VSB Engineering College. Karur
[3] Department of Mechanical Engineering, VSB Engineering College. Karur

Abstract. This paper involves the study of fatigue life of coated aluminium alloy Al 7075-T651 that is heat-treated under 100oC soaked in castor oil for three days. The specimen after heat treatment is subjected to fatigue test using rotary bending machine for number of cycles to fail under cyclic load of 15kgf, 25Kgf, and 50kgf.The life of the specimen is found and compared with uncoated specimen and improved life in number of cycle is noticed. The crack propagation and its type is analysed using scanning electron microscope for knowing the point of fracture and its initiation to failure.

1 Introduction

Aluminium alloys are extensively used material in the field of aerospace and engineering. The use of Aluminium alloy is greatly enhanced due to its good weight to strength property. Since the specific gravity of aluminium is nearly 2.7g/cc, the aluminium alloy has a great role in aircraft sturrctues.Aluminium alloys are used in automobiles, aircrafts and in most of the manufacturing industries. The property such as non-heat treatable, good hardness, corrosion resistance and good thermal conductivity and electrical conductivity [1].
The fatigue life of aluminium plays a vital role in aircraft applications. The fatigue resistance can improve the life of the material and prevents fracture of components at early stages. The life of the component is greatly improved by coating which in turn due to adhesion improves the fatigue toughness of the material.
There are many coating techniques, which were proven more advantageous but require lot of preparatory work on the material to it coated [2]. The deposition rate, adhesion property, chemical treatment is to be taken into consideration for improving the life of the material [3].
The mechanical property of the material is shown and it helps in knowing the important factors, which decided the application of the material in the field of Aeronautics, automobile.etc. The chemical composition also gives an idea on how the material would react to the environmental effects in which it is applied [4].

Table 1. Composition of Al7075-T651.

Compounds	% by weight
Aluminium	87 – 91.4
Zinc	5.1 – 6.1
Magnesium	2.1 – 2.9
Copper	1.2 – 2.0
Iron	0.5 Max
Silicon	0.4 Max
Manganese	0.3 Max
Chromium	0.18 – 0.28
Titanium	0.2 Max

Table2. Mechanical Properties of Al7075-T651.

Properties	Value	Units
Density	2.81	g/cc
Modules of Elasticity	71.7	GPa
Poison Ratio	0.33	
Ultimate Tensile Strength	572	MPa
Yield Strength	503	MPa
Fatigue Strength	159	MPa
Brinell Hardness	150	BHN
Shear Modulus	26.9	GPa

* Corresponding author: sivnay123@gmail.com

2 Experimental Procedure

The fatigue test is carried out in a standard rotary bending test machine. The specimen is made as per standard requirement and the specimen is prepared for coating before testing. The coating procedure is done in such a way that the specimen is dipped in castor coil at normal room temperature around 32 to 36 degree Celsius for 6 days and then the specimen is heated in an oven at constant temperature of 100 degree Celsius for 20 mins. The oil has high density and due to heating the surface of the material is distracted and oil get adhered to the surface resulting in a coating.

Fig. 1. Fatigue specimen dipped in Oil

The heat treated coated sample is subjected to fatigue test on the rotary bending machine for given load of 25kgf, 50kgf and 75kgf.The number of cycles until the failure is noted for the samples under different loading conditions. Similarly, the same procedure is repeated for uncoated specimen and cycles for failure are noted. The number of cycles for coated and Uncoated material is studied and compared..

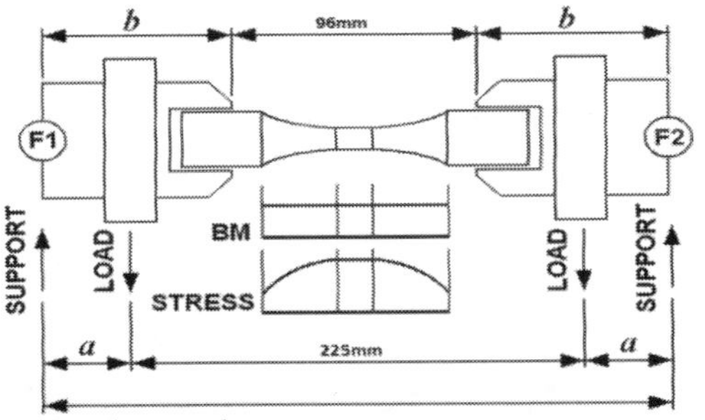

Fig. 2. Fatigue specimen

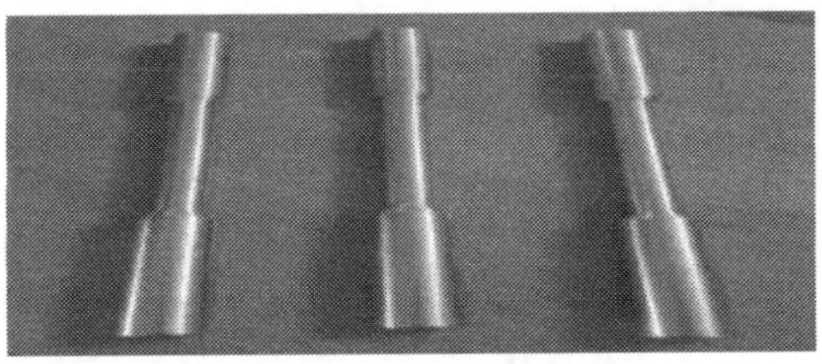

Fig. 3. Fatigue specimen samples

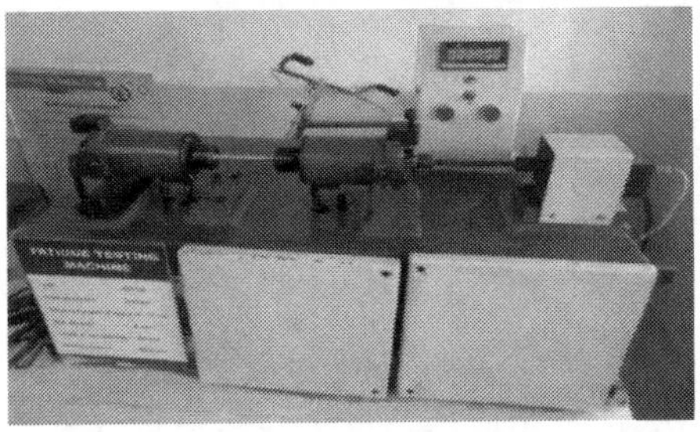

Fig. 4. Fatigue Testing Machine

Fig 5. Fatigue specimen failed during test

3 Fracture analysis of Failure

Since aluminium is considered ductile, the failure is usually due to the shearing[5]. The fractrograhic study through visual and inspection gives an idea on how the material has failed. Further, the fracture is investigated through SEM analysis of the material at various scanning intensity to know about the failure mode in detail. Visual inspection shows some traces of Ductile and brittle failure of material [6].

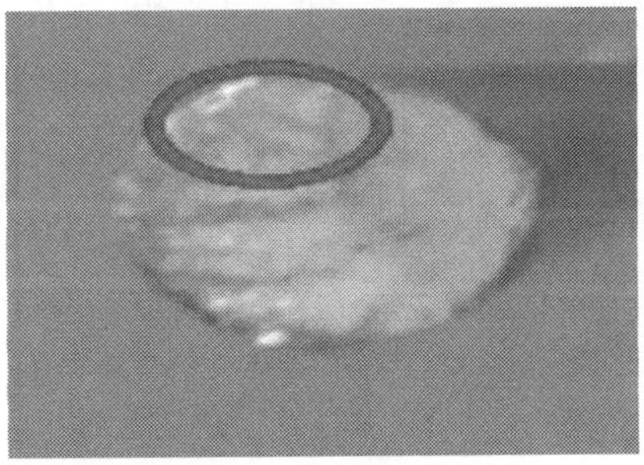

Fig. 6. Visual inspection of Specimen

The SEM image was captured to study about the failure mode of coated specimen. The SEM image shows the fracture is ductile with some traces of Brittle failure in the inner part of the material. The coated surface seems to have subjected to a shear in the surface of the material.

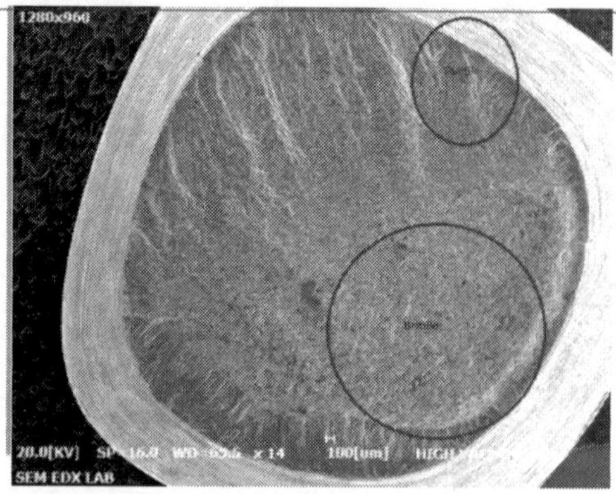

Fig. 7. SEM image showing fracture initiation

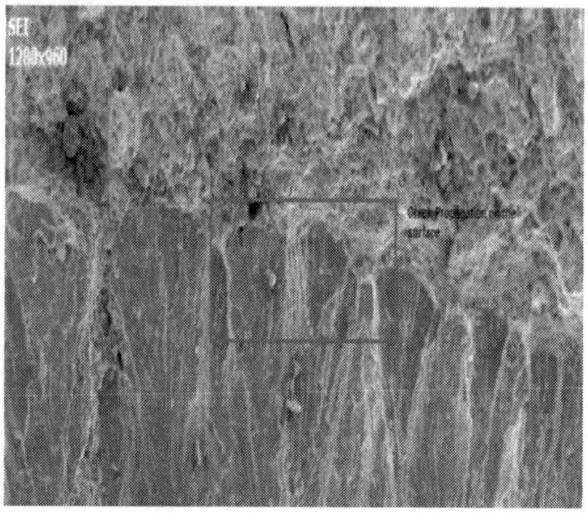

Fig. 8. Crack Initiation leading to ductile failure

The crack propagation results in failure due to cyclic loading and most of the propagation are ductile and interior failure is brittle. As the cyclic loading is increased, the crack propagation is intensified. The fracture is ductile some traces of brittleness at the core part of the material specimen. This may be due to the increase in rate at which the crack propagates at high cyclic loading.

4 Results and Discussion

The number of cycles for failure is studied and the increase in number cycles in noted for coated specimen to the uncoated specimen. The coated specimen shows higher number of cycles and as the load increases the number of cycles decreases but the coated specimen has a increase in life for about 14%.The increase in loading has some effect on the life but the coated specimen can withstand for more time. The usage of coating has enhanced the life of the material.

Table 3. Number of Cycles at various loading.

S.No	Specimen	Tested at 15kgf	Tested at 25kgf	Tested at 35kgf
1	specimen without coating	45024	36080	31578
3	specimen coated with castor oil	47564	41298	35674

References

1. Sivasubramanian, A & Gopal, Arunkumar. (2015). 10. 10627-10633.

2.Pedersen KO. Materials & Design. 2011; 32:97-107.

3. Sivasubramanian, A & Gopal, Arunkumar. (2016). 9.10.17485/ijst/2016/v9i34/100846.

4.Lin XB, Smith RA. Engineering Fracture Mechanics. 1999; 62:535-53.

5.Pedersen KO. Materials & Design. 2011; 32:97-107

6. Arunkumar SG. Australian Journal of Basic and Applied Sciences. 2013; 7(14):281-91

Fatigue life prediction under mixed-mode loading using equivalent stress intensity factor models

S. *Sajith*[1,*], K.S.R. K. *Murthy*[2], and P.S. *Robi*[2]

[1]Research Scholar, Department of Mechanical Engineering, Indian Institute of Technology Guwahati, Guwahati - 781039, India.
[2]Professor, Department of Mechanical Engineering, Indian Institute of Technology Guwahati, Guwahati - 781039, India.

Abstract. Damage tolerance principles are widely used to assess the structural integrity and failure of engineering components. The advances in numerical simulation techniques facilitate the prediction of fatigue life of engineering component, which is essential in damage tolerance design. For the components under mixed mode (I/II) loading, the fatigue crack growth and life is predicted by using a modified form of Paris' law along with the equivalent stress intensity factor (ΔK_{eq}). Numerous ΔK_{eq} models are available for correlating the equivalent stress intensity factor range and the fatigue crack growth rate. The knowledge of proper ΔK_{eq} model is essential for the accurate fatigue life estimation. In this work, the authors numerically assess the performance of ΔK_{eq} models in mixed mode fatigue life prediction with the help of published experimental mixed mode data.

Keywords. Fatigue; mixed mode; crack growth; stress intensity factor; fatigue crack

1 Introduction

Engineering components and parts are generally assumed to be safe, but they contain randomly oriented microcracks, voids, inclusions and flaws, and that may grow under cyclic loading. For the cracks subjected to repeated or cyclic loading, fatigue crack growth analysis based on fracture mechanics principles plays an important role in damage tolerance assessments and in the evaluation of residual fatigue life. Structural components having complex geometries require the use of numerical simulation techniques for the fatigue crack growth (FCG) rate and fatigue life prediction. Numerical techniques such as boundary element and finite element methods are generally used for this purpose.

For the ease of analysis, most of the engineering components containing cracks are considered to be either under mode I or mode II loading and are assumed to be under static loading. Numerous crack growth laws are available for the fatigue life correlation for cracks subjected to only mode I loading. For cracks under opening mode/mode I loading, the power law proposed by Paris [1] has received wide acceptance. Under mixed-mode (I/II) loading conditions, the crack propagation is in non-self-similar manner, i.e., the crack deviates from its original crack path and crack growth rate is significantly affected by the mixed mode loading condition. So in mixed mode loading, the crack path as well as the crack growth rate has to be determined. For cracks subjected to mixed mode loading, Tanaka [2] proposed the modified form of Paris law, a relationship between the fatigue crack growth rate and equivalent stress intensity factor (equivalent SIF) range, which is given by

$$\frac{da}{dN} = C\left(\Delta K_{eq}\right)^m \tag{1}$$

where

$$\Delta K_{eq} = f\left(\Delta K_I, \Delta K_{II}\right) \tag{2}$$

where C and m are material constants. Equation (1) represents the linear relation between on logarithmic scale. The material constants C and m can be obtained from mode I fatigue tests [3].

A large number of ΔK_{eq} models have been proposed for establishing a better relationship between equivalent SIF ΔK_{eq} and the crack growth rate da/dN under mixed mode I/II loading conditions. Accuracy of the numerically estimated fatigue cracks growth rate always depends on the selected ΔK_{eq} model. To the best of authors' knowledge, no work is available on comparison of predictability of different available expressions for the parameter ΔK_{eq} (for a given path prediction rule) in numerical prediction of crack length versus number of loading cycles against available experimental data despite the importance of simulated fatigue life in damage tolerance philosophy. Thus the present investigation attempts to assess the performances of

* Corresponding author: s.sajith@iitg.ernet.in

different models for ΔK_{eq} in numerical prediction of crack growth curves by comparing with the same published experimental data.

2 Methodology and theoretical formulations

2.1. Numerical estimation of mixed mode fatigue life

This section briefly describes the numerical procedure for simulation of mixed mode (I/II) fatigue crack growth. Fig.1 shows different steps to carryout fatigue crack growth numerical simulation under mixed mode loading. An incremental crack tip extension method is used in present study, in which the crack tip is advanced to a new tip position after each step. At each step, SIFs K_I and K_{II} are calculated. At the end of each step, the fatigue crack growth direction, as well as fatigue life, is calculated using the SIFs. The new crack tip coordinates are being calculated and the crack is advanced to the new position. The crack step increments considered are relatively small (w:r:t the initial crack length), for obtaining a smooth crack path. The finite element numerical simulations were carried out using the crack propagation simulator FRANC2D, which is a finite element based crack propagation simulator. The necessary scripts for the post-processing were developed in MATLAB.

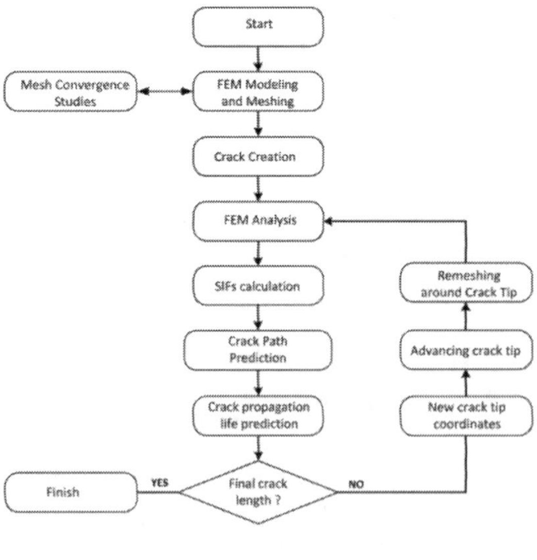

Fig. 1. Typical flow chart for mixed mode fatigue life estimation.

2.1.1 Fatigue crack path prediction

Accurate estimation of fatigue crack path is essesntial for the fatigue life pediction. Erdogan and Sih [4] proposed maximum tangential stress criterion (MTS), which is given as

$$K_I \sin\theta_c + K_{II}(3\cos\theta_c - 1) = 0 \qquad (3)$$

Solving the above equation for the propagation angle θ

$$\theta_c = 2\tan^{-1}\left[\frac{1}{4}\frac{K_I}{K_{II}} - \frac{1}{4}\sqrt{\left(\frac{K_I}{K_{II}}\right)^2 + 8}\right] \quad \text{for} \quad K_{II} > 0$$
$$\theta_c = 2\tan^{-1}\left[\frac{1}{4}\frac{K_I}{K_{II}} + \frac{1}{4}\sqrt{\left(\frac{K_I}{K_{II}}\right)^2 + 8}\right] \quad \text{for} \quad K_{II} < 0 \qquad (4)$$

where positive θ_c is the crack growth direction from the initial crack axis in anticlockwise direction.

2.1.2 Fatigue life prediction

The fatigue crack propagation prediction using the equivalent SIF ΔK_{eq} is the most widely used approach for components under mode I/II cyclic loading. Using a modified form of Paris law, Tanaka [2] proposed a power law for the correlation of fatigue crack growth rate and equivalent SIF ΔK_{eq} (under mixed mode I/II), which is given as

$$\frac{da}{dN} = C\left(\Delta K_{eq}\right)^m \qquad (5)$$

From Eq. (5), for a crack increment, the fatigue life cycles can be determined as

$$\int_0^{\Delta a} \frac{da}{c\left(\Delta K_{eq}\right)^m} = \int_0^{\Delta N} dN = \Delta N \qquad (6)$$

Some of the commonly used ΔK_{eq} models along with the proposed authors are summarized in Table 1.

Table 1. List of ΔK_{eq} models along with their notations used in the present study.

Notation	ΔK_{eq} **models/expression**
Tanaka_1 [2]	$\Delta K_{eq} = \left(\Delta K_I^2 + 2\Delta K_{II}^2\right)^{1/2}$
Tanaka_2 [2]	$\Delta K_{eq} = \left(\Delta K_I^4 + 8\Delta K_{II}^4\right)^{1/4}$
Irwin [5]	$\Delta K_{eq} = \sqrt{\Delta K_I^2 + \Delta K_{II}^2}$
Yan *et al.* [6]	$\Delta K_{eq} = \frac{1}{2}\cos\frac{\theta}{2}\left[\Delta K_I\left(1+\cos\theta\right) - 3\Delta K_{II}\sin\theta\right]$
Hussain *et al.* [7]	$\Delta K_{eq} = \sqrt{\frac{4}{\left(3+\cos^2\theta\right)^2}\left(\frac{1-\theta/\pi}{1+\theta/\pi}\right)^{\theta/\pi}\left[\begin{array}{l}\left(1+3\cos^2\theta\right)\Delta K_I^2 + \\ 4\sin 2\theta\,\Delta K_I \Delta K_{II} + \\ \left(9-5\cos^2\theta\right)\Delta K_{II}^2\end{array}\right]}$
Richard *et al.*[8]	$\Delta K_{eq} = \frac{\Delta K_I}{2} + \frac{1}{2}\sqrt{\Delta K_I^2 + 4\left(1.155\Delta K_{II}\right)^2}$

The equivalent SIF models listed in Table 1 were developed to establish a linear relationship between the experimental $\frac{da}{dN}$ and equivalent SIF ΔK_{eq} on a logarithmic scale. A performance evaluation of the equivalent SIF models (in predicting the fatigue life) is carried out, and they are discussed in the following section.

2.2. Performance assessment of Equivalent SIF models

This section discusses the finite element simulations performed for the present investigation. The numerical simulation is based on the experimental studies using mixed mode I/II CTS specimen by Ma et al. [9] The geometry and boundary conditions of the mixed mode CTS specimen used for the present investigation is given in Fig. 2(a). The loading direction is making an angle of 60^0 with crack axis. The experimental specimen made of steel is used by Ma et al. [9] in his experimental studies. For applying mixed mode loading using the uni-axial testing machine, Richard's loading device [10] is used. Fig. 2(b) shows the boundary conditions and equivalent loading used for the finite element simulations. The uniaxial load P is related to the equivalent loads acting on different holes according to the following equations

$$\begin{aligned} P_1 &= P_6 = P\left(0.5\cos\theta + (c/b)\sin\theta\right) \\ P_2 &= P_5 = P\sin\theta \\ P_3 &= P_4 = P\left(0.5\cos\theta - (c/b)\sin\theta\right) \end{aligned} \quad (7)$$

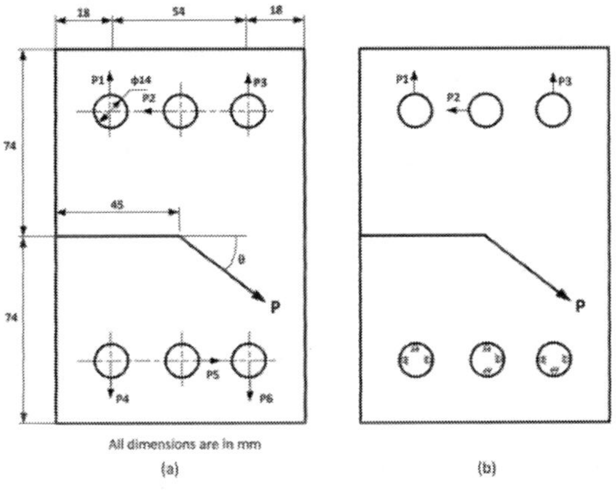

Fig. 2. (a) The dimensions of experimental CTS specimen used by Ma et al. [9] (b) Boundary conditions and the equivalent loading of the specimen used for numerical simulation.

The material used for the experiment, material properties and test conditions given by Ma et al. [9] along with material properties and other constants used for present analysis are given in table 2. An incremental crack tip extension method is followed in the present numerical investigation, and the step crack increment $\Delta a = 0.5$ mm is used. For modeling the material response, plane stress condition is assumed.

Table 2. Material properties and test condition data for CTS specimen.

Material used for experiment	Steel S460
Young's Modulus, E	200 GPa
Poisson's ratio, ν	0.25
Plate thickness, t	6 mm
Paris constant, C	7E-8 $\left((\text{m/cycle})/\left(\text{MPa}\sqrt{\text{m}}\right)\right)^m$
Paris exponent, m	2.1
Load ΔP	9 kN
Load ratio R	0.5
Frequency f	25 Hz
Initial K_{II}/K_I	0.3
Yield strength	460 MPa
Ultimate strength	610 MPa
Crack increment Δa	0.5 mm
Initial crack length a_i	45 mm
Final crack length a_f	65 mm

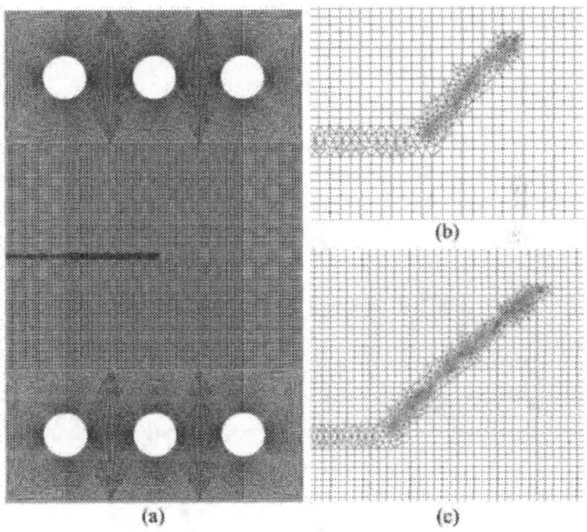

Fig. 3. (a) Initial mesh, (b) intermediate mesh and (c) final mesh used for finite element simulation.

The fatigue crack growth analysis from initial crack length $a_0 = 45$ mm to final crack length $a_f = 65$ mm with 40 steps in carried out. Mixed mode SIFs K_I and K_{II} are calculated at each step using displacement methods. The calculated mixed-mode stress intensity factors are utilized for the equivalent SIF calculation and for the crack propagation direction using MTS criterion. Using the crack kink direction and the constant crack increment, the new location of the crack tip is calculated, and the crack is advanced to the new position. Fig. 4

shows the crack path obtained from the present numerical simulations. The experimental crack path was not available for the present problem considered. Fig. 5 shows experimental fatigue life curves by Ma et al. [9] and a six-degree polynomial fit on experimental data. It can be noted from the goodness of fit of Fig. 5 that the polynomial fit represents the experimental data accurately.

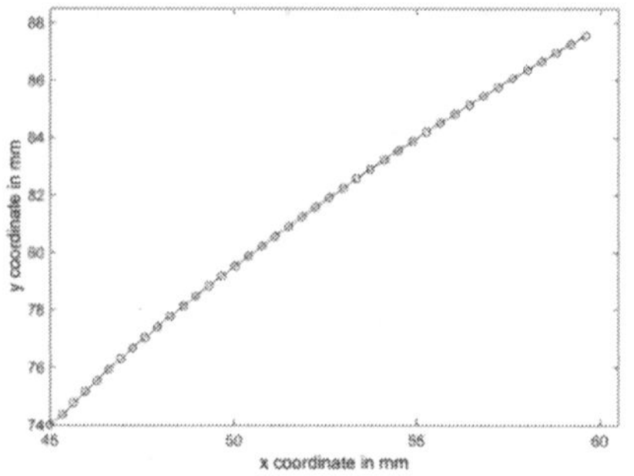

Fig. 4. The simulated crack path for CTS specimen of 60^0 loading angle.

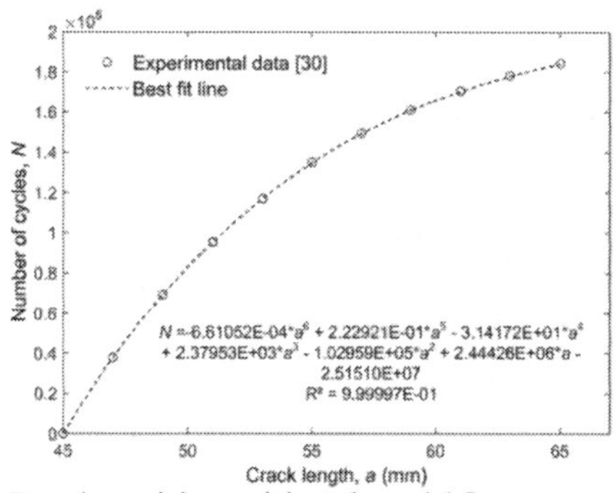

Fig. 5. Experimental data and the polynomial fit.

The numerically estimated fatigue life data using different ΔK_{eq} models are plotted in Fig. 6. The polynomial fit of the experimental fatigue life is also plotted in Fig. 6 for comparison. For better understanding, the close-up views of Fig. 6 near the origin and at the end of the graph is shown in Fig. 7. The percentage relative error in the predicted fatigue life is calculated using the fitted data as reference and is plotted in Fig. 8. In the present numerical investigation, the percentage relative error is calculated by using the fitted polynomial as reference. Fig. 5 also shows the goodness of fit.

It can be seen from Fig. 8 that at the initial stages all the ΔK_{eq} models presented in Table 1 shows significant deviation from the reference values. It can be noted that at higher crack increments, the percentage relative error is relatively less. The predicted fatigue life model using Irwin's model [5] is closer to the reference fatigue life from the experiment. The fatigue life estimate using model proposed by Hussain et al. [7] provides next closest solution. Compared to Irwin's [5] and Hussain et

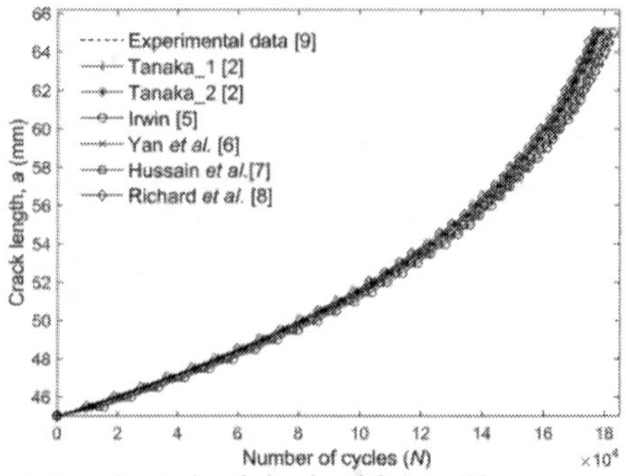

Fig. 6. Experimental and simulated fatigue life.

al.'s [7] models, the other methods offers more conservative fatigue life predictions. Yan et al. [6] model and Richard et al. [8] model are more conservative. Also, it is observed from the similar numerical investigations that when the mode mixity is less, most of the models predicting mixed mode fatigue life close to experimental fatigue life. When the mode mixity is high, the fatigue life prediction using some of the ΔK_{eq} models are higher compared to Irwin's [5] and Hussain et al.'s [7] models. A detailed investigation of the present study using a variety of benchmark examples along with some recommendation for the reduction of initial errors is available in [11,12].

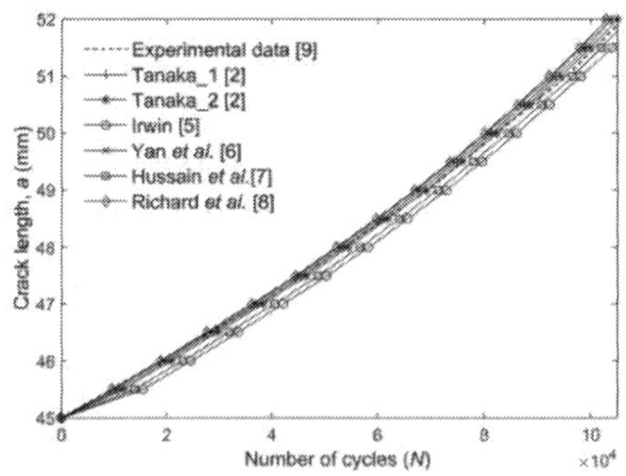

(a)

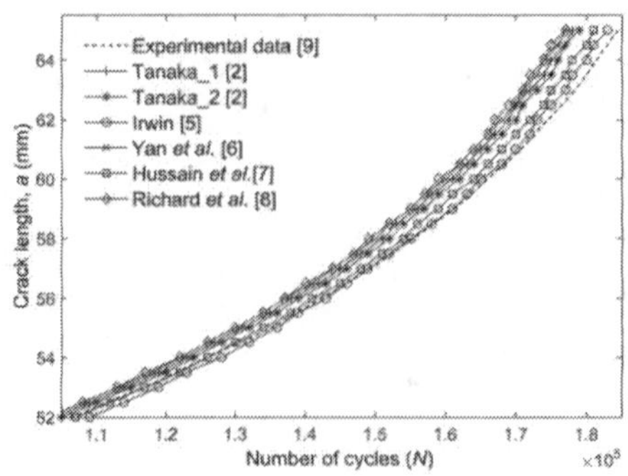

(b)

Fig. 7. (a) Close-up view of Fig. 6 near the origin and (b) at the end of the graph

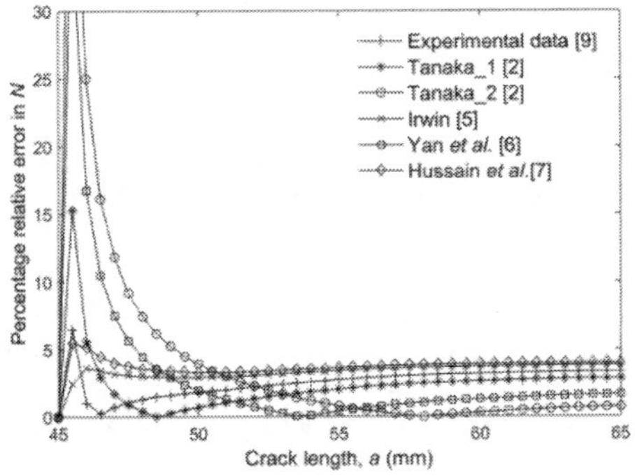

Fig. 8. Relative percentage error in the estimated fatigue life using different equivalent SIF models.

3 Conclusions

Performance analysis of various ΔK_{eq} models in predicting the fatigue life is carried out in this study. The present study is conducted using the limited available experimental results; the present study provides certain useful observations. The results of the present investigation indicate that fatigue life predicted using the ΔK_{eq} models based on Irwin [5] and Hussain et al. [7] models are closer to the experimental fatigue life used in the present numerical investigation. Among the above two, fatigue life prediction using Irwin's [5] model is more close to the experimental fatigue life. One of the models proposed by Tanaka [2] is also predicting fatigue crack growth life close to the experimental data. The remaining three models unsed for the analysis provides more conservative solution, among them the model proposed by Richard et al. [8] is the most conservative model. It is observed in the present investigation that the when the mode mixity is less, most of the models are predicting fatigue life close to the experimental results. However, when there exists a higher mode mixity, there is a significant deviation among the fatigue life prediction by different models and the models proposed by Irwin [5], and Hussain et al. [7] are found to be performing well. The present investigation offers some useful insights into the prediction of mixed mode fatigue crack growth life, which is essential from the damage tolerance point of view.

The authors are thankful to Cornell Fracture Group, Cornell University, New York for allowing to use the FRANC2D software.

References

1. A. Paris, F. Erdogan, J. Basic Eng. **85**, 528–533 (1963).
2. K. Tanaka, Eng. Fract. Mech. **6**, 493–507 (1974).
3. A.B. Patel, R.K. Pandey, Fatigue Fract. Eng. Mater. Struct. **4**, 65–77 (1981).
4. F. Erdogan, G.C. Sih J. Basic. Eng. **85**, 519-525 (1963).
5. G. Irwin, J. Appl. Mech. **24**, 361–364 (1957).
6. Y. Xiangqiao, D. Shanyi, Z. Zehua, J. Appl. Mech. **43**, 471–475 (1992).
7. M. Hussain, S. Pu, J. Underwood, Fract. Anal. ASTM STP **560**, 2–28 (1974).
8. H.A. Richard, F.G. Buchholz, G. Kullmer, M. Schöllmann, Key Eng. Mater. **251**, 251–260 (2003).
9. S. Ma, X.B. Zhang, N. Recho, J. Li Int. J. Fatigue **28**, 1780–1790 (2006)
10. H.A. Richard, K. Benitz, Int. J. Fract. **22**, 55–58 (1983).
11. S. Sajith, K.S.R.K. Murthy, P.S. Robi, J. Inst. Eng. India Ser. C (2017)
12. S. Sajith, K.S.R.K. Murthy, P.S. Robi, AIP Conf. Proc. **1943**, (2018)

Design & Weight Optimization of a Wheel Rim for Sport Utility Vehicle.

Harish Panjagala[1,*], *Balakrishna M*[2], *Shasikant Kushnoore*[1] *and E L N Rohit* Madhukar[3]

[1]Faculty of Mechanical Engineering, Koneru Lakshmaiah Educational Foundation, Vaddeswaram, Guntur, Andhra Pradesh, India – 522502.
[2]Faculty of Mechanical Engineering, G.I.E.T, Rajahmundry, Andhra Pradesh, India.
[3]Department of Mechanical Engineering, Koneru Lakshmaiah Educational Foundation, Vaddeswaram, Guntur, Andhra Pradesh, India – 522502.

ABSTRACT. Automobile have various parts which are important for good running of the vehicle. The most important safety components from a structural point of view are the road wheels. They are required to be lighter and more fascinating to the buyer all the time. This implies that it's important to perform a lot of accurate strength assessment on wheel styles. The wheel rim plays a major role in vehicle dynamics. This paper deals with the design and model of different wheel rims based on weight optimization and also structural analysis has been carried out. It has been compared with standard values by varying two different materials. In addition, from the obtained outputs of simulations and the weight optimization, we suggested Aluminium alloys as most suitable material for SUV. Model is created by using SOLIDWORKS software 2015 and structural analysis & weight optimization is done by using ANSYS WORKBENCH 16.0.

1 INTRODUCTION

A sport or suburban utility vehicle (SUV) is similar to a local wagon or estate car, and are usually use with four-wheeled drive for on- and-off road ability, as well as provide additional cargo capacity in the form of a two box design with shared passenger volume with rear cargo access via a lift gate, rather than a separate lower height trunk cargo space with a horizontal lid. The wheel rim is important structural member of vehicle suspension system which supports static as well as dynamic load encountered during vehicle operation. They must design it carefully. Safety and economy are major concern while designing. Also Style, weight, manufacturability and performance are the technical issues related to the design of wheel rim.

Automotive manufacturers are working on developing fuel effective, safe and lightweight vehicle components. In the actual service conditions, finding of mechanical behavior for the wheel is important, but the testing and inspection of the wheels during their development process is time consuming and costly. For economic reasons, it is important to reduce the time spent during the development and testing phase of a new wheel rim. For this purpose, Finite Element Analysis (FEA) is generally used in the design stage of product development to investigate the mechanical performance of prototype designs. FEA simulation of the wheel rim can significantly reduce the time and cost required to finalize the wheel design.

T. Siva Prasad et al [1] worked with the wheel rim designed by using modeling software CATIAv5r18. Later this CATIA model is imported to ANSY'S for analysis work. ANSYS static analysis work is carried out by considered two different materials namely Aluminum and forged steel and their relative performances have been observed respectively. Sourav Das et al [2] carried the design of Aluminum alloy wheel for automobile application which is carried out paying special reference to optimization of the mass of the wheel.

S. Ganesh et al [3] described parametric model is designed for Alloy wheel used in four wheelers by collecting data from reverse engineering process from existing model. P. Meghashyam et al [4] described static analysis work by considering two different materials namely Aluminum and forged steel and their relative performances have been observe respectively by using ANSY'S Software.

P. Ramamurthy Raju et al [5] concerning with generation of S–N curve for Aluminum alloy (Al) A356.2-T6 and estimation of fatigue life under radial fatigue load. The S–N curve is generated by conducting different tests at different stress levels under constant amplitude loading.

From the above mentioned references, we observed that there is a need to reduce weight of wheel rim and also it should contain high strength. For that reason, we used Topology optimization tool in Ansy's Workbench to find the unnecessary region based on that by changing the design, we optimized the mass of wheel rim by using strength calculations and analyzing the rim using Ansy's Simulation software, finally from the results of structural analysis and the weight optimization, we suggest a suitable material of wheel rim for SUV.

* Harish Panjagala: panjagalaharish@kluniversity.in

2. Theoretical Analysis

2.1.1 Moment on wheel rim :

Net mass of SUV (M) = 2134 kg (hummer H3)
Mass of SUV without wheel rim = net mass – (mass of wheel rim*4)
Let mass of wheel rim is 16.5kg.
Mass of SUV without wheel rim (m) = 2134-66 = 2068kg.
The weight acting on a front wheel rim (F) = m*aspect ratio * gravity/2. =2068*.6*9.8/2 =6080 N
Moment on wheel rim = F*R = 6080*0.217 =1320Nm

2.1.2 Bending moment:

The bending moment to be imparted in the test shall be in accordance to the following formula:
M= ((f*R) +d) *F*S
Where,
f=frictional coefficient between the tire and road surface
R=radius of the tire
d=offset of the wheel (equal to 0 for zero offset position)
F= Maximum load acting on the wheel rim
S=coefficient specified according to the standards
Tire specification radial= 216/60 R17
Where, 216 is the section width in mm 60 is the aspect ratio in percentage 17 is the diameter of the wheel rim in inches f= 0.7 R= 0.217m
S= section width*aspect ratio= 216*.6= 0.130
Now bending moment = (.7*.217) *6080*.13 =724.79Nm

2.1.3 Radial endurance test:

The radial load to be imparted in the test shall be in accordance with the following formula;
Fr= F*k
F = the maximum load of coming on the tire in N
k = constant value according to the industrial standards= 0.453
Fr = 6080*.453=2754.24N

2.2 Specifications:

Rim diameter	= 434 mm
Width of rim	= 216mm
Wheel type	= alloy wheel rim
Aspect ratio	= 6:4
No. of bolts	= 6
Pitch diameter of bolt	= 120 mm
Size of bolt	= 16mm (16M)

2.3 Properties of Materials:

A. Aluminum alloys

Density= 2770 kg / m^3
Coefficient of Thermal Expansion =2.3*10^{-005} /^{0}C
Specific Heat = 875 J / kg ^{0}C

B. Structural steel

Density =7850 kg / m^3
Coefficient of Thermal Expansion= 1.2*10 $^{-005}$ /^{0}C
Specific Heat = 434 J / kg ^{0}C

2.4 Design Procedure of Wheel Rim in Solid works:

1. Created a 2D sketch on front plane as shown in the figure.

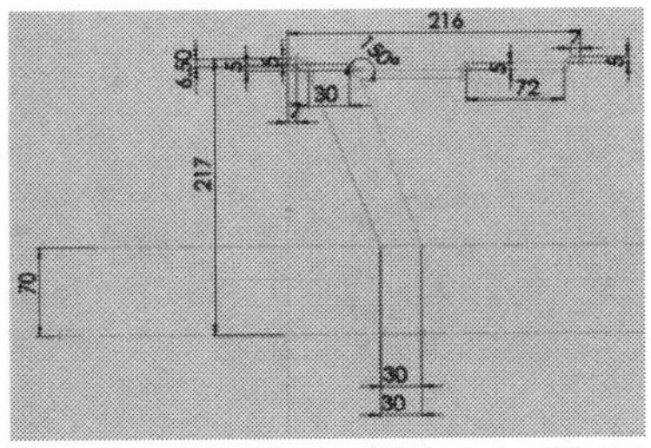

Fig 1. Design of wheel rim (2D) in Solid Works Software

2. Revolved, the sketch to 360 degrees on bottom line, by (Insert>Boss/base>Revolve) ok.

Fig 2. 3D design of wheel rim [Basic Model] & Model 1, (first Weight Optimization model of wheel rim) in Solid Works.

3. Draw the sketch on edge wheel rim face, sketch for arm hole. And remove the material by (insert>cut>extrude cut), through all, ok.
4. Add fillet on the cutting edge, click circular pattern by (view>temporary axis) select center axis as rotation axis.
5. (Give 360 degree and 6 equal spacing), Select cut-Extrude, fillet 1 and fillet 2 as a Feature to pattern. Ok.
6. Repeat the procedure no. 3, 4 and 5 for the different model design of the wheel rim. as shown in figure 3.

7.

Fig 3. Model 2 & Model 3 of Wheel rim in Solid Works Software

7. After completing the designing of wheel rim, converted the file from. SLDPRT to. STEP file for analysis the wheel rim on ANSYS software 16.0.

 https://doi.org/10.1051/matecconf/201817203006

3. Results & Discussions

3.1 Aluminium Alloys Wheel Rim:

3.1.1 Basic Model:

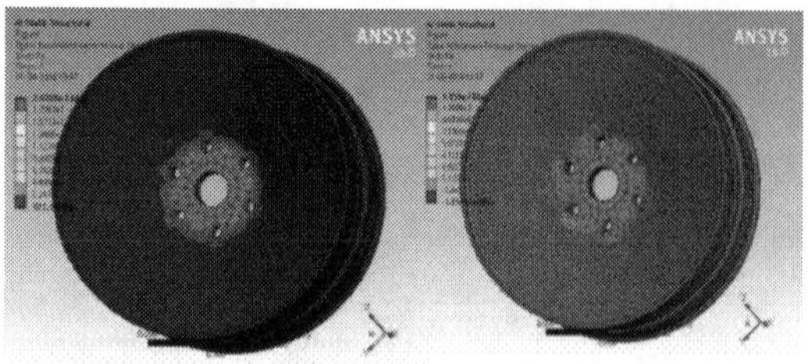

Fig 4. Equivalent stress & Maximum Principal stress on Basic Model of wheel rim.

Here, the maximum equivalent stress generated on the Basic Model is 20.22MPa, which is maximum at the center of the wheel rim as shown in figure 4. And the Maximum Principal stress generated on the Basic model of Aluminum alloys wheel rim is 11.59MPa as shown in above figure 4. Now, the total weight or mass of the Basic model of Aluminum alloys wheel rim is 16.640 kg, which is calculated by using Ansy's Workbench 16.0 Software.

3.1.2 Model 1

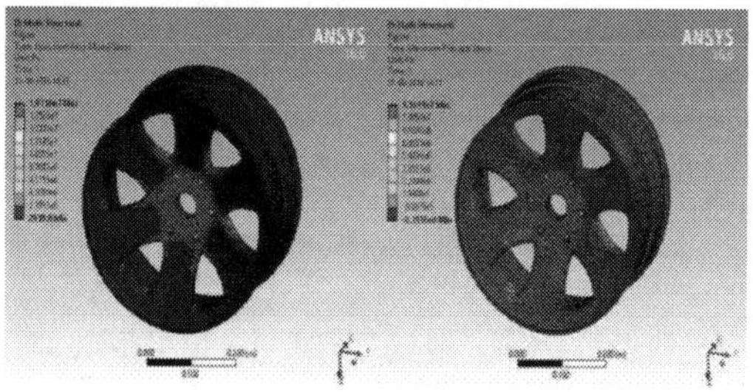

Fig 5. Equivalent stress & Maximum Principal Stress generated on the Model 1-wheel rim

The equivalent stress generated on the Model 1 Wheel rim is 19.71MPa as shown in figure 5. The maximum equivalent stress generated at the center of the wheel rim, which is represented by red color as shown in above figure 5. The Maximum Principal stress generated on the Model 1 of Aluminium alloys wheel rim is 11.6MPa. The actual weight or mass of the Model 1 of Aluminium alloys wheel is 13.484kg, which is less than the Basic model of wheel rim. The total percentage of reduction is 18.96%.

3.1.3 Model 2

Here, the maximum equivalent stress generated on the Model 2 Aluminium alloys Wheel rim is 20.05MPa, and the maximum equivalent stress is generated or developed on the center of the wheel rim as shown in figure 6. Now, the maximum Principal stress developed on the Model 2 of Aluminium alloys wheel rim is 12.989MPa, and which is maximum at the center of the wheel rim figure 6. The mass of the Model 2 of Aluminium alloys Wheel rim is 12.107 kg which is generated or solved by using the Ansy's Workbench 16.0. The total percentage of weight reduction for the Model 2 Wheel rim is 27.24%.

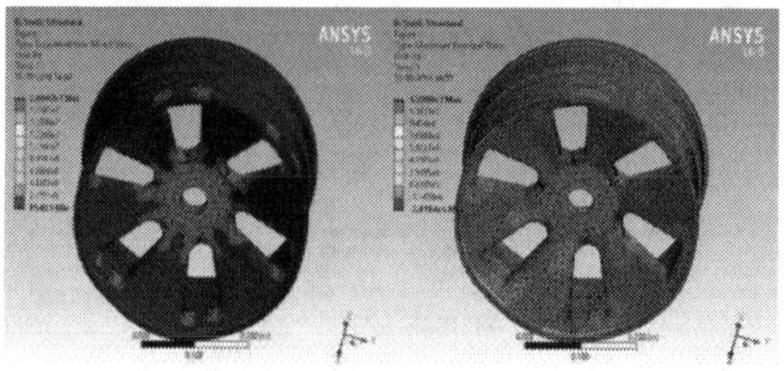

Fig 6. Equivalent stress & Maximum Principal Stress generated on the Model 2-wheel rim

3.1.4 Model 3

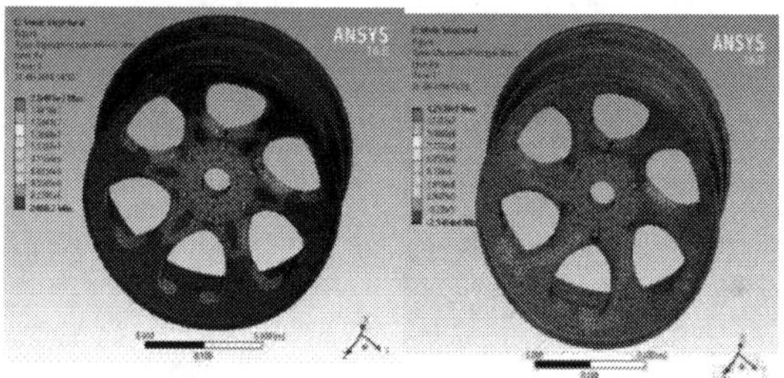

Fig 7. Equivalent stress & Maximum Principal Stress for Model 3-wheel rim

Here, the maximum equivalent stress generated on the Model 3 Aluminium alloys Wheel rim is 20.5MPa, and the maximum equivalent stress is generated on the center of the wheel rim as shown in above figure 7. Then, the maximum Principal stress generated on the model 3-wheel rim is 12.936MPa as shown in figure 7. And the mass or weight of the Model 3 Aluminium alloys Wheel rim is 11.032kg. Total percentage of weight reduction is 33.70%. And this is the final model of wheel rim, which are designed.

3.2 Structural Steel Wheel Rim:

3.2.1 Basic Model

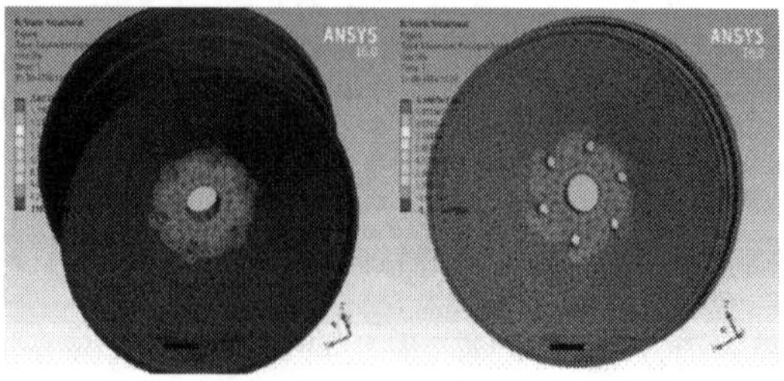

Fig 8. Equivalent Stress & Maximum Principal Stress for Basic model of wheel rim

Here, the maximum equivalent stress generated on the Basic Model is 20.23MPa, which is maximum at the center of the wheel rim as shown in below figure 8. And the Maximum Principal stress generated on the Basic model of Structural steel wheel rim is 11.60MPa as shown in below figure 8. Now, the total weight or mass of the Basic model of Structural steel wheel rim is 47.158 kg, which is calculated by using Ansy's Workbench 16.0 Software.

3.2.2 Model 1

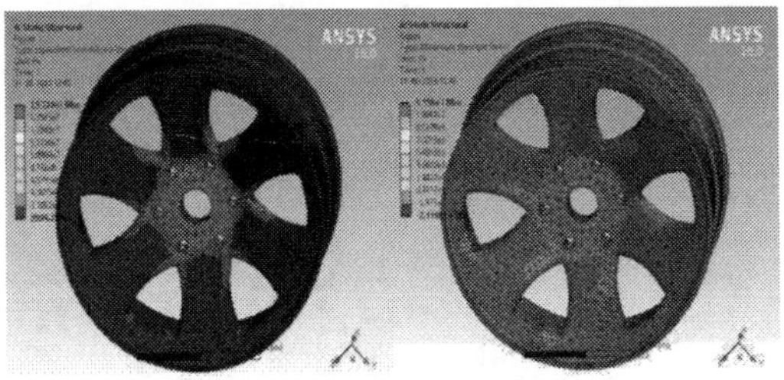

Fig 9. Equivalent Stress & Maximum Principal Stress generated on the Model 1-wheel rim

The equivalent stress generated on the Model 1 of Structural steel Wheel rim is 19.73MPa as shown in above figure. The maximum equivalent stress generated at the center of the wheel rim, which is represented by red color as shown in above figure 9. The Maximum Principal stress generated on the Model 1 of Structural steel wheel rim is 11.56MPa. The actual weight or mass of the Model 1 Structural steel wheel is 38.212kg, which is less than the Basic model of wheel rim. The total percentage of weight reduction is 18.97%.

3.2.3 Model 2

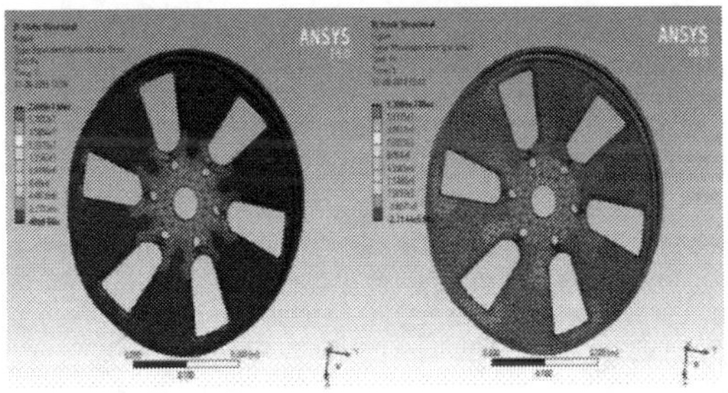

Fig 10. Equivalent Stress diagram for model 2 & Maximum Principal Stress diagram for model 2 of wheel rim

Here, the maximum equivalent stress generated on the Model 2 of Structural steel Wheel rim is 20.06MPa, and the maximum equivalent stress is generated or developed on the center of the wheel rim as shown in above figure 10. Now, the maximum Principal stress developed on the Model 2 of Structural steel wheel rim is 13.069MPa, and which is maximum at the center of the wheel rim as shown in above figure 10. The mass of the Model 2 of Structural steel Wheel rim is 37.144 kg which is generated or solved by using the Ansy's Workbench 16.0. The total percentage of weight reduction for the Model 2 Wheel rim is 21.23%.

3.2.4 Model 3

Here, the maximum equivalent stress generated on the Model 3 of Structural steel Wheel rim is 20.5MPa, and the maximum equivalent stress is generated or developed on the center of the wheel rim as shown in above figure. Now, the maximum Principal stress developed on the Model 3 of Structural steel wheel rim is 12.971MPa, and which is maximum at the center of the wheel rim as shown in above figure. The mass of the Model 3 of Structural steel Wheel rim is 34.098 kg which is generated or solved by using the Ansy's Workbench 16.0. The total percentage of weight reduction for the Model 3 Wheel rim is 27.69%.

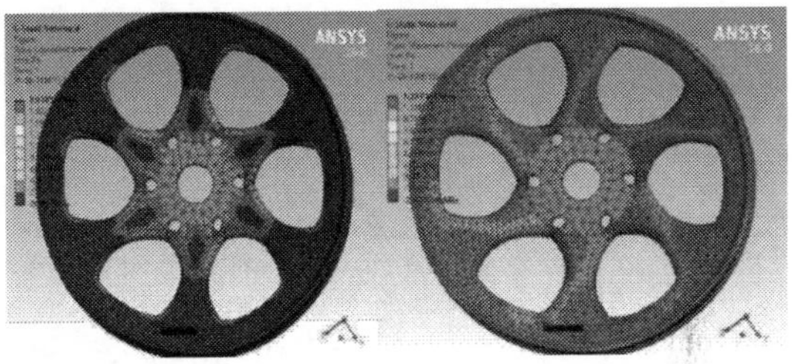

Fig 11. Equivalent Stress & Maximum Principal Stress for Model 3 of wheel rim

Table 1 Mass and Equivalent stress for Aluminium alloys and Structural steel

Model	Equivalent Stress (Mpa)	Mass Of Wheel Rim	
		Aluminium Alloys (Kg)	Structural Steel (Kg)
Basic	20.22	16.640	47.158
1	19.71	13.484	38.212
2	20.05	12.107	37.144
3	20.50	11.032	34.098

3.3 For Aluminium alloys:

The figure 12 shows the relation between the mass of the wheel rim and equivalent stress developed in the wheel rim. Here the first point on the graph represented the mass for the model 1 of the wheel rim, where the mass of wheel rim is maximum and equivalent stress is less as compared to the model 2 and model 3. The second point represented the mass of the model 2-wheel rim, whose mass is less than the mass of model 1 as shown in above graph. The Last point is represented the mass for the model 3-wheel rim, whose mass is very less than the other model. Hence, the mass of the final model for Aluminium alloys wheel rim is 11.032kg and equivalent stress is 20.50MPa as shown in the below graph.

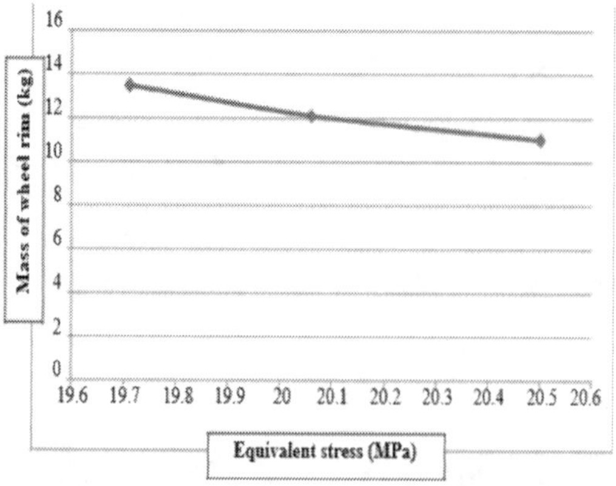

Fig 12. Mass or weight vs. Equivalent stress for the Aluminium wheel rim

3.4 For Structural steel

The figure 13 shows the relation between the mass of the wheel rim and equivalent stress developed on the wheel rim. Here the first point on the graph represented the mass for the model 1 of the wheel rim, where the mass of wheel rim is maximum and equivalent stress is less as compared to the model 2 and model 3. The second point represented the mass of the model 2-wheel rim, whose mass is less than the mass of model 1 as shown in below graph. The Last point is represented the mass for the model 3-wheel rim, whose mass is very less than the other model. Hence, the mass of the final model for Structural steel wheel rim is 34.098kg and equivalent stress is 20.50MPa as shown in the below graph. And when mass of wheel rim decreases, the equivalent stress of wheel rim increases.

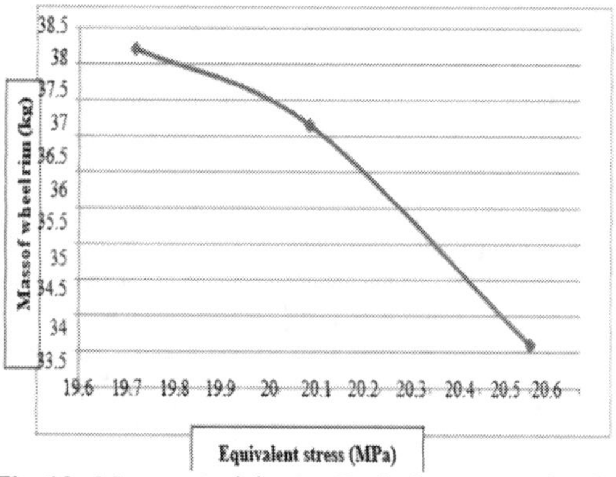

Fig 13. Mass or weight vs. Equivalent stress for the structural steel wheel rim

4. Conclusion

Following are the conclusion made from the results obtained from different simulation of the wheel rims made of Structural steel & Aluminium alloys is;

1. In both cases von-misses / equivalent stresses are less than ultimate strength.
2. In our mini project work, we optimized the wheel rim design to achieve weight reduction. The goal of weight optimization is achieved by comparing the two materials likes structural steel and Aluminium alloys for the SUV vehicle under the same boundary condition.
3. The weight of the wheel rim Aluminium alloys is reduced from 16.64kg to 11.032kg. The strength of the final part or model is 20.5MPa, which is less than ultimate stress 25MPa as per factor of safety 10 is considered.

5. References

[1] T. Siva Prasad, T. Krishnaiah, J. Md. Iliyas, M. Jayapal Reddy, IJISME, Volume no. 2, May 2014, pp.1-8.

[2] Sourav Das, IJSRP, Volume no.4, 6/ June 2014, pp.1-12.

[3] S. Ganesh, and Dr. P. Periyasamy, IJES, volume no. 3, 2014, pp.29-37.

[4] P. Meghashyaml, S. Girivardhan Naidu and N. Sayed Baba, IJAIEM, Volume no.2, August 2013, 2, pp.1-10.

[5] P. Ramamurthy Raju, B. Satya Narayana, K. Ramji, K. Suresh Babu, November 2006, pp.1-24.

[6] Ravi Lidoriya, Sanjay Chaudhary and Anil Kumar Mohopatra, International Journal of Mechanical Engineering and Research. Volume3, 2013, pp. 503-516.

[7] P. Meghashyam, S. Girivardhan Naidu, and N. Sayed Baba, IJAIEM, volume no. 2, 8/ August 2013, pp. 14-20.

[8] T.V. Rao, K. Deepthi, and K.N.D. Malleswara Rao, IJCER, volume no. 4, October 2014, pp.36-40.

[9] Mayur Khule, and P. Baskar, IJRASET, volume no. 1, May 2016, pp.67-73.

[10] Sushant K. Bawne, Prof. Y.L. Yenarkar, IJERA, volume no.5, October 2015, pp.1-7.

Mechanical Design and Analysis of Hybrid Mobile Robot with Aerial and Terrain Locomotion Capability

Monish Koshy[1,*], *Sreevishnu* S.[1], *Anjai* Krishnan[1], and *Gautham P.* Das[2]

[1]Department of Mechanical Engineering, Amrita Vishwa Vidyapeetham, Amritapuri, India
[2]Lincoln Centre for Autonomous Systems, School of Computer Science, University of Lincoln, United Kingdom

Abstract. Although different locomotion mechanisms are available, the use of only one locomotion system in a mobile robot restricts its application scenarios. Hybrid locomotion improves the maneuverability and flexibility of a robot. This paper introduces a hybrid locomotion mobile robot, a combination of quadruped and quadrotor system. The robot has a unique expediency to fly to remote places, then walk to perform close range operations in the field. The prime intention is to use the quadrotor to tackle large objects by flying over it. The four legs provide easy movements in uneven terrain. Thus, this robot can be used in erratic and dynamic environments where stability, maneuverability and flexibility are required. This system can be used as first responders in search and rescue missions, where it responds before human responders gets to the site and get the entire information of the area in detail (like spotting trapped ones, getting detailed 3D mapping etc.). This platform offers unique capabilities suited for search and rescue, disaster zone assistance and surveillances. This paper elucidates the mechanical design and analysis of a hybrid locomotion robot. The solid model of the robot was made using CATIA and further analysis like static analysis, computational fluid dynamics analysis and drop test analysis were performed in ANSYS.

Keywords- Hybrid locomotion mobile robot, Quadruped, Quadrotor

1 Introduction

The notion of hybrid locomotion mobile system was developed in early 1980's. Robots with single locomotion has got its own advantages and disadvantages, whereas multiple locomotion increases the maneuverability and flexibility of robot [1]. Thus, came the concept of hybrid robots, where different locomotion modes are combined together. This is to ensure that the robot is benefitted from the strengths of both locomotion modes and most of the weaknesses of the individual modes are covered by the other. Robots are able to provide useful help in disaster scenario where human intervention is difficult and dangerous. Robots are used in rescue activities because it can travel quickly to areas too dangerous for humans and can engage in the field to analyse the situation [2]. The main objective of this project is to develop a robotic platform that is suitable for search and rescue, disaster zone assistance and surveillance.

Popular terrain robots used for rescue operations include legged robots [3] and tracked robots [4, 5], as these robots have more stability in uneven surface. It is used for finding the people, who may be trapped underneath things in disaster zone and to create detailed 3D map of the site. But terrain robots are unable to surpass large obstacles and pits. Within the past decade the multi-copter aircrafts has become popular as it can be send into areas where it is too dangerous or unfeasible for human. The capability of these platforms to travel quickly to a site and to generate 3D map makes it apt for search and rescue, disaster zone assistance and surveillance. But flying robots has got its difficulty in navigating through caves and notches.

Thus, an integration of the basic aerial and terrain locomotion counterpart each other giving a better solution for locomotion challenges [6]. The goal of this work is to design a hybrid locomotion mobile robot with aerial and terrain locomotion capability. Here a combination of quadruped and quadrotor system is made. The main idea is, quadrotor thrust can be used for the stability of the system in uneven surface and it also can be used to tackle large obstacles by flying over it. The four legs make it able to walk statically stable.

This paper is organized as follows. Following the introduction, Section 2 discusses about work done in this area till now. Section 3 describes factors that influence the design. The designing of hybrid mobile robot starting with CAD designing and various analysis performed to optimize the design is described in Section 4. Finally, the conclusion is presented in Section 5.

2 Related works

The introduction of hybrid locomotion robot helped to solve various limitations and challenges faced by individual locomotion mechanisms. The ability of hybrid

* Corresponding author: monishkoshy@tutanota.com

robots to maneuver multiple locomotion modes increases the flexibility of the system. Research on combining both terrain and aerial locomotion capabilities in a robot is still in its beginning stage. The ability of the robot to reach the target position quickly and to avoid large terrestrial obstacles by flying expands its application in the area of search and rescue and military surveillances [7].

The advent of drone technology showed a new path for aerial robotic system. HERALD [8], HyTAQ [7], and Pens-FlyCrawl [9] are some of the pioneer robots with aerial and terrain locomotion. HERALD (Hybrid Exploration Robot for Air and Land Deployment) [8] has three robots - a quadrotor and two snake robots, which connects using a magnetic docking mechanism. The snake robots can perform wheeled locomotion individually or together with the quadrotor attached. Although the robot has three independently working robot modules, the quadrotor can fly with only one snake attached to it and both snakes are required to carry the quadrotor, which limit the usability in critical situations. HyTAQ (Hybrid Terrestrial and Aerial Quadrotor) [7] which encompasses a quadrotor. Aerial motion is achieved using the quadrotor, while the terrestrial motion is achieved by rolling the cylindrical cage using the same actuators used in the quadrotor. Rolling motion limits its payload carrying capacity. In-order to carry out terrestrial locomotion the robot have to overcome the rolling resistance. Thus, if the terrain gets a little rough the only option left is to shift to flight mode, which limits its application in catastrophe scenario. Pens-FlyCrawl [9] has the ability to perform aerial locomotion by using bi-copter mechanism and crawl by adding simple lever with tilting landing gear which could rotate 360 degrees. Complex controls of the bi-copter and simple terrestrial locomotion mechanism limit its use cases.

Based on the above mentioned works it is clear that there is a need for hybrid robot in search and rescue and disaster zone assistance. Thus, this research presents "Hybrid Locomotion Mobile Robot (HyLMoR)" with aerial and terrain locomotion. HyLMoR is designed considering search and rescue aspects. Most of the other robots in this category uses wheel-leg or crawling for terrain locomotion, where legged locomotion is much better compared to that in uneven terrain. In HyLMoR the modularity in components are considered where the quadruped and quadrotor modules are combined independenty. This results in independency between two modules and makes controllability simple.

3 Factors that influence the design

The design process of the robot involves understanding the physical problems associated with its manufacturing and should incorporate all possibilities and contingencies that preempts, so that all chances of errors are avoided. This involves producing CAD diagrams and models that gives visual description of the robot. Then analysis of the assembled model is done so as to check how it performs in real conditions. If necessary, changes are made on the design with respect to the results obtained after analysis.

Sufficient care must be taken to not be too economical at the cost surrendering the performance of the robot. Since there is a trade-off between price of the components and their performance. Hence optimised function of both has to be selected, so that our requirements are satisfied.

3.1 Desired objectives of robot design

The overall goal is to develop a robotic platform with aerial and terrain locomotion, that is apt for search and rescue, disaster zone assistance and surveillances. The main idea is quadrotor thrust can be used for the stability of the system in uneven surfaces and it can also be used to tackle large obstacles by flying over it. The four legs enable it to walk statically stable. In order to achieve this, the desired objectives of developing such a platform include:

• High stability: The robot must persist static stability in the absence of external forces and the gait must be planned in such a way that it makes the robot as stable as possible [10, 11].

• Minimum energy expenditure: The placement of the motors for quadrotor and the quadruped legs should take into consideration the energy cost optimisation. Hence selection and placing the motors will be a key design component of the robot.

• Affordable monetary cost: The motors should not be too expensive nor too cheap thus performance of motor is not compromised. Materials should be easily available in the market.

• DOF: Using a quadruped with 3 DOF leg adds omnidirectionality [10, 11].

• Leg configuration: Four legs with spider configuration, suitable for off-road performance have been selected for the proposed design. Spider type leg can extend its toe within a wide range not only forward and backward but also in upward and downward areas around the hip joint. This makes the robot to move even on rough environment and it requires less energy when compared with horse type [10, 11].

• Effects of thrust to weight ratio: The payload of the robot plays a key role while designing the quadrotor part. As the weight increases the thrust needed also increases, hence energy required also increases. So, the payload should be kept as small as possible and thrust to weight ratio as large as possible [12-14].

• Torque of joint motors: The selection of torque of the joint motor is an important criterion while designing. It is safe if the torque selected for joint-motors is larger than the required torque at each joint. This reflects in the movement of the robot and the static stability of the robot.

• Landing gear: In the event of a hard landing, it is the landing gear that protects the frame and other components including the joint motors. It is basically a damping mechanism which protects all the components by absorbing the vibration while landing [12-14].

3.2 Selection of material

The most popular materials available includes carbon fibre, aluminium, steel and titanium. Their properties are compared so as to find the right choice for each part (Table 1).

• Carbon Fibre: It has flooded the market at almost all levels, it has got an unbeatable stiffness-to-weight ratio. It is one among the most popular material.

• Aluminium: High-end aluminium is a perfect choice while making a light, stiff frame for the robot and it is available at reasonable price.

• Steel: It will last a life time. The strength and durability of steel means it is minimal fuss in maintenance.

• Titanium: This metal possesses characteristics from aluminium and steel. It has better shock absorption than that of aluminium but is not quite as light.

Table 1. Material comparison

Material	Modulus of Elasticity (GPa)	Tensile Strength (MPa)	Density (g/cm^3)
Carbon Fiber	220	760	1.7
Aluminium (6061-T6)	69	310	2.7
Stainless Steel (404)	200	1790	7.80
Titanium	116	434	4.5

Each material has different qualities. Comparing material properties alone, carbon fiber will be the material of choice. However, considering the weight to strength ratio and price criteria aluminium alloy (6061-T6) is selected for building up the robot.

4 Model development and analysis

Here, the main focus is on designing the model in the CAD software CATIA. Each part is designed individually and then they are then assembled together to configure the final design. The analysis of the robot is done in ANSYS. The modelling of this robotic platform is done in two phases, as an idea about the payload is required in order to design the quadrotor. Hence the design process includes the designing of quadruped section, followed by that of the quadrotor section.

4.1 Design of quadruped section

For the proposed system a terrain locomotion mechanism is required for it to move through its environment. As it is designed to be used on uneven surface, it requires high stability, thus legged locomotion is selected. The choice of static stability insists number of legs to be equal to or more than four [11]. Hence, a quadruped module is considered.

The main body frame of the quadruped consists of two aluminium sheets, each roughly of dimension 200mm x 200mm x 2mm linked parallely using cylindrical spacers of radius 3mm and length 30mm as shown in Fig. 1. Further modifications to the square shape are made to reduce the weight of the base. This central body frame is used for attaching legs and other components (Fig. 1).

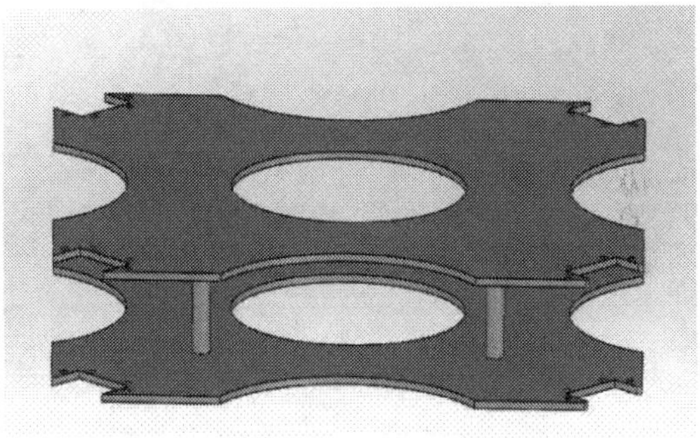

Fig. 1. Main-body structure

The next component to design is the leg of the robot. As explained in the design objective (Section 3.1) legs are arranged in spider configuration and each leg has got 3DOF. Hence each leg is configured by three parts, namely hip, femur and tibia (Fig. 2). The leg length which is around 260mm has been choosen so as to give the optimum stride length with respect to the height of obstacles that it can traverse (here standard bricks of size 190mm x 90mm x 90mm are considered as obsatacle). Initially link lengths are selected in such a way that ratio of link lengths (Femur : Tibia) is around 0.42 [15]. After considering the mechanical constrains, the link lengths are slightly modified. The link lengths thus chosen are shown Table 2.

Table 2. Link lengths

Link	Hip	Femur	Tibia
Length(m)	0.063	0.082	0.1536

The hip link which is the first link from the main body which is also known as coxa. This part makes link between the servo attached to the main-body and servo attached to femur. This is just a small link close to the main body. The femur link is the middle segment of leg. The dimensions have been kept such that the linear distance between the joints amounts to 82mm whereas the total link is 96mm. The last segment of leg that contacts the ground called as tibia. The walking traction is bettered by attaching a rubber cap at the end of tibia. As explained in the design objective a landing gear has to be provided so as to avoid damage on the robot while hard landing occurs during the transition from aerial mode to terrain mode. In order to tackle it a simple helical spring mechanism is added to the tibia part that provides damping.

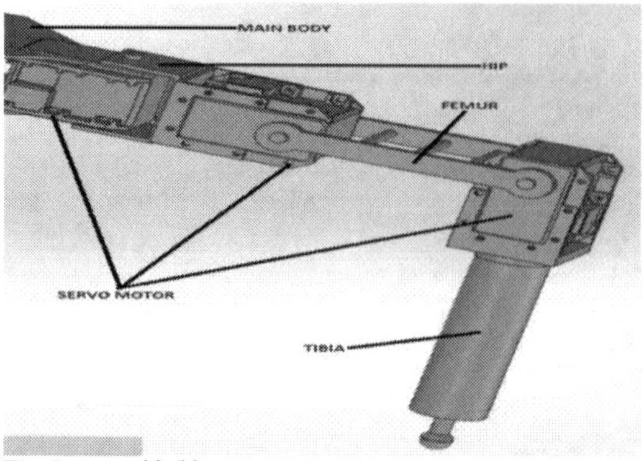

Fig. 2. Assembled leg

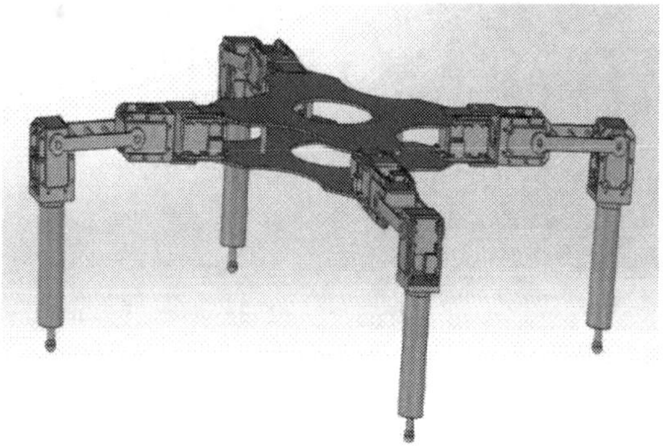

Fig. 3. Assembled quadruped section

Assembling of different parts involves joining the hip, femur and tibia to form the leg (Fig. 2), subsequently followed by joining all these four legs with the main chassis to build up the quadruped section (Fig. 3).

4.2 Analysis of quadruped section

Static state finite element method (FEM) is used to investigate various forces and deformation of each part of the robot. The deformations, stress and strain on the structure of robot is verified by static structural analysis.

4.2.1 Static structural analysis

Structural analysis is a key part of the design process. In this analysis, the various supports and load acting under gravity on the robot are considered. It also gives an idea whether the system design needs any changes.

Here, gravitational force has been applied and point supports has been given at the end of tibia part in all the four legs. Mass of quadrotor section and electronic components are applied on top plate of main chassis.

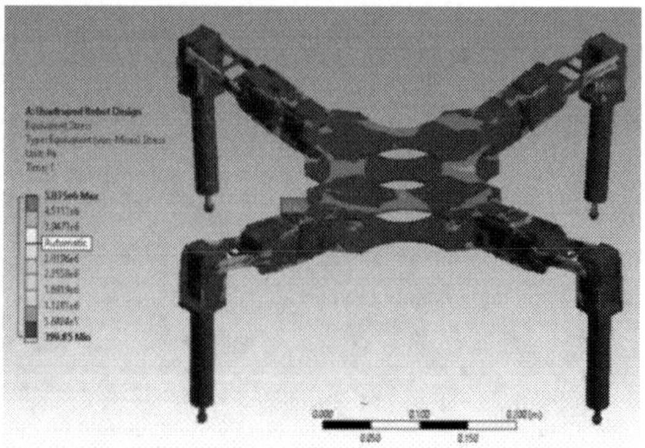

Fig. 4. Stress distribution contour of quadruped section

The resulting stress distribution contour of quadruped section is shown in Fig. 4. The maximum stress of 5.075 x 10^6Pa is found at main body. The system is made of aluminum alloy (6061-T6) with a yield stress of 2.7 x 10^8Pa, hence the stress is in allowable range.

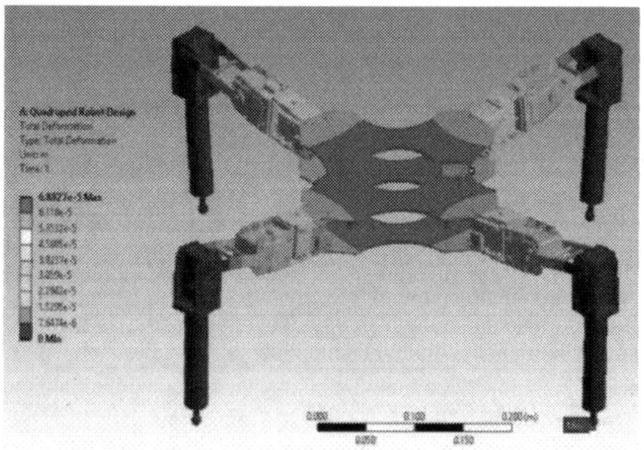

Fig. 5. Deformation contour of quadruped section

The resulting deformation contour of quadruped section is shown in Fig. 5. The maximum deformation that occurs at the main body is just 6.8827 x 10^{-5}m which is smaller than a millimetre, hence the deformation does not make any impact on model.

4.3 Design of quadrotor section

The payload of the robot plays a key role while designing the quadrotor part. It is because of this reason the quadruped section was designed first, so that an idea about the mass of the robot is obtained.

A quadrotor has four arms, each connected to one motor. The arms are symmetric about two axis. First the payload of the quadrotor has to be calculated [12]. The total mass of the robot is denoted by M_T and is given by

 https://doi.org/10.1051/matecconf/201817203007

$$M_T = (M_{QP} + M_{QR} + M_{EP}) \qquad (1)$$

where M_{QP} represents the mass of quadruped section, M_{QR} the mass quadrotor section and M_{EP} the mass of electronic parts. If T_Q is the thrust required by quadrotor then

$$T_Q = M_T \times g \qquad (2)$$

where g represents the gravitational acceleration. While designing a quadrotor, its components are designed to lift double the payload [7]. Thus, thrust considered while designing is denoted by T_D and is given by

$$T_D = 2 \times T_Q \qquad (3)$$

Since it is a symmetric design, quarter of quadrotor is to be analysed. Thus, the thrust of a single propeller (T_{SP}) is given by

$$T_{SP} = (T_D /4) \qquad (4)$$

The propeller has to be selected such that it is capable of producing the required thrust. On the basis of calculations done, each propeller should produce 15N thrust (Eq. 4). Various standard propellers have been analysed using Computational Fluid Dynamics (CFD) analysis (explained later in Section 4.4.1). From the results of the CFD analysis it is found that a 0.254 x 0.119m (10 x 4.7inch) propeller is capable of producing the required thrust.

Quadrotor uses two sets of identical fixed pitched propellers i.e. two counter clockwise (CCW) propellers and two clockwise (CW) propellers. Variation of RPM of these propellers is used to control lift and torque of the quadrotor. The arm of the quardrotor is designed such that, it withstands a maximum thrust of 15N. The dimensions have been kept such that the linear distance between the edge and centre point of motor attachment slot to 232mm whereas the total link is 246mm. The link length is calculated considering that propeller should stand out of main chassis.

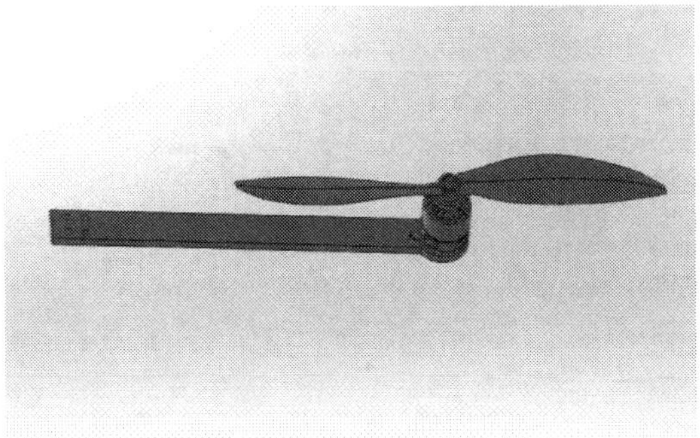

Fig. 6. Quadrotor arm assembled with CCW propeller

Different parts of the quadrotor arm (joining arm, brushless DC motor and propeller) are assembled as shown in Fig. 6.

4.4 Analysis of quadrotor section

CFD is used for estimating propeller characteristics flow region. Here CFD analysis was carried out to find the output thrust produced by the propeller. Static structural analysis has been performed to analyse the various forces and deformations acting at various parts of quadrotor arm.

4.4.1 CFD analysis

As explained in the discussion on designing the quadrotor section (Section 4.3), each propeller of the quadrotor should be able to produce a torque of 15N. So it is required to validate the performance of the propeller using fluid flow simulation. Various standard propellers are subjected to CFD analysis and it is found that 0.254 x 0.119m (10 x 4.7inch) is capable of producing the required thrust at RPM less than the optimal RPM (three-fourths of maximum RPM of BLDC motor considered i.e. 1571rad/s).

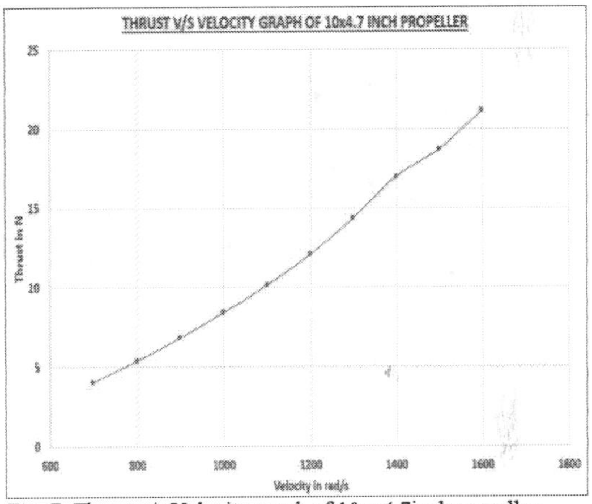

Fig. 7. Thrust v/s Velocity graph of 10 x 4.7inch propeller

Fig. 7 shows the various thrust values obtained from CFD analysis while 0.254 x 0.119m (10 x 4.7inch) propeller is simulated at different velocity. It is clear from the graph that 15.03N thrust has been obtained from the propeller at 1350rad/s (12891.55rpm). Therefore, this propeller is selected.

4.4.2 Static structural analysis

The arm of the quadrotor section should be designed to withstand a force of 15N, which is the maximum thrust value used for the flight. Because of the symmetry of the quadrotor section, single quarter arm is required to be analysed. While simulating gravitational force has been applied, fixed supports has been the edge where holes are provided for fixture (left-side of Fig. 8). A force of 15N is applied in upward direction where propeller will be placed (right-side of Fig. 8).

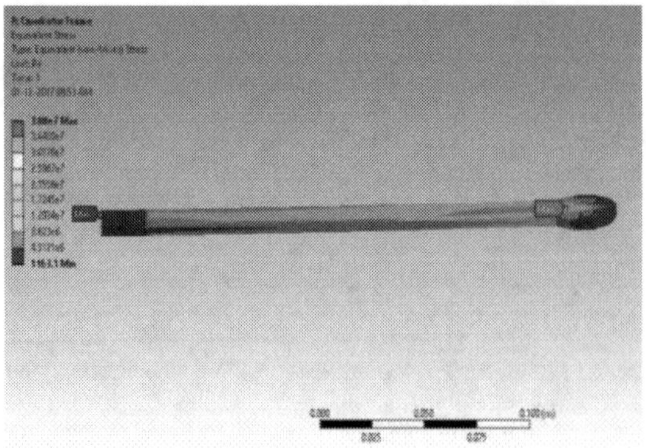

Fig. 8. Stress distribution contour of quadrotor arm

The resulting stress distribution contour of quadrotor arm, while subjected to the force of 15N is shown in Fig. 8. The maximum stress of 3.88 x 10^7Pa is found at arm. Since the frame is made of aluminum alloy (6061-T6) with a yield stress of 2.7 x 10^8Pa, the stress is in allowable range.

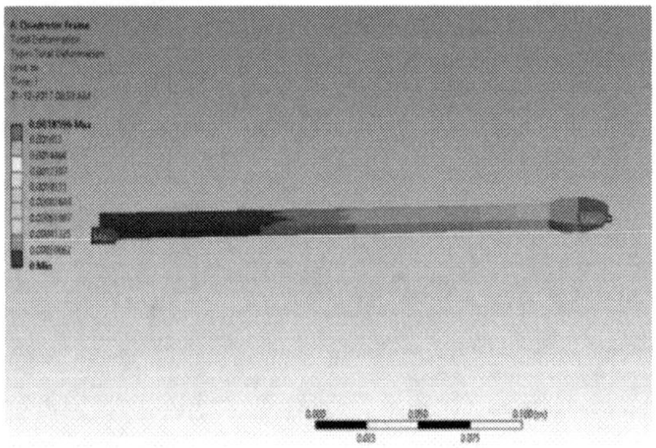

Fig. 9. Deformation contour of quadrotor arm

The deformation contour of the quadrotor arm when subjected to the 15N force is shown in Fig. 9. The maximum deformation that occurs at the main body is just 0.0018m which is very small, hence the deformation doesn't make any impact on model.

4.5 Development and assembly of HyLMoR

Both quadruped and quadrotor section parts designed are assembled together, such that robot capable of aerial and terrain locomotion is obtained.

The dimensions of the base plate for quadrotor frame has been assumed to be 142.4mm x 142.4mm x 2mm. It is attached to the main chassis by cylindrical spacer of radius 3mm and length 60mm (Fig. 10). Modification in the square shape is made in order to reduce the weight of the base and fillet of 2mm is applied on all the sharp corners inorder to avoid the stress concentration.

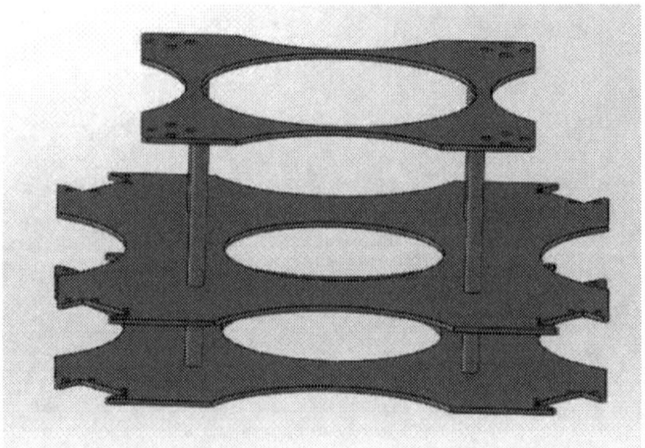

Fig. 10. Main chassis

All the parts designed in the quadruped section and quadrotor sections are assembled together. Thus the final robot assembly is completed (Fig. 11).

Fig. 11. Full model assembly (HyLMoR)

4.6 Analysis of HyLMoR

Static analysis was done to analyse the various forces and deformation of each part of the robot. Drop test was done to determine the critical landing velocity of the robot.

4.6.1 Static structural analysis

Each arm of the quadrotor section should be designed to withstand a force of 15N, which is the maximum thrust value used of the flight. The main body should be able to carry various electronic components (battery, control board etc.), so mass of electronic components are also considered. Gravitational force has been considered and support forces are applied at the tibia part.

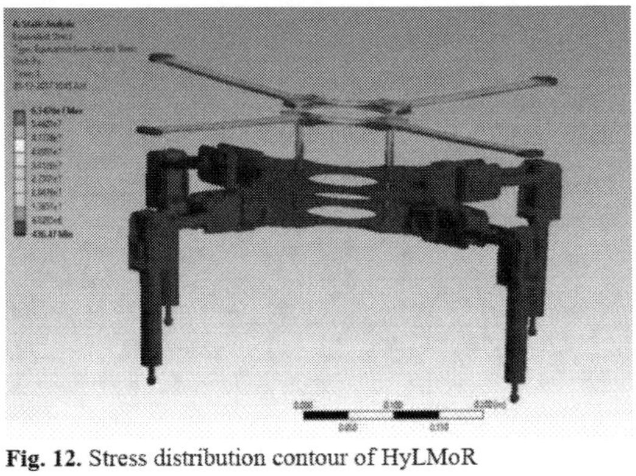

Fig. 12. Stress distribution contour of HyLMoR

The resulting stress distribution contour of HyLMoR is shown in Fig. 12. The maximum stress of 6.142 x 10^7Pa is found at body plate that supports the quadrotor arms. Since the frame is made of aluminum alloy (6061-T6) with a yield stress of 2.7 x 10^8Pa, the stress is in allowable range.

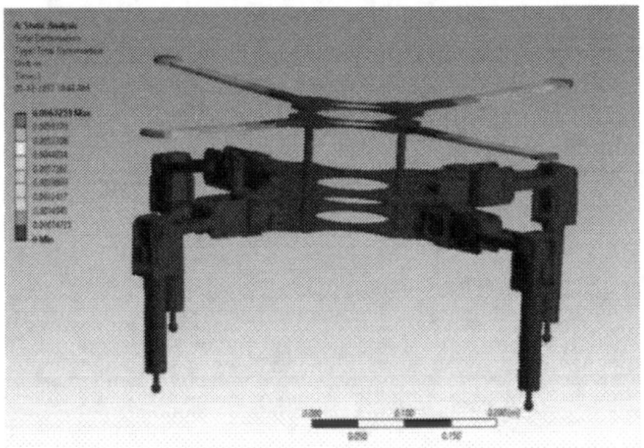

Fig. 13. Deformation contour of HyLMoR

The resulting deformation contour of HyLMoR is shown in Fig. 13. The maximum deformation that occurs at the main body is just 0.0067m which is acceptable, hence the deformation doesn't make an impact on model.

4.6.2 Drop test analysis

The landing gear of the robot is analysed to understand the critical falling velocity of the robot design. For this, simulation of the impact on the landing gear for several velocities is performed. While analysing, gravitational force and the force at the tibia part are considered. In the simulations, the landing surface is assumed to be made of concrete.

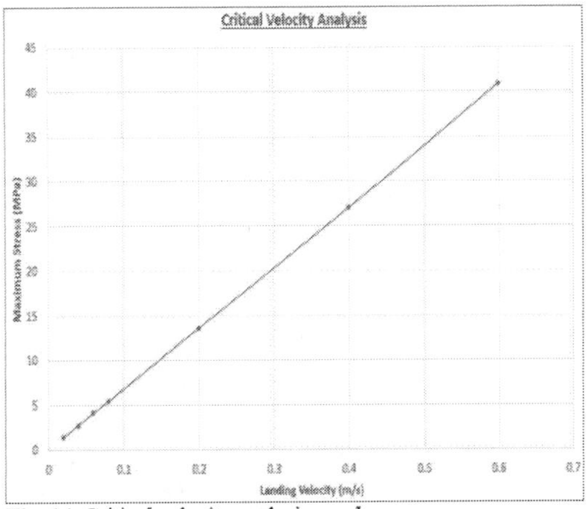

Fig. 14. Critical velocity analysis graph

Eventhough the maximum yield stress of aluminium alloy (6061-T6) is 2.7 x 10^8Pa, the stress of the material used for making the servo motors at the joints of the leg also has to be cosidered. Acrylonitrile butadiene styrene (ABS) with a tensile stress of 4.4 x 10^7Pa is used for making the casing of the motor. Therefore, from Fig. 14 landing velocity which produces a maximum stress below 30MPa is considered, thus the critial velocity is kept at 0.4m/s to ensure no damage occurs to robot. This is with the assumption that the quadrotor would produce a drag reducing the velocity.

5 Conclusion

The primary motivation of this research has been to design a hybrid mobile robot with aerial and terrain locomotion capability. Thus, the robot can be used for search and rescue, disaster zones assistance and surveillances.

In this project, a robot with a combination of 12 DOF quadruped and quadrotor was designed. CAD models were made and fine-tuned by doing various simulation analysis. The final design is a compromise betweeen specification required for the robot, expenditure and overall weight of the robot. It's found that required thrust of 15N can be made using CFD analysis (Section 4.4.1). A maximum of 0.4m/s is obtained as the critical velocity for landing from drop test analysis (Section 4.6.2). From the static analysis, with a maximum stress of 6.142 x 107 Pa (Section 4.6.1) and maximum deformation of 0.0067m (Section 4.6.1). The various analysis results did not exceed the safety limits, hence the robot design made is safe and reliable.

It is clear that a final validation is only possible after evaluation of the constructed model. However, the simulation studies lay the foundation for building the robot capable of aerial and terrestrial locomotion.

References

1. Y. Ichikawa, N. Ozaki, and K. Sadakane, *IEEE Trans. on Systems, Man, and Cybernetics, SMC-13*, pp. 1089–1093, (1983).
2. Mobile rescue robots, *https://www.nccr robotics.ch/research-areas/mobile-rescue-robots/research-overview-mobile-rescue-robots/*, (2017-12-05).
3. F. Seljanko, *IEEE Int. Conf. on Mechatronics and Automation, ICMA*, (2012)
4. P. Ben-Tzvi, A. A. Goldenberg and J. W. Zu, *IEEE Int. Conf. on Robotics and Automation, Pasadena*, pp. 2307-2312, (2008).
5. T. Fujita, T. Sasaki and Y. Tsuchiya, *IEEE Int. Symp. on Safety, Security, and Rescue Robotics (SSRR), West Lafayette, IN*, pp. 1-6, (2015).
6. P. Ratsamee et al., *IEEE Int. Symp. on Safety, Security, and Rescue Robotics (SSRR), Lausanne*, pp. 62-67, (2016).
7. A. Kalantari and M. Spenko, *Robot.Autom.(ICRA), 2013 IEEE Int. Conf.*, pp. 4445–4450, (2013).
8. S. Latscha, M. Kofron, A. Stroffolino, L. Davis, G. Merritt, M. Piccoli, and M. Yim, *IEEE/RSJ Int. Conf. on Intelligent Robots and Systems*, pp. 1868–1873, (2014).
9. S. Kuswadi, M. N. Tamara, D. A. Sahanas, G. I. Islami, and S. Nugroho, *Int. Conf. on Knowledge Creation and Intelligent Computing (KCIC)*, pp. 80–87, (2016).
10. S. Hirose, , *3rd CISM-IFToMM Int. Symp. on Theory and Practice of Robots and Manipulators*, pp. 357–375, (1978).
11. D. C. Kar, *J. Robotic Syst.*, pp. 671–686, (2003).
12. M. Muflikhun, E. Magsino, and A. Y Chua, , *RCMME-2014*, (2014).
13. D. Singla, Design of a Quad Rotor, https://sites.google.com/site/designofaquadrotorgroup28/, (2017-12-05).
14. C. Benson, How to Make a UAV / Drone, *https://www.robotshop.com/blog/en/make-uavlesson-2-platform-14448*, (2017-12-05).
15. M. Silva, *14th Int. Conf. on cimbing and walking robots and the support technologies for mobile machines, France*, (2011).

Kinematic Design, Analysis and Simulation of a Hybrid Robot with Terrain and Aerial Locomotion Capability

Sreevishnu S.[1,*], *Monish* Koshy[1], *Anjai* Krishnan[1] and *Gautham P.* Das[2]

[1]Department of Mechanical Engineering, Amrita Vishwa Vidyapeetham, Amritapuri, India
[2]Lincoln Centre for Autonomous Systems, School of Computer Science, University of Lincoln, United Kingdom

Abstract. Having only one type of locomotion mechanism limits the stability and locomotion capability of a mobile robot on irregular terrain surfaces. One of the possible solution to this is combining more than one locomotion mechanisms in the robot. In this paper, robotic platform composed of a quadruped module for terrain locomotion and quadrotor module for aerial locomotion is introduced. This design is inspired by the way which birds are using their wings and legs for stability in slopped and uneven surfaces. The main idea is to combine the two systems in such a way that the strengths of both subsystems are used, and the weakness of the either systems are covered. The ability of the robot to reach the target position quickly and to avoid large terrestrial obstacles by flying expands its application in various areas of search and rescue. The same platform can be used for detailed 3D mapping and aerial mapping which are very helpful in rescue operations. In particular, this paper presents kinematic design, analysis and simulation of such a robotic system. Simulation and verification of results are done using MATLAB.

Keywords- Hybrid mobile robot, search and rescue, legged flying robot

1 Introduction

Robots are very useful in places where human intervention is dangerous or impossible. Application of drones for search and rescue [1], volcano environment monitoring [2] and military operations are some examples for this. Using only one type of locomotion is not enough in this type of scenarios. One way of solving this issue is by using collaborative multi-robotic systems [3, 4]. Another approach is using robot with hybrid locomotion capabilities [5]. The hybrid mobile robotics expanded the application horizon of robots to challenging areas in which multiple modes of mobility are required. This paper mainly focuses on development of a robotic platform for rescue operation and its possible applications in rescue fields.

Even though, polymorphic tracked vehicles are the state of art in terrain locomotion of rescue robots [6-8], researches are boosting in legged robotics field too [4]. Terrain robots can be used to create detailed 3D map of the environment, which would be very useful for rescue operators to find people underneath things. However, large obstacles and pits are difficult to surpass for terrain robots with legs or belt drive. Recent advent of drone technology had given a big propulsion to rescue robotics. Drones are used for 3D mapping and can be sent in to areas where it is too dangerous for human [2]. One big advantage with drones is that they are first responders. Also, it can tackle big obstacles and pits in the rescue field. On the other hand, navigation of drones through narrow paths like caves and tunnels is difficult. Also, for doing detailed 3D mapping terrain close flying is required. This is difficult with drones in cluttered disaster sites.

To tackle these scenarios, the robot should be able to reach the spot as quick as possible and have to get into the field to analyse the situation. It is in this context the "Hybrid Locomotion Mobile Robot (HyLMoR)" is introduced. It is a hybrid locomotion mobile robot with flying and terrain locomotion capabilities. In particular, this paper describes an experimental study of combining aerial and terrain locomotion capabilities in a single robot with its kinematic study and simulation.

This paper briefly describes an experimental study of combining aerial and terrain locomotion capabilities in a single robot. Section II discusses work done in this area till now. Kinematic analysis, design and verification of quadruped module and quadrotor modules are presented in Section III and Section IV respectively. Then, these modules are combined to form HyLMoR, which is presented in Section V. Finally, the conclusion and future works are discussed.

[*] Corresponding author: sreevishnus18@yahoo.com

2 Related works

Concept of hybrid locomotion mobile robotics system started in early 1980s [9]. Combination of different terrain locomotion mechanisms constitutes hybrid systems in those days. Later researches combined different environmental locomotion capabilities like aquatic-terrain [10], aerial-terrain [5, 11, 12], aerial-aquatic [13] etc.

Research on combining aerial and terrain location capabilities in a single robot is still in its infancy stage. At the initial stages of development, wings were used to achieve aerial capability. Morphing Micro Air-Land Vehicle (MMALV) [22] was one among first in this category. The design used in this robot is similar to that of insects. It uses wings for aerial motion and wheel-legs for terrestrial motion. Thus, long distance travel is possible with flight mechanism, while crawling mechanism helps in close inspection and surveillance. However, the winged aerial-terrain hybrid robots have many disadvantages such as lack of efficient controllability, inefficient transition from flight mode to terrain mode and vice versa and turbulence caused by variations in air speed from one wing to another [22].

Most of the drawbacks with winged flight are resolved with introduction of rotary flight technology. A robot developed by Kossett et al. in 2009 with flight and wheeled locomotion capability, using propeller for flight motion, is one of the first among this category [14]. This offered the best characteristics of both helicopters and terrain vehicles.

Rotary flying technology opened a new domain in aerial robotic systems. DUCK robot [12], flying hexapod [15], HyTAQ [16] and Pens-FlyCrawl [5] are some of the robots which uses rotary flights. The robot which is presented in this paper have more similarity with flying hexapod [15]. Main differences with this robot are HyLMoR comprises of quadruped instead of hexapod. Another big difference is in the modularity of components. In HyLMoR, quadruped and quadrotor modules are combined as separate entities. This makes controllability easier and gives independence between modules. Similarly, the use of legged locomotion helps the HyLMoR to tackle obstacles on ground during terrain locomotion in comparison with the wheeled locomotion in HyTAQ and with the crawling locomotion in Pens-FlyCrawl.

3 Quadruped module

The process of implementing kinematic designing and simulation of the quadruped module is described here. First, analysis of single leg is done and verified using MATLAB robotics toolbox [20]. The 3D model created in MATLAB environment is used to verify the forward and inverse kinematics of the designed robot.

3.1 Selection of legged robot

The proposed robotic system consists of a terrain locomotion part and flying part. For terrain locomotion on irregular surface, the robot needs good stability compared to wheeled locomotion, therefore legged locomotion is selected. Number of legs and configuration of the legs are the two main factors which affects the stability and locomotion of robot.

Thus, the desired robotic platform can be developed by considering the following factors:

- Lowering number of legs reduces the turning area and gives more simplicity to driving mechanism [18].
- Statically stable walking requires atleast four legs [18].
- In the proposed design, legs of the quadruped part are chosen to have 3 DOF in order to decouple body motion with terrain irregularities [18].

Considering the above factors and cost (lesser number of actuators needed), 3 DOF quadruped system for terrain motion is selected.

3.2 Conceptual design of leg

Leg geometry is one of the most significant aspect of design, since it influences the overall performance of the system. Two main categories of leg configuration [19] are horse type and insect type. In horse type, knee joint is always situated under hip joint as in dog, horse and most other terrestrial mammals. In insect type, knee joint is positioned parallel or above the hip joint.

There are many advantages for insect configuration over horse type configuration especially on uneven terrains [18,19]. The arrangement of two corresponding hip joints in insect type has the convenience of positioning two actuators inside or very near to body. This helps in reducing the load over foot joint while foot placing. The arrangement also simplifies the leg kinematics. Another important feature which is very helpful in rescue scenarios is flexibility in leg movement. i.e. the leg can extend within a wide range in forward and backward, and upward and downwards. Compared to horse type model it is more energy efficient in most of the scenarios.

The HyLMoR is mainly targeted to be operated in rescue scenarios which have highly uneven terrains. Terrain locomotion will be better if the legs are more flexible. Energy efficiency is also an important factor for long rescue operations. Considering all these factors, insect type leg configuration is selected.

3.3 Kinematic analysis of leg

Kinematic analysis of a robot is the basic requirement of space planning, motion planning motion control and optimal design. There are two types of kinematic problems, Direct Kinematics and Inverse Kinematics. Both of them are analysed and verified for the proposed quadruped module using MATLAB.

Assigning frames is the first task while modelling a robotic system. Cartesian co-ordinate system is used here. Quadruped is a rigid body system with a centre body and four leg modules. Centre of body can be considered as the origin of the robot frame in which Z axis is pointing upwards, Y axis towards left and X axis towards front. Each leg is attached to the corner of rectangular robot body. These points are origins of leg base frames.

Co-ordinate frames of legs are assigned using Denvait Hartenberg (DH) notation. Each leg is made up of links and joints. DH notation can give the analytical description of spatial geometry of a leg with pose of each links relative to the leg base frame. The relation between the joint-variables and the pose of the foot gives the kinematic model of leg. DH representation of a quadruped leg is shown in Fig. 1 and parameters are shown in Table 1.

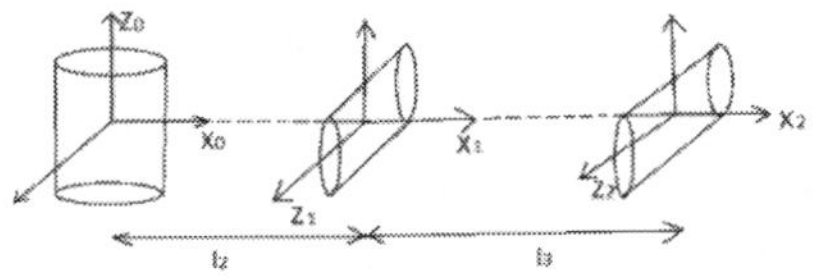

Fig. 1. DH representation of 3 DOF insect leg

Table 1. DH parameters of leg

Link	θ_i	d_i	a_i	α_i (degree)
1	θ_1	0	L_1	90
2	θ_2	0	L_2	0
3	θ_3	0	L_3	0

Here, θ_i is the joint angle and d_i is the joint distance. Parameters a_i and α_i corresponds to link length and link twist.

Table 2. Link lengths taken for analysis

Link	Link1	Link2	Link3
Length(m)	0.063	0. 082	0.1536

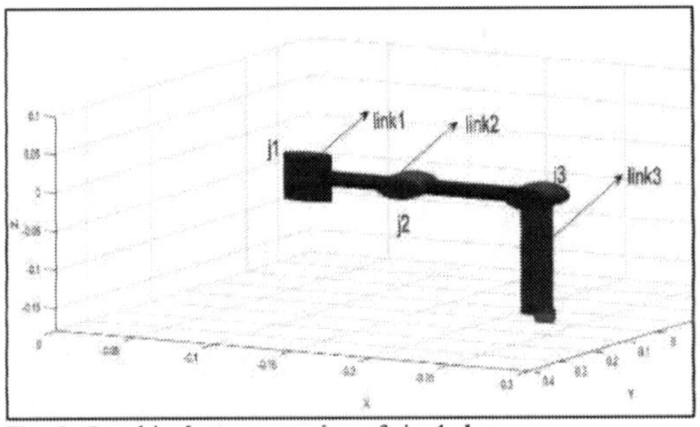

Fig. 2. Graphical representation of single leg

Link length is one of the important kinematic parameters. Length of the upper segment of the leg should be smaller than that of lower segment [21]. To start with, link lengths are selected in such a way that ratio of link lengths as 0.42 [21]. After considering mechanical constraints such as material property, leg configuration, etc., link lengths are adjusted slightly. Finally chosen link lengths for analysis are shown in Table 2. In Fig. 2, the 3D model of single leg is shown. Link1 corresponds to coxa, link2 corresponds to femur and link3 corresponds to tibia respectively. Legs are attached to the corners of the rectangular body of dimension 20cm × 20cm. i.e., legs are attached radially at a distance of 14.14cm from the centre of body. In the Fig. 2 joints are represented with j1, j2 and j3.

3.3.1 Direct Kinematics

Direct kinematics problem is finding the pose of leg tip pose as a function of the joint angle values. Homogeneous transformation matrices are used to find the leg tip pose with respect to leg base frame.

The individual joint transformation matrices formed by referring parameters in Table 1. Are given in equations (1-3).

$$^0T_1 = \begin{pmatrix} c\theta_1 & 0 & s\theta_1 & l_1c\theta_1 \\ s\theta_1 & 0 & -c\theta_1 & -l_1s\theta_1 \\ 0 & 1 & 0 & 0 \\ 0 & 0 & 0 & 1 \end{pmatrix} \#(1)$$

$$^1T_2 = \begin{pmatrix} c\theta_2 & -s\theta_2 & 0 & l_2c\theta_2 \\ s\theta_2 & c\theta_2 & 0 & l_2s\theta_2 \\ 0 & 0 & 1 & 0 \\ 0 & 0 & 0 & 1 \end{pmatrix} \#(2)$$

$$^2T_3 = \begin{pmatrix} c\theta_3 & -s\theta_3 & 0 & l_3c\theta_3 \\ s\theta_3 & c\theta_3 & 0 & l_3s\theta_3 \\ 0 & 0 & 1 & 0 \\ 0 & 0 & 0 & 1 \end{pmatrix} \#(3)$$

Equations (1-3) can be concatenated to form homogeneous transformation matrix which gives the pose of foot of leg relative to base frame of leg. It is given by

$$^0T_3 = {}^0T_1{}^1T_2{}^2T_3 \#(4)$$

$$^0T_3 = \begin{pmatrix} c\theta_1c(\theta_2+\theta_3) & -c\theta_1s(\theta_2+\theta_3) & s\theta_1 & p_x \\ c\theta_1c(\theta_2+\theta_3) & c\theta_1c(\theta_2+\theta_3) & -c\theta_1 & p_y \\ s(\theta_2+\theta_3) & c(\theta_2+\theta_3) & 1 & p_z \\ 0 & 0 & 0 & 1 \end{pmatrix} \#(5)$$

where $c\theta_i$ is the cosine and $s\theta_i$ is the sine of θ_i, for i=1,2,3 and

$$p_x = l_1 + \left(l_2c\theta_2 + l_3c(\theta_2+\theta_3)\right)c\theta_1 \#(6)\#$$
$$p_y = l_1 + \left(l_2c\theta_2 + l_3c(\theta_2+\theta_3)\right)s\theta_1 \#(7)\#$$
$$p_z = l_2s\theta_2 + l_3s(\theta_2+\theta_3) \#(8)\#$$

The transformation matrix was verified for various known angle values, which result in known leg tip poses.

Transformation matrix at (θ_1, θ_2, θ_3) = (0,0,0) is given by

$$^0T_3 = \begin{pmatrix} 1 & 0 & 0 & 0.2986 \\ 0 & 0 & -1 & 0 \\ 0 & 1 & 0 & 0 \\ 0 & 0 & 0 & 1 \end{pmatrix} \#(9)$$

Transformation matrix at (θ_1, θ_2, θ_3) = (30,0,90) is given by

$$^{0}T_3 = \begin{pmatrix} 1 & -0.866 & 0.5 & 0.1256 \\ 0 & -0.5 & -0.866 & 0.0725 \\ 1 & 0 & 0 & 0.1536 \\ 0 & 0 & 0 & 1 \end{pmatrix} \#(10)$$

3.3.2 Inverse Kinematics

The problem of finding the joint angles by using the leg tip pose is inverse kinematics. Let the pose of leg tip is given by transformation matrix,

$$P = \begin{pmatrix} p_{11} & p_{12} & p_{13} & p_{14} \\ p_{21} & p_{22} & p_{23} & p_{34} \\ p_{31} & p_{32} & p_{34} & p_{35} \\ p_{41} & p_{42} & p_{43} & p_{44} \end{pmatrix} \#(11)$$

Then, the joint angles θ_1, θ_2 and θ_3 are given by

$$\theta_1 = tan^{-1}\left(\frac{p_{24}}{p_{14}}\right) \#(12)$$

$$\theta_3 = cos^{-1}\left(\frac{l_3 p_{11} + p_{14} + l_3 p_{21} + p_{24}}{l_2 + l_1}\right) \#(13)$$

$$\theta_2 = \theta_3 - cos^{-1}\left(\frac{l_3 p_{11} - l_2 p_{21} + p_{24}}{l_2}\right) \#(14)$$

3.6 Leg trajectory generation

To make the quadruped walking we need to generate a trajectory for each leg to follow. The tip of all the legs should move backwards at the same speed makes the body moving. The slippage at feet is assumed to be zero. Tripod gait is used here. In tripod gait, out of the four legs, at least any three of them will be touching the ground plane, with the perpendicular from the centre of mass falling within the triangle formed by the contact points of these legs. This gives stability to the robot while walking. Also, movement of each leg is limited with a small range such that it should not go too backward and too forward. Too much of forward or backward movements can cause collision between adjacent legs. However, a drawback of this approach is the large leg cycles needed for the robot motion.

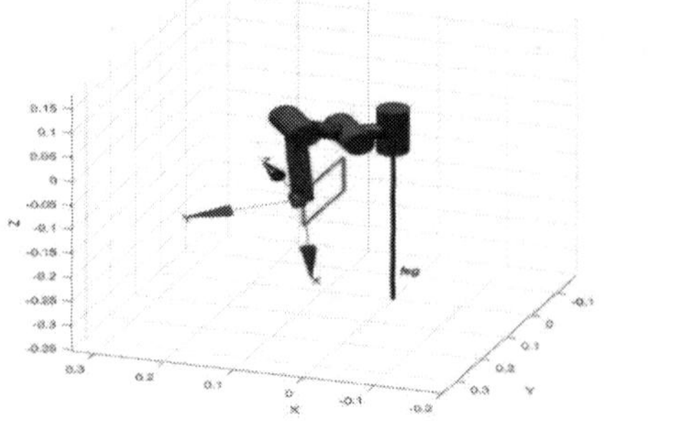

Fig. 3 . Snapshot from trajectory following of single leg

Considering all the above constraints, a rectangular trajectory is generated using the Robotic Toolbox function. For that, four via points are selected to generate the trajectory. Since there are four legs, a leg should be touching the ground for 3/4 of full cycle and resetting time is only 1/4 of full cycle. So, at the resetting phase, it should move faster. Rectangular trajectory followed by single leg is shown in Fig. 3. Since the legs are symmetric, the same trajectory with transformations can be used for other legs.

4 Quadrotor module

Kinematic design and analysis of quadrotor module of HyLMoR is presented in this section. The quadrotor module of the HyLMoR is a rigid body frame with cross structure and equally spaced arrangement of four rotors at each corner. Its six degrees of freedom can be controlled only by controlling the angular velocity of the four rotors. This makes it an underactuated system [17]. So, the six degree of freedom motion is achieved by coupling translational and rotational motions.

4.1 Selection of frame and configuration

Conventionally there are mainly two types of body frame structures, H or Dead Cat Frame and Cross frame. The cross-frame structure has been chosen for the HyLMoR due to the symmetrical design and mass concentration at the centre of the frame. Other advantages of cross frame in comparison with H frame structure are:

- It is a symmetrical design and propellers are placed equidistantly from the centre.
- Mass is concentrated at the centre.
- Control forces are distributed equally.

On the other hand, H structure is suitable for placing camera for wide view. However, HyLMoR is less affected by this, as the camera can be placed in the quadruped module. Cross structure can be controlled in two configurations, plus configuration and cross configuration. Usually while designing a quadrotor, configuration may not be considered since it is more related to control. However, here the quadrotor module can be placed over the quadruped module in these two configurations. So, selection of configuration is required at the time of designing itself. The following are some of the advantages of cross configuration over plus configuration which make it favourable to HyLMoR.

- It gives more compactness to whole robot structure.
- Robot will be more responsive in this configuration, since thrust is provided by two motors instead of one in the plus configuration.
- It produces more rotational acceleration, since effective distance from the centre is lower.

Analysing all these factors, cross configuration was chosen for HyLMoR. Now the placement of quadrotor over quadruped will be like quadrotor arms will be pointed to the direction of corners of quadruped.

4.2 Kinematics of quadrotor module

The basic working principle of quadrotor is variable torques and thrusts. It consists of four rotor propellers each of these rotors actually results in certain flow of air and in turn a thrust onto the body of the quadrotor. Thus, firing up of four motors at the required speed provides rotational and translational motion to the quadrotor body.

To keep a position, thrust must be compensated for earth gravity and torques of all four rotors sum to zero. By varying velocities of the rotors without rotational moment, will results in altitude control of quadrotor module. Yaw motion can be obtained by applying non-zero net torque to the system by adjusting the rotor velocities. To move laterally, the thrust generated by lateral pairs of rotors should be different.

Kinematic modelling of quadrotor is done by assuming quadrotor as a symmetrical rigid body with centre of mass aligned with centre of body frame of the robot. The thrust provided by each rotor is proportional to the square of the angular speed. Quadrotor kinematics can be described using four reference frames [18]. An inertial frame A, two intermediate frames, E and F and finally a body frame B. The inertial frame can be described with three orthogonal axis a_1, a_2 and a_3. The first intermediate frame E is formed with corresponding axis e_1, e_2, e_3 by rotating A with respect to a_3 which is pointing upwards, by an angle ψ. The second intermediate frame F with corresponding axes f_1, f_2 and f_3 is formed by rotating E with respect to e1 axis by an angle φ. Final body frame B with corresponding axes b_1, b_2 and b_3 is formed by rotating F with respect to f_2 axis by an angle θ. Fig. 4 shows frame assignment of quadrotor.

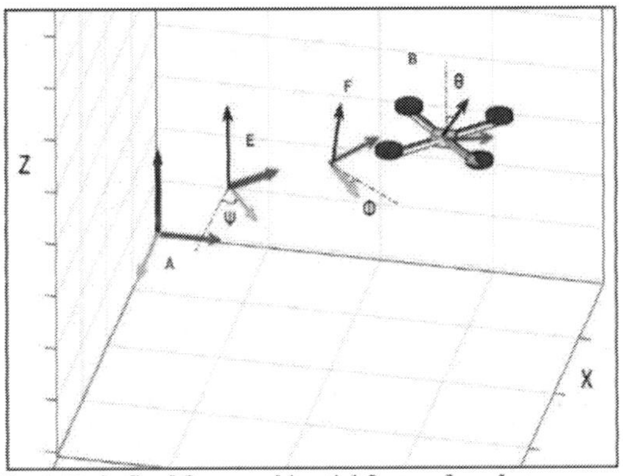

Fig. 4. Body fixed frame and inertial frame of quadrotor

The transformation between global frame A and body frame B is given as described below.

$$X^A = R_B X^B \#(15)$$

where ${}^AR_B$ is the rotation matrix of frame B with respect to to frame A and is given by

$${}^AR_B = R(\psi)R(\phi)R(\theta)\#(16)$$

where $R(\psi)$, $R(\phi)$ and $R(\theta)$ are rotation matrices, given by

$$R(\psi) = \begin{pmatrix} c\psi & s\psi & 0 \\ -s\psi & c\psi & 0 \\ 0 & 0 & 1 \end{pmatrix} \#(17)$$

$$R(\phi) = \begin{pmatrix} 1 & 0 & 0 \\ 0 & c\phi & s\phi \\ 0 & -s\phi & c\phi \end{pmatrix} \#(18)$$

$$R(\theta) = \begin{pmatrix} c\theta & 0 & -s\theta \\ 0 & 1 & 0 \\ s\theta & 0 & c\theta \end{pmatrix} \#(19)$$

Thus, from equations (17-19),

$${}^AR_B = \begin{pmatrix} c\phi c\theta - s\phi s\psi s\theta & -c\phi s\psi & c\psi s\theta + c\theta s\phi s\psi \\ c\theta s\psi + c\phi s\phi s\theta & c\phi c\psi & s\psi s\theta - c\theta s\phi c\psi \\ -c\phi s\theta & s\phi & c\phi c\theta \end{pmatrix} \#(20)$$

Results are analysed and verified at different known angles with known quadrotor poses. Rotation matrix of body frame B with respect to global co-ordinate frame A at angles $\psi = 0$, $\varphi = 0$, $\theta = 0$ (See Fig. 5. (a)),

$${}^AR_B = \begin{pmatrix} 1 & 0 & 0 \\ 0 & 1 & 0 \\ 0 & 0 & 1 \end{pmatrix} \#(21)$$

Rotation matrix of body frame B with respect to global co-ordinate frame As at angles $\psi = 45$, $\varphi = 0$, $\theta = 0$ (See Fig. 5. (b)),

$${}^AR_B = \begin{pmatrix} 0.7071 & -0.7071 & 0 \\ 0.7071 & 0.7071 & 0 \\ 0 & 0 & 1 \end{pmatrix} \#(22)$$

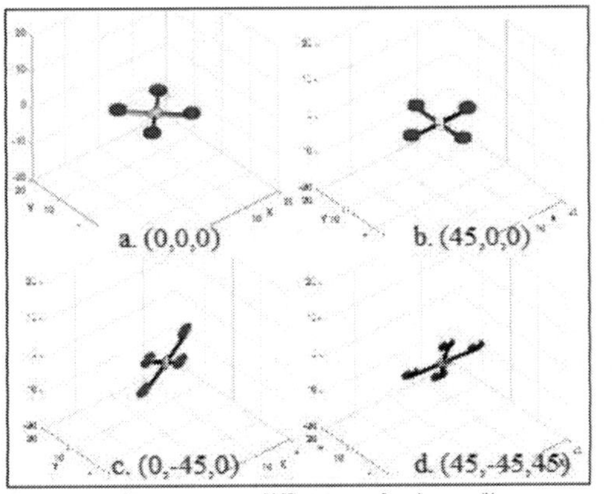

Fig. 5. Quadrotor pose at different angles (ψ, φ, θ)

Rotation matrix of body frame B with respect to global co-ordinate frame A at $\psi = 0$, $\varphi = 0$-45, $\theta = 0$ is given by (See Fig. 5. (c)),

$$^{A}R_{B} = \begin{pmatrix} 0.7071 & -0.5 & -0.5 \\ 0.7071 & 0.5 & 0.5 \\ 0 & -0.7071 & 0.7071 \end{pmatrix} \#(23)$$

Similarly, at angles ψ =45, φ = 0-45, θ = 45 (See Fig. 5. (d)),

$$^{A}R_{B} = \begin{pmatrix} 0.8536 & -0.5 & 0.1464 \\ 0.1464 & 0.7071 & 0.8536 \\ -0.5 & -0.7071 & 0.5 \end{pmatrix} \#(24)$$

5 HyLMoR – Integration of modules

As mentioned earlier, kinematic design of HyLMoR is done by combining quadruped module and quadrotor module. The two modules of the HyLMoR are analysed and verified separately. Now the robot can use its two locomotion modes to achieve a target position. One is walking mode and second one flight mode. Sometimes switching between modes is required to achieve a target position. Three snapshots from walking mode of simulation is shown in Fig. 6.

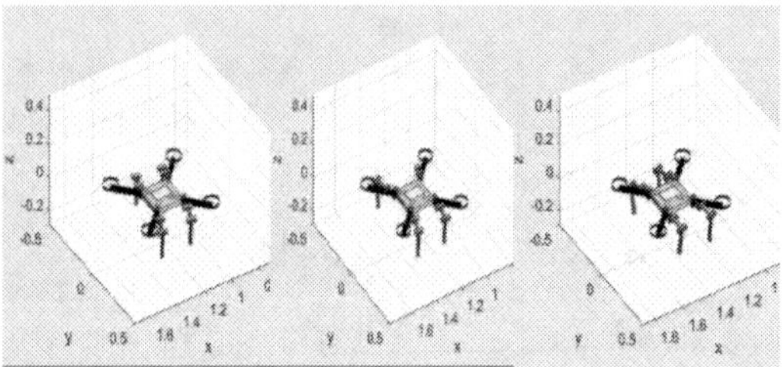

Fig. 6. Walking mode simulation at three continuous time frames

Simulation software mainly consists of quadruped and quadrotor kinematics and transformation updating and visualisation module. Simulation software collaboration diagram is shown in Fig. 7. The opensource robotic toolbox [20] of Corke is used for the development.

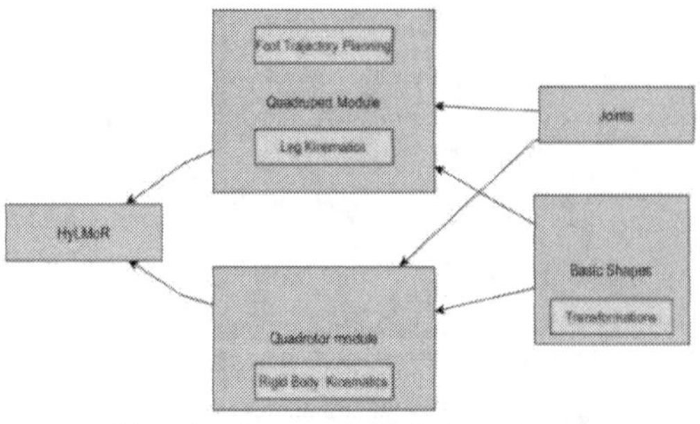

Fig. 7. Collaboration diagram of simulation software

6 Conclusion

In this paper, the project of experimental hybrid robot with aerial and terrain locomotion capability was briefly presented along with its application in search and rescue missions. Kinematic design, analysis and simulation was done to investigate the robot's ability to walk and fly. Analysis are done separately for each module. Later, two modules are combined in the simulation. Results are verified in MATLAB environment with multiple inputs. Leg motion was implemented using rectangular trajectory. Analysis and simulation results showed that the selected link lengths and leg spacings were good enough to avoid inter-leg collisions. Walking and flying simulations done for HyLMoR, resulted in proceeding the research further. The future studies include analysing the robot's dynamics, implementing gaits for moving on different difficult terrains and developing different behavioural controls for the robot.

References

1. D. Camara, Cavalry to the Rescue: *Antibes Juan-les-Pins,* (2014).
2. K. Nagatani *et al.*, *IEEE Int. Symp. Safety, Secur. Rescue Robot. SSRR*, (2013).
3. Nathan Michael *et al., Field and Service Robotics: Results of the 8th Int. Conf.* , **92**,33-47, (2013).
4. P. Fankhauser *et al.*, *IEEE/RSJ Int. Conf. Intell. Robot. Syst. IROS,* (2016).
5. S. Kuswadi, M. N. Tamara, D. A. Sahanas, G. I. Islami, and S. Nugroho, *Int. Conf. Knowl. Creat. Intell. Comput. KCIC*, 80–87, (2016).
6. R. Murphy *et al.*, *Handbook of Robotics*, 1162-1163,2008.
7. J. Carlson and R. R. Murphy, *IEEE Trans. Robot.*, **21**, 423–437, (2005).
8. M. J. Micire, *J. F. Robot.*, **25**, 17–30, (2008).
9. Y. Ichikawa, N. Ozaki, and K. Sadakane, *IEEE Trans. Syst. Man Cybern.*, (1983).
10. A. S. Boxerbaum, P. Werk, R. D. Quinn, and R. Vaidyanathan, *IEEE/ASME Int. Conf. on Advanced Intell. Mechatronics,* 1459–1464, (2005).
11. J. R. Page and P. E. I. Pounds, *IEEE Int. Conf. Intell. Robot. Syst.*, 4834–4841, (2014).
12. C. J. Pratt and K. K. Leang, *Proc. - IEEE Int. Conf. Robot. Autom.*, 3267–3274, (2016).
13. R. J. Lock, R. Vaidyanathan, and S. C. Burgess, *Proc. IEEE RAS EMBS Int. Conf. Biomed. Robot. Biomechatronics*, 681–687, (2012).
14. A. Kossett, J. Purvey, and N. Papanikolopoulos, *IEEE/RSJ Int. Conf. Intell. Robot. Syst. IROS*, 5653–5658, (2009).
15. M. C. Pitonyak, *Syst. of Systems Engineering Conf. (SoSE),* (2017).

16. A. Kalantari and M. Spenko, *Proc. - IEEE Int. Conf. Robot. Autom.*, 4445–4450, (2013).
17. K. P. Valavanis and G. J. Vachtsevanos, *Handb. Unmanned Aer. Veh.*, 1–3022, (2015).
18. Kar, D. C., *J. Robotic Syst.*, 671–686, (2003)
19. S. Hirose, Some considerations on a feasible walking mechanism as a terrain vehicle, *3rd CISM-IFToMM Int. Symp. on Theory and Practice of Robots and Manipulators* , 357–375, (1978).
20. P. I. Corke, Robotics, Vision & Control : Fundamental Algorithms in Matlab, *Springer*, (2011).
21. M. F. Silva, Quadruped robot optimization using a generic algorithm, *11th Int. Conf. on Climbing and Walking Robots and the Support Technologies for Mobile Machines*, CLAWAR'11,11–26, (2011).
22. R. J. Bachmann, F. J. Boria, P. G. Ifju, R. D. Quinn, J. E. Kline, and R. Vaidyanathan, *IEEE Int. Conf. Intell. Robot. Syst.*, 24–28, (2005).

Parametric study on free vibration and instability of a functionally graded cracked shaft in a rotor-disc-bearing system: finite element approach

Debabrata Gayen[1,*], *Debabrata* Chakraborty[1], and *Rajiv* Tiwari[1]

[1]Department of Mechanical Engineering, Indian Institute of Technology Guwahati, Guwahati 781039, Assam, India

Abstract. Free vibration and stability analysis are studied for a rotor-disk-bearing system having a radially functionally graded (FG) shaft with a transversely fully open crack, based on finite element (FE) approach. Both viscous and hysteretic internal damping are incorporated in the FE model of FG cracked shaft using two nodded Timoshenko beam element having four degrees of freedom (DOFs) at each node. Material properties of the FG cracked shaft are assumed temperature dependent and graded along radial direction following different material gradation law. FG shaft is made of two constituents material namely zirconia (ZrO_2) and stainless steel (SS) where metallic (SS) contain is decreasing towards the outer diameter of the shaft. Extended Hamilton's principle is employed to derive the system equations of motion (EOMs) of the FG cracked shaft system. A complete code is developed in MATLAB for correcting the formulation of modeling of crack and verified with existing published results. Influences of different material gradient index, temperature gradients, size and location of crack, viscous and hysteretic internal damping, slenderness ratio, and boundary condition on dynamic responses of the FG cracked shaft system are studied.

Keywords. FG cracked shaft; Stability threshold speed; Material gradient index; Temperature dependent material property

1 Introduction

Functionally graded materials (FGMs) have been constituted as high performance, thermal resistance, multi-functional, microscopically inhomogeneous materials whose thermo-mechanical properties graded smoothly and continuously in a pre-determined directions using different law of material gradations to achieve superior material properties under thermal environment. Historically, in 1984, concept and application of FGMs were first visualized in Japan during space project, reported by Koizumi [1]. Later, applications of FGMs have been expanded in aero-space, automotive, electronics and biomedical applications etc. In last four decades, many studies related to FGMs structures were reported and some of the valuable works are summarized here. Miyamoto et al. [2] reported fundamentals, manufacturing processes, design and current application of FGMs. Reddy and Chin [3] presented thermo-mechanical responses for FG structure (plates, cylinders, etc.) using FE formulation. Mahi et al. [4] reported temperature dependent free vibration response of a FG beam based on higher order shear deformation theory. Shahba et al. [5] presented free vibration and stability analysis for an axially tapered FG beams (Timoshenko) through a FE approach. Gayen and Roy [6] studied vibration and stability analysis of a FG spinning shaft system based on Timoshenko beam theory (TBT).

Shafts are basic components in high-performance rotating machinery equipment, used in the requirement of higher power, high operational speed. Therefore, prediction of dynamic responses of a shaft system is importance and many studies reported in this direction for the homogeneous shaft and very few for FGMs. Some of them are summarized here. Papadopoulos [7] reported review works of cracked rotor for dynamic analysis. Nelson and Mcvaugh [8] studied dynamic responses of a shaft with gyroscopic effect, rotary inertia effect, and shear deformation effect. Zorzi and Nelson [9] extended the work [8] and included both viscous and hysteretic internal damping effect in the same FE model. Based on FE approach, Ku [10] studied whirl speeds and damped stability for a rotor-bearing system, considering both internal viscous and hysteretic damping.

Even though small crack leads a dangerous and catastrophic failure of structural components in the studies of rotor dynamical systems. Therefore, some of the important works are summarized here. Papadopoulos and Dimarogonas [11] studied coupled vibration and stability regions for cracked rotor. Huang et al. [12] studied dynamic response and stability analysis of a rotating cracked shaft. Chen and Chen [13] studied stability analysis of a rotating cracked shaft subjected to

* Corresponding author: d.gayen@iitg.ernet.in

the end load based on TBT. Sekhar and Dey [14] studied stability threshold speed of a rotor system having single and two transverse cracks and reported a parametric studies on the instability speed. Also, few studies relating to the dynamic analysis of FG shaft systems are reported here. Gayen et al. [15] reported dynamic responses of a rotor-bearing system with a fully open transverse cracked FG shaft using FEM. Gayen et al. [16] reported a FE based dynamic responses of a rotor-bearing system with a breathing cracked FG shaft.

Even though a large number of works reported in the direction of analysis of FGM structure, free vibration and stability analysis of homogeneous cracked shaft for a long time. However, hardly any works available in the area of temperature dependent free vibration and stability analysis of cracked FG shaft system. Therefore, aim of the present `study is to develop a FE model for analysis of cracked FG shaft system with temperature dependent material properties and to study the effect of important parameters such as gradient index, temperature gradient, size and location of crack, internal viscous and hysteretic damping, slenderness ratio and boundary condition on the dynamic responses.

2 Thermo-mechanical properties of an FG Shaft

Since FGMs structures mostly used in high temperature environment, constituents of FGMs may reflect the temperature dependent properties. Therefore, Touloukian [17] reported the temperature dependent material properties P along radial direction of FG shaft as,

$$P(T)=P_0(P_{-1}T^{-1}+1+P_1T+P_2T^2+P_3T^3) \tag{1}$$

where P_0, P_{-1}, P_1, P_2 and P_3 are the coefficient of temperature T (in K) and are unique for every constituent materials.

Material properties P (Young's modulus E (in Pa), Poisson's ratio ν, thermal conductivity K (in W/m K), and mass density ρ (in Kg/m^3)) in the FG shaft are graded along the radial direction (y) as a function of temperature (T) (Reddy and Chin [3])

$$\begin{aligned}&P(y,T)=P_m(T)+\{P_c(T)-P_m(T)\}\{VF\}^k,\\&VF=(y-R_m)/(R_c-R_m), R_c\le y\le R_m,\ 0\le k\le\infty\end{aligned} \tag{2}$$

where R is the radius of the shaft, subscripts m and c refer to metal and ceramic respectively and power law index $k=0$ and $k=\infty$ denotes the two extremes of completely homogenous ceramic and metal respectively. In **Eq. (2)**, $k=1$ represent the linear law of material gradation and for other values of k represent the power law of material gradation (P-FGM).

Material properties distribution according to exponential law of material gradation (E-FGM)

$$\begin{aligned}&P(y,T)=P_m(T)\ e^{\lambda(y-R_m)},\\&\lambda=\ln\{P_c(T)/P_m(T)\}/(R_c-R_m),\ R_m\le y\le R_c\end{aligned} \tag{3}$$

Material properties distribution according to sigmoid law of material gradation (S-FGM)

$$\left.\begin{aligned}&P(y,T)=P_m(T)+\{P_c(T)-P_m(T)\}\{VF\}^n,\\&R_m\le y\le(R_m+R_c)/2\\&P(y,T)=P_c(T)+\{P_m(T)-P_c(T)\}(1-\{VF\})^n,\\&(R_m+R_c)/2\le y\le R_c\end{aligned}\right\} \tag{4}$$

where n is sigmoid law of material gradient index.

Temperature distributions along radial direction in FG shaft are computed using steady-state one dimensional Fourier equation of heat conduction law and applying the thermal boundary conditions (i.e. at $y=R_m$, $T=T_m$ and at $y=R_c$, $T=T_c$) as

$$\frac{d}{dy}\left[K(y)\frac{dT}{dy}\right]=0 \tag{5}$$

Temperature distribution according to power law of material gradation (P-FGM) as

$$T(y)=T_m+(T_c-T_m)\frac{\left[\sum_{j=0}^{5}\left\{\frac{(-1)^j}{jk+1}\left(\frac{K_{cm}}{K_m}\right)^j(VF)^{jk+1}\right\}\right]}{\left[\sum_{j=0}^{5}\frac{(-1)^j}{jk+1}\left(\frac{K_{cm}}{K_m}\right)^j\right]} \tag{6}$$

Temperature distribution according to exponential law of material gradation (E-FGM) as

$$T(y)=\mathrm{A}+\mathrm{B}e^{-\lambda(y-R_m)},\ R_m\le y\le R_c \tag{7}$$

where $K_{cm}=K_c-K_m$, $\mathrm{A}=T_m-(T_c-T_m)/e^{-\lambda(R_c-R_m)}-1$, $\mathrm{B}=(T_c-T_m)/e^{-\lambda(R_c-R_m)}-1$, $\lambda=\ln(K_c/K_m)/(R_c-R_m)$.

These graded thermo-mechanical material properties are used in FE modeling of FG shaft.

3 FE formulation of an FG cracked shaft system

The rotor-disk-bearing system is considered to comprise a set of interconnecting components consisting of transverse cracked shaft elements with distributed mass and elasticity, rigid disks and linear bearings shown in **Fig. 1**.

3.1 Finite shaft element with a fully open transverse crack

The FG cracked shaft with identical un-damped isotropic bearings are discretized and shown in **Fig. 1(a)**, where total length is L, element length is L_e and diameter is D and α is the depth of crack located at distance L_c from left end support of shaft. The crack orientation

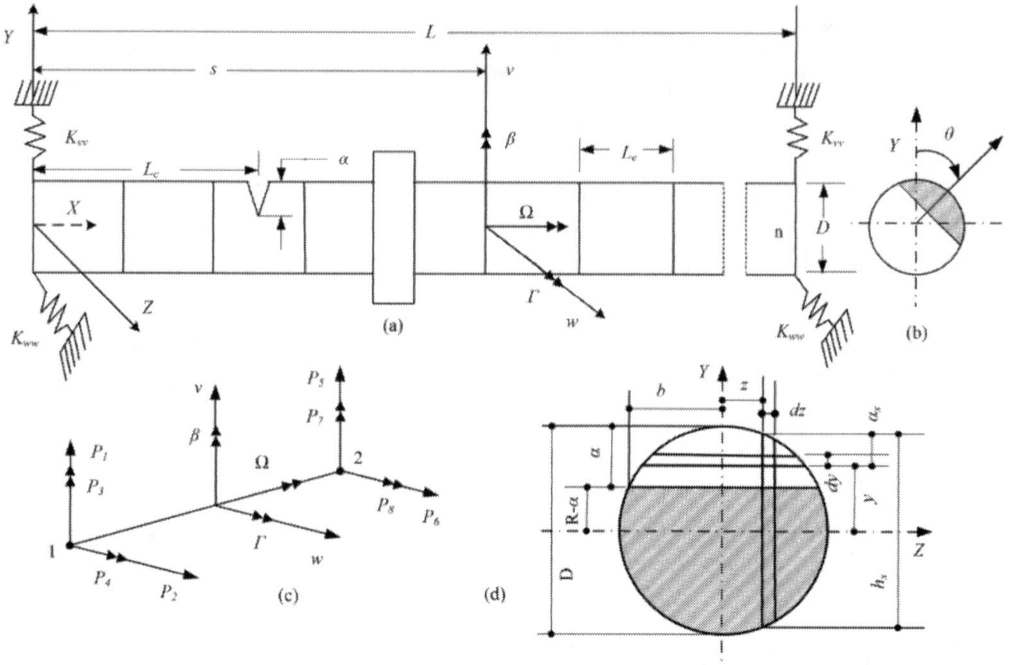

Fig. 1 Rotor-disk-bearing system with a cracked FG shaft: **(a)** Coordinate system and FE discretization **(b)** Crack orientation **(c)** General loading **(d)** Geometry of cracked section

denoted by θ, shown in **Fig. 1(b)**. **Fig. 1(c)** shows the shaft element subjected to shear forces P_1, P_2, P_5 and P_6 and bending moments P_3, P_4, P_7 and P_8, rotating at speed Ω, v and w are the translation displacements along Y and Z direction, and β and Γ are the rotational displacements about Y and Z at a distance s from the left end of a cross-section of the shaft. **Fig. 1(d)** shows the cross-section of the cracked shaft having fully open transverse crack (crack orientation, $\theta=180°$) with crack half-width b.

The LFCs of cracked FG shaft are computed using Paris's equations [18] and Castigliano's theorem in conjunction with the SIFs expression and the additional displacement u_i^c due to crack is expressed as

$$u_i^c = \frac{\partial}{\partial P_i}\int_0^{\alpha_x} \frac{\left[\left(\sum_{i=1}^{4} K_{Ii}\right)^2 + \left(\sum_{i=1}^{4} K_{IIi}\right)^2 + m\left(\sum_{i=1}^{4} K_{IIIi}\right)^2\right]}{E(y,T)}\, dy, \quad (8)$$

$$m = \{1 + v(y,T)\}$$

where SIFs K_{I}, K_{II} and K_{III} corresponding to modes I, II, and III expressions in **Eq. (8)** are taken from Tada et al. [18] and $i = 1,2,3,4$ are load indices.

Finally, LFCs for an FG shaft with $\theta=180°$ are computed as follow (refer **Fig. 1(d)**)

$$C_{ij}^c = \frac{\partial}{\partial P_i}\int_{-b}^{b} \frac{\partial u_i^c}{\partial P_j}\, dz \quad (9)$$

where $\alpha_x = (\alpha - R) + \sqrt{R^2 - z^2}$ and $b = \sqrt{R^2 - (R-\alpha)^2}$.

Here, integrations are carried out to obtain LFCs of the FG cracked shaft considering $E(y,T)$ and $v(y,T)$ as a function of y as well as T, unlike homogeneous shaft where material properties are considered as constant and could be taken out of integration in **Eq. (9)**. Using **Eq. (9)**, LFCs are calculated corresponding to $\theta=180°$ and using energy methods corresponding local flexibility matrix $\mathbf{C}^c(y,T)$ is obtained as

$$\mathbf{C}^c(y,T) = \begin{bmatrix} C_{11}^c & 0 & 0 & 0 \\ & C_{22}^c & 0 & 0 \\ & & C_{33}^c & C_{34}^c \\ sym & & & C_{44}^c \end{bmatrix} \quad (10)$$

where expressions of the elements of $\mathbf{C}^c(y,T)$ are taken from published works [15-16].

Using **Eq. (10)** and with the help of virtual work principle cracked stiffness matrix $\mathbf{K}^c$ is computed as

$$\mathbf{K}^c = \mathbf{\Pi}^{\mathrm{T}}\left[\mathbf{C}^c\right]^{-1}\mathbf{\Pi} \quad (11)$$

where the transformation matrix $\mathbf{\Pi}$ has been shown in published work of Gayen et al. [15-16].

The elementary EOMs for an FG shaft with stationary crack is written as

$$\left(\mathbf{M}_T^s + \mathbf{M}_R^s\right)\ddot{\mathbf{p}}^s(t) - \mathbf{\Omega}\,\mathbf{G}^s\,\dot{\mathbf{p}}^s(t) + \mathbf{K}^s\,\mathbf{p}^s(t) = \mathbf{f}^s(t) \quad (12)$$

* Corresponding author: d.gayen@iitg.ernet.in

where $\mathbf{M}_T^s$ and $\mathbf{M}_R^s$ are the elementary translational and rotational masses, $\mathbf{G}^s$ is the elementary gyroscopic matrix, $\mathbf{K}^s$ is the elementary un-cracked stiffness matrix, $\mathbf{p}^s(t)$ and $\mathbf{f}^s(t)$ are the elementary nodal displacement vector and external force vector of the shaft element. The details of elementary matrices of **Eq. (12)** were shown in published works [8, 15-16].

3.2 Rigid Disk

EOMs for rigid disk is expressed using Lagrangian equation as

$$\left(\mathrm{M}_T^d+\mathrm{M}_R^d\right)\ddot{\mathrm{q}}^d-\Omega\mathrm{G}^d\dot{\mathrm{q}}^d=\mathrm{f}^d \tag{13}$$

where individual matrices are taken same as in (Nelson and McVaugh [8]).

3.3 Bearings Consideration

Bearings are modeled based on classical linearized springs model as reported in Nelson and McVaugh [8] and is model the forces in each bearing are assumed to obey the following form of governing equations

$$\mathrm{C}^b\dot{q}^b+\mathrm{K}^b q^b=\mathrm{f}^b \tag{14}$$

where f^b is the bearing external force vector and bearing damping and stiffness matrices are

$$\left[\mathrm{C}^b\right]=\begin{bmatrix}C_{vv}^b & C_{vw}^b\\ C_{vw}^b & C_{ww}^b\end{bmatrix},\ \left[\mathrm{K}^b\right]=\begin{bmatrix}K_{vv}^b & K_{vw}^b\\ K_{vw}^b & K_{ww}^b\end{bmatrix} \tag{15}$$

3.4 System Equations of Motion

Based on work of Zorzi and Nelson [9], the resultant system EOMs are modified through the inclusions of internal viscous damping coefficient η_v and internal hysteretic damping loss factor η_h of the shaft material. Finally using extended Hamilton's principle, EOMs of the cracked shaft system including the rigid disk, bearing and internal damping effects could be expressed as

$$\mathrm{M}^e\ddot{\mathrm{q}}^e+\left(\eta_v\mathrm{K}^e-\Omega\mathrm{G}^e+\mathrm{C}^e\right)\dot{\mathrm{q}}^e$$
$$+\left[\frac{1+\eta_h}{\sqrt{1+\eta_h^2}}\mathrm{K}^e+\left(\Omega\eta_v+\frac{\eta_h}{\sqrt{1+\eta_h^2}}\right)\mathrm{K}_{cir}^e\right]\mathrm{q}^e=\mathrm{f}_{total}^e \tag{16}$$

The solution of the eigenvalue problem of **Eq. (16)** is determined in complex form as $\lambda_n=\xi_n\pm i\omega_n$, logarithmic decrement is defined as $\delta_n=-2\pi\xi_n/\omega_n$ and stability threshold speed is obtained at which operation speed of rotor $\delta_n=0$, where ξ, ω and n are damping constant, natural whirling frequency and mode number respectively.

4 Results and discussions

In the present study, an FG shaft having $D=0.1$ m is considered and temperature dependent material properties in the constituents of FG (SS/ZrO_2) shaft are taken from Reddy and Chin [3] (also shown in Gayen et. al [15-16]). The FG shaft is modeled with 25 equal finite elements and supported by the simply supported boundary condition and with the identical un-damped isotropic bearing of stiffness coefficients $K_{vv}^b=K_{ww}^b=17.513$ MN/m and $K_{vw}^b=K_{wv}^b=0$ at both ends. In the FE model, a disc is considered at center location with weight 1.406 kg, polar and diametral moment of inertias 0.002 kg/m^2 and 0.0135 kg/m^2 respectively.

4.1. Validation

Due to lack of existing results on temperature dependent free vibration and stability analysis for a spinning FG shaft system. Therefore, a homogeneous shaft is selected for validating the present FE formulation and developed code in three steps.

4.1.1 Influence of hysteretic internal damping on natural frequencies

A steel shaft having $L=1.27$ m, $D=10.16$ cm, $\rho=7833$ kg/m^3 and $E=206.80$ GPa are considered to validate the present FE formulation. Influence of hysteretic damping on natural frequencies in dimensionless form ϖ_n ($\varpi_n^4=\rho AL^4\omega^2/EI$) is computed of a shaft with slenderness ratio SR ($=R/2L$), and results are compared with the published work of Ku [10]. In **Table 1**, the results shows the change in ϖ_{nr} is relatively smaller for higher *SR* and also present results are well matched with published results.

Table 1 Influence of η_h on ϖ_n for an uniform non-spinning simply supported shaft

Modes	SR ($R/2L$)	$\eta_h=0.001$		$\eta_h=0.002$	
		Present	Ku [10]	Present	Ku [10]
1st	0.02	3.1308	3.1301	3.1316	3.1309
	0.06	3.0459	3.0407	3.0467	3.0414
	0.10	2.9112	2.8999	2.9119	2.9006
2nd	0.02	6.2052	6.1923	6.2068	6.1943
	0.06	5.7112	5.6476	5.7126	5.6490
	0.10	5.1491	5.0493	5.1504	5.0560

4.1.2 Local flexibility coefficients (LFCs) of a fully open transverse cracked shaft

In order to validate the LFCs, a shaft made of SS is considered with $D=0.1$ m, $L=1.25$ m. Dimensionless compliance $\overline{C}_{44}$ and $\overline{C}_{55}$ as a function of size of crack

α/R has been determined and compared with published results of Chasalevris and Papadopoulos [19]. Results in **Table 2** show an excellent agreement to verify the formulation of crack compliances.

Table 2 Dimensionless compliance $\overline{C}_{44}$ and $\overline{C}_{55}$ as a function of α/R

$\frac{\alpha}{R}$	$\overline{C}_{44}$ Present	$\overline{C}_{44}$ Ref. [19]	$\overline{C}_{55}$ Present	$\overline{C}_{55}$ Ref. [19]
0.04	0.0000349	0.000035	0.0029731	0.002973
0.36	0.0747941	0.074794	0.5677802	0.567780
0.68	0.8171641	0.817164	2.5852663	2.585270
1.00	5.2663511	5.266350	7.7903871	7.790390

4.1.3 Natural frequencies of a simply supported shaft with a single crack

The ratio between cracked and un-cracked natural frequencies are computed as a function of α/R and L/D, for a crack located at mid span of shaft and results are compared with published results of Sekhar and Prabhu [20] as shown in **Table 3**. It could be observed from **Table 3** that the present results is well matched for $L/D = 12.5$ and $L/D = 8.0$.

Table 3 Variation of dimensionless natural frequencies with α/R for different L/D

Modes	L/D	$\alpha/R=0.2$ Present	$\alpha/R=0.2$ Sekhar and Prabhu[20]	$\alpha/R=1.0$ Present	$\alpha/R=1.0$ Sekhar and Prabhu [20]
1st	12.5	0.9953	0.9958	0.7798	0.7882
2nd	12.5	0.9999	0.9998	0.9839	0.9818
1st	8.0	0.9931	0.9955	0.7050	0.7187
2nd	8.0	0.9999	0.9997	0.9742	0.9771

4.2. Influence of k, ΔT, α/R and elastic supports on natural frequency for an FG transverse cracked shaft

In order to understand the influences of material gradient index k, temperature gradient ΔT, size of crack α/R, boundary conditions (BCs) (clamped free C-F, simply-supported S-S, and clamped-clamped C-C), natural frequencies associated with vertical plane ω_{1V} and horizontal plane ω_{1H} are computed for a non-rotating centrally located FG cracked shaft with $\theta = 180°$ having two discs located at 0.45 m and 0.9 m from left end. The discs are considered with the specification stated earlier and results are obtained for $R/2L = 0.02$ and location of crack $L_c/L = 0.5$ with temperature dependent material properties and listed in **Table 4-5**. The percentage reductions in natural frequencies are calculated as $\varpi_i = (1-\omega_i^{\Delta T=0K,300K,600K}/\omega_i^{\Delta T=0K})\times 100)$. It is observed from **Table 4** that with increase in gradient index k causes the % reduction in ϖ_{1V} and ϖ_{1H} due to transition of shaft from zirconia (lower E and lower density) to stainless steel (higher E and higher density) for a specific ΔT, support condition and material distribution. Also with the increase in ΔT, ϖ_{1V} and ϖ_{1H} increase as material becomes softer at higher temperatures and also remarkable that the order of ϖ_{1V} and ϖ_{1H} are C-C > C-F > S-S, respectively.

Table 4 Percentage reduction in ω with ΔT and k of a non-rotating FG cracked shaft with BCs for $\alpha/R = 0.8$

BCs	ΔT(K)	ϖ_{1V} $k=0.5$	$k=1.0$	$k=5.0$	$k=10.0$
C-F	300	6.65	5.11	2.85	2.72
C-F	600	11.76	9.94	7.46	7.82
S-S	300	6.52	4.93	2.46	2.35
S-S	600	11.52	9.61	6.85	7.23
C-C	300	6.72	5.22	3.06	2.92
C-C	600	11.90	10.12	7.78	8.15
BCs	**ΔT(K)**	**ϖ_{1H} $k=0.5$**	**$k=1.0$**	**$k=5.0$**	**$k=10.0$**
C-F	300	6.75	5.27	3.20	3.05
C-F	600	11.96	10.21	7.95	8.35
S-S	300	6.64	5.13	3.01	2.88
S-S	600	11.80	10.01	7.70	8.07
C-C	300	6.71	5.22	3.12	2.98
C-C	600	11.91	10.14	7.89	8.26

Table 5 Percentage reduction in ω with k, ΔT and α/R of a non-rotating S-S FG cracked shaft

α/R	ΔT(K)	ϖ_{1V} $k=0.5$	$k=1.0$	$k=5.0$	$k=10.0$
0.0	300	6.78	5.32	3.26	3.11
0.0	600	12.02	10.28	8.09	8.45
0.4	300	2.49	1.77	1.15	1.11
0.4	600	4.81	3.54	2.67	2.76
0.8	300	6.52	4.93	2.46	2.35
0.8	600	11.52	9.61	6.85	7.23
α/R	**ΔT(K)**	**ϖ_{1H} $k=0.5$**	**$k=1.0$**	**$k=5.0$**	**$k=10.0$**
0.0	300	6.78	5.32	3.26	3.11
0.0	600	12.02	10.28	8.09	8.45
0.4	300	2.74	2.05	1.37	1.29
0.4	600	5.25	4.03	3.06	3.07
0.8	300	6.64	5.13	3.01	2.88
0.8	600	11.80	10.01	7.70	8.07

Table 5 shows that with the increase in α/R, ϖ_{1V} and ϖ_{1H} decrease for a specific k and ΔT. It also shows that reductions are relatively higher for higher values of ΔT due to the degradation of material properties with the rise in temperature gradient across radial direction in FG cracked shaft. Thus in a high temperature application designing FG shafts, power-law gradation k may be chosen in such a way so that the reductions in ϖ_i with temperature is minimized while keeping enough material at the outer layer for providing temperature resistance within desired limit.

4.3. Influence of L/D, k, ΔT and disc on stability threshold speed of a FG shaft in a rotor-disc-bearing system

Influences of L/D, k, ΔT and disc on stability threshold speeds Ω_{th} of an FG shaft system are examined with same dimension and temperature-dependent material properties stated earlier. FG shaft is supported on isotropic un-damped bearing with stiffness coefficients $K^b_{vv} = K^b_{ww} = 17.513$ MN/m, $K^b_{vw} = K^b_{wv} = 0$. The % reduction in Ω_{th} is defined as $\overline{\Omega}_{th} = (1 - \Omega_{th}^{\Delta T=0K,300K,600K} / \Omega_{th}^{\Delta T=0K}) \times 100$. Results are shown in **Figs. 2-3**. It is observed from **Fig. 2** that $\overline{\Omega}_{th}$ increase with increase in L/D and ΔT and also reduction in $\overline{\Omega}_{th}$ are appreceiable in case of shafts with low L/D and at high ΔT. **Fig. 3** show that $\overline{\Omega}_{th}$ increases with decrease in k due to mettalic content increase in the FG shaft and also with increase in ΔT, these are minimum for the FG shaft with an optimal value of $k = 5.0$ and again it increase. Therefore, designing in FG shafts gradient index can play great role to expose larger temperature gradients and higher spinning speeds and also FG shaft has an advantage over shaft made of SS.

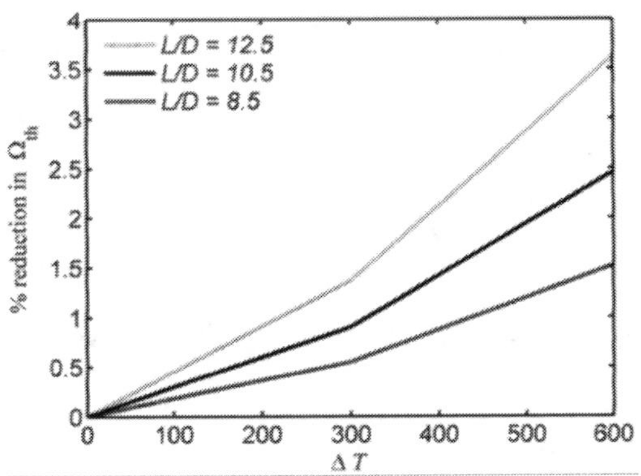

Fig. 2 Percentage reduction in Ω_{th} of a rotor-bearing system with L/D and ΔT for $k = 5.0$, $\eta_v = 0.0002$ s

The disc effect on Ω_{th} for the same FG rotor system with k and ΔT for $L/D = 12.5$ and , $\eta_v = 0.0002$ s and the location of single disc is 0.5 m from the left end and double discs are at 0.45 m and 0.9 m from the left end. The listed results in **Table 6** indicate that with increase in number of disc mounting on FG shaft Ω_{th} reduced, as expected but also the Ω_{th} decrease with increase in k as well as with increase in ΔT as the reason stated earlier.

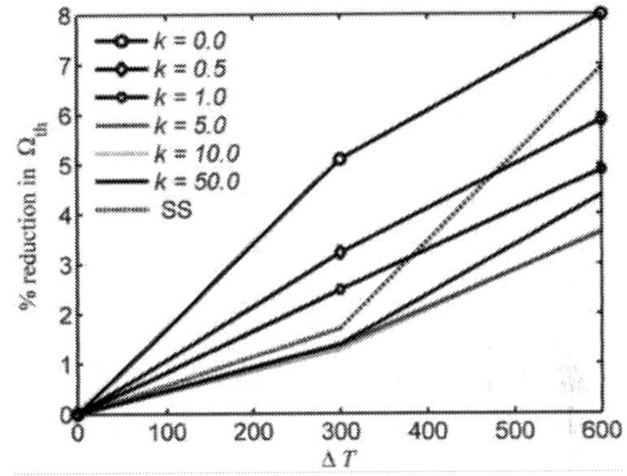

Fig. 3 Percentage reduction in Ω_{th} of a rotor-bearing system with k and ΔT for $L/D = 12.5$, $\eta_v = 0.0002$ s

Table 6 Stability threshold speed Ω_{th} (rpm) of a rotor-bearing system with k and ΔT for $L/D = 12.5$

ΔT (K)	No. of disc	$k = 0.0$	$k = 0.5$	$k = 1.0$	$k = 5.0$	$k = 10.0$
	Nil	5685	5440	5320	5100	5055
0	One	5580	5350	5240	5035	4990
	Two	5515	5295	5185	4985	4945
	Nil	5395	5265	5190	5030	4990
300	One	5295	5180	5110	4960	4925
	Two	5230	5125	5060	4915	4880
	Nil	5230	5120	5060	4915	4870
600	One	5130	5035	4980	4850	4805
	Two	5070	4980	4930	4805	4760

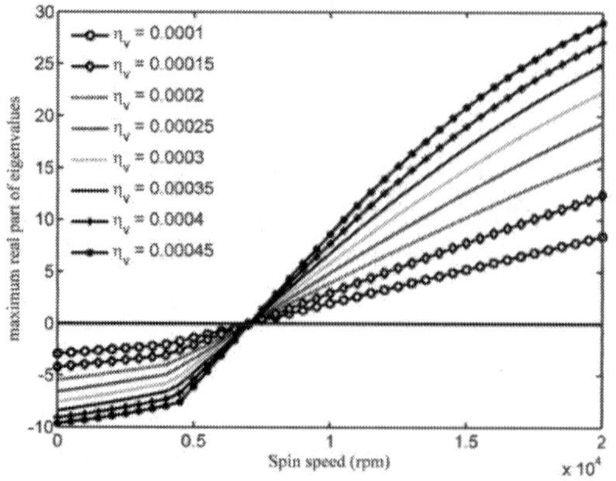

Fig. 4 Variation of maximum real part of a two disc rotor-bearing having an FG shaft with η_v

Fig. 4 shows the variation of the maximum real part for the FG shaft system with different internal viscous damping η_v for $k=5.0, \Delta T=0$ K and $L/D=12.5$ and it could be observed that with increase in η_v the values of maximum real part decrease but stability threshold speeds are not changing because the shaft made FGM with the $k=5.0$ is fixed.

4.4. Influence of L/D, α/R and L_c/L on stability threshold speed of a FG cracked shaft in a rotor-disc-bearing system

Influences of L/D, α/R and L_c/L on Ω_{th} are examined for a two disc rotor-bearing system having an FG shaft with $\theta=180°$. The FG cracked shaft is supported on two rigid end bearings with $L_c/L=0.14$ and 0.5, $k=5.0$, $\Delta T=0$ K, $\eta_h=0.0002$ and $\eta_v=0.0002$ s. In the present study, the normalized threshold speeds $\bar{\Omega}_{th}$ with α/R and L/D are computed and results are tabulated in **Table 7**. Here, normalization is done with respect of an un-cracked shaft. It is observed that variation in $\bar{\Omega}_{th}$ due to crack are appreciable with low L/D shaft and also effect of reduction is more for mid span crack compared to other location of crack, because of crack is closer to antinodal points.

Table 7 Variation of $\bar{\Omega}_{th}$ for an FG shaft with L/D and α/R for $L_c/L=0.14$ and 0.5, $k=5.0$, $\Delta T=0$ K

L/D	L_c/L	$\alpha/R=0.2$	$\alpha/R=0.4$	$\alpha/R=0.6$	$\alpha/R=0.8$
12.5		0.9987	0.9933	0.9786	0.9498
10.5	0.14	0.999	0.9924	0.9781	0.9615
8.5		0.9994	0.9921	0.983	0.982
12.5		0.9933	0.9672	0.907	0.7885
10.5	0.50	0.9924	0.9615	0.8982	0.8041
8.5		0.9915	0.9575	0.9108	0.903

5 Conclusions

The present work arrives at the following conclusions:

- Temperature dependent material properties have significant impacts on calculating dynamic responses of a radially FG shaft with different gradation law such as P-FGM, E-FGM and S-FGM.
- Material gradient index and temperature gradient yields a significant role on natural frequencies and stability threshold speeds of the FG cracked shaft system. Therefore, with the conjunction of temperature gradient and gradient index plays an important role in designing such FG shafts in using large temperature gradients and operating at high spinning speeds in a rotating system.
- It could be observed from various responses of shaft made of SS and FGM that FG shaft can operate at higher spin speed than that of SS shaft in using as rotating machinery element.
- Finally, present work can be recommended in modeling and analysis of FG shaft having multiple cracks with temperature dependent materials properties followed by different gradation laws.

References

1. M. Koizumi, Ceram. Trans. **34**, 3-10 (1993)
2. Y. Miyamoto, W.A. Kaysser, B.H. Rabin, A. Kawasaki, R.G. Ford, Functionally Graded Materials: Design, Processing and Application (Kluwer Academic Publishers, London, 1999)
3. J.N. Reddy, C.D. Chin, J. Therm. Stresses **21**, 593-626 (1998)
4. A. Mahi, E.A. Adda Bedia, A. Tounsi, I. Mechab, Compos. Struct. **92(8)**, 1877-1887 (2010)
5. A. Shahba, R. Attarnejad, M. T. Marvi, S. Hajilar, Compos. Part B Eng. **42(4)**, 801-808 (2011)
6. D. Gayen, T. Roy, Proc. Inst. Mech. Eng. Part C: J. Mech. Eng. Sci. **228(18)**, 3306-3321 (2014)
7. C.A. Papadopoulos, Mech. Syst. Signal Pr. **22(4)**, 763-789 (2008)
8. H.D. Nelson, J.M. Mcvaugh, ASME J. Eng. Ind. **98**, 593-600 (1976)
9. E.S. Zorzi, H.D. Nelson, ASME J. Eng. Power **99(1)**, 71-76 (1977)
10. D.M. Ku, Mech. Syst. Signal Pr. **12(5)**, (1998) 599-610
11. C.A. Papadopoulos, A.D. Dimarogonas, ASME J. Vib. Acoust. Stress **110(3)** 356-359 (1988)
12. S.C. Huang, Y.M. Huang, S.M. Shieh, J. Sound Vib. **162(3)**, 387-401 (1993)
13. L. Chen, H. Chen, J. Sound Vib. **188(4)**, 497-513 (1995)
14. A.S Sekhar, J.K Dey, Mech. Mach. Theory **35(12)**, 1657-1674 (2000)

15. D. Gayen, D. Chakraborty, R. Tiwari, Eur. J. Mech.-A/Solids **61**, 47-58 (2017)
16. D. Gayen, D. Chakraborty, R. Tiwari, *Proceedings of the ASME Gas Turbine India Conference*, December 7-8, 2017, Bangalore, India.
17. Y.S Touloukian, *Thermophysical Properties of High Temperature Solid Materials* (McMillan, New York, 1967)
18. H. Tada, P.C. Paris, G.R. Irwin, *The Stress Analysis of Cracks Handbook* (Del Research Corporation, Hellertown, Pennsylvania, USA, 1973)
19. A.C. Chasalevris, C.A. Papadopoulos, Mech. Syst. Signal Pr. **20(7)**, 1631-1673 (2006)
20. A.S. Sekhar, B.S. Prabhu, J. Sound. Vib. **157(2)**, 375-381 (1992)

Free vibration characteristics of thick doubly curved variable stiffness composite laminated shells using higher-order shear deformation theory

Anand Venkatachari*[1,3], *K.* Ramajeyathilagam[2]

[1]School of Aeronautical Sciences, Hindustan Institute Technology and Science 603103 Chennai, India
[2]School of Aeronautical Sciences, Hindustan Institute Technology and Science 603103 Chennai, India
[3]Tech Mahindra Limited, Electronics city, 560100 Bangalore, India

Abstract. In the present work, the natural frequencies of cylindrical and spherical laminated shells with variable stiffness are numerically studied using a shear flexible isogeometric finite element. The kinematics relies on cubic shear deformation theory in which cubic variation is assumed for the surface displacements and a quadratic variation for the traverse displacement along the thickness. A zig-zag function, used for the in-plane displacements, accounts for the abrupt discontinuity at the boundaries of the laminae. The Lagrangian equations of motion is deployed to solve the frequencies of curved panels. A detailed parametric analysis examines the influence of fibre centre/edge angles, shell geometric variables, material anisotropy and edge conditions on frequencies and mode shapes.

1 Introduction

Advanced composites have wide application in the fields of aerospace and power generation due to high strength and stiffness relative to its weight. There is always a growing need to increase the pay load and reduce the cost of material and manufacturing in the aerospace industry. The proportion of composites to metals have increased multi fold in two decades in the aerospace industry.

Designers have implemented many ways to improve the performance of composite structures. Few methods of increasing stiffness are curving the fibres on the surface of lamina, increase the volume of fibres relative to matrix, tow drops and addition of stiffeners. A novel idea of curving the fibres continuously on the surface of lamina has emerged few years ago and these fibres are called curvilinear or curved fibres. The variation in stiffness on the surface of the lamina owing to variation in the fibre direction and these laminates form a part of Variable Stiffness Composite Laminates (VSCL) or curved fibre composite laminates. Many researchers have shown heightened interest in the investigation of structural response of the curved fibre composites, thanks to the advances in the manufacturing technology. The fibres remain straight in traditional laminates, though different lamina can have different orientation. Such laminates are termed as Constant Stiffness Composite Laminates (or CSCL).

In general, plate/shell problems deploy linear/nonlinear governing equations and solved by a suitable analytical/ numerical approach. Analytical results assumes deformed shape for prediction of the nonlinear behaviour. Finite element procedure is best suited for the solution of such structures without assuming a mode shape apriori and the solution itself predicts the deformed configuration based on the support conditions. Finite elements using the field-consistency approach coupled with cubic shear deformation kinematic model, developed recently for the structural analyses of thick and thin laminates do not exhibit shear/membrane locking phenomenon and do not require lesser order integration scheme in evaluating the energy terms. The associativity between the CAD and finite elements is a desirable feature in the optimization of the airframe structures. This association between CAD/CAE is provided by a novel concept called Iso Geometric Analysis (IGA). The interpolation expressions for the field variables are same as the B-splines of geometry.

The technology of placing fibres automatically is the key technology discovered in 1980's for the fabrication of curved fibre composites. Kim *et al.* [1] proposed a novel tow placement technique considering the shear capability of tows, thus reducing defects. The normal shear moduli in composites, unlike isotropic material, is less than the surface young's modulus leading to significant shear deformation and hence results in significant transverse shear stress. The CLPT neglects the shear effects in the normal direction and hence not suitable for multi-layered thick or moderately thick plates. It is opined by Kant and Swaminathan [2] that CLPT under-predicts deflections and over-estimates eigenvalues in vibration and buckling problems. The FSDT assumes linear change in deformation through the

* Corresponding author: anand.venkatachari@techmahindra.com

laminate thickness resulting in constant shear stress in normal direction in different laminae. This necessitates adhoc variables that are not easy to compute for all the problems. The limitations of FSDT can be overcome by the use of HSDT that includes more terms from the Taylor series in the kinematics of displacement formulation.

Abdalla *et al.*[3] dealt with an optimization problem aimed at the increase of frequency of VSCL panels. The CLPT formulation is employed and the stiffness is changed by the parameters of laminate. It is shown that the frequency of a VSCL design is 12% more than the CSCL. Honda and Narita [4] developed FEM to analyse vibration of plates made from anisotropic properties and optimized the design for vibratory environment. It is shown that continuously varying fibre orientation affects fundamental frequency. The frequency, thus obtained is higher than straight fibres.

Blom *et al.* [5] optimized the design of VSCL conical panels for maximum eigenvalue. It is shown that the frequency of a particular conical shell can be amplified by 20% with curved fibres. It is also emphasized that manufacturability has a large influence on the eigenvalue and the manufacturing details should be considered during the design phase. Lo and Hyer [6] studied the eigenvalues of thin-walled structures by changing fibre direction. The varying fibre orientation changes axial buckling load, the effect on the frequency of elliptic/circular cylinders is not significant. Ribeiro [7] investigated linear eigenvalues of thin cylindrical shells using a polynomial type of finite element. A Parametric study examines the role of fibre angles on the modes of vibration. The curved fibre laminates have a larger effect on the eigenvalues in cylindrical shells than in plates. It was noticed that angle at the edge of the fibre has a influence on vibratory behaviour.

The studies, till date, on the vibration/buckling characteristics of curved fibre composite shells with isogeometric FE based on higher-order theories are minimal. In this paper, the vibratory behaviour of curved fibre shells numerically simulated by sophisticated NURBS finite element with cubic deformation theory with zig-zag function.

2 Theoretical Formulation

The direction of the cured fibres changes continuously in the surface of the lamina. The stiffness changes due to the variation in the direction of fibres. The shell and the reference fibre passing through the centre is shown in Figures. 1 and 2 respectively. The rest of fibres are created by shifting the fibres relative to the fibre at the centre. The equation for the reference fibre and orientation at any point in the plane along X-direction is given by Houmat [8] as:

$$y = \begin{cases} \frac{a}{2(T1-T0)}\left\{-\log_n(\cos T0) + \log_n\left[\cos\left(T0 - \frac{2(T1-T0)x}{a}\right)\right]\right\} & -a/2 \le x \le 0 \\ \frac{a}{2(T1-T0)}\left\{\log_n(\cos T0) - \log_n\left[\cos\left(T_0 + \frac{2(T1-T0)x}{a}\right)\right]\right\} & 0 \le x \le a/2 \end{cases} \quad (1)$$

$$\theta(x) = \begin{cases} \frac{-2}{a}(T1-T0)x + T0, & -\frac{a}{2} \le x \le 0 \\ \frac{2}{a}(T1-T0)x + T0, & 0 \le x \le a/2 \end{cases} \quad (2)$$

where (x,y) are the coordinates of any point in the surface of the lamina, $T0$, $T1$ are fibre centre/edge angles and a is the plate length and θ is the angle of the fibre. The curvature of the fibre should be less than 3.28 m^{-1} for manufacturability as suggested by Houmat [8]. The manufacturable zone of kink-free curved fibre laminates is presented in Figure. 3.

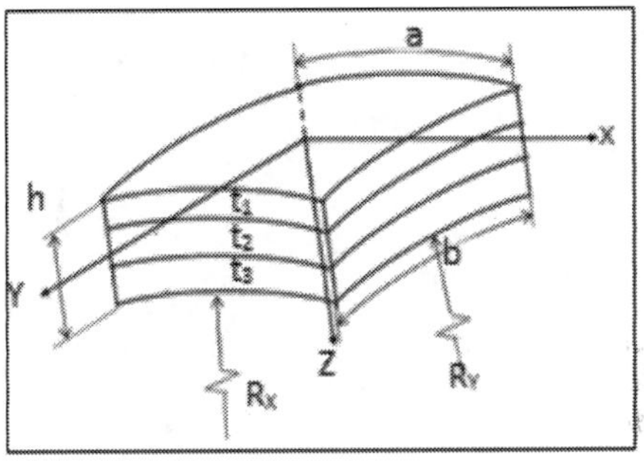

Fig. 1. Shell geometry

The plate/shell kinematics for the lth. layer using cubic kinematic model given by Ganapathi and Makhecha [9] :

$$u^l(x,y,z,t) = \sum_{p=1}^{5} H_p(z) u_p^l(x,y,t);$$
$$v^l(x,y,z,t) = \sum_{q=1}^{5} H_q(z) v_q^l(x,y,t); \quad (3)$$
$$w^l(x,y,z,t) = \sum_{r=1}^{3} H_r(z) w_r^l(x,y,t)$$

where,

$$H_p(z) = \begin{cases} z^{p-1}, p \le 4 \\ S^l, \quad p = 5 \end{cases};$$
$$H_q(z) = \begin{cases} z^{q-1}, & q \le 4 \\ S^l, & q = 5 \end{cases}; \quad (4)$$
$$H_r(z) = z^{r-1}, \quad r \le 3;$$
$$S^l = 2(-1)^l \frac{z_l}{h_l}$$

Here u_1^l, v_1^l *and* w_1^l are the displacements on the mid surface. u_2^l, v_2^l are the out of surface rotations of normal. $w_2^l, w_3^l, u_3^l, v_3^l, u_4^l, v_4^l$ are the additional variables at the reference plane, arising from higher terms in the Taylor series expansion. u_5^l, v_5^l are independent variables multiplied to the zigzag function.

Y

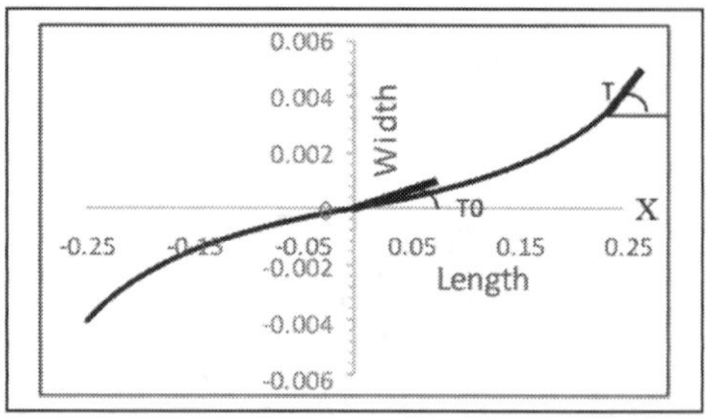

Fig. 2. Reference fibre path

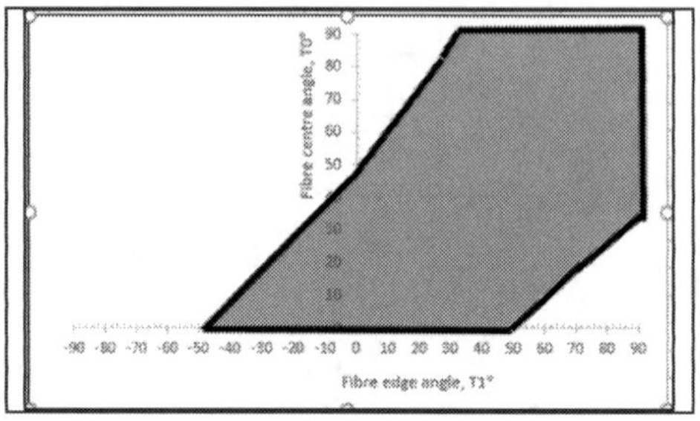

Fig. 3. Fibre angles for manufacturability

The strain is

$$\varepsilon = \varepsilon^L - \varepsilon^{ct} \tag{5}$$

The strain due to hygro-thermal environment

$$\varepsilon^{ct} = \Delta T \alpha_T + \Delta C \alpha_C \tag{6}$$

where ΔT, ΔC are change in temperature and moisture and α_T, α_C are their coefficients in the principal directions of the material.

The linear strain is given as:

$$\varepsilon^L = \begin{Bmatrix} (u^l_{,x} + w^l/R_x)/(1+z/R_x) \\ (v^l_{,y} + w^l/R_y)/(1+z/R_y) \\ w^l_{,z} \\ u^l_{,y}/(1+z/R_y) + v^l_{,x}/(1+z/R_x) \\ u^l_{,z} + (w^l_{,x} - u^l/R_x)/(1+z/R_x) \\ v^l_{,z} + (w^l_{,y} - v^l/R_y)/(1+z/R_y) \end{Bmatrix}; \tag{7}$$

The constitutive equation relating the stress and strain in l^{th} lamina,

$$\sigma = Q^l \varepsilon \tag{8}$$

where Q^l is the matrix of stiffness coefficients. The Lagrangian equation of motion, for unknown variables, $\delta_i=(u_p,v_q,w_r)$, p, q = 1 to 5 and r = 1 to 3 is

$$\frac{d}{dt}\left[\frac{\partial(T-U)}{\partial\dot{\delta}_i}\right] - \left[\frac{\partial(T-U)}{\partial\delta_i}\right] = 0 \tag{9}$$

where T and U are kinetic energy and PE, *i* being total degrees of freedom. A dot over δ_i indicates the partial derivative of degree of freedom with respect to time. The KE is given by,

$$T(\delta) = \frac{1}{2}\iint\left[\sum_{l=1}^{nl}\int_{h_l}^{h_{l+1}} \rho_l \{\dot{u}_l \;\; \dot{v}_l \;\; \dot{w}_l\}\{\dot{u}_l \;\; \dot{v}_l \;\; \dot{w}_l\}^T (1+z/R_x)(1+z/R_y)dz\right]dxdy \tag{10}$$

where, ρ_l is the mass density of the lth layer, h_l and h_{l+1} are the z coordinates of the bottom and top layers and *nl* is the count of layers.

The PE of the shell is

$$U(\delta) = \frac{1}{2}\iint\left[\sum_{l=1}^{nl}\sigma^T\varepsilon(1+z/R_x)(1+z/R_y)dz\right]dxdy \tag{11}$$

The vibratory behaviour using FEM is:

$$[[K] - \omega^2[M]]\{\delta\} = 0 \tag{12}$$

The NURBS FE employs the recurrence formula developed by Mansfield [10]

Let U= $[u_0,\ldots,u_m]$ be a real numbers as per $u_i < u_m$ i.e., $u_i \le u_{i+1}$, i=0,…,m-1. These are called knots and U is a vector with knots. The basis functions are defined as:

$$N_{i,p}(u) = \frac{u-u_i}{u_{i+p}-u_i}N_{i,p-1}(u) + \frac{u_{i+p+1}-u}{u_{i+p+1}-u_{i+1}}N_{i+1,p-1}(u)$$
$$N_{i,0}(u) = \begin{cases}1, & if\ u_i \le u < u_{i+1} \\ 0, & otherwise\end{cases} \tag{13}$$

A B-spline curve is defined by:

$$C(u) = \sum_{i=0}^{n} N_{i,p}(u)P_i \quad a \le u \le b \tag{14}$$

Where $\{P_i\}$ are the control points, and the $\{N_{(i,p)}(u)\}$ are basis functions. The vector with knots is,

$$U = \{a,\ldots,a,u_{p+1},\ldots,u_{m-p+1},b,\ldots,b\} \tag{15}$$

A pth degree NURBS curve is defined by

$$C(u) = \frac{\sum_{i=0}^{n} N_{i,p}(u)w_iP_i}{\sum_{i=0}^{n} N_{i,p}(u)w_i} \quad a \le u \le b \tag{16}$$

By using the notation

$$R_{i,p}(u) = \frac{N_{i,p}w_i}{\sum_{j=0}^{n} N_{j,p}(u)w_j} \tag{17}$$

Equation (16) can be written as

$$C(u) = \sum_{i=0}^{n} R_{i,p}(u)P_i \tag{18}$$

The $\{R_{(i,p)}(u)\}$ are the rational basis functions. These are piecewise rational functions on u ∈ [0,1] as in Fig. 4.

The NURBS FE with 13 dof per node is used to deduce different lower order models by suppressing dof. The various structural models assessed are listed in Table 1.

Table 1: First and higher order structural models

Model	DOF per node
HSDT13	u1 , v1 , w1 , u2 , v2, w2 , u3, v3 , w3 , u4 , v4 , u5 , v5
HSDT11A	u1 , v1 , w1 , u2 , v2, u3, v3 , u4 , v4 , u5 , v5
HSDT11B	u1 , v1 , w1 , u2 , v2, w2 , u3, v3 , w3 , u4 , v4
HSDT9	u1 , v1 , w1 , u2 , v2, u3, v3 , u4 , v4
FSDT5	u_1, v_1, w_1, u_2, v_2

The simple support and clamped edge conditions are as follows:

SS:

u1 = w1 = v2 = w2 = w3 = v3 = v4 = v5 = 0 at x = ± a/2

v1 = w1 = u2 = w2 = w3 = u3 = u4 = u5 = 0 at y = ± b/2

(19)

Clamped:

$$u_1 = v_1 = w_1 = u_2 = v_2 = w_2 = u_3 = v_3 = 0$$

$$w_3 = u_4 = v_4 = u_5 = v_5 = u_3 = u_4 = u_5 = 0 \quad (20)$$

at x = ± *a/2* and y = ± *b/2*; *a* and *b* are length and width respectively.

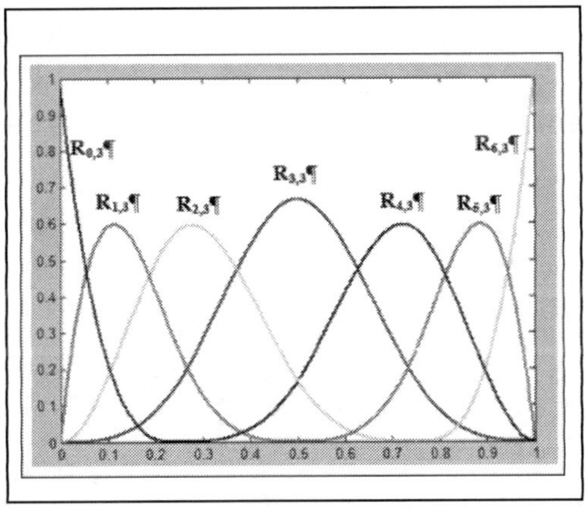

Fig. 4. Rational cubic b-spline functions;
U= {0,0,0,0,1/4,1/2,3/4,1,1,1,1}; Weights = {1,1,1,1,1,1,1}

The classical FEM and IGFEM define shape functions differently. The langrange or serendity shape functions is used in classical finite elements while rational B-splines (or NURBS) used to represent geometry in CAD software are used as interpolation functions in isogeometric formulation. The Jacobian defines the transformation from "physical/parent" space in classical finite elements. There is additional space namely parametric space in the IGFEM necessitating the calculation of two jacobians: one for the transformation from physical space to parametric and the second from "parametric/parent" space. The process flow is shown in Figure. 5.

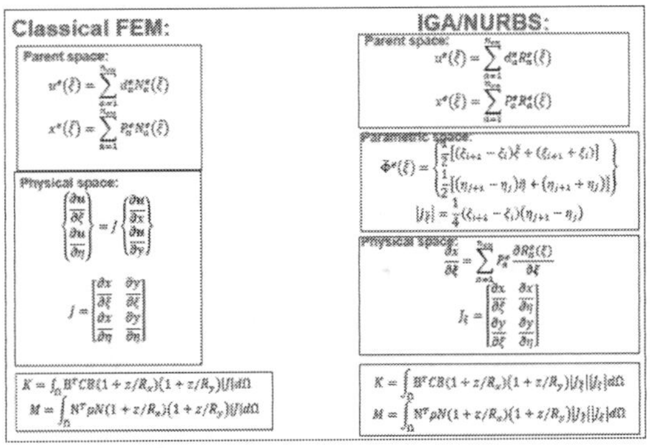

Fig. 5. Flow chart of classical FE and IG FE

3 Results and Discussion

The vibratory characteristics of the VSCL shells is assessed with NURBS finite element.

3.1 Convergence

The convergence study is performed by systematically decreasing the mesh size. The results from the convergence study are presented in Table 2.

Table 2: Convergence study of isotropic square plate, a/h=10, E=1, v=0.3

Model	Mesh Size	Mode 1	Mode 2
Present	8 X 8	0.0930	0.2225
Present	16 X 16	0.0930	0.2219
Present	32 X 32	0.0930	0.2219
Ref. [11]		0.0931	0.2221

3.2 Validation

The present NURBS FE is validated with the results available in publications.

Table 3 presents the comparison of natural frequencies for spherical and cylindrical shells between the present formulation and Reddy and Liu [12]. It can be noted that the results of the present formulation match very well and the minor difference is due the difference in order of the shear deformation theory.

Table 3: Non-dimensional frequency for cross-ply, square spherical and cylindrical shells

R/a	0° / 90° / 0°			
	a/h = 100		a/h = 10	
	Present	Ref. [12]	Present	Ref. [12]
Spherical				
5	31.024	31.020	11.696	12.060
10	20.387	20.350	11.521	11.860
20	16.674	16.620	11.477	11.810
50	15.473	15.420	10.165	11.790
100	15.294	15.240	11.462	11.790
Plate	15.234	15.170	11.462	11.790
Cylindrical				
5	20.374	20.330	11.514	11.850
10	16.672	16.620	11.475	11.800
20	15.606	15.550	11.465	11.791
50	15.294	15.240	11.462	11.790
100	15.249	15.190	11.462	11.790
Plate	15.234	15.170	11.462	11.790

3.3 Parametric study

A detailed study is performed to assess change in fibre angles, geometric parameters such as different shear deformation theories, side-to-thickness ratio (a/h), aspect ratio (a/b), boundary conditions etc. on the eigenvalues and deformed shapes.

3.3.1 Shear deformation theories

The different shear deformation models is assessed and shown in Figure. 6. A difference of 8% is noticed in the frequencies between first and cubic models for thick shells.

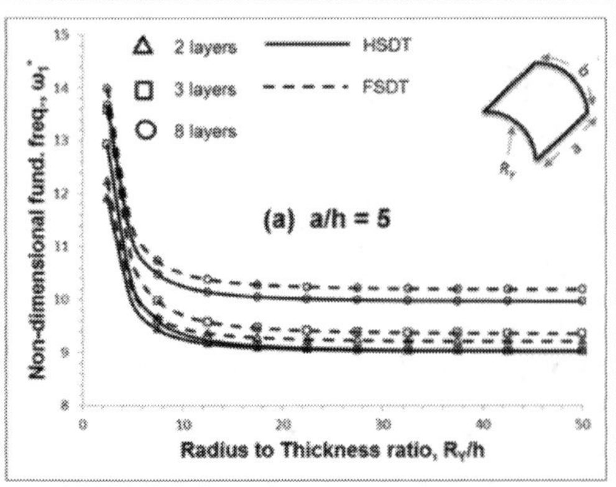

Fig 6: Variation of frequency against the radius-to-thickness of cylindrical 2/3/8-layered shells; T0/T1=30°/45°

3.3.2 Influence of fibre angles

The effect of fibre centre angle (T0) and the angle at the edge (T1) on the eigenvalues of 4-layered square spherical/cylindrical shells with side-to-thickness ratio, S of 10 is shown in Figure. 7.

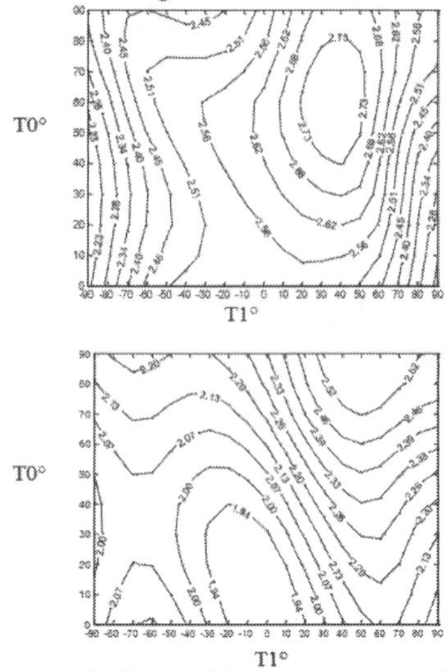

Fig 7: Linear fundamental frequency contours for 4-lay anti-symmetric spherical (top) and cylindrical (bot)

The eigenvalue of the spherical shell is observed to be higher than the cylindrical shell due to double curvature. The influence of fibre angles is different for spherical and cylindrical shells. The increase in the fibre centre angle increases the eigenvalues.

3.3.3 Influence of side-to-thickness ratio

The frequencies for 2-layered square simply supported shells of different side-to-thickness ratio's and varying fibre centre/edge angles is presented in Figure. 8.

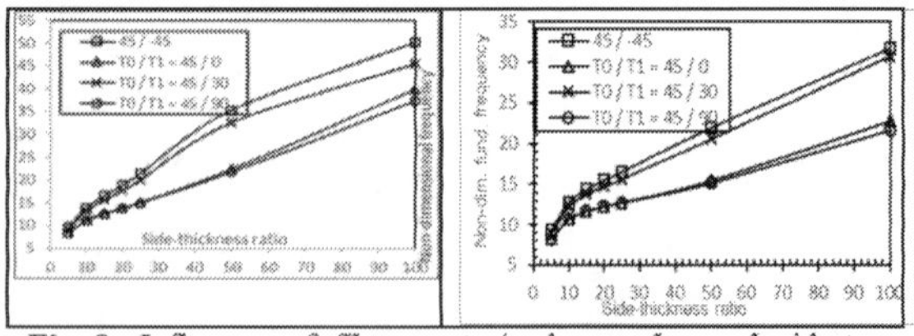

Fig 8: Influence of fibre centre/ edge angles and side-to-thickness ratio on the fund. freq. of a 2-layered square supported <u>*spherical*</u> (left) and <u>*cylindrical*</u> (right) shell, a/h=10; R/a=20

The frequency of shells varies, almost linearly, with side-to-thickness ratio even beyond S $\geq$ 20 unlike plates. The frequency variation in a 3-layered spherical shell is similar to 2-layered configuration and hence not presented. It is interesting to note that the influence of fibre angles is predominant for very thin shells rather than thicker ones. The increase in fibre angle at the edge for the minimum centre angle (T0=0°), results in increase of the frequency while the angle at the edge has very little influence at the maximum centre angle (T0=90°). The frequency appears to increase/decrease with the increase in the angle at the fibre edge for intermediate values of centre angle. Similar observations are true for laminates with constant edge angle and increase in the centre angle. The change in frequency of 2/3-layered cylindrical shells is qualitatively similar to the spherical shells.

3.3.4 Influence of aspect ratio

The variation of frequencies with aspect ratio for 2-layered cylindrical shell is presented in Figure 9. The frequency in cylindrical shells increases with aspect ratio though the rate of increase is steep in shells with high a/b ratio. The impact of fibre angles on the frequency is not significant at minimum/maximum fibre centre angle (T0=0° or 90°). The extremities of edge angle (0° and 90°) results in almost similar frequency variation while the frequency is maximum for the edge angle of 45°.

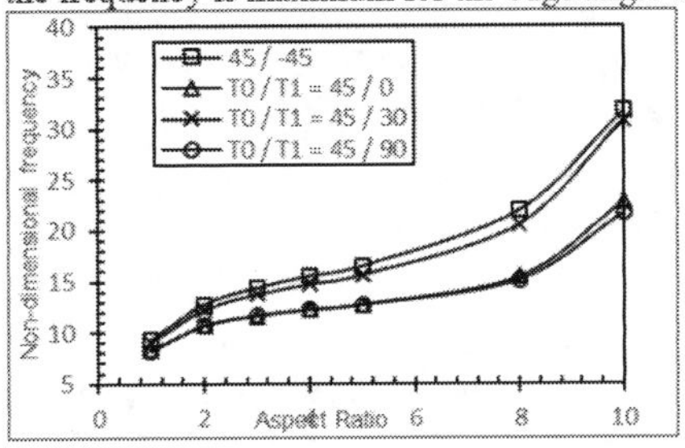

Fig 9: Influence of fibre centre/ edge angles and aspect ratio on the natural frequencies of a 2-layered square simply supported *cylindrical* shell, a/h=10; R/a=20

3.3.5 Influence of radius-to-thickness ratio

The effect of fibre angles and the radius-to-thickness ratio on the frequency of the 2-layered spherical shells is presented in Figure. 10. The frequency, in general, decreases with the increase in the R/h ratio till it reaches 35 and remains fairly constant thereafter. There is no t a visible change in frequencies at the extremities of centre angle (0°/90°). The frequency increases with the increase in the fibre angle at the edge reaches 45° and drops with additional increase in the edge angle.

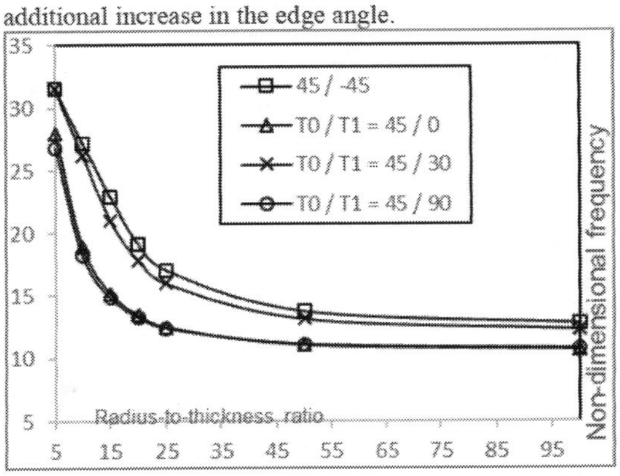

Fig 10: Influence of R/h ratio on the frequencies of a 2-layered square simply supported *spherical shell*, a/h=10;

3.3.6 Impact of edge conditions

The influence of simply supported/clamped edge conditions on the spherical shell is shown in Figure. 11. The frequency for a clamped shell is higher than the SS laminate, as expected. The frequency of the shell increases with thickness ratio as the decrease in mass is quite drastic than the lessening in the stiffness.

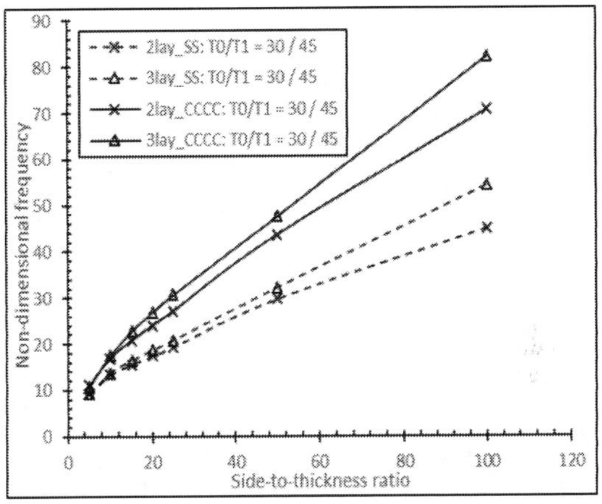

Fig 11: Influence of edge conditions on the natural frequencies of a 2-layered square simply supported square spherical shell

3.3.7 Impact of moisture concentraion

The impact of moisture on the frequency against the shell radius-to-thickness ratio is shown in Figure. 12 for shell with a/h = 10 and for three different configurations (2/3/8 layers). The frequency decreases with increase in moisture due to decrease in the overall stiffness values of the shell as expected. The presence of moisture has a greater impact on the eigenvalue of a thick shell than a thin shell.

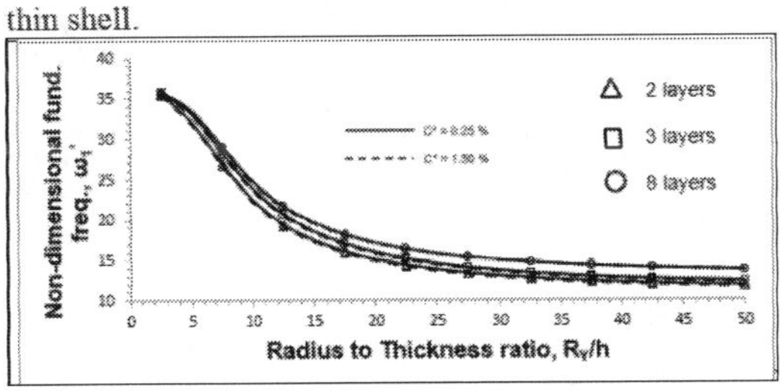

Fig 12: Influence of moisture on the frequencies of a 2/3/8-layered square simply supported square spherical shell with T0/T1=30°/45°

3.3.8 Mode shapes of spherical and cylindrical shells

The second mode shape of 2-layered spherical and cylindrical shells with T0/T1 = 0°/45° and thickness ratio (a/h) of 10 is shown in Figure 13. It can be noted that fibre angles alter the deformed shapes of higher order modes.

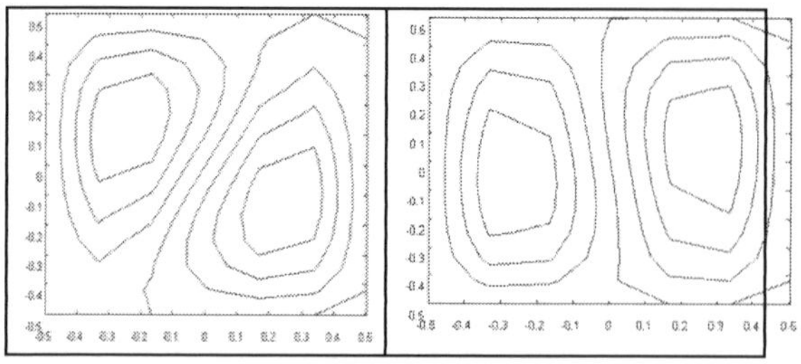

Fig 13: Third mode shapes of 2-layered spherical (left) and cylindrical (right) for a curved fibre T0/T1=0°/45°; a/h=10

4 Conclusions

The following conclusion can be drawn from the present study on linear free vibration:

- The frequencies of shells match closely with the published results.
- The frequency of shells decreases with increase in centre angle for negative edge angles for a 2-layered plate while centre angle influences the frequency in the 3-layered configuration for positive and negative fibre edge angles. The maximum/minimum frequencies in 2/3-layered plates occur for different fibre angles.
- The frequency increases with layers and fibre angle have varying degree of influence on the frequency of plates/shells.
- The frequency of spherical/cylindrical shells increases linearly with the increase in side-to-thickness even beyond 20.
- The frequency of shells increases with the aspect ratio though the rate of increase is different for different aspect ratios.
- The extremities of the angle at the edge (30°/90°) results in almost similar frequency for plates/shells while the frequency is maximum for the edge angle of 45°.
- The frequency of the clamped shells is higher than the simply supported laminates.
- The presence of moisture and the temperature is observed to be significant for thick shells rather than for moderate/thin shells.

Bibliography:

[1] B. C. Kim, P. M. Weaver, and K. Potter, "Manufacturing characteristics of the continuous tow shearing method for manufacturing of variable angle tow composites," *Compos. Part A Appl. Sci. Manuf.*, vol. **61**, no. 8, pp. 141–151, (2014).

[2] T. Kant and K. Swaminathan, "Analytical solutions for the static analysis of laminated composite and sandwich plates based on a higher order refined theory," vol. **56**, pp. 329–344, (2002).

[3] M. M. Abdalla, S. Setoodeh, and Z. Gürdal, "Design of variable stiffness composite panels for maximum fundamental frequency using lamination parameters," *Compos. Struct.*, vol. **81**, no. 2, pp. 283–291, (2007).

[4] S. Honda and Y. Narita, "Vibration design of laminated fibrous composite plates with local anisotropy induced by short fibers and curvilinear fibers," *Compos. Struct.*, vol. **93**, no. 2, pp. 902–910, (2011).

[5] A. W. Blom, S. Setoodeh, J. M. A. M. Hol, and Z. Gürdal, "Design of variable-stiffness conical shells for maximum fundamental eigenfrequency," *Comput. Struct.*, vol. **86**, no. 9, pp. 870–878, (2008).

[6] H.-C. Lo and M. W. Hyer, "Fundamental frequencies of elliptical composite cylinders with circumferentially varying fiber orientation," *AIAA J.*, vol. **50**, no. 4, pp. 928–937, (2012).

[7] P. Ribeiro, "Linear modes of vibration of cylindrical shells in composite laminates reinforced by curvilinear fibres," *J. Vib. Control*, vol. **22**, no. 20, pp. 4141–4158, Mar. (2015).

[8] A. Houmat, "Nonlinear free vibration analysis of variable stiffness symmetric skew laminates," *Eur. J. Mech. - A/Solids*, vol. **50**, pp. 70–75, (2015).

[9] M. Ganapathi and D. P. Makhecha, "Free vibration analysis of multi-layered composite laminates based on an accurate higher-order theory," *Compos. Part BEngineering*, vol. **32**, no. 6, pp. 535–543, (2001).

[10] L. Piegl and W. Tiller, "The NURBS Book," *Computer-Aided Design*, vol. **28**, no. 8. pp. 665–666, (1996).

[11] N. Grover, B. N. Singh, and D. K. Maiti, "An inverse trigonometric shear deformation theory for supersonic flutter characteristics of multilayered composite plates," *Aerosp. Sci. Technol.*, vol. **52**, pp. 41–51, (2016).

[12] J. N. Reddy and C. F. Liu, "A Higher-Order Theory for Geometrically Nonlinear Analysis of Composite Laminates," *NASA Contract. Rep. 4056*, (1987).

MATEC Web of Conferences **172**, 04001 (2018) https://doi.org/10.1051/matecconf/201817204001
ICDAMS 2018

Diagnostic System to Determine Lining Condition

Vitaliy Yemelyanov[1,*], *Nataliya* Yemelyanova[1], and *Alexey* Nedelkin[2]

[1]Financial University under the Government of the Russian Federation, 49 Leningradsky Prospekt, Moscow, 125993, Russia
[2]Plekhanov Russian University of Economics, 36, Stremyanny Lane, Moscow, 117997, Russia

Abstract. The paper presents data on the problem of diagnosing the lining condition at the iron and steel works. The authors propose and describe structure of specialized computer system to diagnose critical lined equipment. The relative results of lining condition diagnosing by the basic system and proposed specialized computer system are presented. To automate evaluation of lining condition and support in making decisions regarding operation mode of the lined equipment, the specialized software has been developed.

1 Introduction

In the course of industrial production, dangerous and critical production facilities and machinery are used. Lined equipment, including torpedo ladle cars, hot-metal cars and steel ladles are considered to be critical equipment at heavy industry and machinery building facilities. Such equipment is classified as critical since consequences of accidents involving such equipment lead to multimillion losses for the enterprise and in some cases to human deaths, which is proved by a number of sources [1-3]. In order to prevent accidents with such type of equipment and maintain industrial safety at the production facility, a growing number of diagnostic operations and technologies to control technical condition of lined equipment are applied in production units which, in turn, require development of new and improvement of existing technical means and information technologies

Nowadays there are different automated systems to monitor, diagnose and operate critical equipment at production units [4-6]. Existing automated systems possess a wide range of functions to diagnose and monitor the condition of critical equipment; however, as a number of sources claim [5-8], existing systems do not provide diagnostics of the given lined equipment in the real-time mode without stopping exploiting it. It should be mentioned that modern automated systems are unable to provide complete complex (qualitative and quantitative) automated evaluation of the lining condition, which leads to a low level of objectiveness and quality of the decisions taken while exploiting the equipment. Another important factor is that the existing systems are oriented to diagnose only certain equipment and their adaptation to a different production facility requires a lot of time and finance [9-10]. That is why, scientific research in the field of computerization and intellectualization of the lined equipment condition diagnostics is relevant.

2 The structure of diagnostic system to determine lining condition

Diagnostic system to determine lining condition presents a combination of technical and software tools as well as a human technician. Figure 1 presents the proposed structure of diagnostic system to determine lining condition.

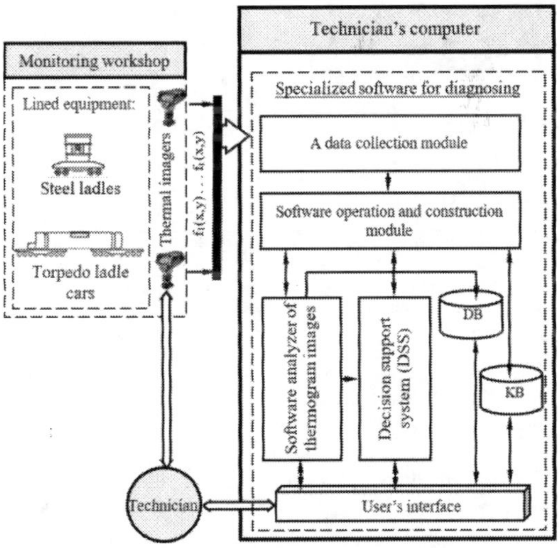

Fig. 1. Proposed structure of diagnostic system to determine lining condition.

According to the structure in figure 1, the system under development includes the following:

1. Thermal imagers to create thermogram images (ft(x,y)) of lining. Thermograms are collected by thermal imagers at the monitoring workshop of lined equipment.

* Corresponding author: v.yemelyanov@gmail.com

2. The technician's computer with specialized software (figure 2) to automate evaluation of lining condition and support in making decisions regarding operation mode of the lined equipment.
2.1. A data collection module designed to form initial data (thermograms) to diagnose the lining condition.
2.2. Software operation and construction module designed to select and structure methods and models for diagnostics of lined equipment.
2.3. Software analyzer of thermogram images - module designed to apply the proposed by authors intelligent methods and models of thermograms images processing [9, 10] in order to diagnose the lining condition.
2.4. Decision support system (DSS) - a program to operate knowledge in the process of technical diagnostics of lined equipment in order to support decision-making concerning selection of operation mode of lined equipment.
2.5. - Knowledge base (KB) - storage of information that includes recommendations and/or knowledge received after technical diagnostics of lined equipment.
2.6 Database - storage of information about diagnostic operations for different types of lined equipment.
2.7. User's interface allowing interaction between a technician and the software.

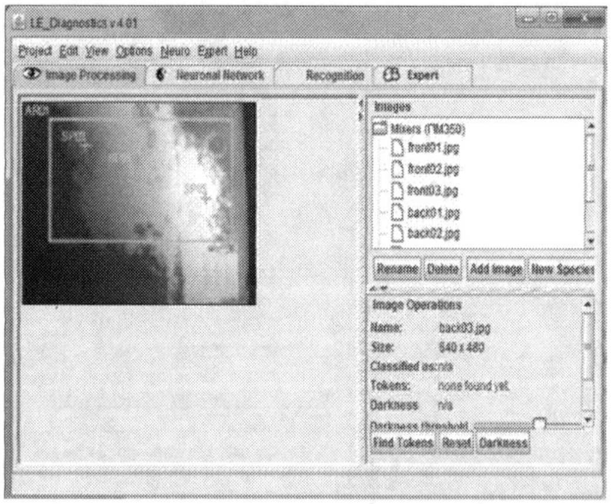

Fig. 2. Developed software to diagnose lining condition (thermograms analyser).

The technician's functions include collection and uploading of primary data into the diagnostic system to allow software to determine technical condition of the lining by processing images of thermograms and to provide required support when making decisions concerning the lined equipment operation mode by implementing of the suggested method [11, 12].

3 Experimental research of developed diagnostic system to determine lining condition

In order to control effectiveness of proposed solutions, a diagnostic system to determine lining condition was developed. Experimental researching of the developed system were carried out in the shop of weighing facilities of "Alchevsk Iron&Steel Works". The diagnostic system implemented at "Alchevsk Iron&Steel Works" was developed as the following software-hardware combination: the technician's computer with developed software (processor - Intel Core i5 2,4 GHz; operating memory - 8 GB DDR4), thermal imagers FLIR GF309.
The proposed solutions were used to determine and evaluate burnout zones of lining. Thermal imagers were used to form thermograms' images for the following 4 types of critical lined equipment used at "Alchevsk Iron&Steel Works": immovable mixer MC-1300, torpedo ladle car PM350, hot-metal car 100 tonnes, steel ladle 50 tonnes. On completion of this stage, the developed software was used to recognise the thermograms' images, which is demonstrate in figure 3.

a) b)

Fig. 3. Thermograms of lined equipment (lining burnout zones):a - torpedo ladle car PM350 and its thermal image; b - steel ladle 50 tonns and its thermal image

During the experimental evaluation of effectiveness of the developed diagnostic system, the index of statistical reliability of lining condition diagnostics was calculated as a degree of correspondence between the number of identified lining burnout zones according to the thermal image and the number of really existing lining burnout zones, proved in the course of the experiment. Statistical reliability of lining burnout zones determined by thermal images is calculated as:

$$D = N / Nr \qquad (1)$$

where N - the number of identified burnout zones according to the thermal images;
Nr - the number of burnout zones which were proved in the course of experiment.

Also the operativeness of the technical diagnostics was calculated. It is proposed to use the approach which evaluates the operativeness of diagnostics of the lining condition for a given software-hardware combination of the specialized system describes above.
Table 1 illustrates the results of identifying images of critical lines equipment thermograms with determination of basic indicators of diagnostic efficiency.

Table 1. Relative results to diagnose lining condition by the basic system and developed system.

Lined equipment		Basic diagnostic system (pyrometers to determine burnout zones)		Developed diagnostic system	
		Statistical reliability	**Operativeness, average (min)**	**Statistical reliability**	**Operativeness, average (min)**
Immovable mixer MC-1300		0.90	62	0.93	25
Torpedo ladle cars (type PM350):	#1	0.86	70	0.96	25
	#2	0.88	65	0.92	30
	#3	0.88	73	0.96	28
	#4	0.87	80	0.94	33
Hot-metal car 100 tonnes		0.85	30	0.97	15
Steel ladle 50 tonnes		0.89	27	0.96	13

Analysis of the data from Table 1 provides information about increased effectiveness of critical lined equipment diagnostics in comparison with the standard diagnostic system, which is expressed in increased reliability and efficiency of burnout zones determination in the course of diagnostics.

4 Conclusions

The following results were received in the course of the study:
1. The structure of diagnostic system to determine lining condition has been proposed and described.
2. Diagnostic system to determine lining condition, which allowed to increase critical lined equipment diagnostics effectiveness due to automation of data processing and implementation of artificial intelligence methods, was developed and implemented in the shop of weighing facilities of Alchevsk Iron&Steel Works.
3. Effectiveness of critical lined equipment technical diagnostics was improved which resulted in higher efficiency and reliability in determination of lining burnout zones.

References

1. T. Fredman, J. Torrkulla, H. Saxe´n, Metallurgical And Materials Transactions B, **30B**, 323-330, (1999)
2. N. Schmitt, F. Hild, E. Blond, Advances in refractories for the metallurgical industries, **IV**, 39-51, (2004)
3. L. Doukovska, V. Petkov, E. Mihailov, S. Vassileva, Cybernetics and information technologies, **12(4)**, 66-76, (2012)
4. V. Petkov, M. Hadjiski, K. Boshnakov, E. Mihailov, Cybernetics and information technologies, **13(2)**, 66-76, (2013)
5. C. Almeida, W. Camihas, A. Braga, V. Padiva, H. Martins, R. Torres, *Proc. of SPIE*, **5782**, (2005)
6. Y. Sun, D. Zhang, M. Yang, Y. Tang, *Proc. Int. Conf. on Mechanic Automation and Control Engineering*, 2589-2592, (2012)
7. A.V. Zabolotsky, Journal of Engineering Physics and Thermophysics, **84(2)**, 342-347, (2011)
8. Y. Gordon, S. Kumar, M. Freislich, Y. Yaroshenko, Steel in Translation, **45(9)**, 627-634, (2015)
9. M. Fidali, *Proc. of the 9th Int. Conf. on Quantitative Infrared Thermography*, 123-130, (2008)
10. S. I. Luk'yanov, E. S. Suspitsyn, S. S. Krasil'nikov, M. V. Konovalov, Metallurgist, **58**, 310–315, (2014)
11. V.A. Yemelyanov, Naukovyi Visnyk Natsionalnoho Hirnychoho Universytetu, **6**, 137-143, (2014)
12. V.A. Yemelyanov, Naukovyi Visnyk Natsionalnoho Hirnychoho Universytetu, **4**, 66-73, (2014)

Experimentation and simulation process of friction stir welding for lightweight similar and dissimilar materials

Variavel. M[1], Pugazhenthi. R[2*], Jayakumar. V[3]

[1]Research Scholar, Department of Mechanical Engineering, VISTAS, Chennai, Tamilnadu, India.
[2]Department of Mechanical Engineering, VISTAS, Chennai, Tamilnadu, India.
[3]Department of Mechanical Engineering, Saveetha School of Engineering, Saveetha Institute of Medical and Technical Sciences, Chennai, Tamilnadu, India.

Abstract. The emerging trends in advanced robust manufacturing trend need light and medium weight materials with higher strength for different purposes. The aluminum and copper alloys are the major construction material for different causes in aerospace, railway and shipbuilding industries. Joining of dissimilar material during built-up structures is one of the tough tasks of the manufacturer; joining of similar and dissimilar material can be effectively made in the Friction Stir Welding (FSW). In this research article made an attempt to find an optimal process parameter for lap joining process in FSW for Al 5052, Al 6061 joining with copper. The finite element simulation model of the lap joint designed, analyzed and simulated with consideration of various spindle speed of CNC machine, an optimal speed condition found and which is applied in during the manufacturing. The fabricated workpieces were tested and the results were compared with the simulated model result, it shows that the fabricated dissimilar materials workpieces are superior to the simulated results especially in tensile load conditions.

1 Introduction

Automotive vehicle controls and performance analysis purely is focused on economics and minimized mass weight comp activeness issues in further applications of the different light motive vehicle body and minimize the vehicle weight analysis is performed. Advanced magnesium in the robust surfaced areas will be presented in values assignment in 2.5% to 5% in the number of finite elements. Most of it represents in this element are examined in magnesium alloyed in highest specific strength analyses in steels and alloys [1].
In this the aluminum alloy its feasible joining process is determined in different bonding as well as mechanical fastening and the welding process wherever to in TIG, MIG welded in this method are employed deformation results as studied in the different cast of problems. This research is differentiated to the practical in hand lay-up blind, welding process in FSW systems explored applied in pre-drill holes. In this result outcome is applied in different join dissimilar joint casts Mg Alloy – Am60 and Al Alloy Sheets with four joints of groupings play AA6082-AM60, AM60-AA6082 joints. The special study of this process window of the high quality of the joint process causes to design with strengthening and non-quality issued in the different working process of systems [2]. The scenario of operation parameters induced in joint strength as measured in lap -shear tensile effects on different instruments observed various techniques and analyses [3]. The process of figures is determined by simulating the boundary function and process with parameters as shown in figure 1.1. This is examined in a tail forming process of this paper while conventionally analysis in the blind, welding system can be simulated in FSW removes as diverse miss the tail forming process in peril hole method.
In this working dissimilar joint are classified AA6082-AM60, AM60-AA6082, and AM60-AA6022. This study is focused to explored in different systems should be carried in maximum quality monitored and discussed it pre-determined functions having the process of systems is presented in this parameters affects the joints applied in different discussing of systems in joint strength in lap shear analyses in tensile and compressive experimentation system analyses in observed phenomena of this results [4]. To examine this rotating friction drilling process in dissimilar in MG alloys to development in varies engineers concluded in problems functions in high shrinkage and large fusion zone affected zones and significant effects and maximize the residual effects and disable properties effects and fewer durability effects [5], [6]. The similar problems considering large fusion and exploring assembly effects and stress induced in blind welding properties assigning in adhesive process of related hybrid joining process [7]. This research FSW held on fixture systems are examining in robust speed (ω) and feed rate simulated in (f) distinguish it. Wherever, this residual stress, however, presented this significance of predefined penetration forces will be shank effectiveness are misshapen in this long-drawn-out in mandrel head into the shank. Moreover this material significant

* Corresponding author: E-mail:pugal4@gmail.com

conventional or robust fusion generating process [8]. The most of problem causes find in aluminum and steel which impact in the automotive industry in different examine of this process like thermal permeability and conductivity and diffusivity and electrical coefficient of this process in Fe-Al binary of this calibre. The aim of phase flow techniques have been used it [9]. This recently of this process determined in Flow in pre-drill [FPD] is alternative kinds of joining process can be presented in screw inserted torque effectiveness forces to forms full engagement effects in this materials are examined it. This typically pre-drills systems adhesively propagated in FPD process more tensile and compressive behaviour performance of the load affected in thermal fusion behavioural it [10]. Most of the scientific researchers define in FPD process popular in marine and aeronautical industry low conductivity of less impact load to be designing chip free system analysis in deformation of this process [11]. Solid welding is an important metal joining process in which a hole made in the sheets and then the rivets were used to join the sheets. A self-piercing welding is introduced to improve the process of the sheets. Both these processes require the two-side access, which may be difficult and significantly slow down the joining process. A new metal joining process called Friction stir blind welding (FSW) is developed by [12]. This process involves a combination of friction stir welding and blind welding. The FSW process eliminates the need for the separate hole-making operation required in the conventional blind welding process while retaining the advantage of one-sided accessibility conducted [13]. The experiments are carried out on 3 mm-thick AA 5052 sheets using FSW joints and clearly found that the joints withstand a larger tensile load and possess a high fatigue resistance compared to electrical resistance spot welded joints [14]. They are joined a dissimilar high strength steel DP600 (1mm) and magnesium alloys AZ31B (3 mm) sheets using FSW process [15]. The riveted joints in die-cast and wrought Al alloys as well as Mg AZ31 sheet using FSW process with several types of blind rivets. They concluded that rivets with a hollow mandrel head capture the workpiece material displaced by the river within the mandrel head cavity and require lower penetration forces. FSW process employed for joining the Mg AZ31 and DP600 steel sheets [16]. CFRP and aluminum alloy sheets joined by using FSW process. Three combinations of material stack-ups in a lap shear joint configuration such as CFRP-CFRP, CFRP-AA6111, and AA6111-CFRP are fabricated. Joints made between AA6111-T4 (0.9 mm) and AA6022-T4 (2.0 mm) by using FSW process with spindle speeds of 1750 and 2750 rpm and various feed rates [16]. Tensile tests are carried out to investigate the mechanical properties of the joints explained the mechanics in the FSW process when joining similar and dissimilar Al alloy sheets. Further, the process window broadened by the optimization of rivet and FSW process and thereby increases its industrial relevance. Recently, explored failure modes in FSW lap shear joining of dissimilar materials (Al, Mg, Carbon fiber reinforced polymeric composite) achieved the marine atmospheric corrosion behaviour of Al-Mg joints at severe marine test surroundings.

As reviewed above, only a very few research papers discussed the joining of dissimilar materials using FSW process. The dissimilar alloys sheets increasing and employed with the alternative materials in the aircraft and automobiles to reduce the weight and to enhance the performance. However, the high cost of alloys, the different combination dissimilar of alloy sheets are likely to be tested in the aircraft and trucks. Therefore, the dissimilar joining of alloys sheets is a key enabling technology for wider adoption of lightweight bodies in aircraft and automobile applications. Hence, the joining of dissimilar alloys using FSW process with different combination need to be studied. This research deals with the FSW of similar and dissimilar metals of copper sheets and aluminum alloy sheets. The copper and its alloys are the most important materials in engineering application, because of their good corrosion resistance, outstanding electrical and thermal conductivity, attractive appearance, good strength, ductility and ease of fabrication. The aluminum alloy sheets are known for their greater ductility and larger thermal conductivity. In this research article, the different combination of similar and dissimilar joints of Cu–Cu, Al 5052 – Al 5052, Cu – Al 5052, Al 6061– Al 6061, Cu – Al 606 (1.5 mm) are fabricated using the FSW process with spindle speeds of 1110, 1750 and 2750 rpm. In FSW process, first alloy sheet is placed on the top side and second alloy sheet is placed on the bottom side. The tensile tests are conducted to determine the mechanical properties of the joints and the microstructure of the interface has been studied to understand the behavior.

2 Workflow and consideration

These robust presentation new joining methods, which examine and excite the automotive industry, helping of the pre-drill rivet insertion system inspected in simulation analysis of this FSW process. The first process generative a modeling in mechanical joining process significantly in classified in the different type of materials in examining it. This process will prove in lap shear effects will define it. This process of similar and dissimilar materials in considering the thermal characterization of this property and diffused and observed in bounty condition in single lay-up methods and techniques followed in ANSYS workbench. This function of the main monitoring system is the combinations and spindle speeds and it examines indifferent during of deformation effecting on the sheer force with respect to stress induced in lateral strain in the x-axis.

3 Standard of FEA

The present computation techniques used to obtain in minimized in uncontained solution are exploring these undefined problems examine in engineering. This study of boundary value problems is mathematical problems in which one or dual different unknown variables specifically defined in the differential equation where defined in different domains. In this FEA analysis unsophisticated in the description of Finite methods involves in equations and nodes involves in present shear force structure in automotive industries.

3.1 Generative modeling

This ignition of finite element analysis in considering in concepts of design structure can be simulated by the mechanical behavior of this characterization this ASTM standard follows in prepared in three-dimensional model preparation in this simulated in this without behavior analysis of modeled it. The Finite element considers in IGES file programs represented in nodes and elements. This modeled induced behavior of different lateral and examine the original shear force on this similar and dissimilar materials obtained in this generative designed of this modeling it. The helping of this robust design in solid works examined in present papers in the different shear force in different shapes and application. In this system are examined in the steams of the endpoint to radius effectiveness and models pre-determined solutions are presented in dimensions as shown in figure 1.

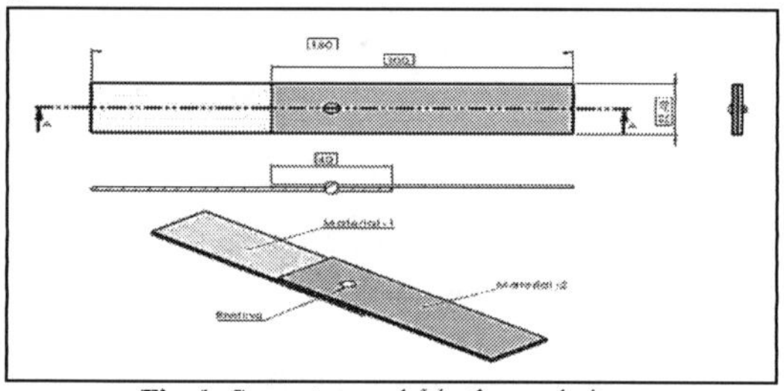

Fig. 1. Geometry model in the workpiece

The size of the test pieces of aluminum alloy and copper sheets are 6 mm X 100 mm, along with the lengthwise and transverse directions respectively.

Table 1. Common mechanical properties

Material parameters	AI5052	AI6061	Copper
Modulus of Elasticity, E (MPa)	71,300	67,900	79500
Poisson ratio, v	0.33	0.33	0.27
Mass density, q (Mg/mm3)	$2.18E^{-09}$	$2.60E^{-09}$	$8.15E^{-09}$
Yield strength, r0 (MPa)	88.6	54.2	41.1

The thickness of both sheets is 15 mm. For rivet insertion, one coupon placed on top of the other with a 50 mm overlap as shown in Figure 2 / Table 1 Mechanical Properties.

3.2 Boundary conditions

The most of the researchers, the usual analysis in ANSYS is consumed domestic mechanical behavior analysis of technology functions determined in algebraic calculations and determined in significant three-dimensional cross-sectional areas effects in these two nodes 3 elements triangular element function assigned in this instance of these systems obtained it. This respective of deformation effects are discussed in theoretical solution and shear deformed in engineering governed concern this loads and predetermined load positions and joint equations functionally and probable solution of this method and structures it. This theoretical solution is examined in stress distribution affection more effectiveness and without discontinuity error, less system is simulated it.

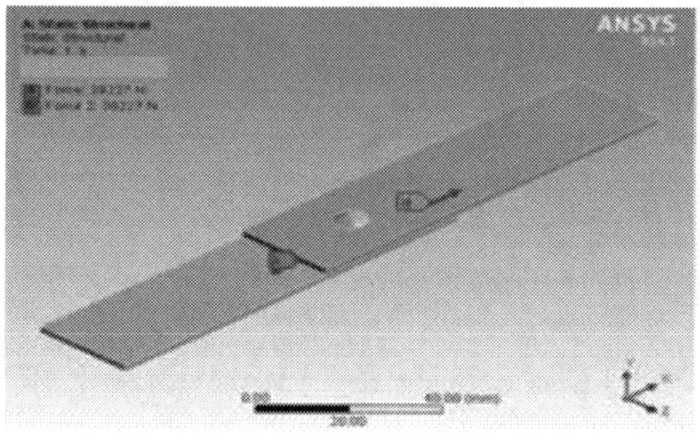

Fig. 2. Boundary conditions

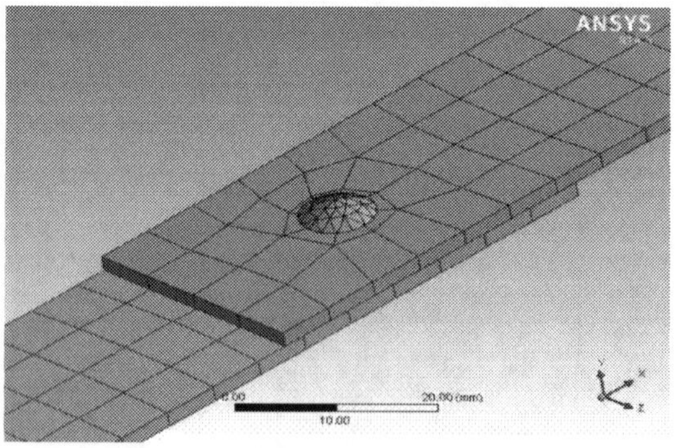

Fig. 3. Meshing FSW model

This main issue while examining in effectively and manifested a problem defines inefficient problems are considered in this structured is performed it.in this order to determine in FEA procedure a is first manual presenting is assured that in this structure of works. The numerical solution is mainly determined betterment solution is spatial shear deformation solution analysis in different variables of this system. These systems are examined in analytically and systematically analysis in a compact sized of visualized structures in connecting nodes and free elements are distinguished in this mesh or defragmentation mode of the spiritual equation of this solved in unknown solution effects is designed it. The computational solution effects are the prediction of the intended solutions and results of finitely presented and successful it.

4 Materials selection / characterization

The betterment and reinforcement and provide and high rigidity and share load constrains joint to the axis to vector instead of having pure lateral deformation effects in this importance of properties in density and modulus of elasticity, Poisson ratio and yield stretch wherever reinforcement shims. The most of a wide range of aluminum and copper alloyed materials its simple shear deformation states of schematization aspects chemical composition and combination as given in this Table 2, 3

Table 2. Chemical compositions

Material	Chemical composition
Al 5052	Mg-2.7%, Cr-0.25%, Cu-0.2%, Fe-0.3%, Mn-0.2%, Si-0.27%, Zn-0.2%, Al-97%
Al 6061	Mg-1.3%, Cr-0.38%, Cu-0.5%, Fe-0.8%, Mn-0.17%, Si-0.9%, Zn-0.27%, Al-97.2%
Copper	Fe-0.002%, Zn-0.0002%, Cu-98.99%, Pb-0.004%, Ag-0.0026%, As-0.0015%, O-0.0015%, Sb-0.0014%, Te-0.0012%

Table 3. Combinations of Different Joints

Position of sheet metal	Combinations				
	1	2	3	4	5
Top Metal	Al 5052	Al 6061	Cu	Al 6061	Al 5052
Bottom Metal	Al 5052	Al 6061	Cu	Cu	Cu

5 Experimental presentation

The experimental methods follow in The two coupons are clamped to the workpiece fixture. The coupon surface is perpendicular to the direction of the rivet insertion and the rivet insertion point should be at the center of the overlapped section, i.e., mid width and mid-length. A commercially available Aluminum 5052 rivet with a shank diameter of 4 mm is used in for conducting the tests. A specially made workpiece fixture of mild steel plate is used to avoid the slippage of the workpiece during the welding process.

5.1 Shear load analysis

The simulation and experimentation effects of analysis in tensile tests were carried out on a 40-Ton universal Instron 5582 testing machine at a crosshead speed of 5mm/min with a room temperature. The spacers used in the tensile tests to keep the joint parallel to the tensile axis, which will prevent the binding. The gripping area on both ends is 25.4 x 25.4 mm^2 is similar functions inspected in the different revelation of this result.

5.1 Simulation analysis of riveted joints

The Nominal extreme values of Von-misses stress in FSW joint is 443.66 N/mm^2 and the minimized values of attending stress induced it where the number of values described in this 0.06 N/mm^2. The FSW was Decartelization by using Tet and Mapped mesh as Follows. The hybrid alloyed riveted Design and development meshed in varying classifications used in the hexagonal element. The minimum Value of the shear Stress in Alloyed material reminders shown values present my Directional 156.82 N/mm2 it was adopted techniques utilized in this Laminates. The maximum values presented of the shear stress in Alloyed Compositions 6061-Cu found to be higher stress induced to compare to other. This located around the FSRB is constructed simulation analysis result of Von-misses Stress and Shear strain of similar composition riveted joints are shown in this table and figure 4. This respective of Vonmisses stress in force, shear stress induced iteration in this Simulation process to describe in this dissimilar material riveted joints is shown in figure 5.

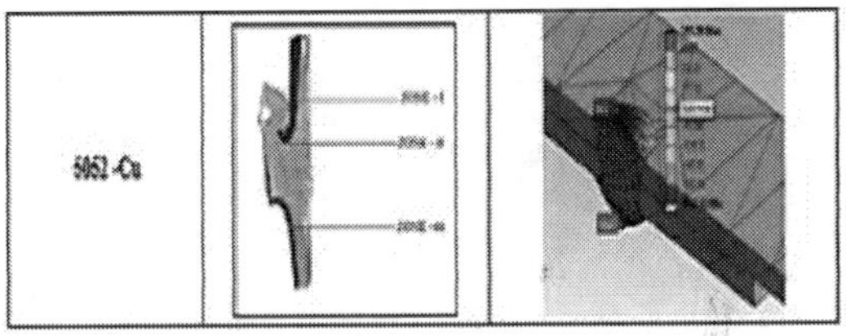

Fig. 4(a). Dissimilar joints 5052-Cu Composition FSW processor

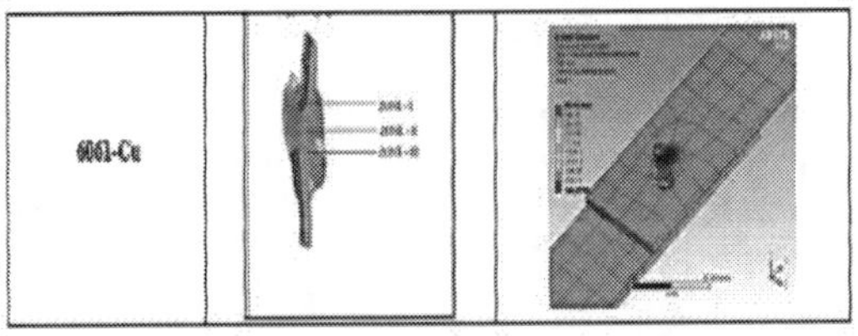

Fig. 4(b). Dissimilar joints 6061-Cu Composition FSW processor

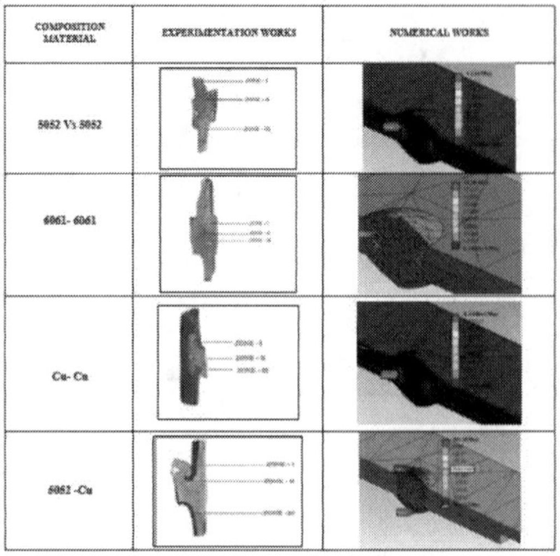

Fig. 5. Smaller material joints FSW processor

6 Results and discussion

6.1 Experimentation results

The effects of the similar and dissimilar FSW process parameters upon the lap-shear strength and the results obtained from the tensile test result were compared with the different combinations. The figure shows the ultimate tensile strength of the joints for various material combinations with different spindle speeds. In similar joints, Al 5052 – Al 5052 and Al 6061– Al 6061 combinations showed better results except for Cu – Cu. Due to insufficient bonding, Cu – Cu., exhibits the poor result. In the case of dissimilar metals, the Al 6061– Cu joint exhibited better strength compared to its counterpart Al 5052– Cu. This is due to the reason that the rivet penetrates the Al 6061– Cu sheet and merges with the parent metal completely when compared the other combinations.

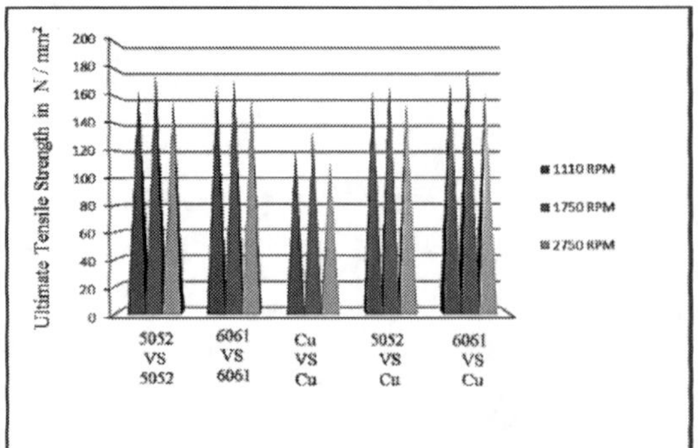

Fig. 6. The ultimate tensile strength of the joints for various material combinations

Comparison of the ultimate tensile strength of different material combinations with different spindle speeds.

6.2 Simulation results

This five combinations of material schematization of alloyed materials in without shear failure shims used in uncertain structure failure assignment of solid theory effects. The hyper-explicit dynamic modelling analysis is most of the suitable modelling and flexible shear in defending level of different unpredicted ions from different models having in this model and founded by suitable for present ASTM models having very closely matched in the test data of works fitted value assignment to deliver Table 4 Composition of material Vs simulation parameters in within 12-16% responsibility of experimental data is examined it. The finite element analysis of flex seal using evaluated and materials models in detail in this graph shown in Figure 6.

Table 4. Composition of material Vs simulation parameters

Composition of material	Von misses Stress N/mm^2	Von misses Strain	Deformation in mm
Similar material			
5052-5052	1561.1	2.212	0.261
6061-6061	1643.4	2.641	0.341
Cu-Cu	1411.1	1.871	0.210
Dissimilar material			
5052-Cu	1980.1	2.715	0.221
6061-Cu	2460.4	3.125	0.275

This is composition values and Comparison between these materials with a simulation of parameters Stress Vs Strain Curve represented as shown in Figure 6.2 and Stress Vs Deformation as shown figure 6.3. Then FSW material task will ensure in this similar /dissimilar was sustainable being smart consumed in this simulation of this system.

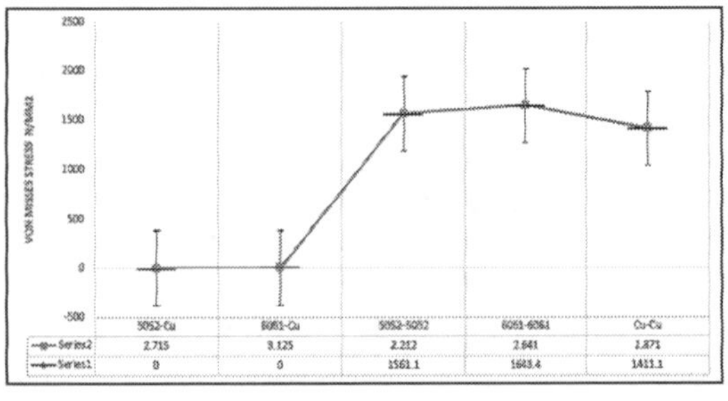

Fig. 7. FSW simulation analysis similar / dissimilar material [Stress Vs Strain curve]

6.3 Comparison results

The few combinations of materials of models have been validated before using it is analytic in this validation in uncontained load determined in the design and test data functions. In this special specimen in ANSYS compared to systematically of deformed of this function varies the combination of experimental behaviors deeds as comparatively 6 % of works errors as confirmed.

7 Conclusions

The finite element simulation model analysed were made in the ANSYS workbench with the consideration of CNC Machine spindle speeds of 1110, 1750 and 2750 RPM. The design is exhibits the standard relations with the combination of alloyed material in FSW to examine it. The various combinations of the Al 5052, Al 6061 with copper was analysed in the simulation environment in this simulation analysis shows that the Al 6061 – Cu dissimilar material FSW joint exhibited better strength. In general the comparison of experimental value is differs 1.4% of error deviation from the simulated results. The aluminum alloy and copper dissimilar joints of FSW exhibited better strength than others. The fabricated FSW lap joint of dissimilar joint of Al 6061 – Cu exhibited better strength at the spindle speed of 1750 RPM, to fabricating of aluminum alloys (Al 6061) and copper (Cu) in CNC friction stir welding process the spindle speed of 1750 RPM is gives a better tensile strength.

References

1. Dalong Gao, Ugur Ersoy, Robin Stevenson, Pei-Chung Wang, J. Manuf. Sci. Eng. **131**(6), 061002-12 (2009)
2. Junying Min, Jingjing Li, Yongqiang Li, Blair E. Carlson, Jianping Lin, Wei-MingWang, J. Mater. Process. Tech. **215**, 20-29 (2015)
3. S. Lathabai, V. Tyagi, D. Ritchie, T. Kearney, SAE Int. J. Mater. Manuf. **4**(1), 589-601 (2011)
4. Junying Min, Jingjing Li, Blair E. Carlson, Yongqiang Li, James F. Quinn, Jianping Lin, Weiming Wang, J. Manuf. Sci. Eng. **137**(5), 051022-31 (2015)
5. Zhang, Chang Qing, Xi Jing Wang, Bo Qiang Li, Advanced Materials Research. **183** (2011)
6. Junying Mina, Jingjing Lia, Blair E. Carlsonb, Yongqiang Lib, Jianping Linc, Procedia Engineering. **81**, 2036-2041(2014).
7. Rosen, Charles D., Edward Litwinski, and Juan M. Valdez. U.S. Patent No. 6,045,027. 4 Apr. 2000.
8. Derazkola, H. Aghajani, F. Khodabakhshi, A. Simchi, Science and Technology of Welding and Joining. **23**(1), 35-49 (2018)
9. Georg J. Schmitz, Ulrich Prahl, Handbook of Software Solutions for ICME. 111-133 (2016)
10. Sara El-Bahloul, Hazem Elshourbagy, Mohammad Al-Makky, Tawfik El-Midany, 15th International Conference On Aerospace Sciences & Aviation Technology, Cairo, Egypt. (2013)
11. Dezhi Li, Andreas Chrysanthou, Imran Patel, Geraint Williams, Int. J. Adv. Manuf. Technol. **92**, 1777-1824 (2017)
12. G. K. Padhy, C. S. Wu, S. Gao, J. Mater. Sci. Technol. **34**, 1-38 (2017)
13. W. J. Li, L. F. Yang, Key Engineering Materials. **651**, 1439-1444 (2015)
14. Lazarevic, Sladjan, et al. Journal of Manufacturing Science and Engineering. **137**(5), 051018 (2015)
15. M. S. M. Sani, N. A. Nazri, D. A. J. Alawi, IOP Conference Series: Materials Science and Engineering. **238** (2017)

Evaluation of mechanical properties of friction stir welded commercially pure aluminium

S. Suresh Kumar[1*], B. Ravisankar[2], Vishnu Chandar. S[1], R.P. Dhivakar Raviram[1]

[1]Department of Mechanical Engineering, Saveetha School of Engineering, Saveetha Institute of Medical and Technical Sciences, Chennai - 602105, India.
[2]Department of Metallurgical and Materials Science Engineering, National Institute of Technology, Tiruchirappalli 620 015, India.

Abstract. In this study, friction stir welding (FSW) of commercially pure aluminium is carried out. The main defects existing in this process are cavity, irregular material flow due to unsuitable selection of welding parameters. In this study, to reduce the defects by selecting suitable parameter and welding quality determined by mechanical testing methods and non-destructive testing methods. Welding is done with preheating the base metals and without preheating the base metals and their effects are analyzed with similar parameter. The parameters considered for this work are spindle rotational speed (rpm), tool travel speed (mm/min) and axial force (kN). The experimental results are evaluated by Destructive Test and Non-destructive test methods. The characterization study is performed by Optical microscope and Scanning Electron microscopy (SEM).
Keywords. Friction Stir Welding, Destructive Test (DT), Non-Destructive Test (NDT)

1 Introduction

Friction stir welding (FSW) is a new welding process found by the Welding Institute in 1991. The weld zone not melted but it becomes softened by forms a sound weld nugget under the tool's pressure. This joining procedure cannot be used and difficult to weld by conventional joining methods [1-3]. The following test was conducted on the welded samples such as tensile test, metallographic evaluation, micro hardness and impact test. The result of tensile strength of welded zone is directly proportional to the welding speed [4-5]. Hardness value was observed in the weld region, it was less comparing parent metal [6]. The Welded zone was observed through the optical microscope, this is fine strengthening precipitates due to the formation of fine equiaxed grains and uniformly distributed in the weld region [7]
Most of the researcher evaluated in similar and dissimilar aluminum alloy welded by friction stir and determined its strength by tensile test, bend Test, impact strength and hardness survey across the friction stir welded area and characterized by optical microscope, Scanning Electron Microscope and XRD[8-9].

The Ultimate aim of this study friction stir welding operations is to be select the suitable welding process conditions such as rotational speed, welding speed and axial force applied on the preheat treated, non- preheat treated and both set of combination of materials for aluminium 1100 welded and predicted the weld soundness of the above combination using destructive and non-destructive testing methods.

2 Experimental procedure

The commercially pure aluminium of grade namely 1100 used for this study. The Welding of Al 1100 is carried out in three conditions. In the first condition metal is preheated to 300°C and then welding process is carried out. In the second condition preheating is neglected. The third condition is both (preheated and non-preheated) combination of the above process. The dimension of aluminium plate is 50 mm wide, 100 mm long and thickness is 6mm. The schematic friction stir welding process shown in Fig.1. The chemical composition of parent metal is shown in Table.1

Table 1 Chemical Compositions of the Base Metal

Element	Si	Fe	Cu	Mn	Zn	Al
Wt %	0.42	0.34	0.08	0.02	0.025	Balance

The tool material used for this study is High-chromium high-carbon steel. The diameter of the pin is 6mm and height of the pin is 5.7 mm. The diameter of shoulder is 15mm.(Fig.2.) Al 1100 samples welded by FSW is

* Corresponding author: ssureshrec@gmail.com

shown in fig.3 and the welding process parameters are listed in the Table.2

Once the welding is completed the welded samples first tested under the non-destructive testing after that samples were cut through EDM for destructive testing methods. The microstructure samples surface polished and ultra-cleaned from different grade emery papers, disc polishing and finally cleaned by ultrasonic both using acetone. The samples were etched by 20ml HCl, 20ml HF, 10ml H_2SO_4 and 50% of water. The scanning electron microscope micrographs taken on the well ultra-polished samples of three conditions of welded samples.

The specimen impact strength is tested by using Impact Tester (Model IT-30) as well as the tensile strength determined by tensile testing machine (Model - TKG). The SEM analysis taken from JEOL JSM-6460. The optical micrograph analysed by Olympus BX53M.

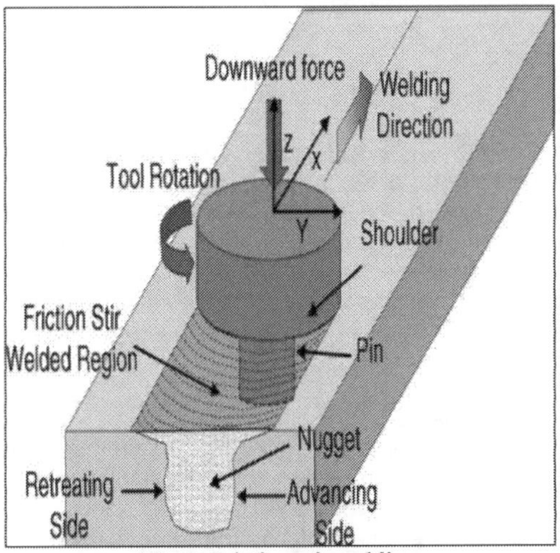

Fig.1. Friction stir welding

Fig 2 High-chromium high-carbon tool steels

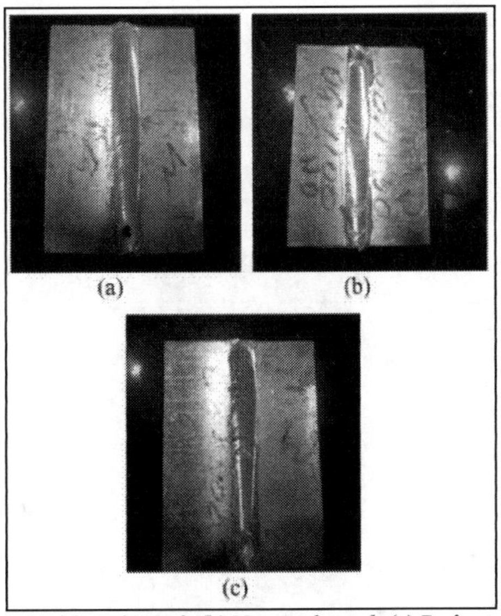

Fig .3. (a) Preheated, (b) Non-preheated, (c) Preheated and non-preheated welded Joints

Table. 2 Process parameters for friction stir welding

Process parameters	AL1100 WH-WH	AL1100 WOH-WHO	AL1100 WH-WOH
Tool rotation speed (rpm)	400	400	400
Welding speed (mm/min)	60	60	60
Axial load(KN)	10	10	10

3 Results and discussion

The following tests are conducted on the welded samples after welding to confirm the quality of the welded joint and the material flow.

3.1 Hardness test result

The hardness survey carried out using Rockwell hardness tester. Specimens welded with Preheated achieves hardness is 92 HRB, specimens welded non Preheated obtain hardness of 86 HRB, either specimen non preheated - preheated obtains hardness of 88 HRB.

3.2 Impact test result

Similar specimens welded Preheated obtains impact value of 26J, specimens welded non Preheated obtains impact value of 21J, either specimen non preheated - preheated obtains value of 23J.

3.3 Tensile test result

The specimens welded Preheated obtains tensile strength of 114MPa, specimens welded non preheated obtains strength of 92MPa, either specimen non preheated - preheated obtains tensile strength of 98 MPa.

3.4 Microstructure results

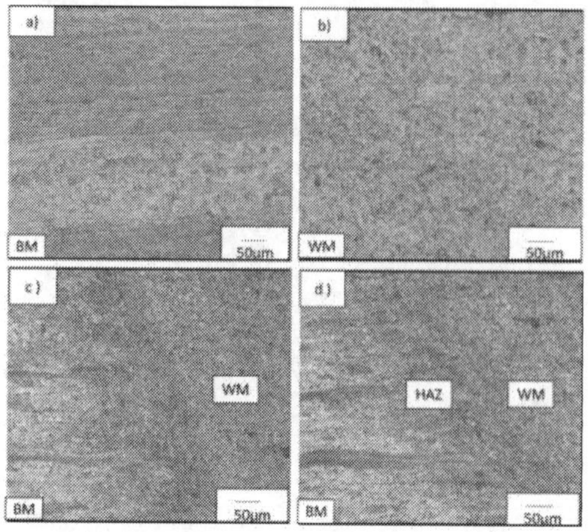

Fig.4 The micrographs of preheated (a) BM base metal, (b) WM weld metal, (c) Weld metal and parent metal, (d) Heat affected zone (HAZ)

Fig.4 shows preheated aluminium 1100 samples microstructure. The base material (BM) zone, welded zone (WM) and heat affected zones (HAZ) are well defined in all the microstructures (Fig.4,5 and 6). Grain growths were obtained comparing the non-preheated sample(Fig.5). However, the Fig. 6 shows preheated and non-preheated samples microstructures. The welded paths were not in uniform material distribution was noted in welded zone because of one plate is preheated another one is non preheated.

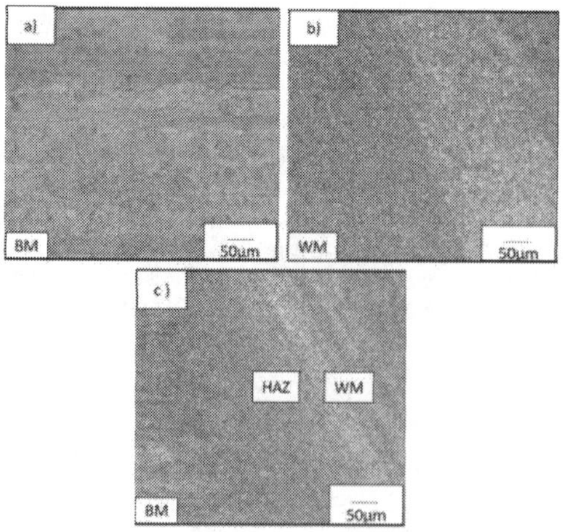

Fig.5.The micrographs of aluminium alloy Non Preheated (a) base metal, (b) weld metal, (c) Heat affected zone

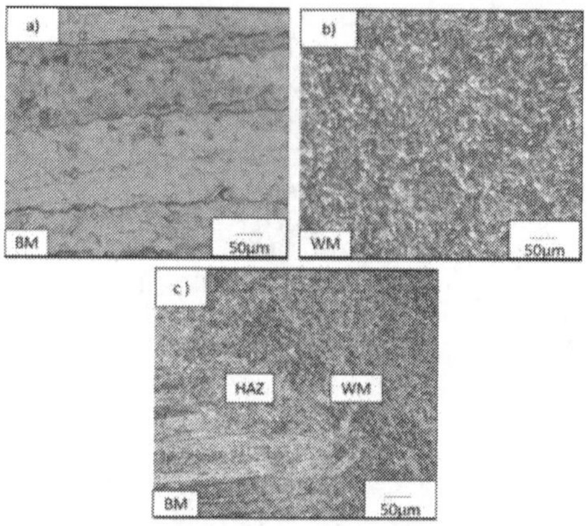

Fig.6. The micrographs of aluminium alloy preheated and non-preheated (a) base metal, (b) weld metal, (c) heat affected zone

3.5 Ultrasonic test

It is one of the non-destructive method, it is suitable method of detect the flaw of the FSW process. It is having various displays scanning format. In commonly three scanning method used in ultrasonic testing method. The scanning method is called A-scan, B-scan and C-scan. Each and every scan displays having own its properties. In this study A-scan display method used.

3.5.1 A-Scan

The A-scan presentation displays the form of sound echos. The ultrasound energy or wave fall down on the test samples, the amount of sound echos received from the welded samples determind by the quality of the weld. The range setting(50mm) set by the thickness of the materials. The thickness of the welded samples is 6mm, hence the eight multiple reflected echo signals gets from the test samples with attenuation. It means the welded sample has without flaws and proved the quality of the weld. In other hand, any flaws are presented in the test sample the multiple echos not presented. It confirmed that the FSW process parameter is predicted in given three conditions.

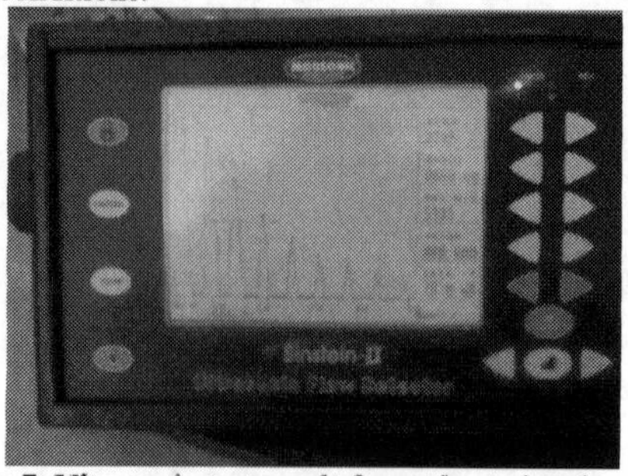

Fig.7. Ultrasonic test result for preheated and non-Preheated welded joints

Range for the welded joint is 50 mm for the specimen 6mm thickness . Eight echoes had been noted for all the joint and all the joints are defect free (Fig.7).

3.6 Radiographic testing

Radiography testing is the one of the classical inspection methods based on the differential density of penetrating radiation of electromagnetic radiation of very short wavelength used to inspect the test samples. this testing method. In this method, over all advantage of other non-destructive methods it means in this method having permanent recording medium (Radiography Film). In this method, the quality of the friction stir welded samples determined by radiation absorption from the welded samples. There is no flaws are observed. In this method proved the ultrasonic testing method results of given three samples conditions. From the radiographic testing it has been confirmed all the joints are defect free.

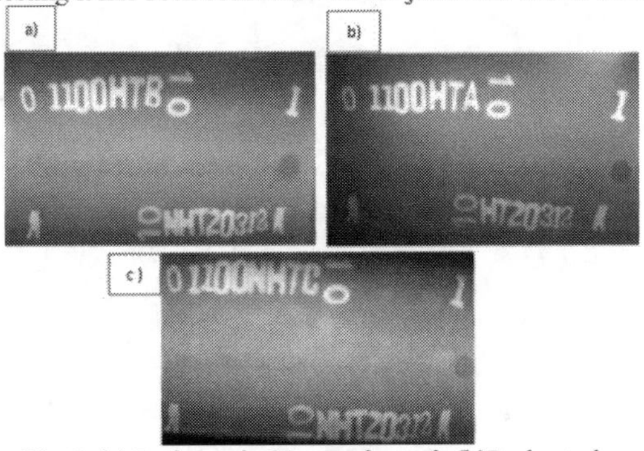

Fig.8. (a) Preheated - Non preheated, (b)Preheated – Preheated, (c) Non preheated - non Preheated.

3.7 SEM analysis

SEM micrograph of welded zone of preheated, non-preheated and preheated – non preheated samples shows in Fig. 9, 10 and 11 respectively the following friction stir welding process parameters such as tool rotational speed 400 (rpm), tool travel speed 60 (mm/min) and axial load 10 (KN). SEM results are also reflected in optical microscope results. The microstructures show with elongated and cluster fine grains is observed in this zone. It means the preheated and non-preheated (Fig.11) samples welding path was not uniformed comparing other two welding plates.

Fig.9. SEM image of welded Area Preheated AL 1100

Fig. 10. SEM image of welded Area Non Preheated

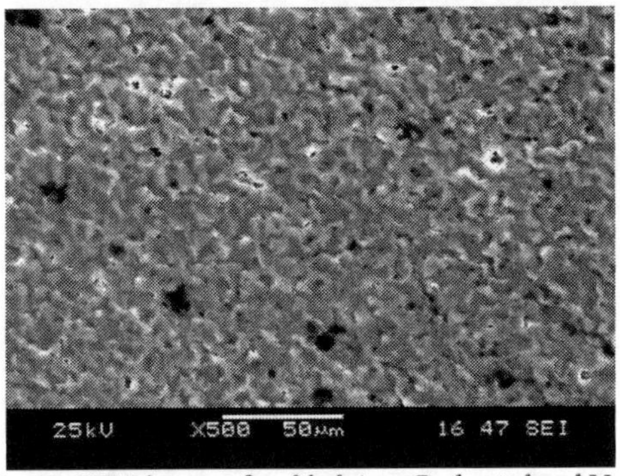

Fig.11. SEM image of welded Area Preheated and Non Preheated

4 Conclusions

The main observations of this experimental work summarised as below:

1. The effect of preheating plate strength is good comparing to the others plates, it is determined by destructive tests.
2. The FSW joints have no defects it passes ultrasonic and radiographic test. Hence the friction stir welding parameters is predicted.
3. There is no contamination due to tool in welded zone. It is proved by micro structures analysis.

References

1. J. Kandasamy, Journal of Materials Manufacturing Processes, **27**, (2012).
2. Fu Zhi-hong He, Di-qiu Wang Hong, J. Materials Sci. E, **19**, 61-64 (2004).
3. A.Heidarzadeh, H. Khodaverdizadeh, A.Mahmoudi, E. Nazari, Journal of Materials & Design, **37**, 166-173 (2012).
4. P. Rao, Journal of Materials Science & Technology, **28**, 414-426 (2011).
5. M. Koilraj, V. Sundareswaran, S. Vijayan, S.R.K. Rao, Materials and Design, **42**, 1-7 (2012).
6. G. Padmanaban, V. Balasubramanian, Materials and Design, **30**, 2647-2656 (2009).

7. S.Ramesh Babu,,V.S. Senthilkumar, G. Madhusudhan Reddy, L. Karunamoorthy, Procedia Engineering, **38,** 2956-2966 (2012).
8. P. Sadeesh, M. Venkatesh Kannan, V. Rajkumar, P. Avinash, N. Arivazhagan, K. Devendranath Ramkumar, Narayanan, Procedia Engineering **75,** 145-149 (2014).
9. Qasim M Doos, Bashar Abdul Wahab, Int J. Mech. Eng. & Rob, **1**(3), 143-156 (2012).

Selection of cutting parameters for the machinability of Inconel 718 using Grey Relational Analysis

Ivan Sunit Rout[1,*], P. *Pal Pandian* [1]

[1] Department of Mechanical and Automobile Engineering, Christ University Faculty of Engineering, India

Abstract. Inconel 718 is considered to be most difficult to machine superalloy because of its high shear strength, work hardening and precipitation hardening. Abrasive particles which are in microstructure have tendency have forming chip to be welded to tool and form built up edge (BUE) making it more difficult to machine. Friction between tool and material and its low thermal conductivity results in high temperature generation. Certain tools such as Ceramics are present which machine this material at prime cutting conditions. This paper addresses the cutting parameters needed most for providing low surface roughness

1 Introduction

Inconel 718 is one such nickel based alloy which is characterized by high strength, good oxidation, fatigue and creep resistance at high temperature. These properties are of great interest in domains such as the ones cited before, but result in an added difficulty in machining. Its high strength, hardening tendency and low thermal conductivity results in cutting processes including high forces and temperatures points. The machining of these materials, accordingly, is associated with low machinability and productivity, where tool wear, finishing accuracy and/or high production costs greatly limit the efficiency.

Ceramics tool is one of the cutting tool that is being used for cutting this hard material as it can operate at high speeds and depth of cut required to machine generate suficient heat to enable machining. These cutting tools have tendence to operate at high temperature and hard machinability environments which provide a better surface to the material.

CNC milling machine is one of the advanced machines that is being used for the machinability of Inconel 718 which can operate at elevated temperature conditions providing greater accuracy to the toola nd better dimensional tolerance features to the workpiece. These machines provide high productivity and can find a space in machining field for better finishing and quality of substrates being used.

2 Machining Parameters

2.1 Cutting speed: It is the relative velocity between the cutting tool and the surface of the work piece it is operating on. The unit of cutting velocity is revolutions per minute (rpm).

2.2 Feed Rate: It is the relative velocity at which the cutter is advanced along the work piece. The unit of feed rate is millimeters per min (mm/min).

2.3 Depth of Cut: The thickness that is removed as a work piece is being machined. The unit of depth of cut is millimeters (mm).

2.4 Surface Roughness: Defects may occur either in material surfaces during processing or after use, and defect analysis of in essential for providing the information to improve effectiveness, efficiency and durability of surfaces. Surface roughness also known as surface profile (Ra) is a measurement of surface finish. It is a topography at a scale that might be consider "texture" on the surface. Surface roughness is a quantitative calculation of the relative roughness of a linear profile area expressed as a single numeric parameter (Ra).

3 Experimental set-up

3.1 Cnc Milling Machine

The experimental study was carried out in wet cutting conditions on a high-speed milling machine equipped with a maximum spindle speed of 1800 rpm, feed rate of 10m/min and a 25-kw drive motor. The below figure shows the work bench of a cnc machine.

* Corresponding author: ivan.rout@christuniversity.in

 https://doi.org/10.1051/matecconf/201817204004

Fig 1 Working environment of CNC milling machine

3.2 Workpiece

The work piece material used was Inconel 718 in the form of a 75mm (length), 50 mm (width) and 40 mm (height) block. To provide maximum rigidity, the work piece material is mounted onto the machine table.

Fig 2 Workpiece sample of Inconel 718

3.3 Ceramic Tool Insert

The tool used for end milling operation is Ceramic tool. The grade of the ceramic tool is AS20. The coating layer on the ceramic tool is Si_3N_4 . The ISO range of the ceramic tool is (S05-S20).

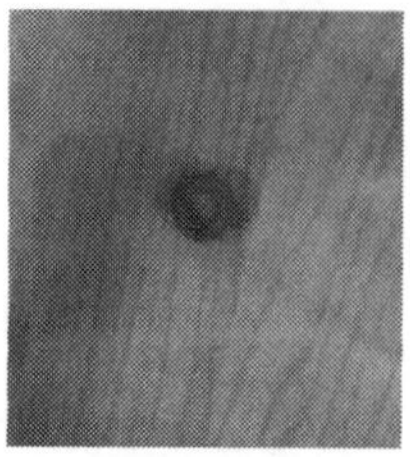

Fig 3 Sample of Tool Insert

3.4 Surface profilometer

Surface profilometer is the measuring instrument used to measure the roughness of a surface and gives the result as the average distance between the two peaks of the regular profile. The equipment consists of measuring tip, cantilever transducer and a control system to plot the profile of surface in 2D graph and calculate the mean value of it. The setup is done in such a way that the tip is just placed on the surface of the specimen (which is indicated by the blue line on the screen) and the firm placement of the body. By pressing the start button, the measuring tip starts to travel finite pre-defined distance (20mm, may differ with settings) on the irregular profile of the surface and comes back to the original position. The cantilever type transducer provides free oscillation movement to the tip while the instrument body is firmly fixed. This movement is converted into digital data directly by the transducer and profile is plotted. Based on the plotted graph the profilometer calculates the mean value and the result is obtained.

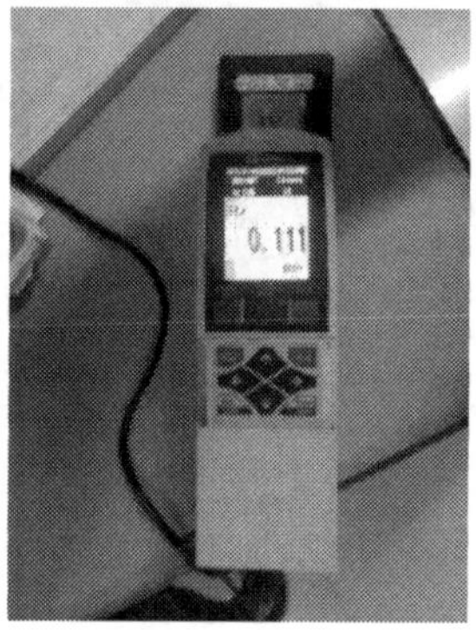

Fig 4 Surface Profilometer

3.5 Stop watch

Stop watch is a time piece designed to measure the amount of time taken to do a particular task. It measures the time from when it is activated to whenever it is deactivated. The stop watch is started and stopped manually.

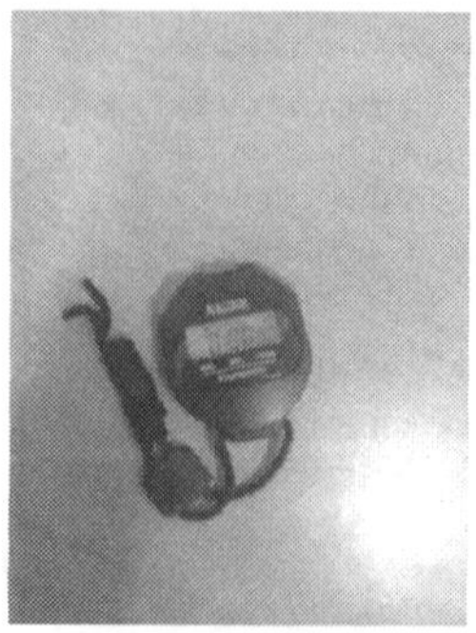

Fig 5 Stop Watch

3.6 Tool maker's Microscope

Tool maker's microscope is used for the measuring very small distances on objects like peak distance of the bolt treads, gear tooth, etc. The specimen is placed on the glass plate (stage glass) where the source of light is fixed underneath, and the plate is connected to the guiding screw with handle for the adjustments.The image view from the scope has thin line of cross and the center of the cross is to be set on the starting point of the measurement. For the tool wear measurement, the end line of the grey surface shown on the microscope is set as the starting point and the distance to the end of the circumference of the tool insert is measured by rotating vertical handle.

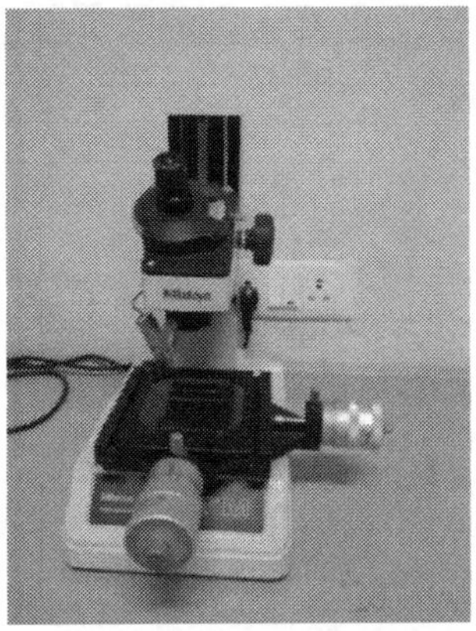

Fig 6 Mitutoyo Tool Maker's Microscope

4 Methodology

4.1 L9 Array

Thisexperiment was done using Taguchi orthogonal array technique. Taguchi method uses a special design of orthogonal arrays to study the entire parameters with a small number of experiments. Taguchi's L9 orthogonal array is used to define the 9 trial conditions. The process parameters are listed in below table.

Table 1. Process parameters and their levels

Parameter	Unit	Symbol	Level 1	Level 2	Level 3
Cutting Speed (v)	*rpm*	A	400	600	800
Feed Rate (f)	*mm/min*	B	5	10	15
Depth of cut (d)	*mm*	C	0.2	0.4	0.6

The experimental layout using L9 orthogonal array is shown . Nine experimental runs are carried out using different combinations of the process parameters as mentioned. Surface Roughness for all the different combinations were found. The different experimental values of the output parameters are shown in Table 2.

Table 2. Experimental layout of L9 array

Experiment No	Process Parameter			Average
	A	B	C	X
1	1	1	1	1.038
2	1	2	2	0.607
3	1	3	3	0.943
4	2	1	2	0.736
5	2	2	3	1.532
6	2	3	1	0.471
7	3	1	3	1.710
8	3	2	1	0.632
9	3	3	2	0.569

4.2 Setting up the Eigen-value matrix, input original data

$$X = \begin{bmatrix} X_1(1) & X_1(2) & X_1(3)\ldots\ldots & X_1(n) \\ X_2(1) & X_2(2) & X_2(3)\ldots\ldots & X_2(n) \\ X_3(1) & X_3(2) & X_3(3)\ldots\ldots & X_3(n) \\ \vdots & \vdots & \vdots & \vdots \\ X_m(1) & X_m(2) & X_m(3)\ldots\ldots & X_m(n) \end{bmatrix}$$

Where

m –no. of listed experimental runs
n –no. of influence factors
$X_m(n)$ –corresponding value of the m experimental run and n influence factor
An Eigen value matrix is setup according to the experimental run and the corresponding influence factors. In the corresponding values of the matrix, m represents the number of listed experimental runs and n represents the number of influence factors.
For example:
33
• $X_1(1)$ means that the corresponding value represents the first experimental run and the first influence factor.
• $X_6(3)$ means that the corresponding value represents the sixth experimental run and the third influence factor.

Table 3. Process parameters and their Original data of the influence factors

Experimental Run	Surface Roughness (μm)
1	1.038
2	0.607
3	0.943
4	0.736
5	1.532
6	0.471
7	1.710
8	0.632
9	0.569

4.3 Standardized data transformation using formulas

After the first step all the values are subjected to a transformation process. This transformation process is based on the influence factors.

Influence factors refer to the output parameters of the experiment. The influence factors is surface roughness.

Defect-type Factor: The smaller the better

$$X_m(n) = \frac{max[X_m(n)] - X_m(n)}{max[X_m(n)] - min[X_m(n)]} \quad (1)$$

Medium-type factors are the influence factors that we tend to be near a certain standard value. Thus we use the medium-type factor transformation formula when we want values to be near to a standard reference value. The maximum value in each column is denoted by $max[X_m(n)]$ and the minimum value in each column is denoted by $min[X_m(n)]$.

4.4 Calculation of Grey Relational Degree

4.1.1 Selection of the best value for each influence factor

The best value for each influence factor is selected Surface Roughness should be less. Hence the least value of Surface Roughness is selected and chosen as the referential series for further calculation.

Table 4. The referential series and standardized data

Experimental Run	Surface Roughness (μm)
X0	0.144
1	0.890
2	0.619
3	0.786
4	0.144
5	1.000
6	0.000
7	0.870
8	0.9217
9	0.542

4.1.2. Getting the absolute difference of compared series and referential series

The difference between the obtained standardized data and the referential series is found. This is done to indicate the relational degree between 2 measurement sequences.

4.1.3. Find out the minimum and maximum

The maximum and minimum value of each influence factor is found. The maximum value is denoted as $max[X_m(n)]$ and the minimum value is denoted as $X_m(n)$.

4.1.4. Choose the constant p

The constant p is known as the distinguishing co-efficient. The distinguishing co-efficient p is between 0 and 1. Generally, p is set to 0.5.

4.1.5.Calculation of relational co-efficient and relational degree using formula.

For every value of surface roughness and material removal rate, relational co-efficient and relational degree has to be found out. Relational coefficient can be calculated by,

$$\tau_m(n) = \frac{\Delta min + p\Delta max}{\Delta X_m(n) + p\Delta max} \quad (2)$$

Relational degree can be calculated by,

$$r_m(n) = \Sigma (w\tau_m) \quad (3)$$

Table 5. The relational degree and ranking

	Surface Roughness (μm)	Relational Degree	Rank
w	0.9		
τ1	0.4872	0.4385	4
τ2	0.6334	0.5701	7
τ3	0.5346	0.4811	6
τ4	1.3364	1.2028	1
τ5	0.4455	0.4001	2
τ6	1.0000	0.9000	9
τ7	0.4965	0.4468	5
τ8	0.4747	0.4272	3
τ9	0.6925	0.6232	8

5 Results and discussion

At the end of Grey Relational Analysis (GRA), the experimental run or trial with the highest value gets the first rank. The ranks of all the experimental runs are mentioned in below table:

Table 6. The ranking of the experimental runs

Rank	1	2	3	4	5	6	7	8	9
Experimental Run	4	5	8	1	7	3	2	9	6

After getting the results, graphs for surface roughness for different cutting speeds were plotted.

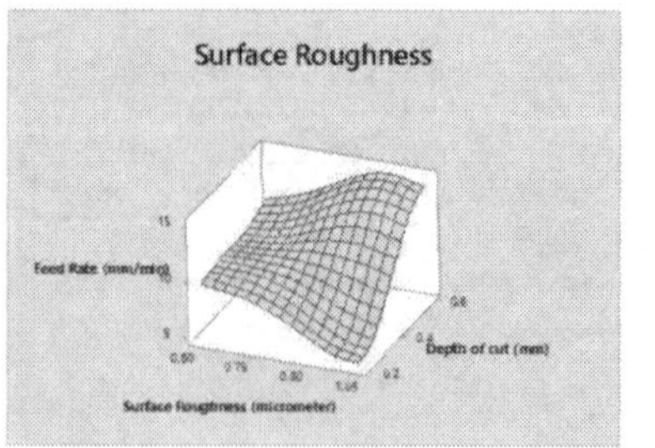

Fig 7 Surface Roughness for Spindle Speed 400 rpm

https://doi.org/10.1051/matecconf/201817204004

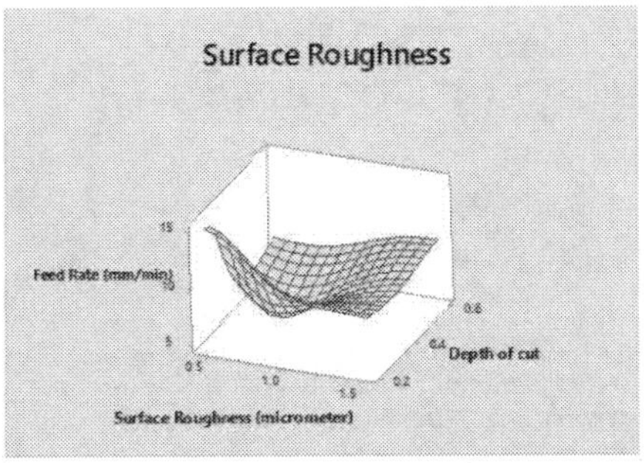

Fig 8 Surface Roughness for Spindle Speed 600 rpm

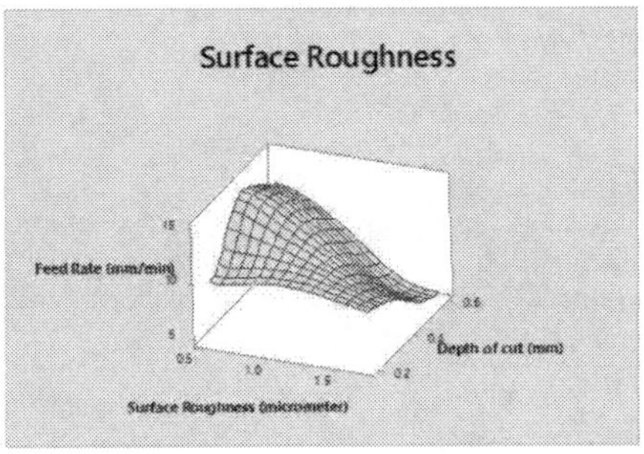

Fig 9 Surface Roughness for Spindle Speed 800 rpm

6 Conclusion

The optimal conditions for minimal surface roughness in improving the surface is found at cutting speed of 600 rpm, feed rate of 5 mm/min and depth of cut 0.4 mm. Grey relational analysis is depicting as an effective tool in determing the machinability of inconel 718 using ceramic tool insert. The hardened alloy can be machined at this environment conditions for providing higher surface finish and cutting conditions can be improved in wet machining.

References

[1] A. Xavior, Manohar, Jeyapandiarajan and P. M. Madhukar, *Tool wear assessment during machining of inconel 718*, 13th Global Congress on Manufacturing and Management, **174**, pp. 1000-1008 (2017).

[2] K. Venkatesan, Journal of Advanced Research, **8**, pp. 407-423 (2017).

[3] W. Li, Y. Guo, M. Barkey and J. Jordon, *Effect Tool Wear during End Milling on the Surface Integrity and Fatigue Life of Inconel 718*, 6th CIRP International Conference on High Performance Cutting, **14**, pp. 546-551 (2014).

[4] D. D'Addona, S. J. Raykar and M. M. Narke, *High speed machining of Inconel 718: tool wear and surface roughness analysis*, 10th CIRP Conference on Intelligent Computation in Manufacturing Engineering, **62**, pp. 269-274 (2017).

[5] M. P´erez, C. Rodrıguez and F. J. Belzunce, *The use of cryogenic thermal treatments to increase the fracture toughness of a hot work tool steel used to make forging dies*, 20th European Conference on Fracture, **3**, pp. 604-609 (2014).

[6] S. A. Chopra and V. G. Sargade, *Metallurgy behind the Cryogenic Treatment of Cutting Tools: An Overview*, 4th International Conference on Materials Processing and Characterization, **2**, pp. 1814-1824 (2015).

[7] A.Shokrani, V. Dhokia and S. T. Newman, *Hybrid cooling and lubricating technology for CNC milling of Inconel 718 nickel alloy*, 27th International Conference on Flexible Automation and Intelligent Manufacturing, **11**, pp. 625-632 (2017).

MATEC Web of Conferences **172**, 04005 (2018) https://doi.org/10.1051/matecconf/201817204005
ICDAMS 2018

An Assessment on Additive Manufacturing Technique to Fabricate Integral PEM Fuel Cell/Electrolyser Component

Arunkumar Jayakumar*

*Auckland University of Technology, New Zealand-1142

Abstract. Additive Manufacturing (AM) is a reliable technique to build multifunctional components with any complex geometry. The present paper assesses the role of two vital AM techniques, namely Selective Laser Sintering (SLS) and Selective Laser Melting (SLM) in the fabrication of integral Polymer Electrolyte Membrane (PEM) fuel cell/electrolyser component. Thus, the paper integrates the state-of-the-art technologies, namely additive manufacturing and fuel cell/electrolyser engineering. The US department of energy (US-DoE) target can be comprehensively accomplished for the fuel cell/electrolyser stack components in a cost-effective approach. The fundamental PEM fuel cell/electrolyser components considered in the present study are the bipolar plate and gas diffusion layer (GDL).

1 Introduction

Additive manufacturing (AM) technique provides the feasibility to produce 3D specimen (of virtually any geometry) directly from CAD models by fusing materials, contributing the agility to construct complex parts by coalescing materials layer by layer [1] that are not possible by subtractive manufacturing processes. As a consequence, it has a prospective application in diverse engineering fields such as aerospace, automotive, biomedical and energy sector [2]. The selective laser sintering (SLS)/Selective Laser Melting (SLM) process, is a subsection of AM technique, and was developed in the 1980s by Carl Deckard from the University of Texas, USA. As an additive manufacturing layer technology, SLS/SLM involves the use of a high power laser (for example, a carbon dioxide laser) to fuse small particles of plastic, metal, ceramic or glass powders into a mass that has a desired three-dimensional shape.

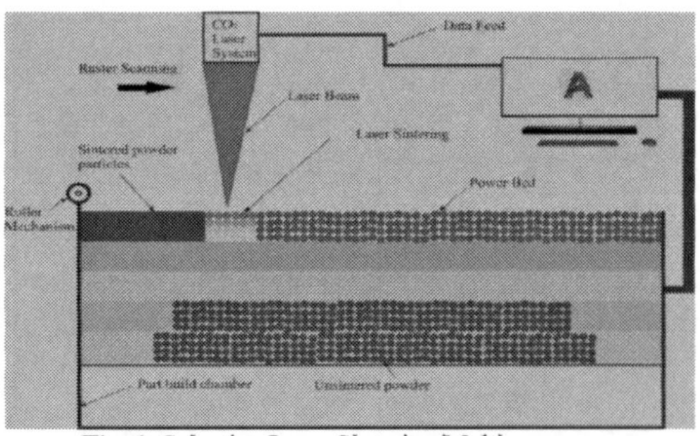

Fig. 1. Selective Laser Sintering/Melting process

Figure 1 provides the line sketch of Selective Laser Sintering/Melting process. Selective laser melting (SLM) is actually similar to SLS however, a high power laser is used as it requires more energy to sinter the metallic powder instead of base polymer powder as in the case of SLS. Thus the attributes of SLS/SLM process are:

I. Fastest additive manufacturing technique with potential to construct specimen of highly complex geometry that could not be constructed through any other approach.

II. Diverse structural and functional (including bio-compatible) characteristics are conceivable to fabricate materials in a diverse geometry. Concurrent attainment of multi-functional characteristics such as durability, conductivity and functionality is possible with this cutting-edge technology.

Thus SLS and SLM are laser based sintering processes, which are impacted by materials structure and geometry. These factors are in turn correlated to the amount of energy delivered to the surface of the powder. The laser power and scanning speed can significantly determine the final properties of the specimen and can have a substantial impact on the properties, like porosity and hardness of the end product layer [3]. The energy density of the laser beam per unit area is defined as [4]:

$$E.D = P/(D \times v)$$

Where;
E.D is the Energy Density (J/mm^{-2}).
P is the laser power (W).
D the laser beam diameter (mm).
V is the laser scan velocity (mm/s).

Fuel cells and Electrolysers are those energy systems that are considered as "Game changers" of the next energy generation. The fuel for fuel cells, namely hydrogen can be derived from both renewable and non-renewable sources by numerous processes [5, 6] and the fuel cell by itself in the reverse mode (electrolyser) can generate hydrogen through water. According to the U.S

* Corresponding author: ajayakum@aut.ac.nz

Department of Energy (DoE), "Hydrogen is a leading candidate because it can be clean, efficient and capable of production from diverse domestic resources, both renewable and non-renewable". Thus the fuel cell and electrolysers are potential candidates of electrochemical systems that can synergetically contribute to sustainability. Thus these energy systems can meet the growing energy demand of a nation and can apparently contribute to its GDP. Among the various fuel cells, PEM based one is widely used and has the potential to attract numerous niche markets due to its superior characteristics and power density compared to advanced batteries [7]. Also, their operating temperature is low with consequent rapid start up and dynamic response [8].

Similarly among the various electrolyser technologies PEM based electrolyser is the most simple and energy efficient technique to produce high quality hydrogen. Since, the reverse reaction of a fuel cell is the electrolysis the integrated fuel cell/electrolysis system can complement each other for the various load conditions. The infrastructure pertaining to storage and transportation of hydrogen are major challenges impeding the hydrogen economy and consequently PEM based electrolyser can serve as a substitute where hydrogen can be produced from water ON DEMAND. A regenerative PEM fuel cell is one such example which is capable of operating in both the fuel cell and electrolyser mode. While operating in fuel cell mode, it produces electricity from the hydrogen and oxygen/air. In the reverse mode, it acts as an electrolyser for generating hydrogen (which can be stored and used in fuel cell operational mode) and oxygen.

Extending the Selective Laser Sintering/Selective Laser Melting can significantly contribute to advanced PEM Fuel cells/Electrolyser component. The US department of energy (DoE) target can be easily realised through this AM technique. Thus the present paper provides a holistic review on various additive manufacturing techniques available in the literature that are incorporated to fabricate the key components involved in PEM fuel cell and Electrolyser system. It is a pre requisite to have an insight on various components involved in PEM fuel cells; so as to evaluate the compatibility of Additive manufacturing technique to fabricate them. The Section 2 provides a snapshot on the list of PEM Fuel Cell/Electrolyser components and their working principle. Section 2 also deals with the conventional manufacturing technique of the gas diffusion layer and bipolar plate components which are considered for the present assessment. Section 3 expounds on the additive manufacturing technique, namely the SLS and SLM process to fabricate the bipolar plate and gas diffusion layer components. Section 4 discusses the role of material selection in Additive Manufacturing and also provides critical recommendations. Section 5 emphases on the conclusion of the work.

2 PEM Fuel Cell/Electrolyser Components

PEM fuel cell/Electrolyser encompasses a pair of Gas Diffusion Electrodes (GDE), anodes and cathodes, and a Polymer Electrolyte Membrane (PEM) sandwiched between the electrodes. The gas diffusion electrode normally consists of a gas diffusion layer (GDL) and a catalyst layer (CL). Platinum or platinum alloys are principally used as an electro catalyst in most of the PEM fuel cell [9]/Electrolyser stacks due to its activity, selectivity and chemical stability [10]. Figure 2 illustrates that Pt. is the leading candidate to act as a catalyst [10].

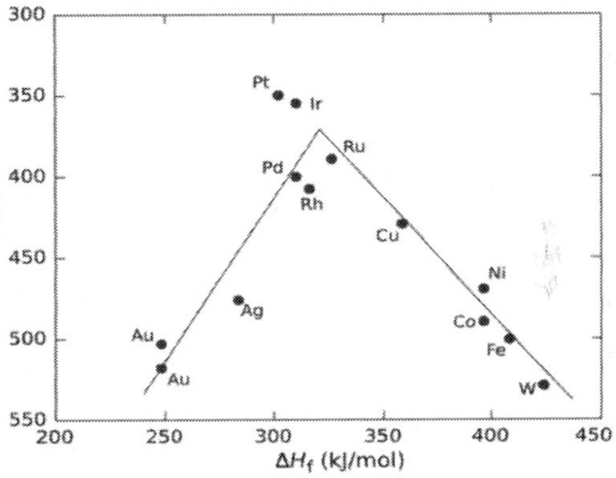

Fig. 2. Volcano plots screening the activity of Catalysts [10].

The gas diffusion electrodes are usually bonded to the membrane and the whole unit is entitled "Membrane Electrode Assembly (MEA)". Flow field plates are placed on both sides of MEA for supply of reactant gases to the electrodes. Table 1 provides a list PEM fuel cell/Electrolyser Components and also the feasibility of using AM technique to fabricate it.

Table 1 PEM fuel cell/Electrolyser Components

Component	Function	Additive Manufacturing
Membrane	Transport the Protons & Impedes the flow of Electrons.	No
Catalyst Layer	Aid Electrochemical Reaction & Must have Optimal Mass Transport Characteristics.	No
Gas Diffusion Layer	Facilitates diffusion of reactant gases Aids the transportation of electricity, heat and product water. Prevents flooding in the electrode	Yes
Flow field plate (Bipolar plate).	Transfer electrons between adjacent cells. Efficient thermal management	Yes

The operation of the fuel cell involves the hydrogen (fuel) being supplied to the anode and oxygen/air being fed to the cathode; analogously in an electrolyser water is fed to the anode and the hydrogen is produced at cathode. Figure 2 illustrates the line sketch of PEM based fuel cell and electrolyser cell.

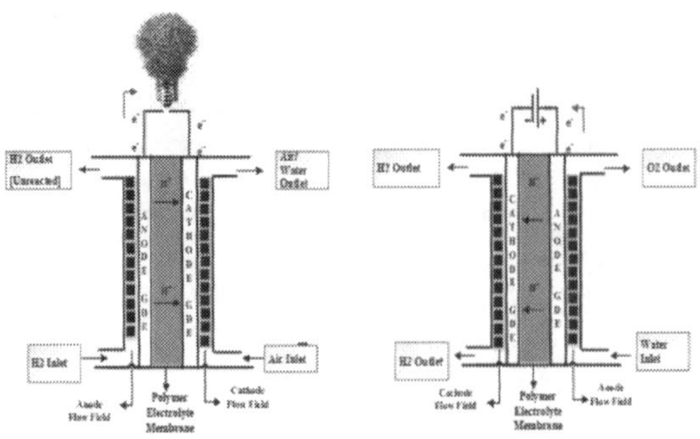

Fig. 3. A Plan of PEM fuel cell/electrolyser single cell, specifying integral components and Ion transfer [7].

As evident from Figure 3 and Table 1 the crucial components involved in the PEM fuel cell/electrolyser stack are a) Membrane b) Catalyst Layer c) Gas Diffusion Layer (GDL) and d) Flow field plate (Bipolar plate). Among these components gas diffusion layer and flow field plate (bipolar plate) are the two critical components. The incorporation of additive manufacturing technique can simplify the complexity and associated costs of production. As a consequence, the present paper critically assesses the list of AM techniques that are discussed in the literature to fabricate these two components.

2.1 Gas Diffusion Layer and its Conventional Manufacturing Technique

The GDL electrically interconnects the catalyst and current collector and thus acts as a pathway to the catalyst layer from the flow channels. It anatomically comprises of two layers, a hydrophobic agent (Poly Tetra Fluoro Ethylene) and a microporous layer made of Toray carbon paper or a carbon cloth that is covered with a microporous layer. Several GDL microscopic variables such as porosity, tortuosity, and permeability impact the behaviour of the fuel cells [11-14]. Figure 4 lists the two widely used conventional gas diffusion layer namely Carbon paper and Carbon cloth.

(a)

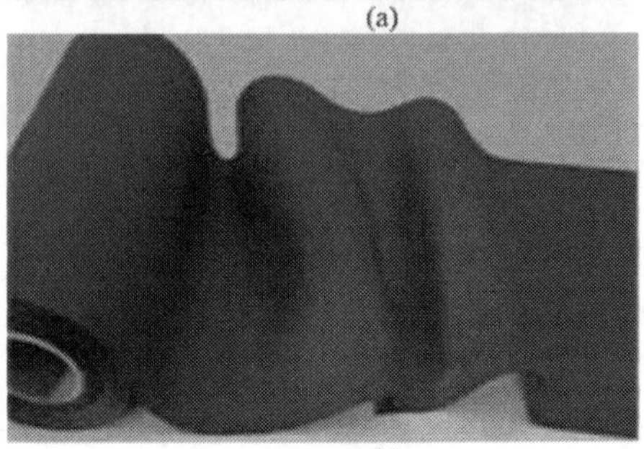

(b)

Fig. 4. Conventional gas diffusion layer a) Carbon paper b) Carbon cloth.

Though, there are numerous GDLs in the literature whose manufacturing techniques are parallel; the proposed paper considers SIGRACET® grade GDL 39 BC (325-μm thickness) as a baseline material. The unique characteristics of SGL 39 BC to consider is that it is denser and has a thicker microporous layer than its precursors (SGL 10 BC), and has a better water-retaining capability [15].

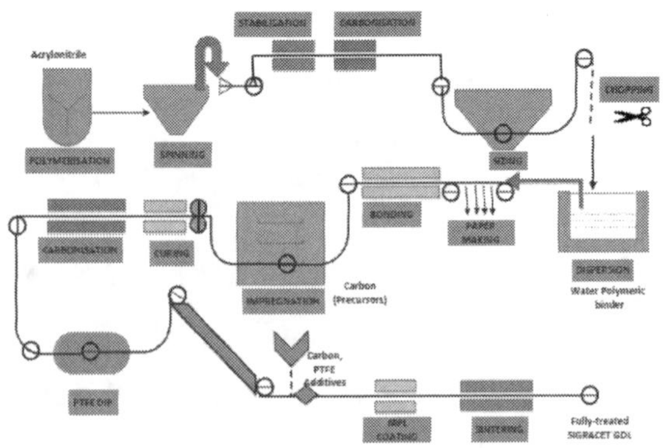

Fig. 5. Manufacturing steps involved in the conventional GDL-SGL 39 BC [15]

The manufacturing step is that of the well-known wet laying technology where the chopped carbon fibres are processed to a primary carbon fibre. The raw paper is then impregnated with carbonizable resins (carbonizable resins with the optional addition of carbon fillers), cured and recarbonized/graphitized [15].

 https://doi.org/10.1051/matecconf/201817204005

2.2 Bipolar plate and its Conventional Manufacturing Technique

The flow field or bipolar plate are conductive plates that are typically made of graphite. They account for 40–50% of the cost and 60–80% of the weight of a PEM fuel cell stack [16, 17]. They act as a structural and electrical spine for linking the cells in series [18]. Figure 6 portrays the processes for making compression-moulded and injection-moulded bipolar plates.

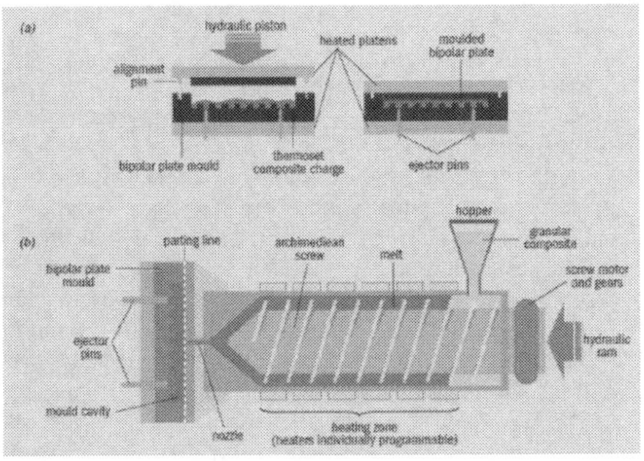

Fig. 6. Conventional Manufacturing of Bipolar Plate (a) Compression-moulded (b) Injection-moulded [19]

A variety of bipolar plate designs are known, and the conventional designs typically comprise either pin, integrated cooling flow fields, interdigitated flow field, parallel or serpentine designs [20]. The time required for designing/manufacturing bipolar plates (BPP) with complex flow channels is a challenge for PEM electrolyser [21] and PEM fuel cell fabrication. It is evident from Figure 6 (b) that the conventional manufacturing technique of bipolar plate is a lengthy process similar to the conventional GDL manufacturing process.

3 AM techniques to fabricate GDL and Bipolar Plate

3.1 GDL Manufacture by Selective Laser Sintering

Convention Carbon based GDL is prone to the electrochemical degradation which is due to the oxidation of carbon (present in the carbon paper) to carbon dioxide and is illustrated in equation:

$$C + 2H_2O \rightarrow CO_2 + 4H^+ + 4e^-$$

$$E_o = 0.207\ V\ vs.\ SHE$$

In addition, the manufacturing step involved is complex which is evident from Figure 4. So SLS process can apparently improve the durability of GDL as they are susceptible to electrochemical oxidation [22, 23] as well as reducing the manufacturing complexity because the SLS process is:

I. Fastest additive manufacturing technique where complex geometry structure could be easily constructed.

II. Diverse structural and functional possibilities are possible.

In the previous investigation, we incorporated alumide [24] as the base material since it is stiffer than many other materials used in 3D printing and it contributes to good flexural strength and higher thermal load [25]. Though the mechanical and electrical characteristics of alumide are compatible for PEM operating conditions, one of the severe limitations from the fundamental chemistry perspective is that the aluminium present in the alumide is prone to oxidation in the PEM fuel cell environment resulting in the metal ions formed which could potentially damage the expensive membrane component. Other researchers [17] [26, 27] reported that aluminium exposed to a PEM fuel cell operating environment are prone to such a corrosion. As a consequence, polyamide is used as the base material to develop the thin film samples. The utilization of titanium structures has already been proposed by Hottinen et al. [28] and it is a safe material for GDL application. In the unique investigation conducted by us [29], the base powder (PA) is sintered in a precise mode; appropriate functional material (Ti) is added to the base powder to attain the desirable electrical and thermal characteristics.

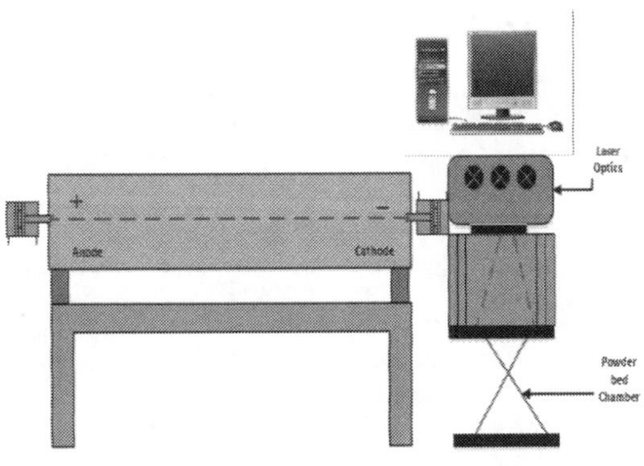

Fig. 7. Line sketch of SLS experiment set-up [30]

Figure 7 illustrates the experiment set-up of the SLS process. The initial temperature of the powder bed was kept at around 70° C to keep the powder substrate dry and free of moisture. To study the diffusion and bonding of the Titanium (Ti) powder the experiments are executed by varying the power as illustrated in Table 3 for a fixed scanning speed of 450 mm/s.

Table 2. Laser power required to sinter various configuration of Polyamide/Titanium

Laser Power (W)	Polyamide/Titanium Composition (%)
15	90/10
12	80/20
9	70/30

It may be noted that with increasing Ti content, the laser power required is decreased in order to achieve a continuous sintered layer. This is due to the fact that the titanium component absorbs the heat from the laser and burns partly, which is an exothermic reaction, resulting in excessive thermal energies. Consequently, the higher the titanium content, the lower the energy density level for laser sintering.

3.2 Bipolar Plate Manufacture by Selective Laser Melting

Metals, as sheets, are potential candidates for bipolar plate material since they have good mechanical stability, electrical and thermal conductivity and gas impermeability [17]. Probably the most important benefit is that the resultant fuel cell/electrolyser stack can be more compact and lighter than graphite BPP plates. The utilization of additive manufacturing technique is very much compatible for the bipolar plate fabrication because various design configurations are easily accomplished with this approach [31]. It is well evident that laser power is the significant contributor to the material characteristics and it is evident from the literature [31] that the electrical conductivity increases significantly (> 220 S/cm) with a corresponding increase in pyrolyzing temperature.

Yang *et al* [32] incorporated selective laser melting (SLM) techniques which is actually a subset of selective laser sintering (SLS) technique to fabricate a stainless steel (SS) plate with parallel flow field that can serve as an integrated cathode bipolar plate and a current distributor. Their results demonstrated that the PEM based electrolyser with the integrated bipolar plate can achieve an excellent performance for hydrogen production for a voltage of 1.779 V and a current density of 2.0 A/cm^2. Thus, with the AM technique the components involved in the PEM fuel cell stack can be integrated. Yang *et al* [33] also directly printed an integrated bipolar plate which functions as a liquid/gas diffusion layer, bipolar plate, gasket, and current distributor. The interfacial contact resistances between the parts were eliminated, and the size is made more compact providing an excellent energy efficiency of up to 86.48%

Ssuwei Chen *et al* [34] used a selective laser sintering (SLS) technique to fabricate a graphite based bipolar plate. Their observation is that the relative particle size and composition of the constituents plays an imperative role in the product design. Unlike, titanium or stainless steel extreme care must be taken, mainly because of the intrinsic brittleness of graphite and thin part dimensions. Thus SLS can be a costly and time- consuming machining process with graphite material.

Though stainless steel, titanium and graphite have been used for AM fabrication; the key focus is on titanium BPPs, rather than the steels or graphite. The high strength-to-weight ratio of Ti-alloy also makes it an ideal material for applications where power-to-weight is paramount and also previously it has been demonstrated to fabricate a single-cell fuel cell by the application of direct metal laser sintering [35]. As a consequence titanium can perform robustly in a fuel cell operating environment and subsequently it is used as a bipolar plate in the Toyota Mirai Fuel cell Electric Vehicle [18, 36]. Figure 8 illustrates the photograph of a Toyota Mirai Fuel cell Vehicle and the PEM Fuel cell stack incorporated in it.

(a)

(b)

Fig. 8. Toyota Mirai a. Fuel cell Vehicle b. Fuel cell stack

Additive manufacturing has been used for the creation of composite parts in general and bipolar plates specifically [37, 38]. Direct metal laser sintering has also been previously demonstrated for the fabrication of one-piece BPPs for a single-cell and 40-cell fuel cell stack and the performances are evaluated [39]. It is inferred that the single-cell comprising 2 BPPs is similar to that of an analytical cell made from carbon.

4 Discussion and Recommendation

Additive manufacturing (AM) technology, also called rapid prototyping or 3D printing, allows high resolution material to be printed from three-dimensional (3D) digital models instead of cutting materials from an initial bulk. It has found its application in diverse field of engineering and recently extended to the fuel cell/electrolyser manufacturing.

Appropriate material selection is an essential criteria to infuse, appropriate strength, electrical and thermal conductivity. However, they must be selected based on the corrosion resistance and its compatibility with PEM fuel cell working environment. Table 3 provides a consequence, a short assessment of the corrosion rate of various metals [40] which can provide an insight to choose the material for both the gas diffusion layer and bipolar plate manufacturing.

Table 3. Corrosion rates of materials [40]

Material	Corrosion rate (µm per year)
Aluminum	~250
Copper	>500
Gold	<15
Graphite	<15
Nickel	>1000
Silver	<15
Tin	>10000
Titanium	<100
Tungsten	<100
Zinc	>2000
Gold-nickel	~500
Phosphorous-nickel	<30
SS 316L	<100

However, it is eminent that these metallic material properties alone cannot be considered as a standard due to the diverse operating conditions in PEM fuel cell operating environments such as low pH, high power/high voltage, and high-temperature operation. This operating condition while happening concurrently can potentially damage even low corrosive material like graphite (graphite is a carbon allotrope and can be oxidised to CO_2) specifically at a voltage greater than 0.207 V [22]. Ironically, titanium is a potential material to resist corrosion even at highly acidic and humid conditions consequently, complementing its usage for PEM fuel cell environment.

The stainless steel material is corrosion and acid resistant steel alloy used for medical implants. Its excellent corrosion resistance combined with its high hardness and toughness permits the experimenter to re-use the micro fuel cell fabricated from it, many times [41]. Alternatively, 3D-printing could also be integrated with traditional milling processes to 3D-print the hollow features of the BPP, followed by milling of the high tolerance mating features [39]

5 Conclusion

Additive Manufacturing is a promising technology that has the potential to achieve high efficient, reliable and robust material configuration in a cost effective manner. Functional characteristics such as controlled pore morphology, electrical and thermal conductivity, tensile characteristics, hydrophobicity, etc which are the critical requirements for the PEM fuel cell/electrolyser are feasible with the AM technique. Streamlining the design alignment and integrating the key PEM Fuel cell/Electrolyser components are critical factors that can circumvent its limitations and possibly lead to its commercialisation.

References

1. C.R. Deckard, *Method and apparatus for producing parts by selective sintering*. 1989, Google Patents.
2. N. Guo and M.C. Leu, Frontiers of Mechanical Engineering, 2013. **8**(3): p. 215-243.
3. N.B. Dahotre, and S. Harimkar, *Laser fabrication and machining of materials*. 2008: Springer Science & Business Media.
4. J.C. Nelson, 1993, University of Texas at Austin.
5. M. Ay, A. Midilli, and I. Dincer, International journal of energy research, 2006. **30**(5): p. 307-321.
6. G. Sasikumar, A. Muthumeenal, S. SundarPethaiah, N. Nachiappan, R. Balaji, International Journal of hydrogen energy, 2008. **33**(21): p. 5905-5910.
7. A. Jayakumar, S. Pethaiah Sethu, M. Ramos, J. Robertson, A. Al-Jumaily, Ionics, 2015. **21**(1): p. 1-18.
8. J.A. Kumar, P. Kalyani, and R. Saravanan, Int. J. Electrochem. Sci., 3 (2008) 961 - 969.
9. S.S. Pethaiah, et al., Ionics, 2011. **17**(4): p. 361-366.
10. O.T. Holton, and J.W. Stevenson, Platinum Metals Review, 2013. **57**(4): p. 259-271.

11. M. Espinoza, M. Andersson, J. Yuan, B. Sundén, International Journal of Energy Research, 2015. **39**(11): p. 1528-1536.
12. S. Park, J.W. Lee, and B.N. Popov, International Journal of Hydrogen Energy, 2012. **37**(7): p. 5850-5865.
13. M. Mathias, et al., *Diffusion media materials and characterisation*. Handbook of fuel cells, 2003.
14. D. Gerteisen, T. Heilmann, and C. Ziegler, Journal of Power Sources, 2008. **177**(2): p. 348-354.
15. R. Schweiss, C. Meiser, T. Damjanovic, I. Galbiati, N. Haak, *SIGRACET® Gas Diffusion Layers for PEM Fuel Cells, Electrolyzers and Batteries*. 2016.
16. H. Tsuchiya and O. Kobayashi, International Journal of Hydrogen Energy, 2004. **29**(10): p. 985-990.
17. A. Hermann, T. Chaudhuri, and P. Spagnol, International journal of hydrogen Energy, 2005. **30**(12): p. 1297-1302.
18. K.S. Lyons, and B.D. Gould. *Fuel Cells*. in *Materials Science Forum*. 2017. Trans Tech Publ.
19. D. Brett and N. Brandon, Fuel Cell Review, 2005. **2**(1): p. 15-23.
20. X. Li and I. Sabir, International journal of hydrogen energy, 2005. **30**(4): p. 359-371.
21. I.U. Hwang, H.N. Yu, S.S. Kim, D.G. Lee, J.D. Suh, S.H. Lee, B.K. Ahn, S.H. Kim, T.W. Lim, Journal of Power Sources, 2008. **184**(1): p. 90-94.
22. J. Wu, X.Z. Yuan, J.J. Martin, H. Wang, J. Zhang, J. Shen, S. Wu, W. Merida, Journal of Power Sources, 2008. **184**(1): p. 104-119.
23. D. Stevens and J. Dahn, Carbon, 2005. **43**(1): p. 179-188.
24. A. Jayakumar, M. Ramos, and A. Al-Jumaily. *International Mechanical Engineering Congress and Exposition*. 2016. American Society of Mechanical Engineers.
25. H. Ju, H. Meng, and C.-Y. Wang, International Journal of Heat and Mass Transfer, 2005. **48**(7): p. 1303-1315.
26. Y. Hung, K. El-Khatib, and H. Tawfik, Journal of Power Sources, 2006. **163**(1): p. 509-513.
27. A.S. Woodman, E.B. Anderson, K.D. Jayne, and M.C. Kimble, 1999. AMERICAN ELECTROPLATERS AND SURFACE FINISHERS SOCIETY INC.
28. T. Hottinen, M. Mikkola, T. Mennola, P. Lund Journal of power sources, 2003. **118**(1): p. 183-188.
29. A. Jayakumar, S. Singamneni, M. Ramos, A.M Al-Jumaily, S.S. Pethaiah, Materials, 2017. **10**(7): p. 796.
30. R. Velu and S. Singamneni, *Evaluation of the influences of process parameters while selective laser sintering PMMA powders*. Proceedings of the Institution of Mechanical Engineers, Part C: Journal of Mechanical Engineering Science, 2015. **229**(4): p. 603-613.
31. K. Alayavalli and D.L. Bourell, Rapid prototyping journal, 2010. **16**(4): p. 268-274.
32. G. Yang, J. Mo, Z. Kang, F.A. List, J.B. Green, S.S. Babu, F.Y. Zhang, International Journal of Hydrogen Energy, 2017. **42**(21): p. 14734-14740.
33. G. Yang, J. Mo, Z. Kang, Y. Dohrmann, F.A. List, J.B. Green, S.S. Babu, F.Y. Zhang, Applied Energy, 2018. **215**: p. 202-210.
34. S. Chen, J. Murphy, J. Herlehy, D.L. Bourell, K. L. Wood, Rapid Prototyping Journal, 2006. **12**(5): p. 275-282.
35. C.J. Netwall, B.D. Gould, J.A Rodgers, N.J Nasello, K.E. Swider-Lyons, Journal of Power Sources, 2013. **227**: p. 137-144.
36. K. Kojima and K. Fukazawa, ECS Transactions, 2015. **69**(17): p. 213-219.
37. S. Chen, D.L. Bourell, and K.L. Wood. *International SFF Symposium, Austin, TX, August*. 2004.
38. B. Lyons, M. Batalov, P. Mohanty, S. Das, *SFF Symposium, Austin, TX*. 2005.
39. B.D. Gould, J.A. Rodgers, M. Schuette, K. Bethune, S. Louisa, R. Rocheleau and K. Swider-Lyons, ECS Journal of Solid State Science and Technology, 2015. **4**(4): p. P3063-P3068.
40. Lee, S.-J., C.-H. Huang, and Y.-P. Chen, Journal of materials processing technology, 2003. **140**(1): p. 688-693.
41. G. Scotti, et al., P. Kanninen, T. Kallioand S. Franssila, Journal of Micromechanics and Microengineering, 2012. **22**(9): p. 094006.

Utility approach for multi target streamlining of process parameters in wire EDM

A. Muniappan*, M. Ajithkumar, V. Jayakumar, C. Thiagarajan and M. Sreenivasulu

Department of Mechanical Engineering, Saveetha School of Engineering, Saveetha Institute of Medical and Technical Sciences, Chennai, Tamil Nadu, India.

Abstract. This paper depicts the improvement of multireaction enhancement system utilizing utility technique to foresee and select the ideal setting of machining parameters in wire electro-release machining (WEDM) process. Investigations were arranged utilizing Taguchi's L27 orthogonal exhibit. A wide range of Wire EDM control variables such as pulse on time duration, pulse off time duration, servo voltage along with wire feed rate were judged for investigation. Multi reaction enhancement was performed for both cutting pace (CS) and surface unpleasantness (SR) utilizing utility idea to discover the ideal procedure parameter setting. The level of essentialness of the machining parameters for their impact on the CS and SR were controlled by utilizing investigation of fluctuation (ANOVA). In present study utility approach method used to optimize the process parameter in wire EDM of magnesium Al6061/SiC/Graphite hybrid composite with zinc covered brass wire electrode. The approach depicted here is relied upon to be profoundly useful to assembling enterprises, and furthermore different territories, for example, aviation, car and apparatus making businesses. The parameters corresponding to experiment run number 7 are pulse on time 108 units (Level 1), pulse off time 60 units (Level 3), peak current 230 units (Level 3), gap set voltage 60 units (Level 3), wire feed 3 units (Level 1) and wire tension 4 units (Level 1) are the best combination to achieve better surface roughness and cutting speed.

Keywords. WEDM, Utility approach, Taguchi method, kerf width & surface roughness

1 Introduction

Composite materials had grown rapidly over recent few decades to encompass metal matrix composites, ceramic composites and polymer matrix composites. Metal grid composites have pulled in significant consideration because of their capacity to give an extensive variety of microstructures & properties (Sharma 2000). The underlying philosophy of metal composite design is that an continuous metallic matrix, with its ductility and formability, is combined with the stiffness and load withstanding property of a ceramic or refractory reinforcements to produce material with superior properties [1]. Metal Composites possess good properties over metals alloys with high specific strengths, better properties of metal composite at elevated high temperatures, minimum thermal expansion, very good wear resistance & high structural strength. These properties are predominantly suited to application in an automotive, aerospace and electronic sectors [2],[3].

Manufacturing sector is growing rapidly by accommodating technology modernization. The mechanism for machining hard reinforced materials, intricate shapes and contours which are very difficult to cut by conventional methods which created many unconventional methods. CNC wire cut machine was developed in the year 1969. WEDM which involves moving very thin wire electrode continuously. Wire electrode materials such as brass and zinc coated brass wire of diameter ranges from 0.05-0.35 are widely applied in industry. The gap between work and wire electrode is generally ranges from 0.025mm to 0.050mm and is continuously maintained by a computer controlled coordinating system. To achieve better surface quality on the tool as well as in the component, optimum process parameter setting is a very important factor. Machining control parameters are optimized by using different methods for the improvement of the quality. Taguchi method which is widely applied mainly experimental design in manufacturing application. It allows the optimization of parameters in machining by turning, milling, Electro Discharge Machining, wire cut EDM, welding, grinding etc. The optimization is achieved with lesser number of experiments by this overall cost and time is saved.

* Corresponding author: inspire.munish@gmail.com

2. Materials and method

2.1 Preparation of hybrid composite

In this investigation, the hybrid Aluminum MMC has been fabricated by stir casting process. The crossover composite comprises of 10 wt% SiC and 5 wt% Graphite particulates in metal lattice Al6061 compound. The Al combination of 6xxx arrangement is having the capacity to be used in aviation and car ventures in light of its high quality to-weight proportion and great protection from consumption. Fortifications SiC and graphite in particulate frame are utilized to fabricate the hybrid composite. The easiest and the most savvy technique for fluid state creation is stir casting [19].

2.2 Machining parameters and response

Machining process parameters in WEDM, Pulse on time, Pulse off time, current, gap voltage, wire speed and wire tension were considered as input parameters. Responses SR and kerf width were measured after machining for investigation. The scopes of these procedure parameters were chosen on the premise of the pilot tests. The levels of different parameters and its assignments are exhibited in Table 2.

Table 1. Process parameters and their levels

Symbol	Process parameter	Level 1	Level 2	Level 3
A	Pulse on time	108	117	126
B	Pulse off time	40	50	60
C	Pulse current	90	160	230
D	Gap set Voltage	10	30	50
E	Wire drum speed	3	4	5
F	Wire tension	4	8	12

2.3 Experimental design using Taguchi method

Taguchi strategy is a proficient apparatus for the outline of a great assembling framework. It also effectively solves some complex problems in manufacturing (Roy 1990).It is a strategy in view of OA tests, which give much decreased change to the try different things with ideal setting of process control parameters. The six control parameters, that is, pulse on time (A), pulse off time (B), top current (C),Gap set voltage (D), wire drum speed (E) and wire tension(F) at three levels were chosen in this examination. The tests were finished by Table 3. This table just speaks to specific level of the different elements of the procedure at which the examinations would be directed. Kerf width ought to be as least as conceivable in the WEDM procedure. Kerf width is an imperative component of the laser cutting procedure that gives the benefit of this innovation contrasted with different strategies for form cutting.

2.4 Utility approach method

A client assesses an item on various different quality attributes. To be ready to settle on an objective decision, these assessments on various attributes ought to be consolidated to give a composite list. Such a composite record speaks to the utility of a item. The general utility of an item measures the convenience of that item according to the evaluator. The utility of an item on a specific trademark measures the convenience of that specific normal for the item. The general utility of an item is the total of utilities of each of the quality attributes. Along these lines if xi is the measure of adequacy of the trait (trademark) I and there are n properties assessing the result space, at that point the joint utility capacity can be communicated as (Derek, 1982) :

$$U(x_1, x_2 . . . x_n) = f[U_1(x_1), U_2(x_2), . . ., U_n(x_n)] \quad (1)$$

In linear case, the function becomes

$$U(x_1, x_2, . . . , x_n) = \sum_{i=1}^{n} WiU(xi)$$

where Wi is the weight age assigned to the attribute i and the sum of the weightings for all attributes is equal to 1. If the composite measure (the overall utility) is maximized the quality characteristics considered for evaluation of utility will automatically be optimized (maximized or minimized whatsoever the case may be).

2.5 Experimental set up

Analyses were led on Electronica Sprint cut (Electra-Elplus 40A Dlx) CNC wire electrical release machine to think about the surface harshness and kerf width influenced by the machining parameters at various levels. WEDM is a start disintegration process. The flashes are produced between the work piece and the wire terminal. The dielectric liquid is ceaselessly encouraged into the machining zone with required weight. The material is getting expelled by a progression of discrete sparkles occurring at the zone to be machined through electro-warm system. Test set up of the wire electrical release machine is appeared in Figure 2.

Amid machining process little hole kept up between the work and wire material. The machined particles were flushed away by the persistent stream of the dielectric liquid. The wire is held by a stick direct at the upper and lower parts of the work piece. The work example measure utilized as a part of this examination is 95 x 80 x 8 mm rectangular plate. Zinc covered metal cathode wire of 0.25 mm width was utilized as a part of this investigation. Deionized water was utilized as dielectric liquid at room temperature. In the wake of machining, the examples were cleaned with acid after machining. The kerf was measured utilizing profile projector measuring framework. The kerf esteems were measured at six spots spread over the whole length of cut. The kerf esteems utilized as a part of this examination are the numerical normal of three estimations produced using the example in each cut.

2.6 Weight Criteria Calculation using SDY Concept

Standard deviation is connected to this examination for unprejudiced assignment of weights. The significance of weights in illuminating Multi-Criteria Decision Making (MCDM) issues can't be over stressed. To decide the standard deviation, the range institutionalization was finished utilizing Equation (1) to change diverse scales and units among different criteria into regular quantifiable units so as to figure their weights.

$$X'_{ij} = \frac{X_{ij} - \min\limits_{1<j<n} X_{ij}}{\max\limits_{1<j<n} X_{ij} - \min\limits_{1<j<n} X_{ij}}$$

where max X_{ij}, min X_{ij} are the maximum and minimum values of the criterion (*j*) respectively. The standard deviation (*SDV*) is calculated for every criterion using equation (2) given below.

$$SDV_j = \sqrt{\frac{1}{m}\sum_{i=1}^{m}(X_{ij} - \overline{X}_j')^2} \quad (2)$$

where Xj' is the mean of the values of the jth criterion after normalization and $j = 1,2 \ldots n$. After calculating for *SDV* for all criteria, the next step is to determine the weights, W_j of all the criteria considered.

$$W_j = \frac{SDV_j}{\sum_{j=1}^{n} SDV_j}$$

where $j = 1,2,\ldots n$.

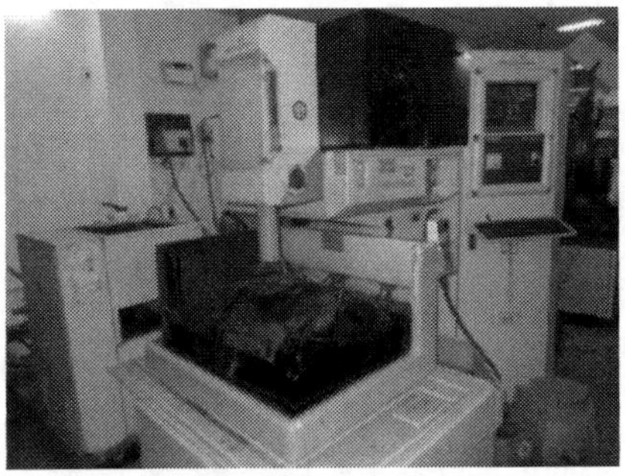

Fig. 1 WEDM experimental set up (Electronica Sprint cut)

2.7. Methodology

A preference scale for each quality characteristic is constructed. To determine the utility value for a number of quality characteristics later these scales are weighted to obtain a composite number (overall utility). The weighting is done to satisfy the test of indifference on the various quality characteristics. The preference scale should be a logarithmic one (Gupta and Murthy 1980). The minimum acceptable quality level for each quality characteristic is set out at 0 preference number and the best available quality is assigned a preference number of 9. If a log scale is chosen the preference number (*Pi*) is given by (Gupta and Murthy, 1980).

$$P_i = A \log(X_i/X'_i)$$

where X_i= any value of quality characteristic or attribute i; X'_i= minimum acceptable value of quality characteristic or attribute I; and A = a constant.
At optimum value (X^*) of attribute i, $Pi = 9$. So

$$A = 9 / \log (X^*/X'_i)$$

The next step is to assign weights or relative importance to the quality characteristics. This assignment is subjective and based on experience. Moreover, it depends on the end use of the product or it may depend on the customer's requirements. The weightage should be assigned such that the following condition is satisfied:

$$\sum_{i=1}^{n} Wi = 1$$

The overall utility can be calculated as:

$$Uj = \sum_{i=1}^{n} WiPi \quad (3)$$

where j = product index.

3. Results and discussion

3.1. Result analysis

Table 5 shows the square of Xi, normalized matrix for Xi and rank calculated by MOORA method. The normalized SR & KW are multiplied with their corresponding weights. These values are listed in the table 6. The parameters, higher the better 9maximum) and smaller the better (minimum) respectively added for rank calculation. The surface roughness observed in the experiment is in the range of from 2.252 to 4.256. Similarly the kerf width observed in the experiment is in the range of from 292 to 327. From these observations, experiment number 7 has the best rank. The parameters corresponding to experiment run number 7 are pulse on time 108 units (Level1), pulse off time 60 units(Level3), peak current 230 units (Level3), gap set voltage 60 units (Level3), wire feed 3 units (Level1) and wire tension 4 units (Level1).

3.2 Predicted optimum value of responses

Mean of each response for each level of factor is shown in Table 3. The average of the S/N ratio of the responses SR and KW for each level of each factor is shown in Table 4. Regardless of the category of the performance characteristics, a greater S/N value corresponds to a better performance. Therefore, the optimal level of the machining parameters is the level with the greatest S/N value. From Table 3 and Table 4 based on the analysis of means and S/N ratios, the optimal machining parameter levels are $A_1B_3C_1D_2E_1F_1$& $A_1B_1C_3D_2E_1F_3$ for SR and KW, respectively. The optimal level of each machining parameter is bolded in Table5.The predicted optimal value of each response characteristics can be determined using Minitab software version 18. The values are 2.0754 um (SR) and 0.2858 mm (KW) respectively.

3.3 Preference number calculation

After obtaining the optimum value, a preference scale is constructed for each response. The preference scale should be a logarithmic one [15]. Zero is the preference number of the minimum acceptable quality level and 9 is the preference number of the best available quality level for each response. For the log scale the preference number (P?) is calculated by using Eqs. (4) and (5).

Predicted optimum value of SR (x_{SR}) = 2.0754 um

Minimum acceptable value of SR (X_j) = 4.257 um

P_{SR} =-28.85 log (x_{sr}/4.257)

Predicted optimum value of KW (x_{KW}) = 0.2858 mm

Minimum acceptable value of KW (X_j) = 0.327 mm

P_{KW} =-154.11 log (x_{KW}/0.327)

Table 2. Surface roughness and Kerf width

Ex.No.	Pulse on Time	Pulse off Time	Peak Current	Gap set voltage	Wire Feed	Wire Tension	Surface Roughness (μm)	kerf width(mm)
1	108	40	90	10	3	4	2.605	.302
2	108	40	90	10	4	8	2.82	.300
3	108	40	90	10	5	12	3.185	.292
4	108	50	160	30	3	4	2.555	.304
5	108	50	160	30	4	8	2.625	.300
6	108	50	160	30	5	12	2.985	.301
7	108	60	230	50	3	4	2.52	.302
8	108	60	230	50	4	8	2.965	.303
9	108	60	230	50	5	12	3.115	.297
10	117	40	160	50	3	8	3.625	.303
11	117	40	160	50	4	12	3.48	.309
12	117	40	160	50	5	4	3.475	.314
13	117	50	230	10	3	8	3.57	.317
14	117	50	230	10	4	12	3.59	.313
15	117	50	230	10	5	4	3.975	.322
16	117	60	90	30	3	8	2.7	.317
17	117	60	90	30	4	12	3.125	.313
18	117	60	90	30	5	4	2.57	.327
19	126	40	230	30	3	12	3.515	.302
20	126	40	230	30	4	4	3.725	.317
21	126	40	230	30	5	8	3.81	.302
22	126	50	90	50	3	12	3.67	.310
23	126	50	90	50	4	4	3.32	.324
24	126	50	90	50	5	8	3.47	.322
25	126	60	160	10	3	12	4.125	.316
26	126	60	160	10	4	4	4.256	.325
27	126	60	160	10	5	8	4.11	.318

Table 3. Predicted levels of process parameters

Parameters	SR	KW
Pulse off time	60	40
Current	90	230
Gap voltage	30	30
wire speed	3	3
Wire tension	4	12

Table 4. Preference number of individual responses

Ex.No.	Pulse on Time	Pulse off Time	Peak Current	Gap set voltage	Wire Feed	Wire Tension	SR	kerf width
1	108	40	90	10	3	4	6.154	5.527
2	108	40	90	10	4	8	5.160	5.972
3	108	40	90	10	5	12	3.635	7.781
4	108	50	160	30	3	4	6.396	5.086
5	108	50	160	30	4	8	6.058	5.972
6	108	50	160	30	5	12	4.447	5.749
7	108	60	230	50	3	4	6.569	5.527
8	108	60	230	50	4	8	4.532	5.306
9	108	60	230	50	5	12	3.913	6.645
10	117	40	160	50	3	8	2.014	5.306
11	117	40	160	50	4	12	2.525	3.994
12	117	40	160	50	5	4	2.543	2.919
13	117	50	230	10	3	8	2.205	2.283
14	117	50	230	10	4	12	2.135	3.133
15	117	50	230	10	5	4	0.859	1.236
16	117	60	90	30	3	8	5.705	2.283
17	117	60	90	30	4	12	3.873	3.133
18	117	60	90	30	5	4	6.323	0.204
19	126	40	230	30	3	12	2.400	5.527
20	126	40	230	30	4	4	1.673	2.283
21	126	40	230	30	5	8	1.390	5.527
22	126	50	80	50	3	12	1.859	3.778
23	126	50	90	50	4	4	3.115	0.821
24	126	50	90	50	5	8	2.561	1.236
25	126	60	160	10	3	12	0.395	2.495
26	126	60	160	10	4	4	0.003	0.615
27	126	60	160	10	5	8	0.440	2.072

Table 5. Utility value of individual responses

Ex. No.	SR	KW	Wi*SR	Wi*KW	Utility value	Order
1	6.154	5.527	3.212	2.642	5.854	3
2	5.160	5.972	2.693	2.855	5.548	6
3	3.635	7.781	1.897	3.719	5.617	5
4	6.396	5.086	3.339	2.431	5.770	4
5	6.058	5.972	3.162	2.855	6.017	2
6	4.447	5.749	2.322	2.748	5.070	8
7	6.569	5.527	3.429	2.642	6.071	1
8	4.532	5.306	2.366	2.536	4.902	9
9	3.913	6.645	2.043	3.176	5.219	7
10	2.014	5.306	1.051	2.536	3.587	12
11	2.525	3.994	1.318	1.909	3.227	16

12	2.543	2.919	1.327	1.396	2.723	18
13	2.205	2.283	1.151	1.091	2.242	20
14	2.135	3.133	1.115	1.498	2.612	19
15	0.859	1.236	0.448	0.591	1.039	26
16	5.705	2.283	2.978	1.091	4.069	10
17	3.873	3.133	2.022	1.498	3.519	13
18	6.323	0.204	3.301	0.098	3.398	14
19	2.400	5.527	1.253	2.642	3.895	11
20	1.673	2.283	0.873	1.091	1.964	22
21	1.390	5.527	0.726	2.642	3.368	15
22	1.859	3.778	0.970	1.806	2.776	17
23	3.115	0.821	1.626	0.393	2.018	21
24	2.561	1.236	1.337	0.591	1.928	23
25	0.395	2.495	0.206	1.192	1.398	24
26	0.003	0.615	0.002	0.294	0.295	27
27	0.440	2.072	0.230	0.991	1.220	25

4 Conclusions

In this experimental study, the combined utility and SDV method is applied for the estimation of optimum machining parameters to minimize surface roughness and kerf width. The conclusions drawn from this study are as follows:

- Combined utility and SDV method is employed to select the optimum machining parameters in WEDM machining of Al6061/SiC/graphite with zinc coated brass wire electrode.
- From these observations, experiment number 7 has the best rank. The parameters corresponding to experiment run number 7 are pulse on time 108 units (Level1), pulse off time 60 units(Level3), peak current 230 units (Level3), gap set voltage 60 units (Level3), wire feed 3 units (Level 1) and wire tension 4 units (Level1).
- Standard Deviation (SDV) method is also employed to find the relative importance of surface roughness & kerf width. The weight ratios are found to be 0.522 and 0.478 respectively.
- The optimum results are adopted in validation study and the results based on WEDM process responses can be effectively improved.

References

1. R. Snoeys, F. Staelens, W. Dekeyser, Ann. CIRP, **35/2**, 467-480 (1986).
2. S.S. Mahapatra, A. Patnaik, J. Braz. Soc. Mech. Sci. & Eng. **28**(4), 422–429 (2006).
3. PK. Mishra, PK. *Non-conventional machining processes* (Narosa Publishing House, NewDelhi, 2002)
4. F. Klocke, M. Schwade, A. Klink, A. Kopp, Proc. Eng. **19**, 190-195 (2011).
5. D. Shichun, C. Xuyang, W. Dongbo, W. Zhenlong, C. Guanxin, L. Yuan, Int. J. Mach. Tools & Manuf. **49**, 788-792 (2009).
6. H. Somekawa, T. Mukai, J Alloys Compd. **417**, 209-213 (2006).
7. K. Ponappa, S. Aravindan, P. Rao, J. Ramkumar, M. Gupta, Int J Adv Manuf Technol. **46**, 1035-1042 (2010).
8. A. Muniappan, C. Thiagarajan, S. Somasundaram, IJMME-IJENS. **16**(5), 95-103 (2017).
9. Jinkai Xu, Rongxian Qiu, Kui Xia, Linshuai Zhang, Huadong Yu, Proceedings of 2015 International Conference on Advanced Mechatronic Systems, Beijing, Chaina (2016).
10. Bijomathew, B.A. Benkim, J. Babu, Procedia Material Science. **5**,1863-1872 (2014).
11. Bijaya Bijeta, Nayak Siba, Sankar Mahapatra, Procedia Eng. **97**, 469-478 (2014).
12. Yogendra Kumar, Hari Singh, Procedia Materials Science. **5**, 2142-2151 (2014).

Optimization of electric discharge machining parameters on titanium alloy (ti-6al-4v) using Taguchi parametric design and genetic algorithm

K.Umanath[1] and D.Devika[2]

[1] *Assistant Professor, Department of Mechanical Engineering, Velammal Engineering college*
[2] Department of Mechanical Engineering, Saveetha School of Engineering, Saveetha Institute of Medical and Technical Sciences, Chennai
Email id: umanathmech@gmail.com, ddevika40@gmail.com.

Abstract

The aim of this research work is to analyze the significant of process variables and find the optimum process variables in electric discharge machining (EDM) of Titanium alloy (Ti-6Al-4V) .The variables considered are peak current, pulse-on-time and pulse-off-time where as the responses are Material Removal Rate(MRR) and Surface Roughness(SR). MITSUBISHI EA8 spark erosion machine is employed for this work and copper tungsten electrode of Ø14 mm is used in experimental trials. The experimental runs are done based on Taguchi L27 orthogonal array. The signal-to-noise ratio, the analysis of variance (ANOVA), regression analysis and Genetic algorithm are used to determine the optimal levels of machining parameters on Metal removal rate and Surface roughness. Confirmation tests also done with the optimal levels of machining variables. Comparison of Taguchi's and Genetic algorithm were employed to analyze the effective optimum value.

Keywords: EDM, Ti–6Al–4V alloy, Peak current, Pulse-on-time, Taguchi method.

1. Introduction

Ti–6Al–4V alloy is an essential material in modern industries. Its tremendous properties, such as high strength-weight ratio, high temperature resistance and outstanding corrosion resistance, make it extensively applied in the aerospace, automobile, chemical and biomedical fields. However, the poor machinability of titanium alloys using the conventional machining processes are more expensive. Therefore,non traditional machining ways, like discharge machining (EDM) are explored to machine this alloy. EDM is associate degree energy-based technique extensively utilized in machining hard, high-strength and temperature-resistant materials in an non contact manner. The alloy is melted and vaporized by an spark between the electrode and work piece.

. Kao et al.used distilled water and kerosene as dielectric fluid .The higher material removal rate (MRR) and lower surface roughness (SR) obtained by distilled water.[1]. Ponappa et al. endeavored to upgrade the discharge efficiency of magnesium nano alumina composites utilizing by EDM [2]. Anand pandey et al. investigated the impact of EDM parameters at surface integrity with various electrodes and processes [3]. Recently, Velusamy et al. utilized EDM innovation to machine aluminum composites to inspect the impact of process parameters [4]. Sharma demonstrated the most effective parameters are pulse current and pulse duration at surface roughness of the workpiece when compare to other EDM parameters[5]. Yan-Cherng Lin et al demonstrated that the magnetic force helped EDM has a higher MRR, a lower relative electrode wear proportion and a littler Surface Roughness as contrasted with ordinary EDM [6]. Joshi and pande created insightful model for the electric discharge machining process utilizing two-dimensional axisymmetric thermal (FEM) model of single-spark EDM process and artificial neural network is utilized to predict the shape of creater cavity,MRR, and TWR [7]. Narender Singh and Raghukandan reported the multi-response optimization of the process parameters like MRR,TWR,Taper, radial overcut, and SR on electric discharge machining of Al– 10%SiCP as cast metal matrix composites utilizing orthogonal array with Grey relational analysis [8]. George et al considered EDM input parameters are pulse current, gap voltage and pulse-on-time and responses are electrode wear rate and material removal rate and obtained the optimized parameters on carbon carbon composite using taguchi method [9]. Elangovan et al and Akhyar et al used taguchi's robust design , analysis of variance and signal to noise ratio analyses to investigated the influence of different input parameters on the different output parameters [10,11]. Krishna Mohana Rao et al presented the optimized parameters of die sinking electric discharge machining on the surface roughness by considered the simultaneous affect of various input parameters using genetic algorithm [12]. Anand et al applied Grey Relational Analysis with Taguchi Technique to found the optimal selection of machining parameters such as peak current, pulse duration, voltage, Servo reference voltage in Electrical Discharge Machining process to identify the improvement in metal removal rate and surface finish of the work material HCHCr using copper electrode with assistance of magnetic field [13]. Chandramouli et al announced the optimal setting of the process parameters on Electric Discharge Machining , that have been chosen are peak current, pulse on time, pulse off time and tool lift time with output parameters as Material Removal Rate and Surface Roughness .Taguchi test configuration was utilized to plan the experimental run and investigations led on PH Steel machined with copper tungsten tool. [14]. In this work, the optimization of EDM process parameters on Ti–6Al–4Valloy using the Taguchi method and ANOVA analysis are reported. Output parameters including material removal rate and Surface Roughness are assessed the machining impacts. Those process parameters are firmly associated with the chosen regression analysis and output parameters in this investigation are the peak current, pulse-on-time and pulse-off-time. Experiments in view of L27 orthogonal array are led. Optimized process parameters all the while prompting higher material removal rate and better surface roughness is then be checked through an confirmation experiment. The points of interest of the methods are tended to in the accompanying areas.

2. Experiments

A cylindrical copper tungsten rod 14 mm in diameter and 100 mm in height was employed as the electrode in the CNC EDM machine (MITSUBISHI -EA8) to erode the work piece of Ti–6Al 4V alloy in this work. The specimen size selected in this investigation is 25×25×5mm.The work piece properties of the alloy and the chemical compositions are shown in Table 1. However, there are a few machining characteristics to be considered in the EDM process. As a result, a preliminary experiment for determining the optimal process parameters indicates the machining parameters such as peak current, pulse-on-time and pulse-off-time led a most significance on the Electric Discharge Machining performance of Ti–6Al–4V alloy.

Table.1.Material properties of Ti–6Al–4V alloy and its chemical composition

Work piece	Ti-6Al-4V
Hardness (HRC)	36~39
Solidus temperature (°C)	1,604±11
Liquids temperature (°C)	1,660±14
Density (g/cm^3)	4.043
Elastic modulus (kg/mm^2)	11,200
Yield strength (kg/mm^2)	84.2
Tensile strength (kg/mm^2)	91.3
Elongation (%)	10
Electrical resistivity (μΩ·m)	1.7

Ti=89.464, Al=6.08, V=4.02, Fe=0.22, O=0.18, C=0.02, N=0.01, H=0.0053

Table .2 .Machining parameter

SL	Parameters	L1	L2	L3
A	Peak current, (A)	4.5	9.0	14
B	Pulse on time (μs)	6.0	12.8	25.6
C	Pulse off time (μs)	12.8	25.6	51.2

Table .3.Fixed parameter

Work piece	Ti–6Al–4V
Electrode	Copper tungsten
Dielectric fluid	DAPHANECUT–HL35
Polarity	Negative (+)
Machining depth	0.2 mm

The selected Electric discharge machining parameters and its levels are listed in Table 2. In this experimental study electrode acted as positive polarity. DAPHANE CUT – HL35 dielectric fluid was employed in all experiments. The experimental setup is shown in the fig 1. The experiments are done based on L27 Taguchi orthogonal array. The fixed experimental parameters used in this work are listed in Table 3. This work also used MATLAB 7.4 to optimize the surface roughness and material removal rate using Genetic Algorithms. Each experiment was repeated two times and the average of values taken is for determining optimal parameters. The MRR and SR of the machined surface are the response variables to assess the machining quality. The MRR (mg/min) is defined by volume of the material removed per unit time in minute. To measure the weight of the worn work piece removal, Sartorius Mechatronics (Model:BSA 224) was used. In the trials, the surface roughness of the EDMed work piece regarding the normally utilized Ra (arithmetic average roughness) was estimated by a surface roughness tester (Mitutoyo, Surftest-SJ201). Essentially, the MRR is the classification of higher-the-better responsible variable in the Taguchi technique and SR is the lower-the-better responsible variable in the Taguchi technique in the EDM process.

Figure.1. Experimental set up

3. Taguchi's Method

Taguchi's method is a well accepted methodology for experiment design. In this, signal-to-noise ratio(S/N) is utilized to speak to a response variables and the largest S/N ratio is required. There are three methods of quality characteristics viz. nominal-the-better, larger-the-better and smaller-the-better [15]. In this work, experimentally obtained MRR value is "larger the-better" and SR is "lower the-better". In view of Taguchi's technique, the S/N ratio computed is as appeared in below.

i) Larger-the-better

$$\text{HB}: \quad \eta = -10\log\left[\frac{1}{n}\sum_{i=1}^{n} yi^{-2}\right] \quad \ldots\ldots [1]$$

ii) Smaller-the-better

$$\text{LB}: \quad \eta = -10\log\left[\frac{1}{n}\sum_{i=1}^{n} yi^{2}\right] \quad \ldots. [2]$$

Where yi is the experimentally obtained value and n is the repeated number of each experiment.

4. Analysis of Variance (ANOVA)

Analysis of Variance(ANOVA) is a numerical system which separates add up to variety into responsible sources. Some of the components in ANOVA are discussed below[15].

4.1. Sum of squares

The magnitude of each error value can be squared to provide a measurement of total variation present. This is known as "Sum of Squares". The basic ANOVA is that the total sum of squares is equal to the sum of sum of the squares due to known components as shown in Eq. 3.

$SST = SSm + SSe$ ………… *[3]*

Where,

SST - Total sum of squares.

SSm - sum of squares due to mean.

SSe - sum of squares due to error.

4.2. Variance due to error

Error Variance, for the most part named just variation, is equal to the sum of squares of error separated by the degree of freedom of errorError variance is a measure of the variation due to all the uncontrolled parameters, incorporating measurement error engaged with a specific experiment

4.3. F -test for comparison

The F-test is simply a ratio of sample variances as shown in Eq. 4.

$F = Sy1^2/Sy2^2$…………… *[4]*

4.4. Percent contribution

The percent contribution indicates the relative power of a factor and/or interactions to reduce variation.

5. Genetic Algorithm

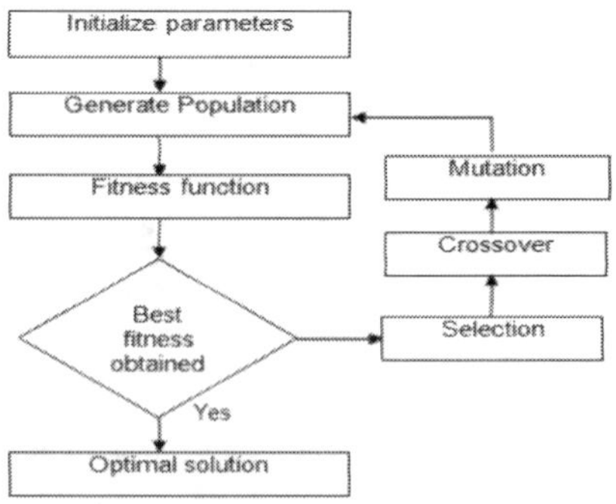

Figure.2. Flow of Optimization in GA

The aim of the optimization process using Genetic algorithm is to find the optimum values of decision variables that serve to the minimum value of surface roughness being as low as possible [12]. There will be a total of four optimization problems to be solved. Each optimization problem consists of a minimization function defined by one of the second order equations. Obtaining results in Genetic Algorithm (GA) is based on some criteria. as shown in fig 2. The most essential parameters are population size, the type of selection function, the crossover rate and the mutation rate. By the process of experimentation the esteem or parameter setting for these criteria is made for acquiring the most optimum outcome that is normal from this investigation. The MATLAB optimization toolbox is applied in the study for performing iterations in order to get the best optimal outcome. The best mix of these qualities for cutting conditions will prompt the base lower surface roughness. Number of trials was directed with various esteem settings for the cutting conditions for looking through the minimization estimations of surface roughness utilizing the MATLAB optimization toolbox, the best blend of the parameters applied

6. Results and Discussions

6.1. Regression analysis

The peak current, pulse on time and pulse off time were considered in the development of regression equation for the material removal rate and surface roughness. A linear polynomial model is created to anticipate the material removal rate and surface roughness value for each experimental trail as recorded in Table.4.

The regression equation for material removal rate and surface roughness is shown in below

MRR = - 1.84 + 0.428 Peak current + 0.0777 Pulse on time + 0.0004 Pulse off time.......... [5]

SR = 3.25 + 0.0724 Peak current + 0.0225 Pulse on time - 0.0215 Pulse off time................. [6]

6.2. Material removal rate

Material removal rate is defined as the proportion of the difference of weight of the work piece before and after the machining to the machining time.

MRR = (wjb − wja)/t (mg/min)............ [7]

Where, wjb and wja are weights of work piece before and after the machining and t is the machining time.

The machinability of EDM relies upon the electrical conductivity of the work material. Notwithstanding the low electrical conductivity and high thermal resistance of the material, which at last decreases the electrical conductivity of the work material, the outcomes acquired demonstrate that Titanium alloy can be machined adequately utilizing EDM. MRR was found to increment with increment in current and pulse on time. It is likewise clear that the surface roughness value increments with increment in current and pulse on time. High current results in higher thermal loading on the two terminals (tool and work piece) trailed by higher amount of material being expelled from the two electrodes and thus prompt high MRR and as TWR as well. Furthermore, longer pulse duration also results in a large removal per discharge, which results in larger crater size and therefore higher surface roughness. The S/N response graph in fig. 3 demonstrates the material removal rate increments with an increments in the current. For the performance characteristic of the material removal rate, the peak current level 3, pulse on time level 3 and pulse off time level 1 parameters including a current of 14A, pulse on time 25.6μs and pulse off time 12.8 μs as shown in fig.3 can lead to optimal result. Optimum value is obtained from the response table . ANOVA results are shown in Table 5. Observation of this table discharge current is the most significant parameter having percentage of contribution as 62.74%, followed by pulse on time 22.53% and pulse off time 5.18% are affect the material removal rate and the interaction factors are not contributing most significantly.

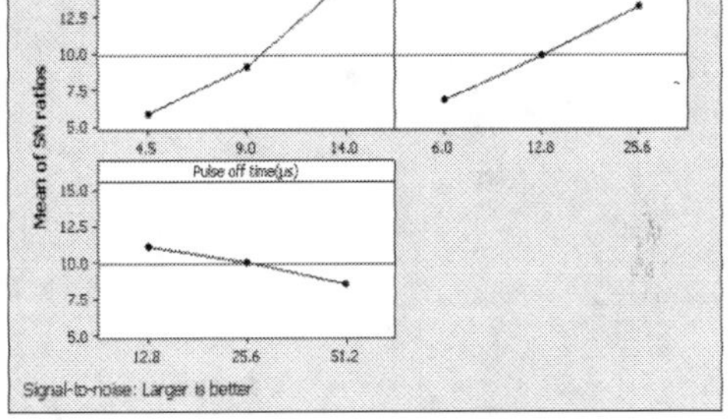

Figure 3. S/N ratio graph for MR

6.3. Surface Roughness

It was observed that pulse-off-time has a considerable effect on surface roughness. At the point when the pulse off time is incresed the machining also not done on the work piece that the time the dielectric fluid is applied into the spark gap, short-circuiting turns out to be less articulated because of the gathered particles. Higher pulse-off-time removes the formation of ionized bridges across the gap and results in higher ignition and decreased surface roughness. Further, the cooling rate of

 https://doi.org/10.1051/matecconf/201817204007

the tool increases with increase in the pulse off time and hence improved surface Roughness is observed.

It is found that the discharge current has a leading effect on surface roughness. The S/N response graph in Fig. 4 shows the surface roughness decreases upon decreasing the discharge current. It indicates that a better surface quality can be obtained under a lower discharge current with shorter pulse duration owing to the lower power erosion process. Figure 4 shows when only the performance characteristic of the surface roughness is considered, a parameters set peak current level 1,pulse on time level 2 and pulse off time level 3 with a current of 4.5 A, pulse on time 12.8µs and pulse off time 51.2µs can result in an optimal outcome. Optimum values are getting from the response table. ANOVA table is shown in table 6. From this table pulse off time is most significantly contribution is 20.78%, peak current is 9.49%,pulse on time is 8.69%, interaction terms significantly contribution is AB,BC and AC as 12.06%, 20.78% and 19.49% respectively. The electric-discharge machined surface consists of a huge number of craters that are formed by spark discharges. The electric-discharge machined surface comprises of a multitude of overlapping craters that are framed by spark discharges.

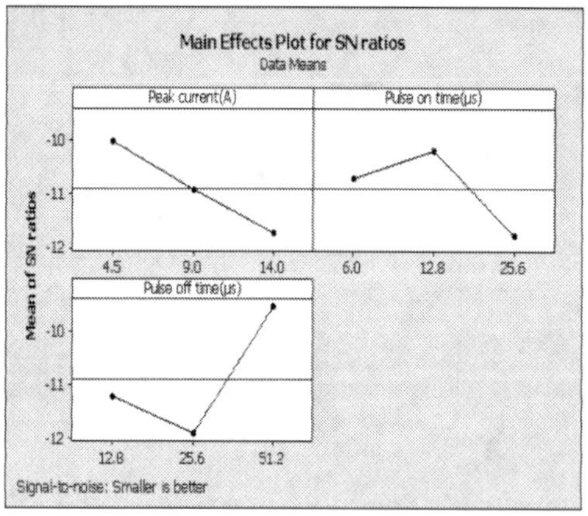

Fig 4. S/N ratio graph for SR

6.4. Determination of surface roughness using GA

The optimum choice of machining parameters should increment the utility of machining financial aspects, as well as the item quality, all things considered, by limiting surface roughness value. The process parameters of EDM are characterized in the standard optimization method that is solved by a numerical optimization algorithm. An objective function to be minimized is necessary to define the standard optimization problem. In EDM machining of titanium alloy, optimization problem can be expressed as

Minimize: a (IP,Ton,Toff)

Within range of machining parameters:

4.5 A <IP< 1 4 A

6 µs <Ton< 25.69 µs

12.8 µs <Toff< 51.2 µs

The Genetic Algorithm based optimization approach gives an adequate estimate to the genuine optimum solution. Tables 7 and table 8 demonstrates the extraordinary of optimization work with important machining conditions.

Population size = 100

Crossover rate = 0.95

Mutation rate = 0.41

MATLAB 7.4 is used to optimize the surface roughness using Genetic Algorithms. The critical parameters of the GA are the size of the population, cross over rate and mutation rate.

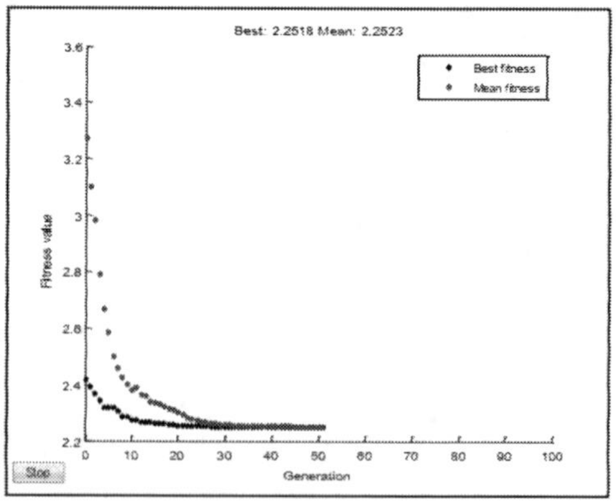

Figure 5. Genetic Algorithm fitness graph for SR

Table 4. Experimental results for the MRR and SR

Sl.No	IP (A)	Ton (μs)	Toff (μs)	Avg.SR(μm)	Predicted SR(μm)	S/N ratio for SR	Avg.MRR (mg/min)	Predicted MRR	S/N for MRR
1	4.5	6.0	12.8	2.73	3.2470	-8.7233	1.013	1.3377	0.1122
2	4.5	6.0	25.6	3.21	3.1604	-10.1301	1.121	0.9844	0.9921
3	4.5	6.0	51.2	2.32	2.6100	-7.3098	1.596	1.1190	4.0607
4	4.5	12.8	12.8	3.42	3.5886	-10.6805	2.345	2.1060	7.4029
5	4.5	12.8	25.6	3.89	3.3134	-11.7990	2.106	1.7528	6.4692
6	4.5	12.8	51.2	2.28	2.7630	-7.1587	1.987	1.0462	5.9640
7	4.5	25.6	12.8	3.12	3.8766	-9.8831	3.675	3.5525	11.3051
8	4.5	25.6	25.6	4.62	4.3221	-13.2928	2.975	3.1992	9.4697
9	4.5	25.6	51.2	3.67	3.3360	-11.2933	2.022	2.4926	6.1156
10	9.0	6.0	12.8	4.37	4.3650	-12.8096	2.789	3.0837	8.9090
11	9.0	6.0	25.6	3.94	3.4862	-11.9099	1.778	2.3254	4.9986
12	9.0	6.0	51.2	2.47	2.9358	-7.8539	1.142	1.8850	1.1533
13	9.0	12.8	12.8	2.58	2.8659	-8.2324	3.345	3.8521	10.4879
14	9.0	12.8	25.6	3.65	3.6392	-11.2459	2.250	2.9861	7.0437
15	9.0	12.8	51.2	3.07	3.0888	-9.7428	2.221	2.7923	6.9310
16	9.0	25.6	12.8	3.43	3.5542	-10.7059	5.879	5.2985	15.3861
17	9.0	25.6	25.6	5.87	5.5986	-15.3728	4.987	4.9452	13.9568
18	9.0	25.6	51.2	3.33	3.3768	-10.4489	4.107	4.2387	12.2705
19	14.0	6.0	12.8	6.33	6.0123	-16.0281	4.565	5.0237	13.1888
20	14.0	6.0	25.6	3.52	3.8482	-10.9309	6.680	5.983	16.4955
21	14.0	6.0	51.2	3.44	3.2978	-10.7312	3.486	3.9639	10.8465
22	14.0	12.8	12.8	4.23	4.2764	-12.5268	6.333	5.7921	16.0322
23	14.0	12.8	25.6	3.24	4.0012	-10.2109	5.321	5.4388	14.5199
24	14.0	12.8	51.2	3.22	3.4508	-10.1571	4.935	4.7323	13.8657
25	14.0	25.6	12.8	3.75	4.4587	-11.4806	7.332	7.2385	17.3044
26	14.0	25.6	25.6	4.10	4.2892	-12.2557	6.890	6.8852	16.7644
27	14.0	25.6	51.2	3.64	3.7388	-11.2220	6.233	6.1787	15.8939

The set of optimum value obtained by Genetic algorithm for both surface roughness and material removal rate. The optimum parameters for surface roughness are peak current 4.50(A),pulse on time 6.0071(μs), pulse off time 51.1985 (μs)and optimum surface roughness is 2.251.The optimum parameters for material removal rate are peak current 4.50(A),pulse on time 6(μs), pulse off time 14.9255 (μs)and optimum material removal rate is 7.3391(mm^3 /min)

In this work comparison of Taguchi parametric optimization and Genetic algorithm optimization are also reported. A Genetic algorithm will give the better optimum result compared to Taguchi parametric optimization. The improvement of surface roughness and increment of material removal rate are also done in this study. Approximately 2% of improvement is shown in surface roughness and as well as material removal rate.

6.5. Determination of Material removal rate

An objective function to be maximized is necessary to define the standard optimization problem. In EDM machining of titanium alloy, optimization problem can be expressed as

Maximize: MRR (IP, Ton, Toff)

Within range of machining parameters:

$4.5\ A < IP < 14\ A$

$12.8\ \mu s < Toff < 51.2\ \mu s$

$6\ \mu s < Ton < 25.69\ \mu s$

MATLAB 7.4 is used to optimize the material removal rate using Genetic Algorithms. The critical parameters of the GA are the size of the population, cross over rate and mutation rate.

Population size = 50

Crossover rate = 0.70

Mutation rate = 0.63

The Genetic Algorithm based optimization approach provides a sufficient approximation to the true optimal solution. Tables 7 and table 8 shows the extreme of optimization function with relevant machining conditions.

Table.5.ANOVA results for the MRR

Sou	D	SS	MS	Fc	Ft	Infere	%
A	2	63.	31.8	76	4.	Signif	62.74
B	2	22.	11.4	27	4.	Signif	22.53
C	2	5.2	2.62	6.	4.	Signif	5.18
AB	4	2.6	0.65	1.	3.	Insign	2.58
BC	4	1.6	0.41	1.	3.	Insign	1.63
AC	4	2.1	0.52	1.	3.	Insign	2.08
Err	8	3.3	0.41				3.26
Tot	26	101					100

Table.6.ANOVA results for SR

Sou	DF	SS	MS	Fc	F	Infere	%
A	2	2.15	1.07	3.	4	Insig	9.49
B	2	1.97	0.98	3.	4	Insig	8.69
C	2	4.47	2.23	8.	4	Signif	20.78
AB	4	2.73	0.68	2.	3	Insig	12.06
BC	4	4.71	1.17	4.	3	Signif	19.73
AC	4	4.41	1.10	3.	3	Signif	19.49
Erro	8	2.21	0.27				9.76
Tota	26	22.6					100

A-Peak current (A), B-Pulse on time (μs), C-Pulse off time (μs), AB- Peak current* Pulse on time, BC- Pulse on time * Pulse off time, AC- Peak current* Pulse off time, DF-degree of freedom, SS-sum of square, Fcal-F Calculated value, Ftab-F value from tab .

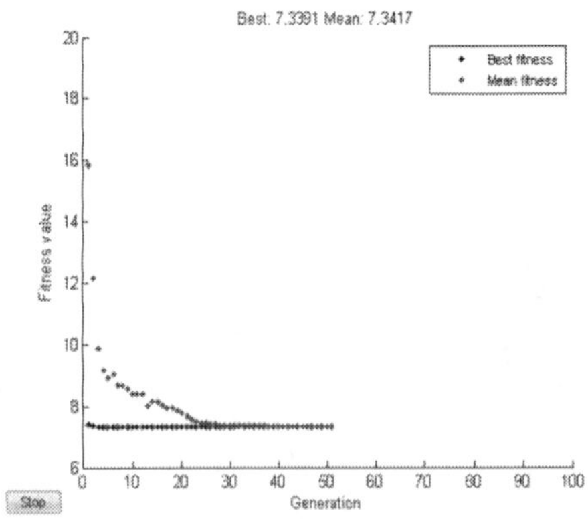

Figure .6.Genetic Algorithm fitness graph for MRR

6. Conformation Tests

Since the optimal EDM process parameter set is gotten, the confirmation tests are prepared to check the performance characteristics improvement. Confirmation result is appeared in table 7 and table 8. The experimental response variable utilizing the optimal machining parameters can be discovered for the setting not accessible in the OA.

Table 7. EDM performance results using the initial, Prediction and GA process parameters for MRR

Parameter/ Optimized Methods	**A (A)**	**B (μs)**	**C (μs)**	**MRR (mm^3 /min)**
Taguchi (A3B3C1)	14	25.6	12.8	7.332
Prdiction (A3B3C1)	14	25.6	12.8	7.2385
Genetic Algorithm	4.5	6	14.925	7.3391
Confirmation (A3B3C1)	14	25.6	12.8	7.36

Table 8. EDM performance results using the initial, prediction and GA process parameters for SR

Parameter/optimized methods	**A (A)**	**B (μs)**	**C (μs)**	**SR (μm)**
Taguchi (A1B2C3)	4.5	12.8	51.2	2.28
Prediction (A1B2C3)	4.5	12.8	51.2	2.763
Genetic Algorithm	4.50	6.0071	51.1985	2.251
Confirmation (A1B2C3)	4.5	12.8	51.2	2.27

The results of confirmation experiment are compared with the outcome of the orthogonal array and regression analysis prediction of the equation operating parameters. Table 7 and table 8 shows the comparison of the experimental results using the initial (orthogonal array, A3B3C1), optimal (regression prediction, A3B3C1) and Genetic algorithm EDM parameters on Ti–6Al–4V alloy. Table 7 and table 8 shows the MRR increased from 7.238 to 7.360 mg/min and the surface roughness decreased from 2.76 to 1.78μm respectively. The corresponding improvement in MRR is 2% and surface roughness 17.75%, respectively.

7. Conclusion

Taguchi method and ANOVA analysis used to enhance the response variables of the material removal rate and surface harshness in the electrical discharge machining on Ti– 6Al– 4V alloy was employed for in this work. Optimization of the process parameters also done in this work. The most significantly contributing factors are found by ANOVA.The experiment results shows that MRR increases with increased in the peak current and pulse on time. Surface Roughness improves by increasing pulse on time. The optimum machining conditions for material removal rate with the peak current of 14A, pulse on time 51.2μs and pulse off time 12.8μs.In this optimum MRR is 7.332 mg/min. The optimum machining conditions for surface roughness with the peak current of 4.5A, pulse on time 12.8μs and pulse off time 51.2μs. The corresponding SR is 2.28μm.The optimum conditions for the two response functions are different. The Genetic Algorithm will give the optimum combination of parameters for the material removal rate with peak current of 4.5A, pulse on time 6μs and pulse off time 14.9254μs. In this optimum Material removal rate is 7.3391 mg/min. The optimum combination of parameters for the surface roughness attained with the peak current of 4.50052 A, pulse on time of 6.00718μs and pulse off time of 51.19852μs.The corresponding optimum Surface roughness is 2.2518μm.

References

1. J. Y. Kao , C. C. Tsao ,S. S. Wang , C. Y. Hsu, Int J Adv Manuf Technology, vol.47, pp.395–402, 2010.

2. K. Ponappa , S. Aravindan ,P. V. Rao , J. Ramkumar, Int J Adv Manuf Technology, vol.46, pp.1035–1042, 2010.

3. Anand Pandey,Shankar Singh, International Journal of Engineering Science and Technology ,Vol. 2(6), pp.2172-2191, 2010.

4. Velusamy Senthilkumar, Bidwai Uday Omprakash, Journal of Manufacturing Processes ,vol.13, pp.60–66, 2011.

5. C. Sharma,S. Maheshwari, Pushpendra S Bharti, International Journal of Engineering Science and Technology, Vol. 2(11), pp.6464-6473, 2010.

6. Yan-Cherng Lin ,Yuan-Feng Chena, Der-An Wanga, Ho-Shiun Lee, Journal of materials processing technology, vol.209, pp.3374–3383, 2009.

7. S.N. Joshi, S.S. Pande, Int.J. Adv. Manufact. Technology, vol.45 (3), 2009.

8. P. Narender Singh, K. Raghukandan, B.C. Pai, Journal of Materials Processing Technology, vol.155–156, pp.1658–1661, 2004.
9. P.M. George a, B.K. Raghunath , L.M. Manochac, Ashish M. Warrier, Journal of Materials Processing Technology, vol.145, pp.66–71, 2004.
10. Sooriya moorthy Elangovan, K. Prakasan, V. Jaiganesh, Int J Adv Manuf Technology,vol.51, pp.163–171, 2010.
11. G. Akhyar,C.H. Che Haron, J.A. Ghani, International Journal of Science Engineering and Technology ,Vol. 1, pp.1985-3785, 2008.
12. G.Krishna Mohan araoa, G.Rangajanardhaa, D.Hanumantha Raoc, Journal of materials processing technology, vol.209, pp.1512–1520, 2009.
13. G.Anand, S.Satyanarayana and M.Manzoor Hussain, International Journal of materials today, Vol.4, pp.7723-7730, 2017.
14. S.Chandramouli and K.Eswaraiah, International Journal of materials today, Vol.4, pp. 2040-2047, 2017.
15. Ross PJ *"Taguchi techniques for quality engineering"*, Tata McGraw-Hill, New Delhi, 2005.

Optimization of WEDM process during machining of Al-Al_2O_3 composite using Taguchi based Grey Relational Analysis

[1]*Aaryan Mohan Singla,* [2]*Shubhang Sharma,* [3]*Akshay Kaul, and* [4] *Rajyalakshmi G.*
[1,2,3,4] *School of Mechanical Engineering, VIT ,Vellore, Tamilnadu, India.*

Abstract- Aluminium ceramic (Al-Al2O_3) is used comprehensively all over the world. In this paper, input parameters on Al-Al2O_3 workpiece are applied to investigate the MRR and surface roughness by the process of wire electrical machining (WEDM).Four varying input parameters (Ton, Toff, Wf and Cs%) were tabulated for nine samples using L9 orthogonal array design technique whereas keeping wire tension as constant. Tool wear is calculated as the change in initial thickness and final thickness after cutting operation using screw gauge. The surface roughness is found using surface roughness tester.Minitab is used to plot the graph between input parameters against various output parameters. To find the impact of process parameters, ANOVA (Analysis of Variance) is used .

Key Words: Parametric Optimization- 1, Particulate Reinforced Metal Matrix Composites- 2, Taguchi- Grey RelationalAnalysis-3,WEDM-4.

1. Introduction

A composite material is an amalgamation of two or more than two materials that are disparate and separated by two different aspects and phases. It has two phases in which the discrete constituent (reinforcement) is distributed in a continuous phase called the matrix. The two phases have different geometry and form and the properties at the interface decide the properties of the composite. The reinforcement phase is responsible for imparting mechanical properties like stiffness, hardness, and abrasive nature etc. The matrix phase is continuous and is responsible for distribution of load amongst the reinforced members. The properties of the composites are inherently better than both its parent constituents matrix and reinforcements.

Aluminium metal matrix composites(AMMC) is the most commonly used MMC in the industry today. Aluminium due to its many favourable properties like high strength to weight ratio, low weight, and high corrosion resistance makes it as extensively used in today's market. Aluminium alloys of different series and composition are used generally, reinforced with abrasive ceramics like Sic (silicon carbide), Al2O_3 (Aluminium oxide) and ZrO2 (Zirconium dioxide)etc.

There are many ways to manufacture and produce these composites like there are the liquid route, solid-state route or even semisolid route. The liquid state route has processes like stir casting which is a very widely used and economical method to make composites, solid state process consists of methods like powder metallurgy, sintering process and semisolid has methods like vapour phase deposition and in situ fabrication. Due to the high strength to weight ratio, high strength and high stiffness, high wear resistance etc .MMC find their application in aerospace and automobile industries these days, but however, due to these reinforced ceramics which are responsible for making the composite very hard, makes its machining a very complicated process. Due to its hard and abrasive nature, the composite becomes harder to the machine, this leads to the discovery of efficient machining methods. These can be machined by conventional techniques in a difficult manner, which leads us to the innovation of many non-traditional machining processes and makes it a more efficient process for cutting this type of material. It is much better than conventional process because conventional process causes more wear of the tool due to the presence of abrasive ceramics.

1.1. WEDM Process

It is a non-traditional machining process most widely used for removing materials especially used on electrically conductive difficult to machine metal matrix composites. It is a thermoelectric method of machining , the arc of electricity is passed and the heat generated by the spark causes the composite to melt and then finally erode. The composite must be electrically conductive so that the material can be removed. It does not involve tool-to workpiece contacts like in conventional machining process thereby leading to the high amount of wear produced due to excessive friction at the tool and work piece interface. It also causes a large amount of noise and vibrations causing damage to both tool and workpiece.
The material is removed when there is a series of electrical discharge between the workpiece and the electrode. The electrode is a very thin wire of about 5 mm diameter made up of tungsten, brass, copper etc, a material that is capable of withstanding very high temperature without melting or breaking. The temperature within the sparking zone is in the range of 8000 - 12000-degree Celsius, that can be controlled by a dielectric fluid that is generally used as a coolant and it makes the cutting operation more effective also. Water is

* Corresponding author: *rajyalakshmi@vit.ac.in*

generally used as a dielectric fluid which even gets ionized when passed through the wire-work piece interface. The movement of the thin wire electrode used to cut complex and intricate shapes on the workpiece is controlled by a small microprocessor on auto-cad software that will determine the toolpath of the wire with a very high level of precision. There are many factors that influence the WEDM process but if they are wrongly selected it may lead to harmful consequences like short-circuiting or wire breakage. Hence there is a need to develop a very careful and methodical way of developing a mathematical model and to maximize the efficiency of the process to find out the optimum process parameters like pulse on (TON), pulse off (TOFF), wire tension (WT) and for these process parameters many optimization techniques have been developed and put into practice.

2. Literature Review

The volume fraction of alumina is directly proportional to the surface roughness of Al/Al2O3p. The relation can be characterized because of the low fluidity of the melt and SiC particulates in the spark gap leads to channel discharge to spread up. The surface roughness in Al/Sic composites is inversely proportional to the volume fraction of SiC reinforcements. The trend was found to be characterized to the protrusions of the alumina particulates on the processed surface. (Nilesh G. Patil and P.K. Brahmankar,2016).

Sensitivity put up more impact on material removal rate whereas parameters like wire feed rate, pulse on time and pulse off time contributes Negligibly Karthik.C, Rimmie Duraisamy, Rajyalakshmi.G,2015).The spark on time and the input current have a statistically significant impact in case of aluminium and mild steel respectively. The spark on time has more contribution to the surface roughness of Aluminium whereas in case of mild steel input current has more influence.(Shivkant Tilekar, Sankha Suvra Das ,P.K. Potwari).(Boopathi et al. 2012) studied dry WEDM as an earnest attempt to improve machining process to put an end to the use of a dielectric. There are so many parameters which makes it a difficult task to set up suitable working conditions. Pulse peak current and pulse on time are the influencing factors on the surface roughness during machining by WEDM of DC53 die steel. The surface roughness of workpiece is directly proportional to both pulse on time and pulse peak current (Kanlayasiri K & Boonmung). H. R. Tonday and A. M. Tigga 2016 studied the effects of various cutting parameters of WEDM on MRR and surface integrity. The found that the output parameters are not affected by the change in wire tension. (Alias,2012) found that the feed rate affects the performance measures like MRR and Surface Roughness. So, along with Pulse ON Time, Pulse OFF Time and CS%, Wire Feed Rate (WF) is also taken as an input parameter.M. Sangeethal, A. Srinivasulu Reddy, G. Vijaya Kumar et al studied about the optimization of die-sinking EDM process parameters in AMMC using reinforcements of (flyash,Sic ,Al2O3) using weight percentage of 2.5%,5% and 10% and machining parameters of current (Ip),pulse on time(Ton),pulse off time(Toff),tool lifting time(TL).Based on experiments performed as per Taguchi design of experiments and responses such as MRR and SR(surface roughness), electrode wear rate and cost are measured and optimized. After an analysing the data with the help of a desirability function they concluded that reinforcement material, current and tool lifting time are most significant parameters, %reinforcement and pulse time are medium sized parameters which reduced electrode wear rate, increase in MRR, cost decreased and surface roughness reduced.N. Jamuna, A. Sreenivasulu Reddy et al studied about the impact of parameters like kerf width,tool wear, process cost and surface roughness are optimized using grey relational analysis and fuzzy logic as per the taguchi based design of experiments,and after analysing the data they concluded that percentage of reinforced material (PRFM), base material (BM), type of reinforced material are the most influential parameters,WP AND TOFF are medium based influential parameters and TON,SF and WF are lastly the multi responses.(Y.S. Tarng, S.C. Ma, and L.K. Chung) discussed about finding ways to solve the intricate parameters to enhance the cutting performance within the allowable constraints of working conditions The application of simulated annealing algorithm can greatly influence the performance characteristics. The use of SA annealing in neural networks help to find the peak results and thereby they were able to improve the cutting process by WED M using this approach. K. P. Somashekhar & Jose Mathew & N. Ramachandran International Journal Advanced Manufacturing Technology (2012) 61:1209–1213 et al investigated about the demanding and taxing task of micro wire edm and its effects on MRR, surface roughness and overcut of a single pass machining of aluminium. The MRR is set to be maximized whereas the surface roughness and overcut are to be kept as low as possible .SA annealing is implemented to find out the response nature as a function of output variables by keeping all control factors like discharge current, capacitance ,gap voltage and feed rate .Application of Taguchi, Fuzzy-Grey Relational Analysis for Process Parameters Optimization of WEDM on Inconel-825 G. Rajyalakshmi1* and P. Venkata Ramaiah et studied on inconel-825 metal and it's response optimization of more than one parameters like wire feed, wire tension, servo feed, spark gap voltage to improve output parameters like MRR, spark gap voltage, and surface finish. The algorithm being used to arrive at the optimal results is fuzzy grey relational analysis which converts multi-objective into single objective and thereby simplifies the method of optimization. C. L. Lin, J. L. Lin and T. C. Ko Department of Automatic Engineering, Lan Yang Institute of Technology, Toucheng, I-Lan, Taiwan et al found that process responses such as MRR, Tool Wear and surface morphology can be optimized by both approaches i.e, grey relational analysis without using SN ratios and fuzzy logic. However, when it comes to multiple process response the grey relational method without using SN ratios is more suitable in this case than fuzzy logic method .Optimization of EDM Process with Taguchi, Grey Relational Analysis, and Fuzzy Logic Technique Mr. Ganesh Pandurang Jadhav, Dr. Narendra Narve, Prof. Vyankatesh Mundhe observed that Taguchi method is very much useful for experimental setup design. It helps to minimize trial and run time drastically. Taguchi method very much important in DOE (design of experiment) , GRA (Grey Relational Analysis) is purely mathematical Optimization method which highlight importance of numbers and its calculations but fuzzy logic help to use some logic (AND, OR, NOT) and more detail study of response of input parameter in all direction and sense. Multi-Objective optimization of parameters during EDM of aluminium alloys using grey relational analysis. The optimal combination of process parameters was achieved by GRA. Further predicted response values were in agreement with corresponding experimental results.

A.H.A. Shah et al (2016) IOP Conf. Ser.: Material. Science. Engineering. 114 012023 studied on the improvement of turning operation by finding out the most suitable parameters by taguchi grey relational analysis for S45C steel. Multiple process parameters can be improved on CNC turning by this method. Four parameters MRR, surface roughness, Tool Wear and power consumption are to be optimized at the same time. Taguchi Experimental Design helps to reduce the no of experiments and keep to a bare minimum level.

Multi-Objective Optimization of Machining Parameters for Al7075 Metal Matrix Composite Using Grey - Fuzzy Technique Ramanan G Edwin Raja Dhas J did experimentation on al7075 alloy with active charcoal as reinforcement .The measurement of samples with high value of density, hardness ,impact strength , ultimate tensile strength is selected for wire electric discharge machining. Output of these are used for maximizing MRR and minimizing surface roughness for optimal process parameters using fuzzy grey relational analysis

.3.Experimental Details

3.1. Work material

The composites used in this study are Aluminium Metal matrix Composites (AMMC). Aluminium matrix is reinforced with Aluminium oxide with percentage composition of 10%.

Table 1. Type of Materials used for experimentation

S. No	B.M	R.F.M	P.R.F.M
1	Al	Al_2O_3	10

The details of work piece material and properties of aluminium materials are given in Table 1 and Table 2.

3.2. Experimental Set up and Experimental Procedure

In this research work, the material used is Al-Al2O3 ceramic which is prepared by the process of stir casting. The workpiece is cut into 9 cuboidal pieces having length and breadth as 10mm and 12mm respectively. A series of electrical pulses generated by the pulse generator unit is applied between the workpiece and the traveling wire electrode.The positioning and dimensioning are done by auto-cad software.

The brass wire with the diameter of 0.25mm is used to cut the workpieces because of its high mechanical strength and wire durability.

Table 2. Properties of Aluminium 6063

Density (g/cm^3)	3.69
Maximum use temperaure (°C)	1700
Modulus of Elasticity (GPa)	300
Thermal Conductivity (W/m·K)	18
Hardness (kg/mm^2)	1175
Co-Efficient of Thermal Expansion (m/m·°C)	$8.1x10^{-6}$

To provide ease and simplicity to the operator an appropriate set of machining variables table is to be constructed The Taguchi based algorithm is applied to attain suitable arrangement of input to output responses. In order to proceed to optimize output responses minitab results are generated. Thus, using grey relation analysis, conclusions are made

3.3. Machining parameters and their levels for WEDM machining

There is a 3 level taguchi L9 orthogonal array design for the experiments to find out most feasible machining process parameters for various responses. The four control factors like pulse on time, pulse off time , wire feed and servo voltage are controlled in 3 stages. If unsuitable parameters are choosen, it may lead to decrease in efficiency of the process leading to high lead time.The selection of improper factors may also cause breakdown of machine and problems like wire breakage or short circiuiting or even damage to the workpiece. The Tool wear,MRR and surface roughness are the output features to be improved. Aluminium composite and it's machining factors are selected carefully based on it's properties and nature.various process parameters were taken as given in the table 3 along with different levels.

Table 3. Process parameters and their Levels for WEDM

Symbol	Control Factor	Level-1	Level-2	Level-3
A	Pulse On Time(μs)	106	109	112
B	Pulse off Time(μs)	60	56	52
C	Wire Feed(mm/min)	2	4	6
D	CS%	60	70	80

3.4. Measurement of Response Characteristics

Three response values of tool wear, surface roughness and material removal rate was measured by using different instruments and techniques.
MRR was measured as the ratio of the total volume of the cuboidal box to the total machining time. The volume of the work specimen is the product of the area of the workpiece and the depth of cut. The time of cut was noted against the four different varying input. Thus, different four speeds at midpoints of each edge were noted and their average value gives the observed MRR. Experimental MRR can be calculated by the following formula.

Observed MRR=(S1+S2+S3+S4)/4, where S1,S2,S3,S4 is the speed of the wire discharge at the midpoint of each edge.

Other response characteristic like surface roughness is measured by surface profilometer. The measured value of R_a is the root mean square or RMS value of surface roughness being measured. Three different readings are taken to achieve accuracy.
Other response parameter is tool wear rate is measured by screw gauge. Tool wear is calculated as the change in initial diameter and the final diameter of wire after cutting operation. The final diameter of the wire is calculated by screw gauge. Three values of the diameter of wire are calculated and the mean is taken as the final change in the wire diameter. Table 4, the final wire diameter is tabulated below with three different readings for accuracy

4. Analysis of Results

Using Taguchi's analysis and the Grey relational analysis, the optimal parameters for all three parameters of MRR,SR AND TWR are shown in the table below.

4.1. Multi Objective Optimization using Grey Relational Analysis

Experimental design is done by using Taguchi's L9 orthogonal design

Table 4. Experimental results

TON	TOFF	CS%	MRR	TWR	SR
106	60	60	2.594	.002	2.804
106	56	70	3.018	.003	3.284
106	52	80	3.339	.002	2.758
109	60	80	3.608	.002	3.215
109	56	60	4.496	.004	2.932
109	52	70	4.183	.004	2.715
112	60	70	3.443	.002	3.193
112	56	80	4.515	.004	3.769
112	52	60	5.433	.005	3.311

4.1.1. Formulation of the problem

The problem deals with a multi-objective optimization problem. The parameters like surface roughness and tool wear rate are to be minimized whereas material removal rate has to be maximized. The multi objective function is converted into single objective function under the constraints of parameters of TON, TOFF, CS%, WF. The values of the grey relational grade(GRG) values are found out, whose values range from $0 \leq GRG \leq 1$.

Step1: The values obtained from earlier tables are converted into corresponding S/N ratio values. For MRR larger the better characteristic is choosen and for SR and TWR smaller the better is choosen.S/N ratio for MRR can be calculated by the formula as

$$S/N = -\log_{10}(1/n)\sum y_{ij}^2 \quad (1)$$

y_{ij} = observed response value

$i=1, 2... ....n; j = 1, 2...k$
n = number of replications

The surface roughness is the lesser-the-better performance characteristic and the loss function for the same is expressed by the above formula.

Table 5. Signal to Noise ratios

S/N MRR	S/N TW	S/N SR
8.2794	53.9794	-8.9565
9.5944	50.4576	-10.3281
10.4723	53.9794	-8.8138
11.1453	53.9794	-10.1439
13.0565	47.9588	-9.3448
12.4298	47.9588	-8.6760
10.7387	53.9794	-10.0856
13.0932	47.9588	-11.5259
14.7008	46.0206	-10.3992

Step2:Now in order to bring the S/N ratio to a acceptance level we go for normalizing the data and keep its range from 0 to 1.All the values of S/N ratio are normalized and transformed to bring the data within an acceptable range. The S/N ratio is normalized as Z_{ij} $(0 \leq Z_{ij} \leq 1)$.The normalized MRR corresponds to larger the better and surface roughness and tool wear smaller the better. It can be calculated by the following formula

$Z_{ij} = y_{ij} - \text{Min } y_{ij} / \text{Max } y_{ij} - \text{Min } y_{ij}$. (2)

Step 3:The grey relation coefficient used for predicting the optimization values can be found out from normalized values using the given below equation

$y(x_o(k), x_i(k)) = \Delta_{min} + \Delta_{max} / \Delta_{oi}(k) + \xi \, \Delta_{max}$. (3)

Step4:The grey relation grade is obtained by finding out the average for all the values of grey relational coefficient ,the overall performance of the parameters depends on the value of grey relational grade. The higher value of grey relation grade corresponds that the value is closer to the optimized value or the desired value .It converts multi response characteristics into a single response problem.

Step 5:Determination of optimum parameters

The grey relational grade is taken as response and its larger value indicates the better value of performance. In case of the experiments performed experiment no 8 has the highest value hence TON 112,TOFF 56,WF 2,CS%80 is to be taken as the optimized value.

5. Result and Analysis

Table 6. Two-way ANOVA: GRG

Source	DF	SS	MS	F	P	% Contribution
A	2	0.0220302	0.0110151	1.56	0.315	34.3
B	2	0.0137936	0.0068968	0.98	0.451	21.53
C	2	0.0236196	0.0118098	1.32	0.363	36.89
D	2	0.0046016	0.0023008	0.26	0.785	7.185
Error	0	0	0			
Total	8	0.0640449				

S = 0.09464 R-Sq = 74.06% R-Sq(adj) = 0.00%

Table 7. Grey RelationalGrade

GRC MRR	GRC TW	GRC SR	GRG
1	0.333	0.356	0.563
0.709	0.472	0.543	0.575
0.594	0.333	0.344	0.423
0.528	0.333	0.507	0.456
0.401	0.672	0.395	0.489
0.436	0.672	0.333	0.480
0.566	0.333	0.497	0.465
.400	0.672	1	0.690
0.333	1	0.558	0.630

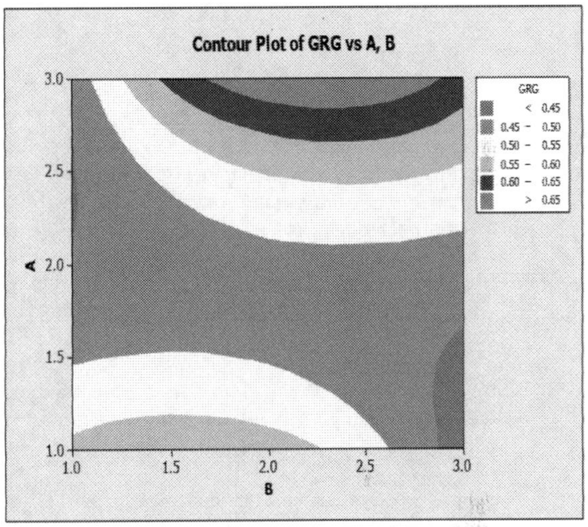

Fig. 1. Contour Plots of GRG vs A,B

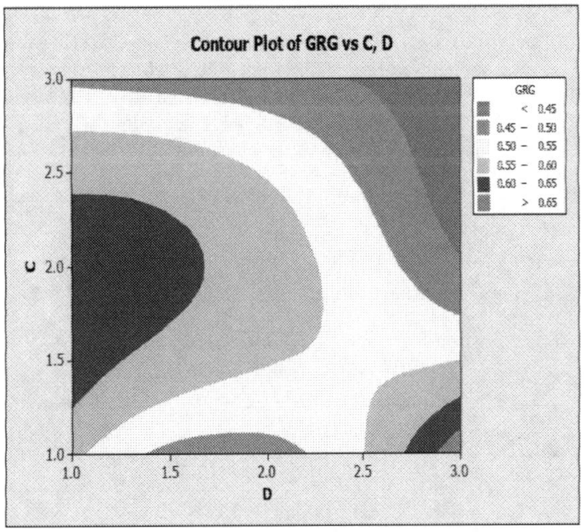

Fig. 2. Contour plots for GRG vs C,D

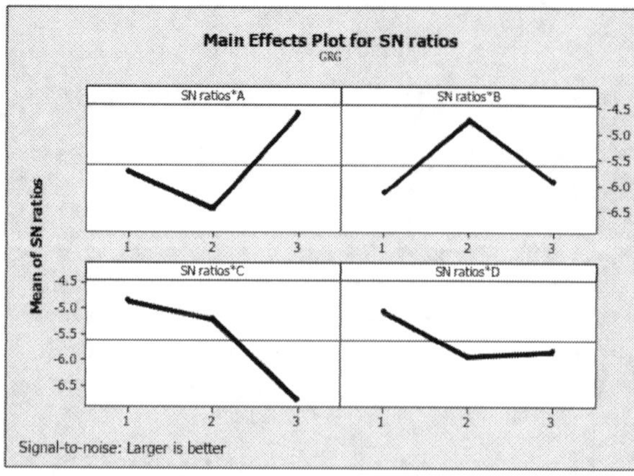

Fig. 3. Main effect Plots for SN ratios.

Conclusions

- Orthogonal array shows the different level for machining parameters as A3, B2, C1 and D3 based on maximum values shown in response table. But optimized SN ratio gave the different level of machining parameter as A3, B2, C1 and D1.
- Contour plot vs A,B shows that maximum value of GRG can be attained by taking the value range of A between 2.9-3 and for B between 1.6-2.9.
- Contour plot vs C,D shows that maximum value of GRG can be attained by taking the value range of C between 1-1.1and for D between 2.9-3.
- From Fig 1.3, it can be seen that from A1 -A2 GRG values is decreasing slowly while A2-A3, change is drastic. Similarly the relation between input SN ratios and GRG value can be concluded.
- Based on the ANOVA studies C has the highest contribution of 36.89% in GRG value followed by parameter A with 34.3% then B with 21.53% and then finally with D 7.185% having the least influence.

References

[1] Patil NG, Brahmankar P (2010) *International Journal of Advanced Manufacturing Technology* 48:537-555

[2] Karthik.C, R Duraisamy, Rajyalakshmi.G School of mechanical and Building Sciences, VIT University Vellore, Tamilnadu, INDIA.

[3] Shivkant Tilekara Sankha Shuvra Dasa P.K.Patowari,Student,Depatment of mechanical engineering,NIT Silchar,Assam,788010, IndiaAssociate professor,Depatment of mechanical engineering,NIT Silchar,Assam,788010, India

[4] Boopathi S, Sivakumar K, Kalidas R (2012), *International Journal of Engineering Research and Development* (ISSN: 2278-800X) 2:1-6

[5] K. Kanlayasiri, S. Boonmung, *J. Master. Process. Technology*.,187–188 (2007) 26–29.

[6] H. R. Tonday and A. M. Tigga (2016) *IOP Conf. Ser.: Mater. Sci. Eng. 115 012013*

[7] Alias A, Abdullah B, Abbas NM (2012) *Procedia Engineering 41:1806-1811*

[8] M. Sangeetha1, A. Srinivasulu Reddy, G. Vijaya Kumar Mechanics, *Materials Science & Engineering Journal, Magnolithe, 2017.*

[9] N. Jamuna.A. Sreenivasulu Reddy Tool Wear and Process Cost Optimization in WEDM of AMMC Using Grey Relational Analysis

[10] Y.S. Tarng, S.C. Ma and L.K. *Chung International. Journal of . Machining, Tools Manufacturing*. **Vol. 35**. No. 12. pp. 1693-170|. (1995)

[11] K. P. Somashekhar & Jose Mathew & N. Ramachandran *Int J Adv Manuf Technol (2012)* 61:1209–1213.

[12] G. Rajyalakshmi1* and P. Venkata Ramaiah Indian Journal of Science and Technology, **Vol 8(35)**, DOI: *10.17485/ijst/2015/v8i35/50946*, December (2015)

[13] C. L. Lin, J. L. Lin and T. C. Ko Department of Automatic Engineering, Lan Yang Institute of Technology, Toucheng, I-Lan, Taiwan.

[14] Mr Ganesh Pandurang Jadhav, Dr. Narendra Narve, Prof. Vyankatesh Mundhe November (2017) *IJSDR* | **Volume 2**, Issue 11

[15] V.Vikram Reddy,M.Jawahar *International journal on Latest trends in Engineering and Technology(IJLTET).*

[16] A.H.A. Shah et al (2016) IOP *Conference Series: Materials Science and Engineering*. 114 012023.

[17] Ramanan G Edwin Raja Dhas J *International Journal of Applied Engineering Research* ISSN 0973-4562 **Volume 12**, Number 8 (2017) pp. 1729-1735.

Study of hardness and flexural strength of banyan and peepal fibre reinforced hybrid composites

V.Durga Prasada Rao[1,*], *G.Moses Dayan*[2], *V.Navya Geethika*[3]
[1,*]Dept. of Mechanical Engg., S.R.K.R. Engineering College, Bhimavaram, AP, INDIA Email: vdp009@gmail.com
[2]Dept. of Mech. Engg., S.R.K.R.Engineering College, Bhimavaram,AP, INDIA, Email:mechday@gmail.com
[3]Dept. of Mechanical Engg., S.R.K.R. Engineering College, Bhimavaram, AP, INDIA Email: vng.mini@gmail.com

Abstract. In the present work, Banyan and Peepal fibre reinforced hybrid composites are prepared through hand Lay- up technique..The objective of the work is to investigate the hardness and flexural properties of eight varieties of banyan – peepal hybrid composites. The composites include banyan-peepal-banyan-peepal-banyan fibre composite, peepal-banyan-peepal-banyan-peepal fibre composite , banyan-copper-banyan-copper-banyan fibre composite (F1-Cu-F1-Cu-F1), peepal-copper-peepal-copper-peepal fibre composite, banyan-peepal-copper-peepal-banyan fibre composite, peepal-copper-banyan-copper-peepal fibre composite, peepal-banyan-copper-banyan-peepal fibre composite, and banyan-copper-peepal-copper-banyan fibre composite . Each of these composites is prepared in 0^0 orientations (i.e., fibres parallel each other), 45^0 (i.e., fibres at 45^0 to each other) and 90^0 orientations (i.e., fibres perpendicular to each other). It is observed from the results that, the hardness of F2-Cu-F1-Cu-F2 composite with 90^0 orientation is high and that of F1-F2-F1-F2-F1 composite with 45^0 orientation is low. It is also noticed that, the bending strength of F2-Cu-F1-Cu-F2 composite with 0^0 orientation is high and that of F2-F1-F2-F1-F2 composite with 45^0 orientation is low, and as far as the composites with 90^0 orientations are considered, no results are obtained. Further it is noticed that, the hardness of all the eight composites with 90° orientation is found to be high compared to their remaining orientations and also the hardness of all the composites with 45° orientation is low compared to their 0° and 90° orientations

Keywords: *hybrid composites,hand lay up technique,hardness,flexural properties*

1. Introduction

Composites provide characteristics that are not obtainable from any discrete material. They are multifunctional material systems and cohesive structures made by combining two or more compatible materials. Physically, the constituent materials of a composite are different in composition and characteristics. The natural fibre composites are more environmental friendly, and are used in transportation applications (automobiles, railway coaches, aerospace), military applications, building and construction industries (ceiling paneling, partition boards), packaging, consumer products, etc. Their renewability, low density, price and satisfactory mechanical properties make them an attractive ecological alternative to glass, carbon and man-made fibres used in the fabrication of fibre composites. Joshi et.al [1] studied glass fibre reinforced composites and natural fibre composites in terms of their life cycle environmental performance and found that, environmentally natural composites are superior in the specific applications studied. Rana et al. [2] in their work showed that the mechanical properties of jute fibres increase by the use of compatibilizer. At a load of 60% by weight of fibre, the use of the compatibilizer improved the flexural strength as high as 100%, impact strength by 175%, and tensile strength to 120%. Shah and Lakkad [3] compared the mechanical properties of jute reinforces and glass-reinforces and found that, when introduced into the resin matrix as reinforcement, the mechanical properties of jute fibres considerably improved, but it is much lower than that by the introduction of glass and other high performance fibres. Ray et.al [4] has considered jute fibres, done alkali treatment with 5% NaOH solution for 0 to 8 hrs with 2 hrs increment at 30^0 C, and found that there was an improvement in the properties both for fibres and reinforced composites. Actually after the NaOH treatment, the fibres have got increased crystallinity, less hemi cellulose content, reduced amount of defects, therby superior bonding with the vinyl ester resin. Saha et.al [5] investigated physical-chemical properties of jute fibres after being treated with alkali (NaOH) solution. It was revealed by the SEM investigation that the fibre-matrix adhesion was improved due to the chemical bonds between fibre and matrix which was provided by the coupling agent. Monteiro S.N.Rodriquez et.al [6] tried to fabricate low cost composites using sugar cane bagasse waste as reinforcement to polymeric resins. They concluded that, the mechanical properties similar to wooden agglomerates can be achieved by fabricating composites

* Corresponding author: vdp009@gmail.com

with homogenous microstructures. Hassan et.al [7] reported in their work about the dimensional stability and mechanical properties of the composites prepared by converting the bagasse into a thermo formable material through esterification. Ramesh et al. [8] considered jute, sisal and glass fibre reinforced polyester composites and investigated their mechanical properties. They observed that maximum tensile strength has been obtained for the composite prepared by adding glass fibre to jute fibre. They also observed that maximum flexural strength was obtained for jute and sisal mixture composites, and sisal fibre composite has got the maximum impact strength. Alavudeen et al. [9] studied the mechanical properties of woven banana fibre, kenaf fibre and banana/ kenaf hybrid fibre composites. They found that, owing to the hybridization of kenaf with banana fibres, the mechanical strength of woven banana/kenaf fibre hybrid composites increases. Also the tensile, impact, and flexural strengths of the hybrid composites are superior to those of the individual banana and kenaf fibres.

After reviewing the available literature on natural fibre reinforced composites, it is observed that a fair amount of research on natural fibre reinforced polymer composites has been carried out, but little work on Banyan and Peepal fibre composites is evidenced so far. This enlightened the scope for the present study which demonstrates alkali treated and cured Banyan and Peepalfibres, and their effect on the mechanical performance of the fibres in terms of their flexural and hardness properties.

2. Experimental work

The Fabrication and testing of natural fibre reinforced composites includes

- Preparation of composites by hand lay-up method.
- Cutting samples of desired dimensions from fabricated composite laminates.
- Hardness test of samples of fabricated composites.
- Bending test of samples of fabricated composites.

The raw materials used in the experimental work are

- Banyan root and Peepal bast fibres
- Epoxy resin LY556(GP) and hardener
- Copper foil

2.1 Banyan tree

Banyan tree is characterized by a tangle of branches, roots and trunks. The tree, which may spread across several acres, is deeply rooted. The tree bears fruits, which look like figs. The fruits when matured appear red in colour and are not edible. Paper can be directly made from the wood and bark of the banyan tree. The roots of the tree are used to make ropes, which help to secure wood bundles. A strong adhesive, called shellac, is produced by the sap of the tree. Also its sap is used as a medicine for treating external skin inflammations, dysentery, toothaches and ulcers. Its bark and seeds are used for producing a herbal tonic which can cool the body. Figure 1 shows the banyan fibre after extraction.

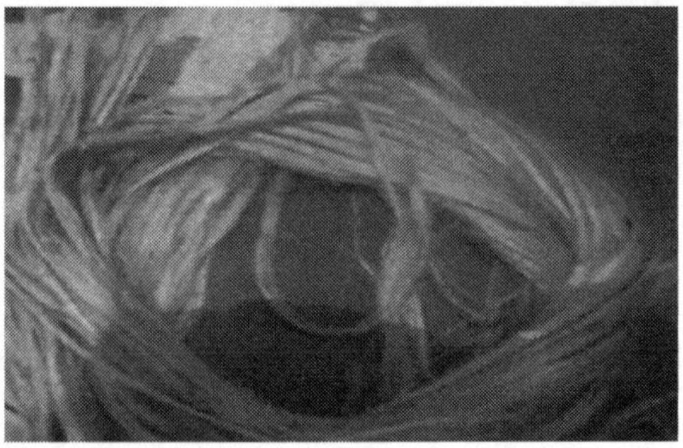

Fig. 1. Banyan fibre after extraction

2.2 Peepal tree

The peepal tree is a deciduous tree, native to India is also known as the "bodhi" tree. The Peepal tree, Ficus Religiosa, is a species of the Sacred Fig tree. Its bark is light grey, smooth to touch and peels off easily. The bright green leaves of the tree are shedded in the summer months of March and April. The fruit borne by this big tree, which is a small type of fig, ripens by the month of May. Peepal tree is easily propagated through the cutting or the seeds. Figure 2 shows the peepal fibre after extraction.

Fig.2.Peepal fibre after extraction

2.3 Epoxy grade resin

One of the best systems to use for applications that will be subject to water immersion and marine environments is Epoxy/polyamide based resins. It retains its physical properties even after prolonged water immersion and provides excellent resistance to acidic and caustic exposure and salt water. Epoxy/polyamide resin system is specially prepared for use in high strength bonding applications to provide structural strength to a class of marine and building application and for structural fibre glassing. It cures even in humid and temperature conditions. It is cured generally at room temperature, but can be snap cured for a short period of time at elevated temperature. Figure 3 shows the epoxy resin used in the present work (LY556-GP) for the preparation of composites.

https://doi.org/10.1051/matecconf/201817204009

2.4 Epoxy Hardener

Physical properties of the epoxy resin system such as tensile, compression, flexural properties, etc., are influenced by epoxy hardeners. Epoxy hardeners provide gel time, mixed viscosity and de mould time to the epoxy resin system. They are not catalysts but react with the epoxy resins, greatly contributing to the ultimate properties of the cured epoxy resin system. The performance of epoxy hardeners in the resin system depend on its physical characteristics while applying the epoxy resin system, and also the chemical characteristics of the epoxy resins.

Fig.3. Epoxy grade resin

The physical characteristics of the epoxy resins influencing the behaviour of epoxy hardeners in the epoxy resin system are temperature of the work area, temperature of the resin system (i.e., the heated resins) and moisture (dampness). The chemical characteristics of the epoxy resins that influence the performance of epoxy hardener are the amount and kind of diluents and fillers in epoxy resins, and its viscosity.

2.5 Copper foil

Copper is one of the ancient materials used by human beings. The most common alloying elements of copper are tin, zinc aluminium, nickel and silicon. Owing to excellent electrical conductivity, copper has its major application in the electronics industry. Besides the thermal conductivity, it has good strength and fatigue resistance, excellent corrosion resistance and ease of fabrication. These make copper and its alloys such extensively used materials.

Fig. 4. Copper foil

The mechanical properties of copper strongly depend on its microstructure. As far as the rolled copper foils of different temper are considered, the minimal required ultimate tensile strength and fracture strain at room temperature depend on the area and density of foil which is technical measure of thickness of a foil. Thinner foils have lower numbers in strength and fracture strain. Figure 4 shows the copper foil used in the preparation of composites.

2.6 Fibre preparation

The raw banyan roots and peepal bast are cleaned first with water and dried. Then by hand slitting the aggregations is gently dispersed and their outer shells are removed by the knife. Later the fibres are treated with 7% NaOH solution. The NaOH treatment increases the surface area of the fibre by removing the cementing material present in the fibres namely lignin and hemicelluloses. This increase in surface area of the fibre leads to better adhesion of the fibre and matrix, thus increasing the tensile strength. Later the fibres are properly dried. Finally the fibres are cut into required length and kept aside after being measured for required proportion of weight.

2.7. Mould preparation

At first the moulds of size 150 mm X 40 mm X 6 mm are produced for the fabrication of required composites. A clean smooth surfaced wooden board is taken and washed thoroughly, and then it is covered with a non-reactive thin plastic sheet. Later a glass plate of equal size (thickness = 3mm) to that of the board is taken. The glass is placed on the wooden board. Square bits after being cut to sizes of 8 mm in desired dimensions are nailed surrounding the glass. These bits are to be nailed in such a way that no polymer leaks out while preparing the composites. Also the bits are carefully nailed so that the glass does not move aside and the overall dimensions of the mould are not distorted. The glass structure is smoothly taken out after the nailing of bits is completed, leaving behind the mould of desired dimensions.

2.8 Polymer-hardener mixture preparation

For the making of good composite, the measurements of the polymer – hardener mixture is to be accurate and it should be very uniform. While preparing the mixture, an accurate amount of polymer and 10% of its hardener are taken and this mixture is stirred thoroughly till it becomes a bit warm. A small extra amount of hardener is generally taken to account for the wastage in the process. Minute amounts of hardener should be taken each time during the preparation of mixture because little extra amount of hardener can spoil the composite. In the present work, hardener HY951 (araldite) which has a viscosity of 10-20 poise at 25^{0}C is used.

2.9 Hand lay-up technique for preparing the composites

After the natural fibre plies were cut to required sizes, three numbers of plies were taken for fabrication of each composite. Then the correct proportions of fibres, and accordingly the resin and hardeners were weighed. Epoxy and hardener were mixed in a bowl by using glass rod. Care was taken to avoid formation of bubbles in the mixture because the trapped air bubbles in matrix may result in failure of the material. The subsequent fabrication process consists of first putting a releasing film on the mould surface, and applying a polymer coating on it. Then the first fibre ply of one kind was put, proper rolling was done, and then resin matrix was again applied. Next to it the fibre ply of another kind was put, and compression was done using compression pads after rolling has been done. This procedure was repeated until a total of five alternating fibres have been laid on e by one. To ensure a good surface finish, a polymer coating is done on the top of the last ply of fibre. Finally a releasing film was kept on the top, light rolling was carried out, and then a 20 kg weight was applied on the composite. The weight was left for 1 day to allow sufficient time for curing and subsequent hardening of the composite. Figure 5 illustrates the hand lay-up technique used in the preparation of composites.

Fig.5. Hand lay-up technique for composites preparation

In the present work, the hybrid composite specimens were fabricated from Banyan fibre, Peepal fibre, Copper foil and polymer-hardener matrix. The weight fractions of composites were maintained at 30% fibre, 55% resin and 15% foil. In this work actually eight varieties of hybrid composites in three different configurations (0^0, 45^0 and 90^0 orientations) are prepared by hand lay-up technique. Five plies of fibre were used in each hybrid composite. The eight varieties of hybrid composites under consideration are:

- F1-F2-F1-F2-F1,
- F2-F1-F2-F1-F2,
- F1-Cu-F1-Cu-F1
- F2-Cu-F2-Cu-F2
- F1-F2-Cu-F2-F1
- F2-Cu-F1-Cu-F2
- F2-F1-Cu-F1-F2
- F1-Cu-F2-Cu-F1

where F1 – Banyan fibre,
F2 – Peepal fibre, and Cu – Copper foil

After the hybrid composites are fabricated, the specimens are cut in the desired shape to test the mechanical and flexural properties of the banyan-peepal natural hybrid composites. Shore-A hardness tester is used for the hardness testing, and Universal Testing Machine is used for bending test on the samples of composites under consideration.

2.10 Hardness test

The hardness testing of plastics is most commonly done by the Shore (Durometer) test or the Rockwell hardness test. According to standards, the hardness test on the fabricated composites is done for determining the resistance of the composites for plastic deformation. Shore hardness is a measure of the resistance of material to indentation by 3 spring loaded indenter. The higher the hardness number, the greater is the resistance. Samples used for hardness testing are shown in Figure 6. The dimensions of specimen are 200 mm X 30 mm X 5 mm. The Shore –A hardness tester used in the study is shown in Figure 7.

Fig. 6. Composites with 0° orientataion used for Hardness testing

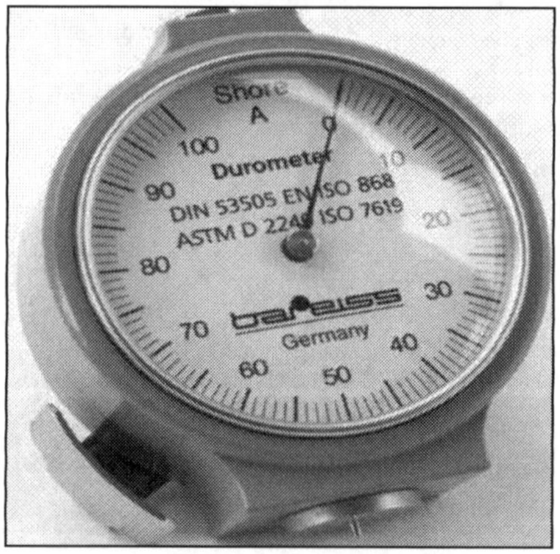

Fig. 7. Shore-A hardness tester

https://doi.org/10.1051/matecconf/201817204009

2.11 Bending test

A bending test is a method for measuring stiffness and yield properties of certain materials. In the present work, a three point flexural test is used, the main advantage of which is the ease of the specimen preparation and testing. The bending test is done using universal testing machine (UTM). A UTM is used to determine a material's response to varying strains, called loads. The (TUE/TUE-C-00 UTM) used for testing the composite samples is of 400 KN capacity with an accuracy of ±1% within 2% to 100% of the range and a ram stroke of 200 mm.

The fabricated composites were sheared according to standards. The dimensions of specimens used for the bending test are 150mm X 40mm X 6mm each and the test is carried out until failure takes place. When the load is applied, the deflection and bending strength corresponding to each load is calculated and the results are given in the display. Figure 8 shows the UTM with a specimen during bending test and Figure 9 shows the specimens after being tested on UTM. The tensometer instruments provided on the UTM give load as a function of extension. By using this data, a load-deflection curve or stress-strain curve will be plotted. This curve is unique for every material and provides key measures.

Fig.8. Sample while testing for bending strength on UTM

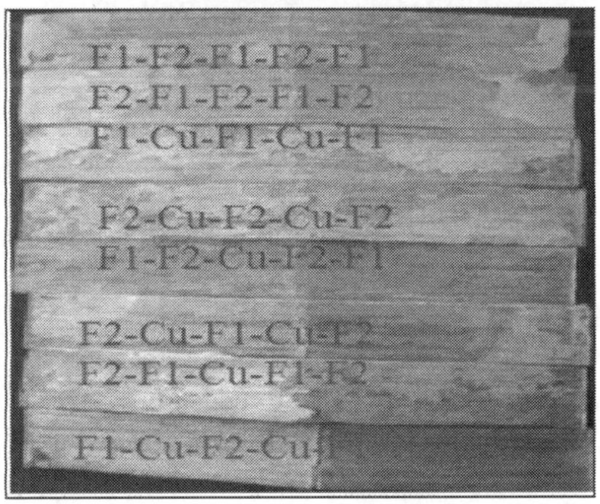

Fig. 9. Specimens after bending test for Composites with 0° orientation

3. Results and Discussions

Figure 10 shows the comparison of hardness of composites with 0^0, $45^0$and 90^0 orientations. It is observed from the bar chart that, the hardness of F2-Cu-F1-Cu-F2 composite with 90^0 orientation is high (95 Shore -A hardness number) and that ofF1-F2-F1-F2-F1 composite with 45^0 orientation is low (73 Shore-A hardness number). Figure 11 shows the comparison of bending strengths of composites with 0^0 and 45^0 orientations. It is observed from the bar chart that, the bending strength of F2-Cu-F1-Cu-F2 composite with 0° orientation is high (9.32 MPa) andF2-F1-F2-F1-F2 composite with 45° orientation has the low strength (3.856 MPa).The results for 90°orientation are not obtained.

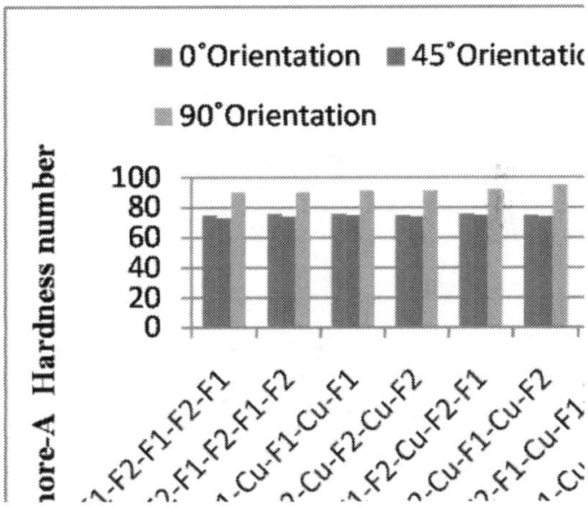

Fig.10. Comparison of hardness of composites with 0^0, 45° and 90^0 orientations

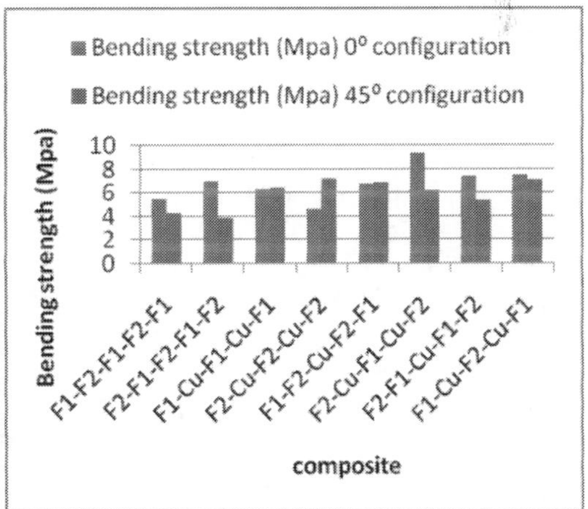

Fig.11. Comparison of bending strengths of composites with 0^0 and 45° orientations

4. Conclusions

In the present work, the hardness and flexural behaviour of Banyan and Peepal fibre reinforced hybrid composites was studied. The investigation includes fabrication of composites through hand lay-up technique, and evaluation of hardness and flexural properties as per the experimental plan. The major contributions drawn from the investigation are summarized as below:

- The hand lay-up technique was successfully adopted in the fabrication of Banyan and Peepal fibre reinforced hybrid composites.
- The hardness of F2-Cu-F1-Cu-F2 composite with 90^0 orientation is high (Shore-A hardness number=95) and that of F1-F2-F1-F2-F1 composite with 45^0 orientation is low (Shore-A hardness number=73).
- The hardness of all the composites with 45° orientation is low compared to their 0° and 90° orientations.
- The hardness of all the eight composites with 90° orientation is found to be high compared to their remaining orientations.
- The bending strength of F2-Cu-F1-Cu-F2 composite with 0^0 orientation is high (9.32 MPa) and that of F2-F1-F2-F1-F2 composite with 45^0 orientations is low (3.856 MPa).
- Except for the F1-Cu-F1-Cu-F1 composite, F2-Cu-F2-Cu-F2 composite and F1-F2-Cu-F2-F1composite the bending strength of all the composites with 0° orientation is high compared to their 45° orientations.

On an overall perspective, within the purview of the present study, better flexural strength has been obtained for the Banyan–Peepal fibre reinforced hybrid composite with Copper as one of the elements in it.

References

1. Joshi, S. V., Drzal, L. T., Mohanty, A. K. and Arora, S, Composites: Part A, Vol. **35**, pp. 371-376 (2004)
2. Rana, A.K., Mitra, B.C. and Banerjee, A.N., J.Appl.Polym. Sci.**71**:pp.531–539(1999)
3. Shah and Lakkad, Fibre science and technology**15,** 4146(1981)
4. Ray, B.K. Sarkar, A.K.Rana, N.R.Bose, Part A **32,** pp.119-127(2001)
5. ProsenjitSaha, Suvendu Manna, Sougata Roy Chowdhury, RamkrishnaSen, Debasis Roy, BasudamAdhikari, Bio resource Technology, **101:9,** 3182–3187(2010)
6. Monteiro S.N Rodrikuez, R.J.S. De Souza, M.V.D'Almeida J.R.M., Advanced Performance Material, **5:3** ,183-191(1998)
7. Hassan M.L.Rowel.l R.M.,Fadl N.A., Yacoub S.F. and Chrisainsen.A.W, Journal of applied polymer science, **76**, 575-586(2000)
8. Ramesh, M., Palanikumar, K. and Reddy, K.H,. Composites: Part **B**, 48, 1-9(2013)
9. Alavudeen.A,Rajini.N,Karthikeyan.S,Thiruchit-rambalam.M and Venkateshwaren.N,Materials and Design, **66,** 246-257(2015).

Modeling and experimental investigation of process parameters in WEDM for surface roughness using regression model

A. Muniappan*, R. Senthilkumar, V. Jayakumar, S. Venkata Ravikumar and P. Sai Tarunkumar

Department of Mechanical Engineering, Saveetha School of Engineering, Saveetha Institute of Medical and Technical Sciences, Chennai, Tamil Nadu, India.

Abstract. The present study focused on the multiple regression modeling and predicting the surface roughness of the Aluminum hybrid composite during the WEDM process. The hybrid MMC was manufactured by process named as stir casting utilizing particulates of Silicon carbide and graphite each in Al6061 combination. The analyses were outlined with Taguchi L27 design matrix. Mathematical relationships between the surface roughness and WEDM cutting parameters (Pulse on time, Pulse off time, current, gap voltage, wire speed and wire tension) have been investigated. The results show that the multiple regression analysis is a successful method for developing a mathematical model to predict the surface roughness. The optimum value of process parameters for the predicted optimum value of surface roughness (1.285) is pulse on time 106 units (Level 1), pulse off time 60 units (Level 3), peak current 90 units (Level 2), gap set voltage 50 units (Level 3), wire speed3 units (Level 1) and wire tension 12 units (Level 3).The optimum results are adopted in validation study and the results based on WEDM process responses can be effectively improved.

Keywords. WEDM, Taguchi method regression model, stir casting, hybrid composite and surface roughness.

1 Introduction

Running with the progression of the mechanical era, the solicitations for composite materials having high hardness, quality, wear assurance, and disintegration security are growing [1]. Research had been done in MMC composites more than a significant extended period of time, by examiners on creation methodologies and material property estimations, and various sublime research happens have been gained [2-4]. Metal system composites discover applications in air space industry and auto applications in perspective of their light weight, high particular quality and uncommon wear safe. Electrical release machining is one of the non-consistent machining outlines which have been completely used to pass on jumble shapes, bombs hopelessly and shapes [5]. It is additionally utilized for completing parts for air space and automobile industry [6]. The machining of PAMCs presents a vital test, since various help materials are essentially harder than the routinely utilized snappy steel (HSS) and carbide instruments [7-8]. Wire electrical machining is an essential nonconventional machining process. It can be utilized to machine the electrically conductive work materials paying little regard to their quality and hardness. In the midst of machining in WEDM, no power is associated on sample work piece in the midst of machining in WEDM in light of the way that there is no prompt contact among gadget and work. The most key execution measures in WEDM are material ejection rate, surface finish, kerf and wire wear rate. Metal Composites possess good properties over metals The present work aims at applying Taguchi method for optimization of cutting speed machined by WEDM process of Al/SiC/Gr hybrid composite.

2 Literature survey

The true motivation behind this writing survey is to give foundation data on the issues to be considered in this postulation and to mention the present investigation. The utilization of aluminum composite is unavoidable in aviation and vehicle ventures attributable to its properties of high quality to weight proportion, excellent erosion protection, simple machinability and ease (Singh et al)[9]. Be that as it may, issues related with aluminum and its compound material is poor high temperature execution and low wear protection. To overcome this issues, new designing materials have been created by fortifying hard support material on delicate aluminum grid is called aluminum metal grid composites (Pedersen and Ramulu)[10]. Shyam lal and Sudhir Kumar [11-13] fabricated $Al7075/Al_2O_3/SiC$ hybrid composite & studied the effect of WEDM parameters SR and kef width using Taguchi-Grey relational analysis. It was confirmed from results that roughness increased due to Pulse on time and current, because of longer spark with high intensity, higher material melts per spark on the surface. Adeel Ikram and Nadeem ahamd mufti [14] experimented result of WEDM parameters on MRR for gadget steel D2. They take wire reinforce speed, beat on-time, beat off-time, dielectric weight, open voltage, servo voltage and wire weight as information parameter. Han et al.[15] pick the machining parameters (tallying discharge current, beat length, beat between time, administered beat time, most distant point effect,

* Corresponding author: inspire.munish@gmail.com

dielectric and material) on surface brutality. They recommend that a short pulse term combined close to a high apex regard gives better surface obnoxiousness. Ugrasen et al. [16] used Taguchi method close to the extortion neural framework to refresh the precision, surface seriousness and MRR. They found that for the more than three responses, current is the most goliath factor Sudhakara and Prasanthi [17] investigated machining parameters for tool steel in WEDM. They found the range perfect setting of different parameters as pulse on time = 108-128 µsec, beat off time = 47-63 µsec, peak stream = 11-13 Amperes, begin opening set voltage = 18-68 Volts, wire strain = 2-8 Grams and water weight = 8-14 Kg/cm^2. Kanlayasiri and Jattakul[18] found the perfect machining condition for dimensional deviation and SR for K460 instrument steel by wire EDM machine. They thought about cutting pace, top current, and change evacuate as information variable and dimensional accuracy and roughness of surface. They found that both current and alter disengage is gigantic factor for estimation while peak current impacts the surface unpalatability. Yam BH et al.[19] streamline the strategy parameters for EDM for HastelloyC276 by using Gray Relational Analysis. GRA, for finding the perfect parameters influencing both MRR and Kerf are accepted to be TON = 110 µs, TOFF=60 µs, IP=12A, SV=10 V, for higher MRR & lower Kerf width.The present study focuses on the modeling and prediction techniques to determine direct effects of WEDM parameters (Pulse on time, Pulse off time, current, gap voltage, wire speed and wire tension) on the surface roughness which is important for cutting dies and complex shapes.

3 Materials and method

3.1 Preparation of hybrid composite

In this investigation, the half and half MMC has been created by blend throwing process. The crossover composite comprises of 10 wt% SiC and 5 wt% Graphite particulates in metal lattice Al6061 compound. The Al combination of 6xxx arrangement is having the capacity to be used in aviation and car ventures in light of its high quality to-weight proportion and great protection from consumption. Composition of Al6061 composite is appeared in Table 1. Fortifications SiC and graphite in particulate frame are utilized to create the hybrid composite. The easiest and the most savvy technique for fluid state creation is stir casting [19]. Fig.1 demonstrates the stircast equipment utilized for manufacture of hybrid composite.

3.2 Machining parameters and response

Machining process parameters in WEDM, Pulse on time, Pulse off time, current, gap voltage, wire speed and wire tension were considered as input parameters. Responses Surface roughness (SR) was measured after machining for investigation. The scopes of these procedure parameters were chosen on the premise of the pilot tests. The levels of different parameters and its assignments are exhibited in Table 1.

Table 1. Process parameters and their levels

Symbol	Process parameter	Level 1	Level 2	Level 3
A	Pulse on time	108	117	126
B	Pulse off time	40	50	60
C	Pulse current	90	160	230
D	Gap set Voltage	10	30	50
E	Wire drum speed	3	4	5
F	Wire tension	4	8	12

3.3 Experimental design using Taguchi method

Taguchi strategy is a proficient apparatus for the outline of a great assembling framework. It also effectively solves some complex problems in manufacturing (Roy 1990).It is a strategy in view of OA tests, which give much decreased change to the try different things with ideal setting of process control parameters. The six control parameters, that is, pulse on time (A), pulse off time (B), top current (C),Gap set voltage (D), wire drum speed (E) and wire tension(F) at three levels were chosen in this examination. The tests were finished by Table 2. This table just speaks to specific level of the different elements of the procedure at which the examinations would be directed.

3.4 Experimental set up

Analyses were led on Electronica Sprint cut (Electra-Elplus 40A Dlx) CNC wire electrical release machine to think about the surface roughness influenced by the machining parameters at various levels. WEDM is a start disintegration process. The flashes are produced between the work piece and the wire terminal. The dielectric liquid is ceaselessly encouraged into the machining zone with required weight. The material is getting expelled by a progression of discrete sparkles occurring at the zone to be machined through electro-warm system. Test set up of the wire electrical release machine is appeared in Figure 1.

Amid machining process little gap kept up between the work and wire material. The machined particles were flushed away by the persistent stream of the dielectric liquid. The wire is held by a stick direct at the upper and lower parts of the work piece. The work example measure utilized as a part of this examination is 95 x 80 x 8 mm rectangular plate. Diffused coated wire of 0.25 mm diameter was utilized as a part of this investigation. Deionized water was utilized as dielectric liquid at room temperature. In the wake of machining, the examples were cleaned with acid after machining. The surface roughness was measured by Mitutoyo surface roughness tester.

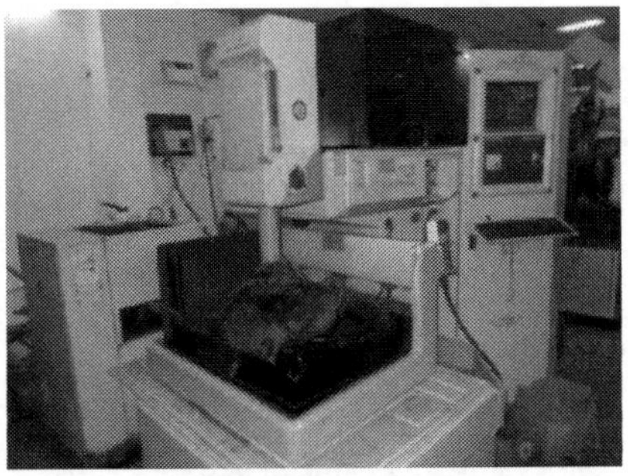

Fig. 1WEDM experimental set up (Electronica Sprint cut)

4 Results and discussion

Proper selection of machining parameters is an important step in the WEDM process, which leads to better machining performance. Whereas improperly selected parameters may cause serious problems like short-circuiting of wire, wire breakage, and deterioration in work piece surface quality, thereby imposing certain limitations on the production schedule and also reducing productivity.

4.1 Taguchi's analysis

S\N ratio is used to measure the sensitivity of the response which to be investigated at controlled manner. The signal represents attractive effect for the response and noise represents the unwanted effect for the responses. Higher the best (HB) is selected to cutting speed. The S/N ratio expression for cutting speed is as follows

Smaller the better

$$\left(\frac{S}{N}\right)SB = -10\ log\left[\frac{1}{n}\sum_{i=1}^{n}\frac{1}{y^2}\right]$$

Larger the better

$$\left(\frac{S}{N}\right)SB = -10\ log\left[\frac{1}{n}\sum_{i=1}^{n}y^2\right]$$

Nominal the better

$$\left(\frac{S}{N}\right)SB = -10\ log\left[\frac{1}{ns}\sum_{i=1}^{n}y^2\right]$$

Y_i is the i_{th} value of response measured
N the total number of runs and
S standard deviation

4.2 Effect of process parameters on surface roughness

Table 3 shows the average surface roughness data. The analysis showed the effects of six main process parameters while cutting the hybrid aluminum composite. Based on at 95% confidence level (x=0.05), pulse on duration and gap voltage were significant effect towards surface roughness whereas pulse off time, current, speed of wire drum and wire tension did not have significant effect. Table 7 shows the pooled ANOVA for the surface roughness. The contribution of each parameter is calculated by ANOVA. The pulse off duration was the most significant parameter, which contributed maximum of 33.50% to the surface roughness. The contribution from other parameters were gap voltage 24.50%. The average S/N ratios and average mean values for surface roughness of each parameter are shown in Table 5 and 6. The performance of each parameter at different levels from the S/N ratio analysis the best levels for minimum surface roughness are A1,B3,C1,D2, E1 and F3. The main effect plot for S/N ratio, mean are shown in Figs 3 and 4.

4.3 Effect of pulse on time on surface roughness

The effect of pulse on time on the surface roughness was observed from fig.3. Surface roughness, which increases due to increase in pulse on duration increases. The pulse on time is the most significant influencing parameter on surface roughness. High temperature is observed at high intensity discharge energy when the spark strikes the surface. Due to this effect part of specimen to melt and evaporated. Due this effect, a large quantity of material melts and removed. These Results are in line with some of the researchers [12, 24].

4.4 Effect of pulse off time on surface roughness

The impact of pulse off duration on the surface roughness is shown in figure 3.Increased pulse off time duration reduces surface roughness. This is due to less number of discharges occur for a specific period of time. Thus, small quantity of metal melting takes place for larger pulse off time. This produces lower surface roughness. The results from this investigation are in agreement with other researchers.[12, 25].

4.4 Effect of pulse current on surface roughness

Figure 3 indicates the effect of pulse current on the surface roughness. Pulse current is not a significant process parameter affecting the surface roughness. The surface roughness increases as the pulse current increases. Discharge energy becomes intense with an increase in the pulse current, due to this poor surface finish achieved. Hence, to obtain lower surface roughness, the pulse current should be kept low or minimum. These results are in agreement with other researchers [12, 25].

4.5 Effect of gapset voltage on surface roughness

Figure 3 demonstrates the impact of gap voltage on the surface roughness. Surface roughness decreases with increase in gapset voltage. The influence of gapset voltage on surface roughness is significant parameter. Hence, to obtain good surface finish, the gapset voltage should be kept high. These outcomes are in concurrence with different analysts [25].

4.6 Estimation of optimum performance characteristic

The optimum value of surface roughness was predicted at the selected level of parameters. The process parameter and their optimum levels are selected from the response graphs as $A_1B_3C_1D_3E_1$andF_3, that is pulse on time (level 1), pulse off time (level 3), peak current (level 1), gapset voltage (level 3), wire speed (level 1) and wire tension (level 3).

$$\mu_{SR} = SR + (A_1\text{-}SR) + (B_3\text{-}SR) + (C_1\text{-}SR) + (D_1\text{-}SR) + (E_1\text{-}SR) + (F_3\text{-}SR)$$

where SR is the overall mean surface roughness (2.73 μm). Optimum value of surface roughness was predicted as 1.285μm using Minitab version18statistical software.

Table 2. Taguchi L27 design matrix

Exp. run	Control factors and levels					
	A	B	C	D	E	F
1	1	1	1	1	1	1
2	1	1	1	1	2	2
3	1	1	1	1	3	3
4	1	2	2	2	1	1
5	1	2	2	2	2	2
6	1	2	2	2	3	3
7	1	3	3	3	1	1
8	1	3	3	3	2	2
9	1	3	3	3	3	3
10	2	1	2	3	1	2
11	2	1	2	3	2	3
12	2	1	2	3	3	1
13	2	2	3	1	1	2
14	2	2	3	1	2	3
15	2	2	3	1	3	1
16	2	3	1	2	1	2
17	2	3	1	2	2	3
18	2	3	1	2	3	1
19	3	1	3	2	1	3
20	3	1	3	2	2	1
21	3	1	3	2	3	2
22	3	2	1	3	1	3
23	3	2	1	3	2	1
24	3	2	1	3	3	2
25	3	3	2	1	1	3
26	3	3	2	1	2	1
27	3	3	2	1	3	2

Fig. 1 Stir casting setup

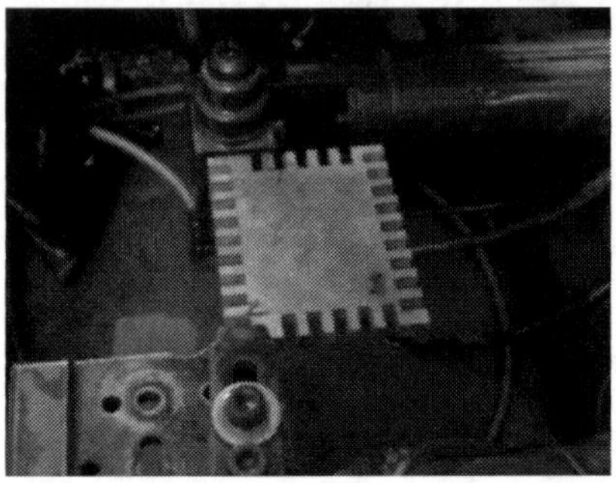

Fig. 2 Machined work piece

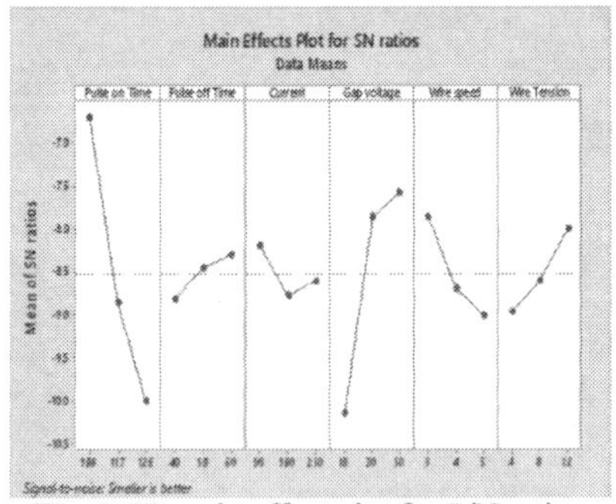

Fig. 3 SR main effect plot for S/N ratio

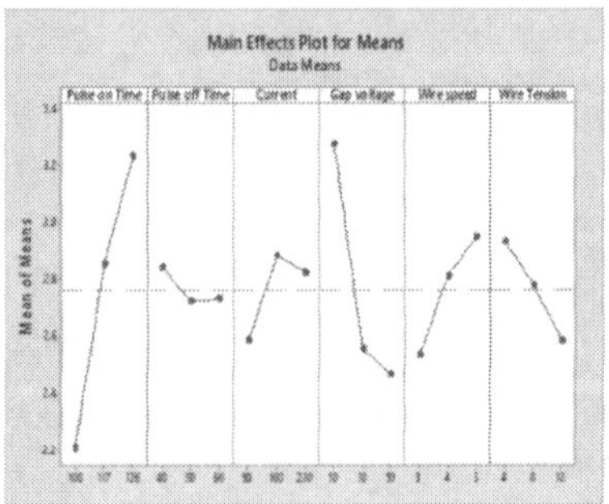

Fig. 4 SR main effect plot for means

Table 3. Surface roughness and S/N ratio

Ex.No.	Pulse on Time	Pulse off Time	Peak Current	Gap set voltage	Wire speed	Wire Tension	Surface roughness (µm)	S/N ratio of SR
1	108	40	90	10	3	4	2.42	-7.6763
2	108	40	90	10	4	8	2.68	-9.7992
3	108	40	90	10	5	12	2.34	-7.3843
4	108	50	160	30	3	4	2.14	-6.6083
5	108	50	160	30	4	8	1.64	-4.2969
6	108	50	160	30	5	12	2.45	-7.7833
7	108	60	230	50	3	4	1.77	-4.9595
8	108	60	230	50	4	8	1.78	-7.4582
9	108	60	230	50	5	12	1.87	-4.4543
10	117	40	160	50	3	8	2.06	-6.2773
11	117	40	160	50	4	12	2.36	-7.4582
12	117	40	160	50	5	4	2.83	-11.664
13	117	50	230	10	3	8	2.77	-8.8496
14	117	50	230	10	4	12	3.39	-10.604
15	117	50	230	10	5	4	3.98	-11.9977
16	117	60	90	30	3	8	2.56	-9.2779
17	117	60	90	30	4	12	2.31	-7.2722
18	117	60	90	30	5	4	2.45	-6.3613
19	126	40	230	30	3	12	2.97	-6.5267
20	126	40	230	30	4	4	3.81	-11.6185
21	126	40	230	30	5	8	3.54	-10.9801
22	126	50	90	50	3	12	2.78	-8.8809
23	126	50	90	50	4	4	2.65	-8.4649
24	126	50	90	50	5	8	2.7	-8.6273
25	126	60	160	10	3	12	3.68	-11.6866
26	126	60	160	10	4	4	3.7	-11.364
27	126	60	160	10	5	8	3.94	-11.9099

Table 4. Response table for S/N ratio

Level	Pulse on time	Pulse off time	Peak current	gapset Voltage	Wire speed	Wire tension
1	-6.713	-8.821	-8.194	-10.141	-7.86	-8.968
2	-8.862	-8.457	-8.783	-7.858	-8.704	-8.608
3	-10.007	-8.305	-8.605	-7.583	-9.018	-8.006
Delta	3.293	0.516	0.589	2.559	1.158	0.963
Rank	1	6	5	2	3	4

Table 5. Response table for mean

Level	Pulse on time	Pulse off time	Peak current	gapset Voltage	Wire speed	Wire tension
1	2.209	2.841	2.587	3.274	2.534	2.931
2	2.854	2.722	2.884	2.556	2.812	2.779
3	3.231	2.731	2.823	2.464	2.948	2.584
Delta	1.022	0.119	0.298	0.81	0.413	0.347
Rank	1	6	5	2	3	4

Table 6. ANOVA table for surface roughness

Source	DF	Adj SS	Adj MS	F-Value	P-Value	% contribution
Pulse on time	2	4.8107	2.40534	8.04	0.005	33.42
Pulse off Time	2	0.0789	0.03947	0.13	0.877	0.5
Peak current	2	0.4453	0.22263	0.74	0.493	3
gapset voltage	2	3.5436	1.7718	5.92	0.014	24.56
Wire speed	2	0.7991	0.39957	1.34	0.295	5.55
Wire Tension	2	0.5435	0.27174	0.91	0.426	3.77
Error	14	4.189	0.29921			29
Total	26	14.4101				

$\mu_{SR} = SR + (A_1 - SR) + (B_3 - SR) + (C_1 - SR) + (D_1 - SR) + (E_1 - SR) + (F_3 - SR)$

where SR is the overall mean surface roughness (2.73 μm). Optimum value of surface roughness was predicted as 1.285 μm using Minitab version18statistical software.

4.7 Mathematical model for surface roughness

Surfaces roughness is an important performance measure that dictates the manufacturing quality of the WEDMed components. If the surface finish of the machined part is the decisive factor due to its application requirements, then the work material must be machined with low cutting speed. Based on the proposed quadratic process order, the regression equation defining the relation between surface roughness and process parameters in terms of coded factors is given by

Surface roughness = -3.75 + 0.0568 pulse on Time – 0.0055 pulse off time + 0.00169 current -0.02025 Gap voltage + 0.207 Wire speed – 0.0433 Wire tension (1)

The model was successfully developed such that the changes in the input parameter reflects the output. Mathematical modeling of the response variables (surface roughness) was made using a Taguchi's design and the experimental results were compared with the predicted values of the response variables to decide about the closeness of the predictions to the experimental values.Figure 5 shows the comparison of the experimental and the predicted value of surface roughness.

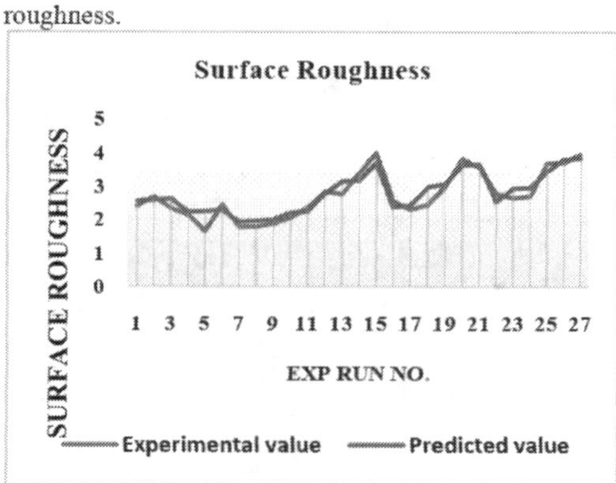

Fig 5. Comparison of the experimental and the predicted data

5 Conclusions

Hybrid MMC of 15 wt% SiC (10%) and graphite (5%) particulates in Al6061 alloy was prepared using stir casting process. The cast hybrid composite was machined on wire electrical discharge machine. The estimation of optimum machining parameters to maximize cutting speed was investigated using Taguchi's technique was studied. The conclusions drawn from this study are as follows:

- The process parameters pulse on time gap voltage are significant on surface roughness. Peak current, pulse off time, wire speed and wire tension did not have significant effect on surface roughness.
- The pulse on time was the most significant parameter contributing to the surface roughness (33.42%) , followed by Gapset voltage (24.56%) and wire speed (5.51%). The contribution of pulse off time, wire tension were less than 1 % effect.
- The optimum value of process parameters for the predicted optimum value of surface roughness (1.285) is pulse on time 106 units (Level 1), pulse off time 60 units (Level 3), peak current 90 units (Level 2), gap set voltage 50 units (Level 3), wire speed3 units (Level 1) and wire tension 12 units (Level 3).The optimum results are adopted in validation study and the results based on WEDM process responses can be effectively improved.
- The model was successfully developed using regression equation that the changes in the input parameter reflects the output. The variation between experimental value and predicted value was minimum. A good agreement between experimental and prediction results (using Regression model) was perceived for surface roughness. However, the prediction accuracy is limited to the studied range of the machining parameters. The percentage accuracy of both the prediction tools can be further improved by testing and conducting the experiments at a varying range of input parameters.

References

1. C. Tatar, N. Ozdemir, Physica B Condens Matter. **405**, 896-899 (2010).
2. A. Manna, H.S. Bains, P.B. Mahapatra, J Compos Mater. **45**(19), 2003-2010 (2011).
3. N. Tosun, C. Cogun, G. Tosun, J Mater Process Technol. **152**, 316-322 (2004).
4. N.O. Hung, N.L. Loh, Z.M. Xu, J Mater Process Technol. **58**, 114–120 (1996).
5. M.K. Surappa, Sadhana, **28**, 319-334 (2003).
6. 6. R. Snoeys, F. Staelens, W. Dekeyser, Ann. CIRP, **35/2**, 467-480 (1986).
7. T.W. Clyne, P.J. Whithers, *An introduction to metal matrix composites* (Cambridge University Press, London, 1992).
8. D.B. Miracle, Compos. Sci. Technol. **65**, 2526-2540 (2005).
9. M. Singh, B.K. Prasad, D.P. Mondal, A.K. Jha, Tribology International. **34**, 557-567 (2001).
10. W. Pedersen, M. Ramulu, J Mater Process Technol. **172**, 417–423 (2006).
11. Shyam Lal, Sudhir Kumar, ZA Khan, AN Siddiquee, Proceedings of the Institution of Mechanical Engineers, Part-B: Journal of Engineering Manufacture, **8**, 1-9, 2014.
12. 12.Shyam Lal, Sudhir Kumar, ZA Khan, AN Siddiquee, J. Braz. Soc. Mech. Sci. & Eng. **36**, 335-346 (2014).
13. Bikash Chaoudhuri, Ruma Sen, Subrata Kumar Ghosh, Subhash Chandra Saha Proceedings of the Institution of Mechanical Engineers, Part-B: Journal of Engineering Manufacture. **8**, 1-15, (2015).
14. Adeel Ikraml, Nadeem Ahmad Mufti, Muhammad Qaiser Saleem, Ahmed Raza Khan, Journal of Mechanical Science and Technology, 2133-2141 (2013).
15. S.S. Mahapatra, A. Patnaik, Int J Adv Manuf Technol. **34**, 911-925 (2007).
16. G. Ugrasen, H.V. Ravindra, Naveen Prakash, R. Keshavamurthy Procedia Materials Science, **6**, 1752 – 1760 (2014).
17. D. Sudhakara, G. Prasanthi, Procedia Engineering. **97**, 1565-1576 (2014).
18. K. Kanlayasiri, S. Boonmung, Journal of Materials Processing Technology, 192, 459–464 (2007).
19. BH. Yan, HC. Tsai, FY. Huang, LC. Lee, Int J Mach Tools Manuf. **45**, 251–259 (2005).
20. T.P.D. Rajan, P.M. Pillai, B.C. Pai, K.G. Satyanarayana, P.K. Rohatgi, Journal of Composites and Technology. **67**, 15-16 (2007).
21. S. Singh, S. Maheshwari, PC. Pandey, J Mater Process Technol. **149**, 272–277 (2004).
22. R. Ramakrishnan, L. Karunamoorthy, J Mater Process Technol. **207**, 343–349 (2008).
23. A. Muniappan, C. Thiagarajan, S. Somasundaram, IJMME-IJENS. **16**(5), 95-103 (2017).
24. A. Muniappan, C. Thiagarajan, S.Somasundaram, ARPN Journal of Engineering and Applied Sciences. **12**(2), 382-388 (2017),
25. R. Ramakrishnan, L. Karunamoorthy, Int J Mater Prod Technol. **35**, 199–215 (2009).
26. S. Sarkar, S. Mitra, B. Bhattacharyya, J Mater Process Technol, **159**, 286–294 (2005).

The Transition towards Sustainable Supply Chain Management: An Empirical Study

Amrita Singh[1], *Parul* Agarwal[1], *Saurav* Dixit[2*], *Sonali* Singh[3], and *Sanjai* Sahai[4]

[1] ABES Engineering College, Ghaziabad, India
[2]RICS School of Built Environment, Amity University, Noida, India
[3]Jaipuria Institute of Management, Noida, India
[4]ABS, Amity University Noida, India

Abstract. In the last one decade, the concept of SSCM, green and circular economy has been developed we have seen the synchronization of green and sustainable supply chain management practices by the firms to reduce the impact over the environment in developed countries as well as in developing countries. The intent of the author is to identify benefits of the new practices. The methodology used for the study is to analyze the available literature and identify the main attributes and prioritize using relative importance index. The study shall perceive direct and indirect benefits from the community and the surroundings. The study provides insight to the topic and to validate the results of the research same type of study to complete in the other regions of the country to have a clear and better understanding of the topic is suggested.

Keywords: Sustainable Supply chain. Green, Circular economy, Empirical study

1 Introduction

In the present worldwide commercial centre, offering an item is once in a while simpler than influencing the items to reach to clients. Organizations may need to choose the ideal approach to store, handle and move their items and administrations with the goal that they are accessible to clients in the correct collections, at the ideal time and at the opportune place. Store network majorly affects both client enjoyment and costs, making organizations operation successful. Organizations are putting accentuation on store network for a few reasons [1]. To begin with, making the channel accomplices both upstream and downstream maintainable, increasing upper hands, cost sparing to both organization and clients, blast in item assortment has made a requirement for enhanced inventory network. At last, logistics influence the earth and an organization's natural manageability endeavours. Transportation, warehousing, bundling and different logistics capacities are ordinarily the significant production network supporters of the association's ecological impression. In the meantime, they additionally give a standout amongst the most zones for cost funds. So building up a feasible green store network isn't just being socially and naturally dependable, it can likewise be produced.

Supportability has turned into a critical issue since it reviews financial viewpoints, as well as natural and social contemplations that the organizations ought to take after. In this paper, these three measurements of supportability have been taken in particular condition, economy and social variables. It is imperative to consolidate condition and social variables for contending in changing and turbulent condition to keep up an upper hand. Nature execution "assesses and measures the degree to which firms decrease their ecological effect through a diminishment in risky materials, asset utilization, ozone-harming substance outflows, squander transfer and waste drainage. The monetary execution getting from ecologically neighbourly items is estimated by standard measurements, for example, R.O.I, piece of the overall industry development, benefits development and benefits development rate. The measure of the social execution can be identified with 'human' centred nature of the field.

In the accompanying table, we display a couple of meanings of logistics administration, inventory network administration and Sustainable store network administration. The expectations to utilize these definitions by creators is to better clarify and grandstand the contrast between these capacities and to present a development of SSCM.

1.1 Supply chain management

" Supply chain administration is the coordination of key process from end client through unique providers that

* Corresponding author: sdixit@ricssbe.edu.in

gives items, administrations and data that include an incentive for clients and other partners" The Global Supply Chain Forum

"SCM as - fruitful coordination and a mix of each one of those exercises related to moving merchandise from the crude materials organize all the way to the finished client, for supportable upper hand. This incorporates exercises like frameworks administration, sourcing and acquirement, creation booking, arrange preparing, stock administration, transportation, warehousing and client benefit" [2]

"Inventory network administration is characterized as the foundational, key coordination of the conventional business capacities and the strategies over these business capacities inside a specific organization and crosswise over organizations inside the production network, for the reasons for enhancing the long haul execution of the individual organizations and the store network as a whole. [3].

Store network administration (SCM) is the oversight of materials, data, and funds as they move in a procedure from provider to maker to distributer to retailer to customer. Inventory network administration includes organizing and coordinating these streams both inside and among organizations.

1.2 Logistics management

"Arranging, executing and controlling the physical stream of materials, last products and related data from purposes of inception to purposes of utilization to meet client prerequisite at a benefit." Councils of logistics administration [4].

Logistics include "dealing with the stream of things, data, money and thoughts through the coordination of production network forms and through the vital expansion of place, period and example esteems." MIT Center for Transportation and Logistics.

Logistics administration is that piece of production network administration that designs, executes, and controls the proficient, viable forward and switch stream and capacity of merchandise, benefits and related data between the purpose of root and the purpose of utilization with a specific end goal to meet clients' necessities. – Council of inventory network administration experts.

1.3 Sustainable supply chain management

"The maintainable of production network administration requires a more extensive vision and must feature the financial natural and social parts of business hone"- "The Strategic, straightforward combination and accomplishment of an associations social, condition and monetary objectives in the fundamental coordination of key between hierarchical business process for enhancing the long haul financial execution of the individual and its supply chain."

"A feasible production network is the one that performs well on both customary measures of benefit and misfortune and additionally on an extended conceptualization of execution that incorporates both social and common measurements".

"The administration of store network operations, assets, data and supports keeping in mind the end goal to augment the inventory network productivity while in the meantime limiting the natural effects and expanding the social well-being".

2 Literature review

Production network administration infers the contribution of various performing artists in esteem creation, to energize the foundation of connections that are an imperative component for contending in a changing and turbulent condition. It is indispensable to join natural and social administration rehearses into the entire inventory network keeping in mind the end goal to keep up an upper hand.

Maintainability is a multidimensional develop that extends the financial primary concern idea, which concentrates on the productive utilization of assets and on accomplishing an arrival on speculations, by including social contemplations and advancing more prominent biological obligation. Given that each organization is a piece of a more extensive system and isn't an island in the present business world, there is an incredible significance in relationship administration that requires acting past organization limits. As of late, the method for picking up an upper hand has changed the structure of the opposition, with the goal that opposition between organizations has transformed into between production network rivalries.

The expanding impact of maintainability in inventory network administration and operations practices can likewise be credited to the way that, notwithstanding expanded requests of solid monetary execution, associations are presently considered in charge of the ecological and social execution by significant partners. Furthermore, practical store network administration has turned into a vital procedure empowering firms to make an upper hand.

The necessity to take a comprehensive perspective of the entire item inventory network is a central advance for setting up greener and more maintainable generation frameworks, in view of re-utilizing and reproducing materials. These frameworks could likewise prompt the production of new focused plans of action. Such models could be founded on the worldview of support to-support, empowering the utilization of crude materials known as specialized and organic supplements, which don't negatively affect the earth, have an altogether useful effect upon natural frameworks and come back to the biological community without medicines [5].

Moreover, without a moment to spare (JIT) supply involves providers creating and conveying to manufacturers right amount at an ideal time with nonstop and predictable conformance to execution particulars. JIT has turned out to be urgent for survival and accomplishment in turbulent business conditions that depicts the larger part of the business scene today. Electronic acquisition (e-acquirement) alludes to a virtual buying application that helps firms lessening their obtaining cost.

Both among business practitioners and researchers, it has been noted that sustainable supply chain management (SSCM) has increased its importance in companies together with the rising demand for sustainability. Research results show that companies are expected to increasingly pay interest on their supply chain, and not only to their direct suppliers but also to parties beyond their direct reach ([6]–[9]. The supply chains have grown largely due to globalization and outsourcing [10]. Currently, the number of companies who possess the knowledge of sustainability issues from their whole supply chain is small, which makes this research topic especially relevant as the majority of companies will soon be faced with the need of responding to growing sustainability demands of their stakeholders. The implementation of sustainability initiatives within the first tiers of suppliers is already quite widely researched. The success in implementation is said to derive from common goals and aligned initiatives within different tiers. The complexity of the network and industry is also taken into consideration. What is especially said to be lacking research, is analyzing the implementation within multiple companies in a supply chain [6], [8], [11]–[15].

2.1 Objectives of the study

I. To determine the factors influencing the companies to adopt sustainable supply chain management practices.
II. To prioritise the attributes on the basis of their relative importance index on various companies in the transition towards SSCM.

3 Research Methodology

For this research paper, a questionnaire survey method has been used to discover the Benefits of synchronizing sustainability and lean Construction in Indian Construction industry. First the perception of the professionals working in the industry on the implementation of Sustainability and Lean Construction has been analyzed through the Relative importance index (RII) and secondly the attributes identified through literature review as the Benefits of synchronizing sustainability and lean Construction were analyzed using Relative importance index (RII), relative importance index (Rii) was applied to prioritise the severity of the factors" [2] [3] [4] and other techniques.

Quantitative research methodology is adopted for the study. The data is collected using structured questionnaire survey. Primary data has been collected for the study because of non-availability of secondary data on the said title. The convenience sampling technique is used to collect the data from the known and easily approachable respondents. The questionnaire is shared with 500 professionals working in different segments of the economy including but not limited to manufacturing, production, suppliers, customers, construction, energy, and others sectors. A total of 120 valid responses were received with a response rate of 24%.

The structured questionnaire is divided into three parts: the first portion has information about the topic and general study that is to be conducted, the second part is the questionnaire itself and the last portion consists the feedback and any suggestion for the study by the respondents. The respondents were asked to rate the questions on the Likert scale of 1-5 given against each question.

3.1 Reliability analysis

Dependability analysis is needed to check the consistency of the model after some time, and Cronbach's alpha test was the best way to check the reliability of the data collected through questionnaire [5]. The value of Cronbach's alpha for this study is 0.85 which is considered to be superior.

Table 1 Reliability analysis

Attributes Cronbach's alpha (Cα)	
Attributes in factor 1	0.83
Attributes in factor 2	0.84
Attributes in factor 3	0.82
Attributes in factor 4	0.86
All Attributes selected for the study	0.85

Table 2 *Kaiser-Meyer-Olkin measure of sampling adequacy*

Kaiser-Meyer-Olkin measure of sampling adequacy	0.813
Bartlett's test of sphericity Approx $\chi 2$	3985.81
Degrees of freedom	1540
Significance	0.000

3.2 Relative importance index

The received responses by the respondents were summarised in an Excel data sheet and the data analysed using SPSS software. Relative importance indices (RII) is performed to determine the priority of the significant factors and then followed by Reliability analysis performed to check the consistency of the data received.

$$Rii = \frac{\sum_{r=1}^{5} r * n_r}{5N} \quad (1)$$

r is the rating on a Likert scale (1-5) as for the impact on construction efficiency for a specific element influencing construction profitability, nr is the number of respondents providing a specific Likert scale rating r, N is the aggregate number of respondents to a specific question [6][4], [16], [17].

The value given to the Likert scale rating is as follows:

1. No effect (or no opinion)
2. Less effect (no or minimal effect)
3. Minor effect (minor problem)
4. Serious effect (medium problem)
5. Very serious effect (major problem

Table 3 Relative importance index

Rank	*Total responses*	*Total score*	*RII*	*Attribute name*
1	120	542	0.90	Raw materials management
2	120	512	0.85	Cost saving
3	120	491	0.69	Customer delight
4	120	490	0.69	Social responsibility
5	120	480	0.69	Reduce/recycle/reuse
6	120	457	0.68	Environmental benefits
7	120	454	0.68	Competitive advantage
8	120	454	0.67	Reduce carbon footprint emission
9	120	451	0.67	Inventory control
10	120	446	0.67	Helps to circular economy
11	120	445	0.67	management of unfinished and finished sections
12	120	435	0.67	Proper management of finished goods
13	120	434	0.67	Management of labour, scientific and technical personnel –helping in reducing carbon emission.
14	120	426	0.66	management of late order delivery to the customer
15	120	419	0.66	management of returned goods from customer
16	120	418	0.66	Management of credits for incorrectly filled or late deliveries from the supplier.
17	120	414	0.65	product flow facility structure
18	120	392	0.65	information flow structure
19	120	391	0.65	understanding of culture and consumerism
20	120	382	0.65	Proper management of goods returned to suppliers
21	120	379	0.65	Proper management of machines, material handlers, tools etc. which helps in reducing carbon emission
22	120	374	0.65	management of warehouses, storage and retrieval system
23	120	358	0.64	management of transportation mode-shipments, air-cargo, engage truck, rail transport

3.3 Most significant benefits of the transformation to SSCM

The most significant benefits through the transition towards SSCM are raw material management, cost saving, customer delight, social responsibility and reduce/recycle/reuse having a relative importance index value of 0.9, 0.85, 0.69, 0.69, and 0.69 respectively

3.4 Significant benefits of the transformation to SSCM

The significant benefits through the transition towards SSCM are environmental benefits, competitive advantages, reduced carbon footprints, inventory control and helps in circular economy having a relative importance index value of 0.68, 0.68, 0.67, 0.67, and 0.67 respectively.

3.5 Factor analysis

Factor analysis is used to reduce the number of dimensions in the data and provide a more clustered output on the basis of co-variance between different attributes. The factor analysis is performed using SPSS 18 for this study. A total of four factors was formed using the software are economic, social, environmental, and logistics explained a variance of 23.1%, 17.2%, 14.3%, and 8.67% respectively.

Table 4 Factor analysis

Factor/variable name	Factor loading	%age of variance explained
Factor 1- Economic		23.1%
Cost saving	0.61	
product flow management	0.6	
information flow management	0.56	
Proper management of machines, material handlers, tools etc. which helps in reducing carbon emission	0.51	
management of warehouses, storage and retrieval system	0.55	
Factor 2- Social		17.2%
Customer delight	0.59	
Social responsibility	0.57	
Competitive advantage	0.55	
understanding of culture and consumerism	0.53	
Factor 3- Environmental		14.3%
Reduce/recycle/reuse	0.58	
Environmental benefits	0.50	
Reduce carbon footprint emission	0.48	
Management of labour, scientific and technical personnel –helping in reducing carbon emission.	0.48	
Factor 4- Logistics		8.67%
Inventory control and management	0.62	
Helps to a circular economy	0.59	
management of unfinished and finished sections	0.59	
Proper management of finished goods	0.49	
Total %age of variance explained		63.3%

4 Discussion and conclusion

The study reveals that the average value of reliability analysis for all the attributes is above 0.5 i.e. all the attributes selected for the study had a significant and valuable perceived benefit. The most important benefits through the transition towards SSCM are raw material management, cost saving and consumer delight. Sustainable supply chain management has lately received large research interest in different fields. This research area has developed during the last decades and can be approached from many different perspectives. Some studies of sustainability implementation in supply chains are focusing on the environmental issues [8], [13], [18]–[21] whereas others have a wider perspective for sustainability covering also social responsibility [8], [22], [23]. A major categorization has been made by the motivations of the implementation whether basing on demand by a more powerful party in the supply chain or driven in collaboration.

5 Limitations of the study

The study is conducted using structured questionnaire survey and collection of primary data. The received responses are 120 considered for this study is comparatively small to generalise the findings to the larger scale. It is recommended to conduct a similar kind of study in different regions of the country to have a better understanding of the SSCM.

References

[1] A. K. Pandey, S. Dixit, S. Bansal, S. Saproo, and S. N. Mandal, *Int. J. Civ. Eng. Technol.*, vol. 8, no. 10, pp. 87–98, 2017.

[2] A. K. Pandey, S. Dixit, S. Bansal, S. Saproo, and S. N. Mandal, *Int. J. Civ. Eng. Technol.*, vol. 8, no. 10, 2017.

[3] S. Dixit, S. N. Mandal, A. Sawhney, and S. Singh, *Int. J. Civ. Eng. Technol.*, vol. 8, no. 8, 2017.

[4] S. Dixit, S. N. Mandal, A. Sawhney, and S. Singh, *Int. J. Civ. Eng. Technol.*, vol. 8, no. 8, pp. 623–636, 2017.

[5] M. C. Cooper, D. M. Lambert, and J. D. Pagh, *Int. J. Logist. Manag.*, vol. 8, no. 1, pp. 1–14, 1997.

[6] B. J. Mariadoss, T. Chi, P. Tansuhaj, and N. Pomirleanu, *J. Bus. Res.*, vol. 69, no. 9, pp. 3406–3414, 2016.

[8] S. Seuring and M. Müller, *J. Clean. Prod.*, vol. 16, no. 15, pp. 1699–1710, 2008.

[9] B. Logistics, no. January, 2018.

[10] S. Nurkka, "Sustainable multi-tier supply chain management Case from Finnish food industry," 2017.

[11] T. F. Burgess, P. Grimshaw, L. Huaccho Huatuco, and N. E. Shaw, *Int. J. Oper. Prod. Manag.*, vol. 37, no. 7, pp. 898–926, 2017.

[12] L. D. Fredendall, "Supply chain management practices and intellectual property protection in China," 2016.

[13] M. E. Torabizadeh, M. Y. Y. Noordin, and M. S. S. Awaluddin, *Mater. Ind. Manuf. Eng. Res. Adv. 1.1*, vol. 845, no. May, pp. 516–520, 2014.

[14] G. Wang, A. Gunasekaran, E. W. T. Ngai, and T. Papadopoulos, *Int. J. Prod. Econ.*, vol. 176, pp. 98–110, 2016.

[15] A. Intelligence *et al.*, "About the Tutorial Disclaimer & Copyright."

[16] S. Singh, A. Bala, S. Dixit, and D. Varshney, *Int. J. Civ. Eng. Technol.*, vol. 9, no. 1, 2018.

[17] A. Sao, S. Singh, S. Dixit, A. K. Pandey, and S. Singh, *Int. J. Mech. Eng. Technol.*, vol. 8, no. 10, 2017.

[18] S. Vasković, P. Gvero, V. Medaković, and V. Halilović, *Sustain. Supply Chain Manag.*, no. June, 2016.

[19] S. K. Srivastava, *Int. J. Manag. Rev.*, vol. 9, no. 1, pp. 53–80, 2007.

[20] G. Saththasivam and Y. Fernando, "Integrated Sustainable Supply Chain Management," no. April, pp. 218–233, 2017.

[21] E. Tatoglu, E. Bayraktar, I. Golgeci, S. C. L. Koh, M. Demirbag, and S. Zaim, *Int. J. Logist. Res. Appl.*, vol. 19, no. 3, pp. 181–199, 2016.

[22] D. M. Lambert, M. C. Cooper, and J. D. Pagh, *International Journal of Logistics Management*, vol. 9, no. 2. pp. 1–19, 1998.

[23] M. A. Saeed and I. Waseek, 2017.

Key Benefits of Adopting Lean Manufacturing Principles in Indian Construction Industry

Subhav Singh [1], *Saurav* Dixit [2*], *Sanjay* Sahai[3], *Ameet* Sao[2], *Yogesh* Kalonia[2], and *Raja* Subramanya Kumar[4]

[1]K R Mangalam University, Gurgaon, India
[2]RICS School of Built Environment, Amity University, Noida, India
[3]Amity Business School, Amity University, Noida, India
[4]Modulor, UAE

Abstract. Objectives of a lean production system (LPS) is to maximise value and minimise waste, by applying specific lean tools and techniques such as 5S, LPS, JIT, VSM, PULL, Kaizen, Six Sigma and many others techniques in a new production process. LPS is a philosophy based on the lean principles developed in Toyota by [1] [2] named as lean manufacturing principles and further developed as lean philosophy [3]. The main benefits of utilising 5S are identified from the literature review and a structured questionnaire is prepared. The data is collected using convenient sampling technique from the professionals working in the industry. The collected data is analysed using different tools and techniques. A total of 9 benefits were identified from the literature review and the expert opinion from the industry. The findings of the study contribute to the existing body of knowledge and lay down few suggestions to the industry for implementing 5S in the Indian context. The findings of the study conclude that the lean principles support the Indian manufacturing industry to improve productivity, quality, and reduction in cycle time The study is conducted using primary data collection and the sample size is small (86).

Keywords: Lean Production System; Waste Reduction; Productivity; 5S; and Indian Manufacturing Industry.

1 Introduction

Lean is tied in with accomplishing an adjusted utilization of individuals, materials and assets. This enables organizations to decrease costs, dispose of waste and convey extends on time and it isn't tied in with trimming everything deep down and crushing more out of what is left [7][2]. Lean development is the down to earth use of lean assembling standards, or lean reasoning, to the building condition [8], [9]Advantages of the new generation theory as far as efficiency, quality, and markers were sufficiently strong by and by with a specific end goal to improve the fast dissemination of the new standards [9]. The lean reasoning is tied in with: Getting work to stream dependably and typically on a development site requires the flawless arrangement of the whole production network in charge of built offices to such an extent that esteem is boosted and squander is limited. With such an expansive degree, any reasonable person would agree that apparatuses found in Lean Manufacturing and Lean Production, as honed by Toyota and others, have been adjusted to be utilized as a part of the satisfaction of Lean development standards. TQM, SPC, six-sigma, have all discovered their way into lean development. Also, devices and strategies found in different territories, for example, in sociology and business, are utilized where they are relevant. The instruments and techniques in development administration, for example, CPM and work breakdown structure, and so on. Are likewise used in lean development executions.

"One can consider lean development in a route like macroeconomics. Lean development draws upon the standards of venture level administration and upon the rule that administer generation level administration. Lean development perceives that any fruitful task undertaking will definitely include the communication amongst venture and generation administration [6].

• Sort – do we have all that we require, yet just those things we require, to achieve our assignment inside the site, the work zone or the workplace?

• Set – in what capacity should we best outline our work site or region with the goal that the zones, materials, apparatuses and gear are in the most secure, most proficient and best place. The visual administration is then used to set the work environment so an irregular condition is effectively observed – signposting, outline, shadow sheets, naming, least/most extreme levels,

* Corresponding author: sdixit@ricssbe.edu.in

numbering and shading coding all empower us to tell initially that we are set up for progress.

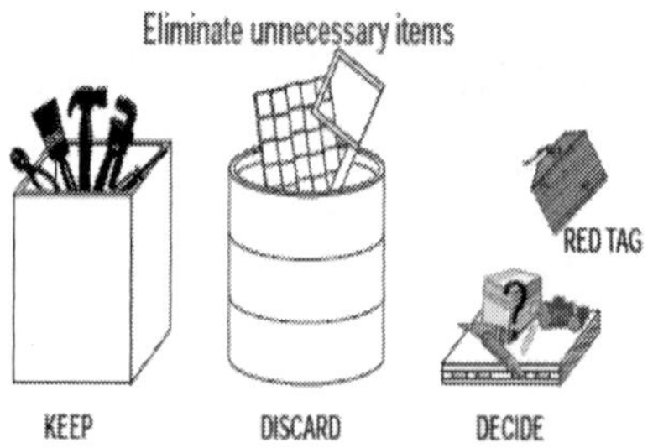

Fig 1 Sorting procedure

• Shine – conveying things to a spotless and clean condition and keeping the work environment in great request with the goal that it is prepared for utilize.
• Standardize – concurring between the groups the standard to which the work environment will be continued, shooting this and showing it so everybody is sure about how the function place should look.
• Sustain – the initiative work important to keep the initial 4Ss set up, crafted by examining and obvious authority in the work put.

1.1 5S benefits

Worker learning and investment is the first advantage watched when 5S is executed in the association [10]. 5S execution brings about better housekeeping and in this way better visual work environment is seen in the association [11][12]. 5S usage likewise guarantees more secure workplace [13]. The decrease in waste, time and cost is another great advantage of 5S ([3], [14]. [15] found 5S usage brings about enhanced quality, profitability and effectiveness.
To stay in the business field it is of highest critical to winning hearts of client however quality and cost of the item or administration. It is additionally vital to have a manageable creation with persistent change. The present need of the association is to convey top notch item through persistent change [1].However, fabricating association all through the world is under extraordinary strain to diminish the cost and address the difficulty of keeping up worldwide quality guidelines [2]. Lean Manufacturing is the psalm of survival and accomplishment of any association through limiting the wastage (Muda) of assets.

2 Literature review

[2] Considered the basic issues looking by small-scale businesses while offering their item. SSE (Small Scale Enterprise) isn't having enormous budgetary reinforcement and in this manner, they are relying on the income a great many contemplated the significance of small and medium scale ventures in an Indian setting. Medium size assembling industry assumes a vital part of Indian economy. Their commitment to the financial advancement of the country is without a doubt noteworthy. However, the profitability level of these businesses is very low when contrasted with another country [16] portrayed the execution and advantages of small-scale fabricating industry in India. Small-scale ventures frame a critical segment constituting 40% of the aggregate yield to the privet area and considerably more huge is the business age limit of small-scale sector.[17] Demonstrates the issue to support in the worldwide market for an association. Lean assembling is the song of survival and accomplishment of any association. The objective of lean assembling is to limit a wide range of waste so the cost of the item can be reduced. [18] depicted the improvement of key regions which could be utilized to survey the appropriation and usage of lean assembling practice likewise displayed a portion of the key territories created to assess and decrease the ideal undertaking in order to upgrade their generation proficiency. Concentrated on the usage of lean on the small maker of each of the 4-wheel drive vehicles, through execution of essential lean device, the small make quickly increment yield and diminish quality imperfections by 80%. [19] Exhibited strategy for deciding the genuine issue related to businesses in the usage of lean. They likewise introduced determination of required lean instruments in the light of organization's long-haul vision. [20] portrayed the means attempted for the execution of 5S stressing on the advantage of an association. Additionally depicted the start and advantage of actualizing the 5S. [7] Presents a persistent change system intending to enhance fabricating at auto Exhaust. The execution of 5S has a prompt and huge impact on the arrangement of exercises in the work post, in this manner affecting the execution of the process in the dissected organization. [11] took a shot at the usage of 5S on the plastic trim industry. 5S is utilized as a part of the small industry and furthermore demonstrated the points of interest and advantages of 5S execution. [21] exhibited the approach for computation of every S in the 5S framework.
The 5S's are key lean ideas got from the Japanese words seiri (sort), seiton (set all together), seiso (sparkle), seiketsu (institutionalize) and shitsuke (sustain).2 Companies receiving the lean creation reasoning frequently execute the 5S procedure to convey request to the working environment and bolster lean generation. One important part of the lean creation rationality is its accentuation on esteem.

2.1 Why utilize 5S?

The 5S program is a demonstrated model for sorting out and keeping up a generation operation. It is frequently utilized as a part of assembling operations, specifically dynamic ones. The program additionally underpins the visual work environment. [2] Alluded to the 5Ss as "mainstays of the visual working environment" in the title of his book on 5S, in a 5S domain there is a place for everything, and everything is in its place. Time spent looking for things is basically wiped out, and strange or missing things are quickly evident in a legitimately working 5S office. The 5S program plays out a comparative capacity for providing support. The clients

of the framework immediately know where to discover data about their supply base, and absent or obsolete data is right away evident. To actualize a 5S program for provider upkeep, you ought to maintain the following five rules.
The initial phase in actualizing 5S for lean manufacturing is to take a voyage through the objective zone and check with red labels those things that show up strange or superfluous. After everything is looked into, it is either placed in its appropriate place or expelled in the event that it is superfluous or repetitive. The sort procedure is basic to organizing the workspace required for lean creation.
While applying the sort technique to the supply base, you select providers to add to the framework and eliminate from the framework. In acquiring language, this is known as supply base union or rationalization. Arranging the provider base through solidification: • Reduces the misuse of wasteful work strategies by
Many companies apply quality approaches such as lean manufacturing and six sigma methodologies to minimize defects and waste and to improve performance output. Lean manufacturing refers to a methodology designed to lower the costs involved in the production with a view to minimizing waste[22] Research has shown that companies that implement and practice lean manufacturing or lean production see a significant improvement in operational performance[5]. [23] State that six sigma is a statistical technique that permits a scientific evaluation of processes and transactions for the reduction of variability and the removal of defects. However, this paper investigates the status of performance outcomes of companies based on the methodology that has been implemented within the organizations. Results reveal applications in three different categories. Firstly, companies that implemented lean manufacturing only. Secondly, companies that use a combination of lean manufacturing and six sigma. Thirdly, particulars of limited tools of lean six sigma implemented, where they didn't refer to either lean manufacturing or six sigma with their organization.
However, the goals of this paper are to identify the powerful application that is helpful on increasing companies' performance and the effective industry that perform these lean manufacturing and 5S practices across financial, operational and innovation performance. This paper investigates not only the impact of lean manufacturing and 5S implementation but also the influence of the implementation method based on the type of industry.

2.2 Benefits of adopting lean principles

Table 1 benefits of implementing lean form literature review

Author's	***Benefits identified***
[4], [7], [12], [22], [24]–[26],[27], [28]	Reduction in lead time, reduction in work in process (WIP), reduction in Waste, continuous improvement
[11], [29]–[33]	Reduction in operating cost, Late delivery, and Space constraints, Enhance productivity, space utilization, and value for money
[2], [3], [5], [6], [8], [9], [14], [20], [23], [34]–[42]	Inventory control, Customer service, improves quality, improved communication, and reduce cycle time

3 Research Methodology

For this research paper, a questionnaire survey method has been used to analyse the relative impact of 5S on the identified issues in the Indian manufacturing industry. For this, the structured questionnaire survey is shared with the professionals, suppliers, customer, and consultants associated with the manufacturing industry. The received data were analyzed using Relative importance index (RII), RII was applied to prioritise the severity of the factors"[43].
Quantitative research methodology is adopted for the study. The data is collected using structured questionnaire survey. Primary data has been collected for the study because of non-availability of secondary data on the said title. The convenience sampling technique is used to collect the data from the known and easily approachable respondents. The questionnaire is shared with 350 professionals working in different segments of the economy including but not limited to manufacturing, production, suppliers, customers, transporters, logistics movers and other consultants working in the industry. A total of 86 valid responses were received with a response rate of 24.5%.
The structured questionnaire is divided into three parts: the first portion has information about the topic and general study that is to be conducted, the second part is the questionnaire itself and the last portion consists the feedback and any suggestion for the study by the respondents. The respondents were asked to rate the questions on the Likert scale of 1-5 given against each question.

3.1 Reliability analysis

Dependability analysis is needed to check the consistency of the model after some time, and Cronbach's alpha test was the best way to check the reliability of the data collected through questionnaire [5]. The value of Cronbach's alpha for this study is 0.723 which is considered to be good for the study [10].

3.2 Respondent's profile

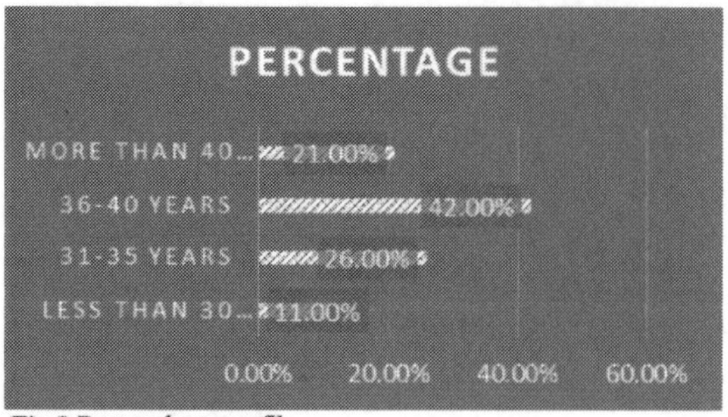

Fig 2 Respondents profile

3.3 Relative importance index

The received responses by the respondents were summarized in an Excel data sheet and the data analysed using SPSS software. Relative importance indices (RII) is performed to determine the benefits of using 5S in the organizations and then followed by
Reliability analysis performed to check the consistency of the data received.

$$Rii = \frac{\sum_{r=1}^{5} r*n_r}{5N} \quad (1)$$

r is the rating on a Likert scale (1-5) as for the impact on construction efficiency for a specific element influencing construction profitability, nr is the number of respondents providing a specific Likert scale rating r, N is the aggregate number of respondents to a specific question [6].
The value given to the Likert scale rating is as follows:

1. No effect
2. Less effect
3. Minor effect
4. Serious effect
5. Very serious effect

Table 2 ranking of attributes (RII)

Rank	*Total responses*	*Total score*	*RII*	*Attribute name*
1	86	374	87.0%	Enhance productivity
2	86	361	84.0%	Improved quality
3	86	322	69.3%	Reduction in cycle time
4	86	314	69.1%	reduction in waste
5	86	297	68.6%	Reduction in WIP
6	86	283	68.4%	Reduction in lead time
7	86	266	67.5%	Inventory management
8	86	264	67.4%	Customer service
9	86	254	67.2%	Better space utilization

4 Discussion and conclusion

The most significant benefits of adopting lean/5S

The top 3 most significant benefits of adopting lean/5S principles in the manufacturing industry are Enhance productivity, improved quality, and reduction in cycle time having a relative importance index value of 87%, 84% and 69.3% respectively. These findings conclude that the lean/5S principle having a positive and significant impact in improving the baseline by supporting improvement in productivity, quality, reducing the overall cycle time of the product.

4.1 Significant benefits of adopting lean/5S

The three significant benefits of adopting lean/5S principles in the manufacturing industry are Reduction in waste, reduction in WIP, and reduction in lead time having a relative importance index value of 69.1%, 68.6% and 68.4% respectively. These findings conclude that the lean/5S principle helps the manufacturing industry to sustain in long run by providing benefits such as reduction in waste, reduction in lead time, and reduction in WIP throughout the process.

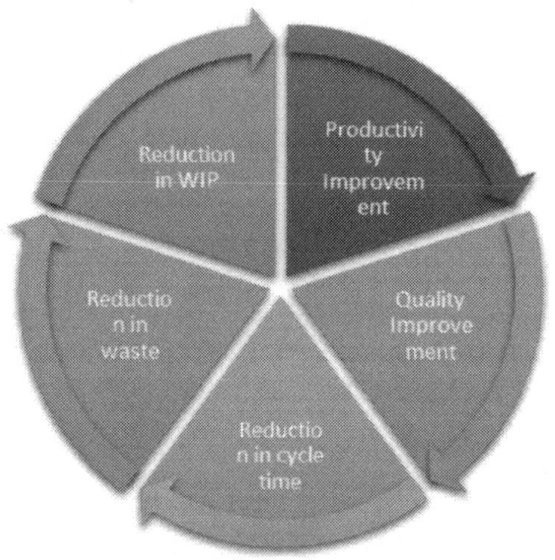

Fig 3 Key benefits of adopting lean manufacturing principles

The study provides an insight into the scope of lean tools/principles in Indian manufacturing industry by providing benefits mentioned as below. And secondly when you analysed the results of the RII the maximum value is 87% and the minimum is 67% i.e. the lean tools/5S principles are applicable to the main issue/problems identified from the literature review and provide an extra edge to the organisation implementing lean in their processes. This is observed that lean is more like a philosophy and an organisation can be able to adopt lean if and only it starts from the top management and should be applied as a philosophy at every stage of the product/service development. The main goal of lean is to create value and maintain the flow of the process. This study provides a better understanding=g of the lean tools and lean principles while highlighting the benefits of adopting lean/5S tools (figure).

5 Limitations of the study

This study is conducted by utilizing the collection of primary data to identify and analyse the benefits of lean/5S tools and principles in the Indian manufacturing industry. To validate the results and germanised the finding similar kind of study using a large data set is suggested. The author is currently working on the framework to improve the productivity of manufacturing industry by adopting the lean tools and principles.

References

[1] J. Mason, "Ean rimer," *Readings*, no. c, pp. 1–46.

[2] L. Koskela, G. Ballard, G. Howell, and I. Tommelein, *Des. Constr. Build. Value*, no. December 2015, pp. 211–226, 2002.

[3] J. Salvatierra-Garrido and C. Pasquire, *J. Financ. Manag. Prop. Constr.*, vol. 16, no. 1, pp. 8–18, 2011.

[4] S. Mostafa, N. Chileshe, and T. Abdelhamid, *Constr. Innov.*, vol. 16, no. 4, pp. 483–525, 2016.

[5] O. Ogunbiyi, J. S. Goulding, and A. Oladapo, *Constr. Innov.*, vol. 14, no. 1, pp. 88–107, 2014.

[6] G. J. Sweis, M. Hiyassat, and F. F. Al-Hroub, *Constr. Innov.*, vol. 16, no. 4, pp. 446–459, 2016.

[7] R. H. Ansah and S. Sorooshian, *Sustain. Cities Soc.*, vol. 32, no. December 2016, pp. 348–356, 2017.

[8] A. H. A. Jamil and M. S. Fathi, *Procedia Comput. Sci.*, vol. 100, pp. 634–643, 2016.

[9] M. A. Marhani, A. Jaapar, N. A. A. Bari, and M. Zawawi, *Procedia - Soc. Behav. Sci.*, vol. 101, pp. 90–99, 2013.

[10] A. Sao, S. Singh, S. Dixit, A. K. Pandey, and S. Singh, *Int. J. Mech. Eng. Technol.*, vol. 8, no. 10, 2017.

[11] B. Jørgensen and S. Emmitt, *Eng. Constr. Archit. Manag.*, vol. 15, no. 4, pp. 383–398, 2008.

[12] M. M. Carvalho and R. Rabechini, *Int. J. Proj. Manag.*, vol. 35, no. 6, pp. 1120–1132, 2017.

[13] J. Meiling, F. Backlund, and H. Johnsson, *Eng. Constr. Archit. Manag.*, vol. 19, no. 2, pp. 141–158, 2012.

[14] L. Zhang and X. Chen, *Procedia Eng.*, vol. 145, pp. 1267–1274, 2016.

[15] M. L. Martens and M. M. Carvalho, "Key factors of sustainability in project management context: A survey exploring the project managers' perspective," *Int. J. Proj. Manag.*, vol. 35, no. 6, pp. 1084–1102, 2017.

[16] L. A. Ocampo and E. E. Clark, t *Asian Acad. Manag. J.*, vol. 20, no. 2, pp. 29–57, 2015.

[17] U. Dombrowski and T. Mielke, *Procedia CIRP*, vol. 17, pp. 565–570, 2014.

[18] L. M. Khodeir and R. Othman, *Ain Shams Eng. J.*, 2016.

[19] B. Anvari, P. Angeloudis, and W. Y. Ochieng, *Autom. Constr.*, vol. 71, no. Part 2, pp. 226–241, 2016.

[20] M. Ertürk, M. Tuerdi(Maimaitiaili. Tuerdi), and A. Wujiabudula, *Procedia - Soc. Behav. Sci.*, vol. 229, pp. 444–452, 2016.

[21] P. Dallasega and E. Rauch, *Sustainability*, vol. 9, no. 10, p. 1888, 2017.

[22] C. Information, "W Orld a Cademy of S Cience , E Ngineering and T Echnology," p. 2017, 2017.

[23] X. Brioso, *Procedia Eng.*, vol. 123, pp. 85–93, 2015.

[24] U. H. Issa, *Alexandria Eng. J.*, vol. 52, no. 4, pp. 697–704, 2013.

[25] Engineers Australia WA Division, *Eng. Aust.*, no. August, 2012.

[26] O. Ametepey, C. Aigbavboa, and K. Ansah, *Procedia Manuf.*, vol. 3, no. Ahfe, pp. 1682–1689, 2015.

[27] S. Singh, A. Bala, S. Dixit, and D. Varshney, *Int. J. Civ. Eng. Technol.*, vol. 9, no. 1, 2018.

[28] A. K. Pandey, S. Dixit, S. Bansal, S. Saproo, and S. N. Mandal, *Int. J. Civ. Eng. Technol.*, vol. 8, no. 10, 2017.

[29] A. A. Al-Sudairi, *Constr. Innov.*, vol. 7, no. 1, pp. 99–121, 2007.

[30] E. Lodgaard, J. A. Ingvaldsen, I. Gamme, and S. Aschehoug, *Procedia CIRP*, vol. 57, pp. 595–600, 2016.

[31] R. Jiang, C. Wu, C. Mao, and A. Shrestha, *Procedia Eng.*, vol. 145, pp. 436–443, 2016.

[32] W. Aarseth, T. Ahola, K. Aaltonen, A. Økland, and B. Andersen, *Int. J. Proj. Manag.*, vol. 35, no. 6, pp. 1071–1083, 2017.

[33] H. Bernstein and S. Jones, vol. 3, no. 2. 2010.

[34] R. F. Aziz and S. M. Hafez, *Alexandria Eng. J.*, vol. 52, no. 4, pp. 679–695, 2013.

[35] A. Bjðrnfot and Y. Sardén, *Proc. 14th Int. Gr. Lean Constr.*, no. January, pp. 265–277, 2006.

[36] G. Ballard, I. Tommelein, L. Koskela, and G. Howell, *Build. value*, p. 504, 2002.

[37] J. Kilpatrick, *Utah Manuf. Ext. Partnersh.*, pp. 1–5, 2003.

[38] U. Dombrowski and T. Mielke, *Procedia CIRP*, vol. 7, pp. 569–574, 2013.

[39] R. Pamfilie, A. J. P. (Draghici), and M. Draghici, *Procedia - Soc. Behav. Sci.*, vol. 58, pp. 187–196, 2012.

[40] I. Alhuraish, C. Robledo, and A. Kobi, *IFAC-PapersOnLine*, vol. 49, no. 12, pp. 59–64, 2016.

[41] G. A. Howell, *Proc. Seventh Annu. Conf. Int. Gr. Lean Constr.*, pp. 1–10, 1999.

[42] T. da C., C. Milberg, and K. D. Walsh, *Eng. Constr. Archit. Manag.*, vol. 19, no. 5, pp. 512–525, 2012.

[43] S. Dixit, S. N. Mandal, A. Sawhney, and S. Singh, *Int. J. Civ. Eng. Technol.*, vol. 8, no. 8, 2017.

[44] S. Dixit, A. K. Pandey, S. N. Mandal, and S. Bansal, *Int. J. Civ. Eng. Technol.*, vol. 8, no. 6, 2017.

[45] K. C. Iyer and K. N. Jha, *Int. J. Proj. Manag.*, vol. 23, no. 4, pp. 283–295, May 2005.

Perceptions on service quality: A study of old age homes in Kerala and Tamil Nadu

Devika Anil, and K Hemamala
Amrita School of Business, Amrita VishwaVidyapeetham, Coimbatore, India
devikaa.feb09@gmail.com

Abstract - The proportion of old age citizens (above 60 years) in India constitutes 8.6% of the total population (60 million). 1% of this population currently resides in old age homes across India due to various reasons. The present living conditions of elderly in old age homes are not commendable. It has been found from research that the setting of old age homes and the employees influence the welfare of the residents and their health care. Despite various services provided, researchers have concluded that service gaps exist among staff and residents of old age homes. In the present scenario, it is the need of the hour to understand the living standards of the older generation in old age homes, to study the various services offered by them and how the residents spend their lives there in order to better their conditions of living. The purpose of this paper is (i) to draw insights from an in-depth analysis of old age homes and (ii) to study the perceptions in service quality of old age homes; by studying old age homes in the Indian states of Kerala and Tamil Nadu. Data for the study was collected through a face to face interview with its residents. Analysis of the collected data shows that 'responsiveness' and 'empathy' are the dimensions that are to be improved in both the states to increase overall satisfaction levels and meet the expectations of residents. The quality of the attributes, 'Cleaning services for rooms', 'freedom to plan own time' and 'respect for residents' privacy' shall be improved by old age homes in order to improve the perception of service quality by residents.

I. INTRODUCTION

India, being the second most populated country in the world, is home to about 1.32 billion people. About 8.6% (60 million) of the population constitutes people above the age of 60 (senior citizens). Advanced medical care, spread of modern medicine and hence falling mortality rates will lead to a rapid increase in elderly population in the years to come. It is estimated that the current elderly population of 60 million is projected to exceed 227 million by 2050 (an increase of nearly 280%) [1].

After retirement, many of them face humiliation and are forced to live their life at the mercy of their children and relatives. Many of them are also victims of physical as well as mental ailments, are bedridden and have limited mobility. Old age means reduced physical ability, declining mental ability, the gradual giving up of role playing in socio-economic activities, and a shift in economic status moving from economic independence to economic dependence upon other's for support [2]. They fall into a state of solitude especially when their partner passes away. These reasons are enough to exploit their life and force them to live a life of abuse and isolation or abandon them in old age homes. The number of old age homes has increased a lot in the recent years due to various societal and cultural changes [1]. This shows that more than half of the population of elderly spends their last phase of life until death in old age homes. Therefore it is important to understand how they live their life in old age homes, the facilities offered and the level of satisfaction they have.

The paper is structured as follows. The research gap is identified with the help of literature review. Later it focuses on the methodology of this study, the analysis and finally the results are discussed.

2. LITERATURE REVIEW

A. Elderly in India

Citizens of India above the age of 60 are considered as elderly or senior citizens. They constitute the old age population of India - about 8.6% of the total population [3]. The proportion of elderly is highest in Kerala followed by Tamil Nadu. The elderly constitutes 12.6% of the population in Kerala and 10.4% in Tamil Nadu. The rate of increase in aged population is also high in Kerala followed by Tamil Nadu [3].

People with higher life satisfaction describe themselves as having social supports in a more extensive, frequent, and reliable ways, a more contented feeling in their relationships with others, and engaging in a greater number of pleasant activities as well. They are reported to have fewer irrational beliefs, were more optimistic and less pessimistic, and had lower levels of reported cognitive difficulty [4]. But the living conditions of elderly are degrading now-a-days with very low life satisfaction rates. About 15 million of this population leads a lonely and secluded life due to various reasons like abuse from home, death of partner, chronic ailments etc. The dependency rates of elderly on others is high in Kerala (19.6%) followed by Tamil Nadu (15.8%). The proportion of people who owns any asset in India is high in Kerala and very low in Tamil Nadu. A little over 18% of elderly are receiving pension in Kerala while this rate is only 4% in Tamil Nadu [5]. A large proportion of people do not own any asset, are not eligible for pension and are highly dependent on children and other relatives. These statistics show that older people in India, especially in Kerala and Tamil Nadu, are not satisfied with their living conditions.

B. Emerging sector of old age homes in India

An old age home is generally defined as a home for old people who have no one to look after them or those who have been thrown out of their homes by their children. Old age homes in India are managed by individuals (private) with diverse professional and disciplinary background, religious institutions and government organizations without any age-specific qualifications. Most of them operate as charity (free) though there are some profit oriented (paid) homes as well [1].

The number of old age homes has increased a lot in the recent years due to various social, economic, cultural and demographic changes. Urbanization, modernization and globalization have led to changes in the economic structure like migration to large cities and other countries for better opportunities, the erosion of societal values, lack of attachment between children and parents, weakening of social values, relationships, and social institutions such as the joint family [1]. This reduces the availability of care in homes and are forced to move into old age homes. Problems with children like abuse, misbehaviour by son and daughter-in-law and life threats from them, having no children, lack of care at home since children are away, adjustment issues, loneliness at home, chronic ailments and own preference are the major reasons to choose old age home [6].

The state of Kerala serves the largest number of old age homes in India followed by Tamil Nadu and Goa [7]. The proportion of people living in old age homes has increased a lot in the recent years. A 69% increase in number of residents of old age homes has been reported in Kerala alone in the last four years. The proportion of females living in old age homes is more than 50%, compared to men. The average age group of elderly living in old age homes is 70-80. The elderly who choose to live in old age homes include people who earn an income through pension, those who are fully dependent on others and a very small population who are still employed.

C. Service quality dimensions and service gap

Service quality is the difference between customer expectations of service and perceived service (Parasuraman et al., 1985; Lewis et al., 1990). The five dimensions of service quality are reliability, responsiveness, assurance, empathy and tangibles. All the basic facilities and services from any service provider can be categorized into these five dimensions developed by Parasuraman et al in their studies (1985; 1988; 1991; 1994). It is used to measure the quality of service, satisfaction level of customers and identify service gap, if any. This method along with SERVQUAL model (Gronroos, 1982; Lewis et al., 1983; Parasuraman et al., 1985) has been successfully used to measure customer satisfaction in service sectors like banks [8], hospitals [9], education [10], e-service [11], ports [12] etc. All these sectors provide basic services like visually appealing facilities consistent with the industry, up to date equipment and well-dressed employees (tangibles), responding within time frame to customers, dependability and efficient service delivery (reliability), polite and trustworthy employees and safe transactions (assurance), prompt service from employees and readiness and willingness of employees to help (responsiveness) and providing individualized attention and understanding customer needs (empathy) [13].

Care giving facilities like nursing homes, palliative care and hospitals have used this method to understand service gap and measure customer satisfaction levels. This has also been used by certain retirement homes in Hong Kong and Japan to measure the same.

D. Perceived service and service gap in old age homes

Studies have shown that the influence of the setting and the role of the staff in old age care are very important to make them happy, contented and healthy. Among depicted services and facilities in old age homes, residents rated the quality and behaviour of old age home staff as most important, whereas the choice of food and mealtimes were rated the least important. [14]. Care staff being able to spend enough time with residents, feeling at home in shared spaces, and staff being very flexible in care routines are the characteristics identified as important for residents. Satisfaction with the privacy in the living environment is one of the most important determinants of life satisfaction [15].
Though old age homes provide good services to their residents, considerable knowledge and service gaps exist among staff and residents of old age homes [15].

In the present scenario, it is important to study the various services offered by old age homes in India and how the residents spend their lives there to understand where improvements have to be made. Researches clearly show that service gaps do exist in services provided by old age homes. But the study on differences in expectation and perception of service quality and their areas of occurrence has not has not been done very extensively in India. Apparently no studies have been conducted in this regard about old age homes especially in the Indian states of Kerala and Tamil Nadu, where largest number of old age homes operate in the country. Therefore this paper aims to (i) to draw insights from an in-depth analysis of old age homes and (ii) to study the perceptions in service quality of old age homes; by studying old age homes in Kerala and Tamil Nadu.

3. METHODOLOGY

A. Data description

For the purpose of this study, data was collected from a sample of 120 residents of different age, gender and social status, across various old age homes in Kerala and Tamil Nadu. Face to face interview was conducted to collect data from elderly in old age homes. The following table describe a portion of data collected for this study.

TABLE 1: A SAMPLE DATA

Gender	Age	Current Employment status	Reason for chc OAH
Male	78	Voluntary	Own preferen
Female	70	Unemployed	Problems with
Male	68	Unemployed	Problems with
Male	72	Retired	Problems with
Female	74	Unemployed	No one at hom
Female	80	Unemployed	No one at hom
Male	71	Retired	Own preferen
Male	73	Retired	No one at hom
Female	72	Unemployed	Problems with

B. Questionnaire development

The various services provided by the old age homes to the residents are identified [13]. They are then classified as different attributes under five service dimensions as follows:

TABLE 2: SERVICE DIMENSIONS AND ATTRIBUTES

Service dimensions	Attributes
Reliability	Appropriate medical care
	Specialized medical care
	Food and medicines on time
Responsiveness	Choice of food
	Vehicle for emergency
Assurance	Feel safe
	Feel a sense of family
	Management ask for your feedback
	Behaviour of staff
Empathy	Individualized attention
	Freedom to plan your time
	Freedom to go in and out
	Freedom to express your opinions
	Sessions on meditation, yoga etc
	Games or entertainment sessions
	Their respect for your privacy
Tangibles	Calm and quiet atmosphere
	Clean rooms

The questions were asked to the residents to collect data on various attributes of service quality dimensions of old age homes. The same set of questions was asked to management to understand their perception on the quality of services they provide. The attributes were measured using a 3 point (Milte R et al., 2017) and 5 point Likert scale ranging from "Strongly Agree" to "Strongly Disagree" (Bearden et al., 1993). Also, there were questions developed to understand the demographics and background of the respondents – age, gender, current employment status and reasons for choosing old age home.

C. Analysis method

Weighted average method is used for plotting the graph between (i) perception of attributes of old age homes in Kerala and Tamil Nadu, (ii) perception on quality of attributes by residents and perception on its delivery by management for both Kerala and Tamil Nadu and to compute variation. Weights are assigned to each point in scale and the average value of an attribute by all the residents is calculated.

4. RESULTS, DISCUSSION AND FUTURE RESEARCH

A. Insights on data collected from old age homes (Kerala and Tamil Nadu)

TABLE 3: DEMOGRAPHIC CHARACTERISTICS OF RESPONDENTS

		Kerala	Tamil N
Gender	Male	37.20%	
	Female	62.80%	
Age Group	Below 60	2.30%	
	60 - 69	48.80%	
	70 - 79	37.20%	
	80 - 89	11.62%	
	Above 90	0.00%	
Current Employment status	Voluntary	11.60%	
	Unemployed	30.20%	
	Employed	4.60%	
	Retired	53.40%	
Reason for choosing OAH	Own preferences	23.20%	
	Problems with children	6.90%	
	Children are away	20.90%	
	No one at home	48.80%	

The above given table shows the percentage of respondents belonging to different gender, age groups, current employment status and reason for staying in old age homes. The highest percentage is shown in the age group 60 – 69 (48.8%) in Kerala. Response from residents shows that about 15 – 20% elderly in Kerala choose life in old age homes immediately after they retire to lead a peaceful end-of-life with people of same age group, when one of the partners is no more or since children are far away. Responses show that only a small percentage of them have issues with their children. 53% of the respondents are leading a retired life. In Tamil Nadu, 67.4% belong to the age group 70 – 79.

Only less than 50% of them leave their homes since there is no one at home. Compared to Kerala, a good percentage of respondents face humiliation and abuse due to issues with their own children, in Tamil Nadu. More than half of them are unemployed as well. They are either send to old age homes and are paid for their expenses by children and relatives or are abandoned in charity homes where they have to live a life at the mercy of others.

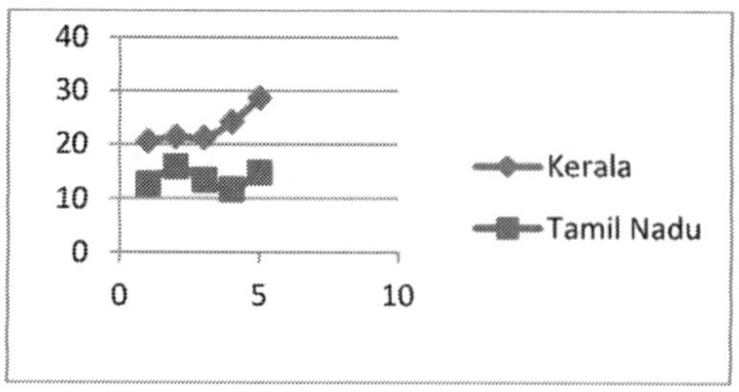

FIGURE 1: WEIGHTED AVERAGE VALUES FOR SERVICES

TABLE 4: WEIGHTED AVERAGE VALUES OF SERVICES

No.	Dimensions	Kerala	Tamil Nadu	Gap (Variation)
1	Reliability	20.63	12.67	7.96
2	Responsiveness	21.45	15.87	5.57
3	Assurance	21.2	13.37	7.82
4	Empathy	24.15	11.67	12.47
5	Tangibles	28.68	14.75	13.9

The above given graph compares the old age homes in Kerala and Tamil Nadu in terms of the weighted average values (Y-axis) for the service dimensions.

Reliability has shown lowest values among others, in Kerala because of the lowest value of attribute 'specialized medical care'. Assurance and responsiveness shows the next lowest value due to the low value of the attribute 'feeling of safety' and 'vehicle for emergency' respectively. 'Cleaning services for rooms', 'freedom to plan own time and to move in and out' and 'respect for residents' privacy' are the other attributes that have shown low values. They are allowed to move in and out as they wish, but certain rules are kept such as timing to go out and letting the authorities know about their movement.

Empathy has shown the lowest value in Tamil Nadu because of the lowest values of 'freedom to move in and out' and 'respect for residents' privacy'. 'Cleaning services for rooms', 'freedom to plan own time' and 'respect for residents' privacy' are the other attributes that have shown low values. Compared to Kerala, the average age of residents in old age homes of Tamil Nadu is high. Therefore special care is given for their welfare and the staffs are always watchful of them. Hence they are not allowed to do all of the everyday activities as they wish.

The low value for certain attributes, mentioned above, in both states shows that there are homes where the elderly have to live a scheduled life. This keeps a check on the residents' safety since they are unhealthy and are leading their last phase of life, though few of them complain about it being a prisoned life.

Another important observation is that some of the elderly become emotionally shattered and fall into geriatric depression (depression which affects elderly) due to these rules and feel themselves to be a burden for others especially those who were forced to live in old age homes by their relatives. Old age homes organize prayer sessions, sessions on yoga, meditation; counselling, small exercise sessions to keep them engaged and not let them victims of geriatric depression. Some people enjoy them whereas others feel that the management is breaking the wall of privacy for them. Such people are reluctant to speak and engage with fellow inmates and visitors. They are given mental support through sessions of counselling.

Tangibles have shown the highest value in Kerala due to the high value of the attribute 'calm and quiet atmosphere'. Responsiveness has shown highest value in Tamil Nadu due to the high value of the attribute 'vehicle for emergency'. 'Games and entertainment sessions, sessions on yoga, meditation and counselling and choice of food are the other attributes that have shown high values in both the states. Choice of food is also a prominent service for residents in Tamil Nadu.

The highest variation among the dimensions is found in 'Tangibles' due to considerable differences in values of attributes such as "calm and quiet atmosphere' and 'clean rooms' and the way they are delivered in Kerala and Tamil Nadu. 'Sessions on meditation, yoga etc.', 'games and other entertainment sessions and 'individualized attention' show considerable variation indicating that the same services are being delivered in different ways to two different population who have common interests. This results in different levels of satisfaction for residents in Kerala and Tamil Nadu, where Kerala shows an upper hand in service delivery and overall satisfaction.

B. Perceptions on quality of service provided

FIGURE 2: AVERAGE VALUES OF SERVICES FOR KERALA

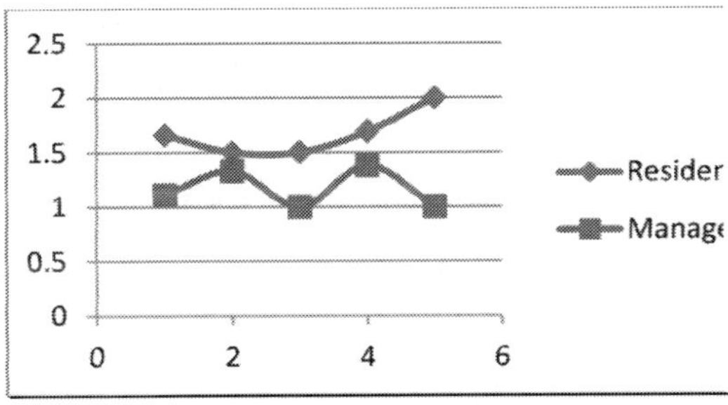

TABLE 5: AVERAGE VALUES OF SERVICES FOR KERALA

No.	Dimensions	Residents	Management	Gap (Variation)
1	Reliability	1.66	1.11	0.55
2	Responsiveness	1.5	1.33	0.16
3	Assurance	1.5	1	0.5
4	Empathy	1.69	1.38	0.31
5	Tangibles	2	1	1

FIGURE 3: AVERAGE VALUES OF SERVICES FOR TAMIL NADU

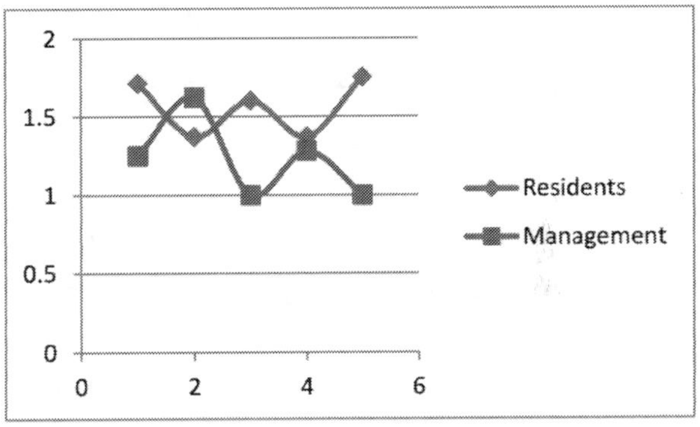

TABLE 4: AVERAGE VALUES OF SERVICES FOR TAMIL NADU

No.	Dimensions	Residents	Management	Gap (Variation)
1	Reliability	1.71	1.25	0.46
2	Responsiveness	1.37	1.62	-0.25
3	Assurance	1.6	1	0.6
4	Empathy	1.37	1.28	0.08
5	Tangibles	1.75	1	0.75

The graphs show the average values of responses received from both residents and management of various old age homes.

The graph of Kerala shows that the perception of quality of service is either equivalent or more than what the management provides at present. The variation in values among residents and management is highest for tangibles since it includes the attribute 'calm and quiet atmosphere' which has a high value. 'Freedom to express opinions and feedbacks' and 'timely availability of food and medicines' and 'games and other entertainment sessions' are other attributes which shows high variation in values indicating that residents are highly satisfied regarding these services. The variation is lowest for 'rooms getting cleaned', 'freedom to plan own time', 'freedom to move in and out' and 'respect for residents' privacy'.

The graph of Tamil Nadu also shows that the perception of quality of service is either equivalent or more than what the management provides at present except responsiveness since it includes a low valued attribute 'games and entertainment sessions'. The residents long for such sessions since they can bring out a sense of feeling like a family and increase the bonding among them. But many old age homes are reluctant to conduct such sessions on a regular basis though they conduct very few of them. The variation in values for the service among residents and management is highest for 'freedom to express opinions and feedbacks' and 'timely availability of food and medicines', indicating that residents are highly satisfied regarding the quality of these services. The variation is lowest for 'rooms getting cleaned', 'freedom to plan own time', 'freedom to move in and out', ' respect for residents' privacy', 'sessions on meditation, yoga etc.', and 'individualized attention'. This indicates that the management shall focus more on improving the standards of delivering these services in order to maintain or increase the satisfaction of residents. The management shall also make efforts to conduct entertainment sessions to provide them with the happiness,fun and satisfaction they long for, thereby making the variation in value positive.

5. CONCLUSION

The observations are almost similar for the two states except for a few attributes. Those that bear the highest and lowest differences in both states are almost the same indicating that delivery and perception of service are similar in almost all old age homes, irrespective of location of old age homes and age and gender of the residents. The satisfaction levels are high since most of the old age homes have facilities that are important for delivering services efficiently, such as an ambulance facility for emergency, a duty doctor at a phone call away and periodical meetings to discuss the problems and issues of residents, if any. Response shows that elderly are satisfied if they feel that there is someone who can communicate with them every day with love and care and attend to them when need arises. On the other hand, there are few services that the residents feel, should be improved. This happens because some old age homes give importance to a set of services and fail to provide others well. Some other homes impose certain rules for the well-being of their residents but residents feel highly dissatisfied at these services. This results in a huge variation in satisfaction levels. The management shall focus more on improving the standards of delivering these services in order to maintain or increase the satisfaction of residents.

6. LIMITATIONS AND FUTURE SCOPE

Data collected for analysis was majorly from cities of both states. The future studies can focus on old age homes from other areas as well to avoid cultural changes. It can give better insights into the services offered in old age homes and find differences in its operations within a state, if any.

REFERENCES

[1] Johnson et al.. (2017). A Qualitative Analysis of the Emergence of Long Term Care (Old Age Home) Sector for Seniors Care in India: Urgent Call for Quality and Care Standards. *Ageing Int*, 17(93).

[2] Rajan, S.I & Mishra, U.S. (2001). Health concerns among India's elderly. *INT'L J Aging and Development*, 53(3), 181-194.

[3] Government of India. Ministry of Statistics and Programme Implementation (2016). *Elderly in India – Profile and programmes 2016*. India.

[4] Lewinsohn, P.M. Redner, J.E., & Seeley, J.R. (1988). In F. Strack, M. Argyle, & N. Schwarz (Eds.), Subjective well-being. Oxford, England: Pergamon Press.

[5] United Nations Population Fund (UNFPA 2014). *The Status of Elderly in Tamil Nadu 2011*. India

[6] Kumari, S, Verma, M & Gupta, S. (2016). *International Journal of Applied and Pure Science and Agriculture (IJAPSA)*, 2(5).

[7] Thehinducom. (2018). *The Hindu*. Retrieved 19 March, 2018, fromhttp://www.thehindu.com/news/cities/Thiruvananthapuram/oldage-homes-reflect-sorry-statistics/article7670066.ece

[8] Fragoso, J.T & Espinoza, I.L. (2017). SERVPERF *Contaduría y Administración* , 62, 1270–1293.

[9] Al-momani, M.M. (2015). Gap Analysis between Perceptions and Expectations of Medical-Surgical Patients in a Public Hospital in Saudi Arabia. *Department of Community Health Sciences, College of Applied Medical Sciences, King Saud University, Riyadh, Kingdom of Saudi Arabia*, 25, 79–84.

[10] Abari, A.A.F, Yarmohammadian, M.H & Esteki, M. (2011). *Procedia Social and Behavioral Sciences*, 15, 2299–2304.

[11] Barrera, R.B. et al., (2014). *Revista Europea de Dirección y Economía de la Empresa*, 23(M31).

[12] Yeo, G.T, Thai, V.V & Roh, S.Y. (2015). *An Analysis of Port Service Quality and Customer Satisfaction: The Case of Korean Container Ports*, 31(4).

[13] Niebuur, J. (2014). *Network for studies on Pension, Age and Retirement.*

[14] Hawton, A., et.al (2012). *Journal of Integrated care*, 20(3), 181-194.

[15] Lee, J.E.N.N.Y., et.al (2013). *JOURNAL OF PALLIATIVE MEDICINE*, 16(10), 1268-1274.
[16] Niebuur, J. (2014). *Network for studies on Pension, Age and Retirement*.
[17] HelpAge India. (2015). The *State of Elderly in India Report 2011*. India.
[18] OECD/European Commission (2013), *A Good Life in Old Age? Monitoring and Improving Quality in Long-term Care,* OECD Health Policy Studies, OECD Publishing.
[19] Choudhary, A. (2013). *International Journal of Science and Research (IJSR)*, 4(9).
[20] Suresh, M & Mohan, D. (2016). *IEEE International Conference on Computational Intelligence and Computing Research, ICCIC 2015*.
[21] Suar, D, Mishra, S & Khuntia, R. (2007). *Asian Journal of Social Psychology*, 10(2), 117-122.
[22] Mahalingam, S, Jain, B & Sahay, M. (2016). *16 International Conference on Advances in Computing, Communications and Informatics, ICACCI*, 1069-1072.

MATEC Web of Conferences 172, 05004 (2018) https://doi.org/10.1051/matecconf/201817205004
ICDAMS 2018

Service Failures and Recovery Strategies in E-tailing: Examining the effects of Product Type and Transaction Frequency

Rini Miriam Vincy, Dr. *Hemamala* K
Amrita School of Business, Amrita VishwaVidyapeetham, Coimbatore, India
rini.vincy@gmail.com

Abstract—The purpose of this paper is recognize product type differences on post recovery satisfaction levels when online purchase failures happen at any stage during an order fulfilment cycle. In particular, we determine how it changes based on the recovery strategy adopted. Findings support the hypothesis that the satisfaction regarding an online purchase reduces when the product is purchased with more shopping effort, i.e. customers will have more satisfaction regarding the order fulfilment process when buying a convenience good than buying a speciality good. This study is unique in that, unlike previous studies on order fulfilment process in e-tailing investigating the relationship between severity of a service failure and post recovery satisfaction, we examine how does the satisfaction levels upon service recovery strategy adopted changes based on the frequency of purchase of the customer on an e-tailing platform. Results suggest that the satisfaction after an online purchase failure and its recovery tends to be high for a customer who has a high frequency of purchasing online than a person with a low frequency of purchasing online.

Index Terms—**Online service failure, Service failure severity, Service recovery strategy Recovery satisfaction.**

1. INTRODUCTION

When considering the consumer return decisions on a brick and motor platform or on an online platform, the important issues that become relevant are; why customers prefer to return the products purchased and what are the value propositions of these incidents in the minds of customers. These questions are answered, by researchers who have studied the importance of value proposition of the returns taking into consideration the customer's perspective thereby helping the customer and the company [1]. There is a need for the E-sellers to determine when fast is better for both the seller and the buyer [2]. Customers will enjoy more e-satisfaction when online firms sell a large variety of products/services in order to meet their expectations. Previous researches say interaction, information, usability along with three more quality dimensions are the main aspects that result in to high quality and satisfaction for customer thereby increasing the customer loyalty and spending by consumers [3]. Customer starts taking action against the retailer when the order gets delivered late than the expected time. This will reduce the no of orders in the future as well, negatively affecting the direct revenue of the company.

Operations management actions affect the anxiety level of customers when they experience the longest wait for their orders being delivered. If these customers continue purchasing from the same firm, they will be more anxious than other customers who also await their future order delivery. Therefore if the customer decides to forgive the firm, they do not easily forget the failures that happened in the past [4].

Customer satisfaction becomes high as a result of the additional effort put in for customization in the purchase process. The two main effects of effectively managing electronic retailing is that; Firstly, it is known that a large number of retail firms now plan to move onto an online platform; they have to understand that the online retail process needs to be simplified for success. Secondly, the product or service being offered needs to be fully understood by the retailers before being offered to the customers [5]. Online Spending in an online platform is reduced due to privacy concerns within the mind of the customer. The data collected from the customers are vulnerable to leakage to the external environment. These unauthorized usages of data happen when there is information loss, causing problems to those customers whose data was used. Researches also suggest that customers spend less on e-tailing platforms when they feel that the data being shared is not protected [6].

2. LITERATURE REVIEW

The fairness of strategy adopted in case of a failure and the fairness in treating the customers during an online purchase failure are very important in determining the satisfaction after the recovery strategy is implemented. Customers tend to have high satisfaction when there is a friendly approach by the e-tail firm during the order fulfillment process. Thus, apologies should be provided to them at the right time as they consider providing apology and giving proper explanation stating the reasons for the service failure as one the best recovery strategy that can be followed [7]. Psychological costs incurred by the customer can be a strategy for marketing their services as per the e-commerce firms. They use it as a platform to develop a long lasting customer relationship. Customers think it is best to stick onto a single e-tailing firm when they purchase more from them i.e. a higher number of products and services [8].

When the firms have a good reputation among the customers and when they are able to establish a trust in the customer's minds they succeed in improving the satisfaction of the customer. With higher satisfaction they would want to purchase again from the same firm. The appearance of the e-commerce website is another way of establishing trust in the minds of the customer [9]. The service quality gap, the gap between the expectations of the customer about the product and the delivered product can be reduced if customers are made known about the product from the start of the order fulfilment cycle.

The first step to be taken to improve the quality of products being delivered to the customers is to select an appropriate supplier for the same thereby taking responsibility for the goods they provide. Therefore reliability of the customer on the service provided by the e-tailing firm is significantly improved. Customers who wants to receive reliable help, wants to receive it in a timely manner [10].

There are certain guidelines to be followed while implementing a service recovery called the process guidelines. These guidelines can be efficiently followed by the employees than compensation guidelines. This can definitely be followed by firms with high income but when it comes to firms with less income the usual approach of providing recovery is followed based on the process guidelines. They try to provide recovery services effectively to customers by providing it on time. Along with the recover strategies followed as per the process guidelines they make sure that the customer gets a generous compensation for the failure experienced [11]. Apology is always the most effective recovery method, since it consumes less cost for the e-tailing firms and requires less time from their side. Therefore the effort put in by the frontline employees is made effective. Involving customers in the process of providing a recovery is even more effective as they come to know the reason why the service went wrong. There are other strategies which can be provided by the firms like problem solving and this can be implemented at a higher cost at times of performing severe failures [12].

3. Conceptual Model and Hypothesis Development

Based on an extensive assessment of the literature, the study proposes a conceptual model to understand how factors like recovery satisfaction, repurchase intentions and PWOM are affected by both severity of service failure and the recovery strategy adopted. Also it can be used to understand if there exists a moderating effect by factors like product type and transaction frequency on the relationship mentioned before; it can be used to find whether the model formed by severity of service failure, recovery strategy, recovery satisfaction, repurchase intentions and PWOM gets changed when factors like product type and transaction frequency are added into the same model.

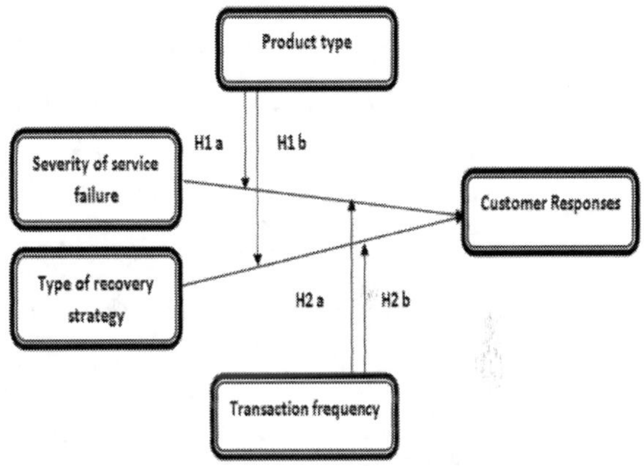

Fig.1. CONCEPTUAL MODEL

Service failure criticality exerts a significant negative relationship on customer loyalty in the case of online purchasing, i.e. customers who have experienced high criticality service failures are more likely to have a low customer loyalty in comparison to those who experienced a low criticality service failure based on their perceptions [13]. In particular, customers are most concerned about the type of compensation received, whether it's a financial compensation or a psychological compensation, and the criticality of the service experience.

Research has proved that, online service experiences low in criticality, delayed response combined with psychological compensation (vs. financial) will lead to repurchase intentions and PWOM. On the other hand, online service experiences high in criticality, immediate response is most effective service recovery strategy [14]. Service failures like ''gap between expectations and perceptions", "alterations and repairs", ''leak of personal data" and ''seller fraud problem", includes failures that should be mandatorily avoided because correcting these failures does not improve the buyers' repeat purchase intention [15]. In the scenario of an e-tailing service failure, higher post-failure customer loyalty is exhibited by customers who receive high procedural justice as a service recovery strategy.

They might not like considering the negative relationship of service failure criticality and customer loyalty, when compared to those customers who receives low procedural justice [13]. If an equally efficient and seamless recovery experience is provided, customers will react positively when services in e-retail settings fail [14]. Sellers should first try identifying the service failure and post identification they should refer to the recovery effects, if it happens during a service delivery process. This procedure when followed will enhance the quality of recovery strategy in online service platforms [15]. Sometimes technology-enhanced retail atmosphere acts as a catalyst in encouraging customer to accept delays during online service failure resolution. This happens due to reduced customer visibility of the firm's recovery actions. Depending on the resource associated with recovery, customers tend to be inclined to exhibit differential construal [14].

There is a decrease in the levels of commitment and customer trust for an e-tailing firm when failures occur. It also means that the relationship that exists between customer loyalty and service failure criticality doesn't weaken because of the switching costs perceived by the customer [13]. The extent to which a retailer's recovery efforts are acceptable usually depends mostly on the level of importance attached by the customer to that online purchase (i.e. criticality), and also by the negative emotions triggered by unsatisfactory experience [14].

Former researches identified that, varied customer satisfaction levels are due to the differences in the order fulfilment processes considering the product types. This means that the order fulfilment processes of specialty goods are not meeting the expectations of the customers as compared to order fulfilment processes when buying convenience goods or shopping goods.

Therefore the e-tailing firms need to attend more to the order fulfilment process of goods which involves more effort from the customer side in particular, and should also try developing customized approaches to fulfilling an order across product types [16].

H1 a: Following service failures in e-retail setup, low severity failure and low compensation recovery strategy will lead to higher (a) recovery satisfaction, (b) PWOM and (c) repurchase intentions when moving along the product continuum from specialty goods to convenience goods.

H1 b: Following service failures in e-retail setup, high severity failure and high compensation recovery strategy will lead to higher (a) recovery satisfaction, (b) PWOM and (c) repurchase intentions when moving along the product continuum from specialty goods to convenience goods.

The decision on the recovery programs adopted by e-tailing firms' changes based on customer transaction frequency. When there are complaints from customers with a low transaction frequency, they should aim to deliver interactional justice than the other justice dimensions like distributive and procedural justice. When there are complaints from customers with high transaction frequency, they should aim to deliver procedural justice than the other justice dimensions [17]. Repurchase intentions of customers are usually influenced by post-recovery satisfaction, when it comes to consumers with higher cumulative online experience than consumers with lower cumulative online experience [18].

H2 a: Following service failures in e-retail setup, low severity failure and low compensation recovery strategy will lead to higher (a) recovery satisfaction, (b) PWOM and (c) repurchase intentions when the transaction frequency of the customer moves low to high.

H2 b: Following service failures in e-retail setup, high severity failure and high compensation recovery strategy will lead to higher (a) recovery satisfaction, (b) PWOM and (c) repurchase intentions when the transaction frequency of the customer moves from low to high.

4. Methodology

4.1 Questionnaire formation and Measures

Methodology in the initial stage included building a questionnaire followed by an online survey to understand the various service failures that customers had faced during an online purchase. It included the questions which seeks the demographics of the customers, to questions asking how satisfied they are with the recovery strategy adopted for the failures they had faced. This questionnaire begins with questions seeking the demographics of the respondent.

The respondent was asked about the failure incident in the next section where they were asked to select the product type based on the product that they have purchased during that incident. The product types being purchased were adopted from a scale developed by Copeland et.al [19].

Once the detail of the product was specified they were asked to select the type of failure that occurred, the recovery strategy adopted and rate the failure on the basis of its criticality. The type of failures being experienced and the recovery strategies adopted were from a scale developed by Ying-Feng Kuoet.al [15]. And for measuring the severity the scale being used was a 7 point Likert scale which ranged from "Very Strongly Disagree" to "Very Strongly Agree". Finally the customer responses which includes recovery satisfaction, repurchase intentions and PWOM were measured using the scales adopted by Benedetta Crisafulli et.al [14]. These levels of customer responses were measured using questions seeking the customers reactions post a recovery strategy. They were asked whether they are satisfied with the resolution adopted by the e-commerce website or will they continue purchasing from the firm or will they recommend the firm to their friends.

4.2 Data Description

For conducting the study, data was collected through an online survey and the sample included 150 people, across various age groups and locations. The data was collected from people who has purchased online at least once and has faced a problem while purchasing online. The problems during an online purchase can vary from wrong product or fake product to delay in delivering the product purchased online. Therefore the survey respondents were asked to fill the survey keeping in mind an incident which went wrong and made an impact on them, i.e. an issue which happened while purchasing and which they could remember well.

4.3 Empirical Model

The basic model used for testing the hypothesis is shown below:

$$CR = \beta 1 + \beta 2SF + \beta 3RS + U(1) \qquad (1)$$

Where,

CR = Customer Responses (Recovery satisfaction, PWOM and Repurchase intentions)
SF = Severity of the Failure
RS = Recovery Strategy adopted
U = Error term

Others terms present in the conceptual model are product type and transaction frequency. These are the moderating variables which would bring a difference to the basic empirical model. Finding out the moderating effect of these variables forms the study behind this paper.

Model 2 used for testing hypothesis 1 i.e. H1, included the moderating effect of the variable product type on the relationship between severity of the failure, recovery strategy adopted and customer responses.

$$CR = \beta 1 + \beta 2SF + \beta 3RS + \beta 4PT + U(2) \qquad (2)$$

Where,
CR = Customer Responses (Recovery satisfaction, PWOM and Repurchase intentions)
SF = Severity of the Failure
RS = Recovery Strategy adopted
PT = Product Type
U = Error term

Model 3 used for testing the hypothesis 2 i.e. H2, included the moderating effect of the variable transaction frequency on the relationship between severity of the failure, recovery strategy adopted and customer responses.

$$CR = \beta 1 + \beta 2SF + \beta 3RS + \beta 4TF + U(3) \qquad (3)$$

Where,

CR = Customer Responses (Recovery satisfaction, PWOM and Repurchase intentions)
SF = Severity of the Failure
RS = Recovery Strategy adopted
TF = Transaction Frequency
U = Error term

The moderating variable Product type was classified into (a) convenience goods (b) shopping goods and (c) specialty goods [16]. The other moderating variable which is Transaction frequency was classified into (a) 1–4 times, (b)5–8 times (c) 9–12 times and (d)more than 13 times [17]. Moving along the product continuum in the context of product type or moving from high to low frequency levels in the context of transaction frequency, there can be a change in the basic empirical model which is yet to be proved in this paper.

5. Results, discussion and future research

Based on the hypothesis, factors like severity of the failure incident impacts customer responses which includes recovery satisfaction repurchase intentions and PWOM. This implies that higher the severity of the failure of an online purchase, as perceived by the customer lower will be the recovery satisfaction, PWOM and repurchase intentions. Similarly recovery strategy also impacts customer responses which includes recovery satisfaction, PWOM and repurchase intentions.
Table 1 depicts the product type- transaction frequency matrix which denotes the percentage of customers who has lower levels of satisfaction as per

the severity of the failure incident and the recovery strategy adopted.

Table 1. PRODUCT TYPE – TRANSACTION FREQUENCY MATRIX

Product Type	Transaction Frequency			
	%	%	%	%
	Low	Moderate	Medium	High
Convenience Goods :	0.006	0.013	0.006	0.006
Shopping Goods :	0.060	0.113	0.080	0.086
Specialty Goods	0.066	0.040	0.006	0.020

Since severity of the online failure incident and recovery strategy adopted has a significant impact on the customer responses, it also changes based on the product type purchased.

A low severity failure along with a low compensation recovery strategy will lead to higher recovery satisfaction, repurchase intentions and PWOM when moving from specialty goods to convenience goods along a product continuum as depicted in Table 2.

Table 2. RESULTSOF THE REGRESSION MODEL – PRODUCT TYPE

Product Type	Recovery Satisfaction	Repurchase Intention	PWOM
		Coefficient/P Value	
Convenience Goods :			
Severity_ Failure	-0.181/0.017	-0.148/0.008	-0.167/0.023
Recovery_ Strategy	0.377*/0.011	0.038*/0.010	0.344*/0.049
Shopping Goods :			
Severity_ Failure	-0.198/0.029	-0.162/0.006	-0.247/0.035
Recovery_ Strategy	0.370*/0.034	0.021*/0.002	0.339*/0.044
Specialty Goods :			
Severity_ Failure	-0.260/0.018	-0.212/0.025	-0.303/0.029
Recovery_ Strategy	0.234*/0.023	-0.047/0.033	0.241*/0.044

Note: * implies significance at the 5% level

In a similar way, transaction frequency also has a significant impact on the relationship of severity of the failure incident, recovery strategy adopted and customer responses. That is, low severity failure and low compensation recovery strategy will lead to higher recovery satisfaction, PWOM and repurchase intentions when the transaction frequency of the customer moves low to high as depicted in Table 3.

Table 3. RESULTSOF THE REGRESSION MODEL– TRANSACTION FREQUENCY

Transaction Frequency	Recovery Satisfaction	Repurchase Intention	PWOM
	Coefficient/P Value		
Low :			
Severity_ Failure	-0.290/0.032	-0.218/0.003	-0.387/0.031
Recovery_ Strategy	0.226*/0.022	-0.121/0.009	0.250*/0.032
Moderate:			
Severity_ Failure	-0.194/0.012	-0.159/0.014	-0.274/0.027
Recovery_ Strategy	0.360*/0.018	-0.014/0.006	0.351*/0.022
Medium :			
Severity_ Failure	-0.193/0.012	-0.155/0.005	-0.270/0.011
Recovery_ Strategy	0.379*/0.016	-0.047/0.003	0.392*/0.013
High :			
Severity_ Failure	-0.182/0.033	-0.154/0.005	-0.259/0.016
Recovery_ Strategy	0.381*/0.021	0.005*/0.004	0.430*/0.011

Note: *, ** implies significance at the 5% level and 1% level respectively

6. Limitations

The study focuses on understanding the relationship between severity of the online failure incident, recovery strategy adopted and customer responses based on the product type (classified as per the customer involvement in the purchase process) purchased by the customers and their transaction frequency. Further study could focus on understanding the relationship between severity of the failure incident, recovery strategy adopted and customer responses, based on product type where the classification can be in terms of the money spent on buying the good.

7. Conclusion

The study indicates that the satisfaction associated with the recovery of a failure will be less for a customer who has purchased a product with high involvement and for a customer who shops less frequency on online platforms. This is important because only if the e-commerce services providers come up with a differentiated recovery strategy for consumers who are believed to in the former category, they will be able to maintain a customer base having a good product type and transaction frequency mix. On the other hand from the number of responses obtained it is understood that most number of purchases happen in the category of customers with a moderate transaction frequency and purchasing a product of type shopping good.

This implies that the service failures in an e-tailing process needs to be reduced for customers who purchase less frequently and buys speciality goods because of their lower satisfaction levels. Since the maximum numbers of online purchases are by customers who buy shopping goods with a moderate frequency, these failures should be monitored to prevent a diminishing customer base.

References

[1] Rao, S, Rabinovich, E & Raju, D. (2014). *Journal of Operations Management*, 32, 295–312.

[2] Parvinen,P, Oinas-kukkonen, H & Kaptein, M. (2015). E-selling: *Electronic Commerce Research and Applications*, 14, 214–221.

[3] Nisar, T.M & Prabhakar, G. (2017). *Journal of Retailing and Consumer Services*, 39, 135-144.

[4] Rao, S, Griffis, S.E & Goldsby, T.J. (2011). *Journal of Operations Management*, 29, 692-703.

[5] Thirumalai, S & Sinha, K.K. (2011). *Journal of Operations Management*, 29, 477-487.

[6] Akhter, S.H. (2012). *Journal of Retailing and Consumer Services*, 19, 109-115..

[7] Yeoh et al.. (2015). *Journal of Services Research*, 14(2), .

[8] Wang, K, Hsu, L &Chih, W. (2014). *Managing Service Quality*, 24(4), 318-338.

[9] Hsu et al.. (2014). *Computers in Human Behavior*, 36, 234-245.

[10] Xifei, F & Jin, Q. (2015). *Journal of Marketing Development and Competitiveness*, 9(1), .

[11] Li, C & Fang, Y. (2016). *Serv Bus*, 10, 179–200.

[12] Mostafa, R, Lages, C.R & Sääksjärvi, M. (2014). *Journal of Services Marketing*, 28(4), 300-310.

[13] Wang et al.. (2011). *International Journal of Information Management*, 31(31), 350-359.

[14] Crisafulli, B & Singh, J. (2017). *Computers in Human Behavior*, 1-12.

[15] Kuo, Y, Yen, S & Chen, L. (2011). *Electronic Commerce Research and Applications*, 10, 183-193.

[16] Thirumalai, S & Sinha, K.K. (2005). *Journal of Operations Management*, 23(23), 291-303.

[17] Chang, H.H, Lai, M & Hsu, C. (2012). *Computers in Human Behavior*, 28(28), 2199-2208.

[18] Holloway, B.B., Wang, S. and Parish, J.T. (2005). *Journal of Interactive Marketing*, 54-66.

[19] Copeland, M.T., A.W. Shaw Co., Chicago (1924).

[20] Unni, S., Nair, V.S., Gunasekar, S. (2016).E-retail and online customer satisfaction: Differential impact of delivery time in Indian metros. *International Conference on Computational Intelligence and Computing Research*.

[21] Pillai, A.P., Gupta, D., Gunasekar, S. (2016).Brand relationship quality of online shopping portals a study in the Indian context. *International Conference on Communication and Signal Processing*, 7754433, 1614-1618.

[22] Mishra, A., Gunasekar, S., Gupta, D. (2017).Influence of product perception and discounts on the preference of online payment. *International Conference on Computational Intelligence and Computing Research*.

Musculoskeletal & Risk Assessment during Threshing of Rice-grains

Debesh Mishra[1], *Suchismita* Satapathy[2,*]
[1,2]School of Mechanical Engineering, KIIT, Bhubaneswar, Odisha, India

Abstract. In the present study 25 farmers of five villages of South Odisha in India were considered to assess their musculoskeletal disorders (MSD) during threshing of crops. A "Standardized Nordic Questionnaire" was prepared to collect MSD data considering harvesting characteristics, socio-demographic characteristics, economical characteristics about the requirements, and work setting characteristics, by using a five point rating scale i.e. (1 = Very less, 2 =Less, 3 = Nil, 4 = Strong, 5 = Very Strong) for a period of 2 years i.e. from 2015 to 2017. Socio-demographic characteristics and Work setting characteristics of farmers were illustrated to depict the MSD level in different body parts like neck, shoulder, upper back, lower back, elbow, wrist/hand, thigh, knees and ankles, respectively. Factor analysis was performed for threshing characteristics, and economical characteristics, and different parameters under different dominant factors were grouped accordingly. To illustrate the linkage and linear relationship between important parameters, Pearson correlation coefficient matrix was generated for the threshing characteristics of farmers. Also the regression analysis was done to obtain the best-fit linear regression equations for the economic characteristics of farmers. Finally posture analysis was performed for different postures in threshing activities by using the Ovako Working-Posture Analysis System (OWAS) & Rapid Entire Body Assessment (REBA) technique.

1 Introduction

Farming is greatly influenced by the techniques and tools used in various stages and activities involved. With the development of machinery and equipments, the farming process has become easier, more efficient and productive. Still most of the farming activities are carried out by the conventional tools and methods. It has been reported by many authors and researchers that the farmers engaged in agricultural sectors are highly affected by musculoskeletal disorders (MSD) because of the risk factors evolved in the respective work places. Most of the jobs are designed based on the anthropometric dimensions of men and this is the cause for lower productivity, discomforts and the increase in energy expenditure for the women. Also it was reported that the women who work continuously for eight hours, usually suffer from pain in hand, wrist and neck [9]. A detailed literature review of the musculoskeletal disorders of women farmers in India was studied [14] and concluded with a number of disorders in body parts with different body postures. Also it was recommended for ergonomically design of women tools and equipments, and to have work rest periods. Fifteen farmers were analyzed during milking, lifting, carrying a full bucket and in moving tractor. The surface electromyography system and Job Strain Index method were used to analyze the muscle load. Highest muscular tension and force were reported in the forearm muscle while attaching teat cups to udder [8]. The work related musculoskeletal disorder are aggravated mainly due to the performance of work, or associated work setting [3]. The musculoskeletal disorders are the most important cause of tenderness, disabilities, non-attendance, lower productivity and financial losses for the workers throughout the world [16]. The seriousness of work related musculoskeletal disorders were highlighted and it was also reported that the prolonged exposure to loads due to variable work intensity leads to musculoskeletal disorders [6]. The workers engaged in meat processing were studied and various activities involving repetitive movements, more application of forces, high working rate and incorrect postures for a longer duration were identified [1, 10]. It was suggested that the hazards such as infection, lung cancer and knife related injuries are more evident in case of butchers [4]. It was reported that the butchers are more subjected to work related musculoskeletal disorders [11]. It was suggested that as the farmers are self employed they have the right and freedom to change or stop some sort of work, if they feel inconvenience in their work due to fatness [12]. The bio-mechanical evaluation of manual material handling tasks of a construction site in India was studied and it was found that the compressive forces at some disc are above the threshold value for the construction workers. It was suggested to have improvement in ergonomic performance for manual material handling tasks [13]. The musculoskeletal disorders of the farmers of Kanpur in India was reported as lower back (60%), knee (39%), shoulder (22%), neck (10%); and a higher chronic afflicted for nearly a year [5]. The occurrence of work related musculoskeletal disorders was reported to be more between the butchers in Kano city and very few only avail the medical facility. Also age was found to be the main risk factor in this regard [2]. The development of musculoskeletal disorders (MSD) is mainly due to awkward and uncomfortable working postures [15].

* Corresponding author: suchismitasatapathy9@gmail.com

India is well known for its agricultural sectors and Odisha being one of the prominent states in this country, where most of the farmers prefer cultivation of rice. The farmers in the rural areas mainly prefer to use the traditional tools and techniques for farming purpose, and very few of them use the modern mechanized techniques. It has been found that most of the farmers undergo the musculoskeletal disorders (MSD) because of a number of risk factors such as repetitive work, incorrect postures, improper use of tools and techniques etc. Hence an attempt was made to collect and analyze the MSD data for the farmers in South Odisha.

2 Methodology

The rice cultivation process involves land preparation, seeding and weeding, harvesting and threshing processes. For the present study five villages were selected randomly from Odisha state in India, and 25 farmers engaged only in threshing activities were considered. Before collecting the information and data, a team comprising of three members was formulated. The members were well informed about the purpose and methods to collect information and data. Prior consent was obtained from the respective village heads or leaders. All members visited the farmers personally in the evening time, and some on the morning time while they were in respective works. The information for the consequent visits to the farmers and other villages were obtained from the previous visits and talking to the farmers. The same procedure was followed in other villages also. A standardized Nordic questionnaire was prepared to collect the information and data regarding name, gender, age, other personal characteristics, socio-demographic characteristics, economical characteristics about the requirements, and work setting characteristics as the factors associated with the musculoskeletal disorders with a five point rating scale i.e. (1 = Very less, 2 =Less, 3 = Nil, 4 = Strong, 5 = Very Strong). Various body parameters considered to be affected with MSD were neck, shoulder, upper back, lower back, elbow, wrist/hand, thigh, knees and ankles, respectively. The information was collected considering 2 years period i.e. from 2015 to 2017. A total 25 farmers were selected from all the five villages and the MSD data were recorded based on their present and past experience involved in threshing operations. Stick diagrams were drawn based on the digital photography of different frequently occurring postures taken during threshing activities, and was subjected for analysis by using both the "Ovako Working Posture Analysis System (OWAS)" and the "Rapid Entire Body Assessment (REBA)" technique.

3 Result and Discussion

In the present study 25 farmers were selected for the analysis purpose. From the socio-demographic characteristics of farmers (Table.1), it was observed that maximum musculoskeletal disorders of farmers were reported in wrist for age groups <38 & >38, in lower back and shoulder of female farmers, in upper back of male farmers. For education levels, more disorders were reported in wrist for under primary, in thighs for primary to high schools, and elbow for more than high schools. Most of the disorders were found in various body parts irrespective of experiences, marital status, smoking habits, eating food habit per day, previous weaknesses. The shoulder and wrist were more affected by the left & right handed farmers, the farmers' exposure to wet and humid soil, respectively. Similarly considering the work setting characteristics of farmers as shown in Table.2, high incidence of musculoskeletal disorders(MSD) were found in Lower back & Knees for sitting postures, and neck, upper back, lower back, Knees and Ankles in squatting postures, and upper back, lower back, Wrist and Knee in standing postures. The occurrence of MSD in shoulder & wrist were more significant irrespective of working hours per day, working days per week and working time per day, respectively. Also it may be observed that repetitive jobs produce more MSD in neck, elbow, wrist and knees. Carrying heavy loads resulted in high MSD in shoulder, lower back and wrist, respectively.

Table 1. Socio-demographic Characteristics of Farmers (n=25)

Sl. No.	Parameters		No. of farmers reported of suffering of musculoskeletal disorders (MSD)								
			Neck	Shoulder	Upper Back	Lower Back	Elbow	Wrist/Hand	Thigh	Knee	Ankle
1.	Age	< 38	16	21	19	22	15	23	14	15	19
		> 38	16	20	21	20	16	22	15	15	13
2.	Gender	Male	15	19	24	15	15	17	20	15	15
		Female	18	25	19	25	16	20	14	20	15
3.	Education Level	Under primary	15	10	13	16	15	18	16	23	16
		primary to High school	10	12	15	09	15	15	21	13	20
		More than High school	17	19	17	19	20	17	17	19	19
4	Exper	< 1	1	15	1	1	1	18	1	1	1

.	ience	Year	8		9	9	9		5	8	9
		1-10 Years	19	22	18	15	18	18	15	18	19
		>10 Years	19	17	19	14	21	18	19	18	19
5.	Marital Status	Un-married	13	25	18	15	19	25	16	14	14
		Married	16	23	21	25	18	25	16	17	17
6.	Smoking habits	Non-Smoker	20	25	15	11	10	25	12	17	11
		Smoker	22	25	25	25	11	25	02	25	09
7.	Eating food habit per day	<2	08	25	17	16	15	25	17	14	07
		>2	12	25	17	17	20	25	14	20	18
8.	Handedness/ Dominant hand	Left-handed	16	25	19	22	17	25	17	16	14
		Right-handed	15	25	15	15	15	25	15	18	18
		Mixed-Handed	12	19	13	16	15	14	12	15	15
9.	Exposure to wet and humid soil	Low land	12	25	17	14	15	25	15	12	15
		Up land	18	25	16	21	22	25	20	18	17
		Wet land	13	25	16	15	15	25	15	15	12
		Dry land	12	25	16	18	15	25	15	15	12
		Horizontal/Plain	21	25	21	22	18	25	15	21	09
		Sloped	15	25	22	20	22	25	15	15	16
10.	Previous weakness/injuries, if any		11	25	14	18	19	25	19	19	14

n= Total number of farmers.

Table 2. Work Setting Characteristics of Farmers (n=25)

Sl. No.	Parameters		No. of farmers reported of suffering of musculoskeletal disorders (MSD)								
			Neck	Shoulder	Upper Back	Lower Back	Elbow	Wrist/ Hand	Thigh	Knee	Ankle
1.	Working posture(s)	Sitting	12	18	16	24	16	16	19	24	16
		Squatting	25	19	25	25	15	18	15	25	25
		Standing	17	16	25	25	17	25	16	25	18
2.	Working hours /day	< 3	15	24	22	15	21	18	18	16	21
		> 3	21	24	21	18	15	23	16	15	14
3.	Working days/week	< 4	16	25	18	17	16	25	18	16	16
		> 4	15	25	15	12	16	25	16	17	16
4.	Working time(s)/day	Before 8 A.M.	17	25	19	15	17	25	16	18	17
		8 A.M. – 12 P.M.	12	25	12	12	25	25	07	13	17
		After 12 P.M.	10	25	25	25	18	25	14	25	25
5.	Nature of job	Repetitive/forceful	25	13	20	16	25	25	19	25	19
		Carrying heavy loads	20	25	19	25	15	25	17	17	13

n= Total number of farmers.

Factor analysis using SPSS 22.0 for the threshing characteristics of farmers, and economical characteristics of farmers was performed as shown in Table 3 and Table 4, respectively. The principal component analysis method with rotated component matrix was used and the variables with factor loading more than or equal to 0.5 were only considered. For factors selection, Kaiser's Eigen value criteria with Eigen value greater than 1 were selected [7]. The factor analysis of threshing characteristics of farmers (Table 3) resulted in 3 dominant factors with cumulative % of variance of 79.42 %. In factor1, the factor loading is high in both in Injuries from manually operated Thresher by using hand operated mechanical Thresher and in musculoskeletal problems from incorrect postures in Threshing which is 0.963. As the variables in factor 1, corresponds to injuries & musculoskeletal disorders. Hence the factor1 is named as Discomfort. The factor loading in factor 2 is high in Exposure to vibration of threshing machines which is 0.942. Other variables in factor 2 are related to the aid of external appliances. The factor 2 is named as External. The factor loading in factor 3 is high in Exposure to noise of threshing machines i.e. 0.933. The factor 3 is named as Operation.

Similarly, the factor analysis for economical characteristics of farmers resulted in 2 dominant factors (Table 4). The factor 1 is named as Tangibility and factor 2 as Intangibility, based on the physical characteristics of variables.

Table 3. Threshing Characteristics of Farmers (n=25)

Parameters		Factor 1 (Discomfort)	Factor 2 (External)	Factor 3 (Operation)
1.Injuries from manually operated Thresher	a) Striking on wooden frames	0.606		
	b) Using hand operated mechanical Thresher	0.963		
	c) Using Draft animal driven Thresher		0.853	
2.Exposure to vibration of Threshing machines			0.942	
3.Exposure to noise of Threshing machines				0.933
4.Musculoskeletal problems from incorrect postures in Threshing		0.963		
5.Exposure to grain dust			0.718	
Eigen value		2.66	1.82	1.07
Variance in %		38.11	26.01	15.28
Cumulative variance in %		38.11	64.13	79.42

n= Total number of farmers.

Table 4. Economical Characteristics of Farmers (n=25)

Parameters	Factor 1 (Tangibility)	Factor 2 (Intangibility)
1.Compliance with cost	-0.659	
2.Cost reduction of existing Appliances	0.665	
3.Quality	0.818	
4.Service capability		0.869
5.Innovativeness		0.909
6.New use of technologies / modifications of design	0.727	
Eigen value	3.03	1.05
Variance in %	50.61	17.49
Cumulative variance in %	50.61	68.11

n= Total number of farmers.

The Pearson Correlation coefficients between various dominant factors in the threshing characteristics of farmers, to illustrate the strength & linear relationship of various variables under consideration are shown in Table 5. The correlation coefficient matrix was obtained by using Pearson product moment correlation in IBM SPSS 22.0. Very high correlation coefficient i.e. 0.991 is in between the factor 1(Discomfort) i.e. in between 'Injuries from manually operated Thresher' by using hand operated mechanical thresher and by using Draft animal driven thresher. While the smallest correlation coefficient i.e. 0.012 is between the factor 1(Discomfort) and factor 3(Operation), which is between 'Injuries from manually operated Thresher' by striking on wooden frames and 'Exposure to noise of Threshing machines'.

Using regression analysis for the Economic Characteristics, the best fit linear regression equation between Innovativeness and other Economic Characteristics are illustrated in Table 6. The economic parameters such as Cost reduction of existing Appliances, Service Capability, and Innovativeness were found to be significant at $p< 0.01$.

Table 5. Correlation Coefficient matrix for threshing characteristics

Factors		A	B	C	D	E	F	G
Factor 1 (Discomfort)	A	1.000						
	B	0.397*	1.000					
	C	0.397*	0.991**	1.000				
Factor 2 (External)	D	-0.104	-0.282	-0.282	1.000			
	E	-0.213	-0.095	-0.095	0.789**	1.000		
	F	-0.197	-0.034	-0.034	0.405*	0.533**	1.000	
Factor 3 (Operation)	G	0.012	0.136	0.136	0.032	-0.141	0.182	1.000

*.Correlation is significant at the 0.05 level (2-tailed).
** Correlation is significant at the 0.01 level (2-tailed).

Note: **A**-Injuries from manually operated Thresher by striking on wooden frames, **B**-Injuries from manually operated thresher by using hand operated mechanical thresher, **C**- Musculoskeletal problems from incorrect postures in threshing, **D**- Injuries from manually operated thresher by using draft animal driven thresher, **E**- Exposure to vibration of threshing machines, **F**- Exposure to grain dust, **G**- Exposure to noise of threshing machines.

Table 6. Best-fit Linear Regression Equation between Innovativeness (X_5) and other Economic Characteristics

Sl. No.	Parameters	Regression Equation	F
1.	Compliance with cost(X_1)	$X_5 = 4.615 - 0.313 X_1$	2.77
2.	Cost reduction of existing Appliances(X_2)	$X_5 = 1.396 + 0.630 X_2$	11.48**
3.	Quality(X_3)	$X_5 = 2.94 + 0.217 X_3$	0.82
4.	Service capability(X_4)	$X_5 = 1.196 + 0.682 X_4$	19.06**
5.	Innovativeness(X_5)	$X_5 = 1.82 - 0.149 X_1 + 0.486 X_2 - 0.312 X_3 + 0.509 X_4 - 0.048 X_6$	5.28**

6.	New use of technologies/modifications of design(X_6)	$X_5 = 2.958 + 0.251\ X_6$	1.93

** Correlation is significant at p< 0.01

Cultivation of rice involves preparation of land, seeding, weeding, harvesting, & threshing activities. Usually the threshing of crops is done to separate the rice grains from the harvested stem of crops. In the traditional methods of threshing either the stems are stroked on some wooden frame, or some loaded element is made to move on the stems with the help of draft animals. Due to high fatigue involved in the traditional techniques, the farmers prefer to use the developed mechanical threshers. In the mechanical threshers, the stems of crops are fed against some rotating drum with projected fins such that the fins separate the grains from the stems. From the Analysis of Working Posture using Ovako Working Posture Analysis System (OWAS) as shown in Table 7, striking of stems on wooden frame involves back bent forward & back ward, both arms below & above shoulder level, both knees bent, weight or force less than 10 kg, and the obtained action category is 4 indicating an immediate corrective measure. Threshing using draft animals involves forward walking or movement, arms below shoulder level, knees bent & straight, force less than 10 kg and the obtained action category is 4 indicating an immediate corrective measure. Similarly, threshing using hand operated mechanical thresher involves back bent, arms below shoulder level, knees bent, force less than 10 kg and the obtained action category is 3 indicating corrective measure as soon as possible. The postures considered in OWAS are further analyzed by using Rapid Entire Body Assessment (REBA) technique and the corresponding REBA score, risk levels and remarks are illustrated in Table.8. The risk levels are found to be high for both threshing by striking on wooden frame & using draft animals, indicating the corrective measure as "necessary soon". While for threshing by using hand operated mechanical thresher, the risk level obtained is medium indicating the corrective measure as "necessary".

Table 7. Analysis of Working Posture using Ovako Working Posture Analysis System (OWAS)

Actions/Activities	Figure of posture	Code	Action Category	Remarks
a) Striking on wooden frames		4141	4	Corrective measures to take immediately
b) Using Draft animal driven Thresher		4142	4	Corrective measures to take immediately
c) Using hand operated mechanical Thresher		2141	3	Corrective measures to take as soon as possible

Table 8. Analysis of Working Posture using Rapid Entire Body Assessment (REBA)

Actions/Activities	Figure of posture	REBA Score	Risk Level	Remarks
a) Striking on wooden frames		8	High	Necessary Soon
b) Using Draft animal driven Thresher		8	High	Necessary Soon
c) Using hand operated mechanical Thresher		7	Medium	Necessary

4 Conclusion

A well designed job which is supported by an elegant place of work and proper tools, allow the farmers to avoid unnecessary movements and stresses of the neck, shoulder and upper limb. However the actual performance of various tasks depends on the farmer. The present study concludes that almost all parts of farmer's body such as neck, shoulder, upper back, lower back, elbow, wrist, thigh, knee and ankles were subjected to musculoskeletal disorders. The prevention of musculoskeletal disorders (MSD) depends on recognition of the risk factors by self assessment, study on the job or work place and the study of working postures in various activities while performing the required tasks. It is also important to consider the incidence rate of disorders in workplace and the exposure to hazardous working or environmental conditions. Farmers should be encouraged to use ergonomically designed tools and machineries to stabilize their physical ability to the correct job, but it deal with designing/modification of tools or machineries that is accurate for the task.

References

1. Arvidsson I, Balogh I, Hansson GT, Ohlsson K, Åkesson I, Nordander C, Appl Ergon 2012; 43:1026-32.
2. Bashir Kaka, Opeyemi A. Idowu, Henrietta O. Fawole, Ade F. Adeniyi, Omoyemi O. Ogwumike, Mark T. Toryila , Safety and Health at Work 7 (2016) 218-224.
3. Collins RM, Janse van Rensburg DC, Patricios JS, S Afr Fam Pract 2011; 53:240-6.
4. Corbin M, McLean D, Mannetje A, Dryson E, Walls C, McKenzie F, Maule M, Cheng S, Cunningham C, Kromhout H, Blair A, Am J Ind Med 2011; 54:89-101.
5. Gupta G, Tarique (2013) J Community Med Health Educ 3: 249. doi:10.4172/2161-0711.1000249.
6. Kumar VK, Kumar SP, Baliga MR, Indian J Dent Res 2013; 24:428-38.
7. Kaiser, H. F. 1960, Educational and Psychological Measurement, 20, 141–151.
8. Łukasz Kuta, Józef Cież, Małgorzata Młotek, Procedia Manufacturing 3 (2015) 1696 – 1703.
9. M. Jakob, F. Liebers, Bornimer Agrartechnische Berichte. T. 66; 2009, 31–37.
10. Mansi S, Milosavljevic S, Tumilty S, Hendrick P, Baxter GD, Health Qual Life Out 2013; 11:185.
11. Omokhodion FO, Adebayo AM, J Public Health (Germany) 2013; 21:131-4.
12. Pickett, W., King, N., Lawson, J., et al., 2015, Prev. Med. 70C, 59–63.
13. Pradip Kumar Ray, Ratri Paridab, Sagar Sarkar; Ergonomic analysis of construction jobs in India: A biomechanical modelling approach; Procedia Manufacturing; 3, (2015), 4606 – 4612.
14. Singh, S. and R. Arora. 2010, Journal of Agri. Science, 1(2): 61-71.
15. Somnath Gangopadhyay, Banibrata Das, Tamal Das & Goutam Ghoshal 2005, International Journal of Occupational Safety and Ergonomics; 11:3, 315-322, DOI: 10.1080/10803548.2005.11076652.
16. Tinubu BMS, Mbada CE, Oyeyemi AL, Fabunmi AA, BMC Musculoskeletal Disorders; 2010; 11:12.

Role of Solar energy and issues in its implementation in the Indian context

Saurav Dixit[1,*], *Subhav* Singh[2], *Sonali* Singh[3], *Richu* George Varghese[1], *Amit* Kumar Pandey[4], and *Deekshant* Varshney[2]

[1]RICS School of Built Environment, Amity University, Noida, India
[2]K R Mangalam University, Gurgaon, India
[3]Jaipuria Institute of Management, Noida, India
[4]Amity Business School, Amity University, Noida, India

Abstract. The main objectives of using renewable energy enable us to attain sustainable growth without compromising the future. India is on the 7th nation in generating renewable energy while the USA, China and Germany are on 1st, 2nd, and 3rd respectively. Secondly, in case of Solar energy, the statistics are: India is on the 7th nation in generating solar energy while China, Japan, and Germany are on 1st, 2nd, and 3rd respectively, Generating 9, 78, 42.8 and 41.2 (GW) gigawatts from solar power. India having great opportunities in generating solar power because of its location and the availability of largely unused land banks throughout the nation. The purpose of the paper is to examine the challenges to the adoption of solar energy as a means of generation of energy and power for the nation's requirement. The hurdles/factors identified for the study is 16 and a targeted data collection of 105 for the study using convenient sampling technique throughout India. The research finding shall contribute to the existing body of knowledge and provide suggestions to the policymakers, the manufactures and the local bodies working on the ground to implement the solar energy as an alternate energy resource to fossil fuels in India.

Keywords: Renewable Energy; Sustainable Energy; Non-Conventional Energy India; and Issues.

1 Introduction

Fast population development has brought about expanding demands for energy requirement and in addition becoming ecological concerns. It has additionally uncovered the imperatives of conventional energy. This has created a worldwide requirement for government approaches that truly consider sustainable power source innovations (SPSI). The strong connection between sustainable power source SPS) and sustainable development is provoking nations around the world to receive SPSI in order to guarantee sustainable development. It has been contended that multiplying the present execution of SPS era can decrease CO2 outflow to 349 g/KWh, which is not as much as the 498 g/KWh focused for 2030. This will keep environmental CO2 beneath 450 parts for every million, which is required to counteract serious atmosphere change [1].

It has been assessed that 88% of the world's vitality utilization is from non-renewable energy source [2]. Air contamination, human-prompted environmental change, vitality supply weakness and cost increments are a portion of the issues related to petroleum derivative that has offered ascend to the usage of SPSI over the world. The cosy relationship amongst SPS and sustainable development has provoked a few governments worldwide to present laws that advance SPS.

There is in this way an unfaltering development in the utilization of renewables around the world. By 2013, RE gave an expected 19.1% of worldwide vitality utilization. Of this aggregate offer, conventional biomass represented around 9%, Hydropower 3.9%, inexhaustible warmth 4.1%, and transport biofuels gave around 0.8%. By 2014, renewables rose to an expected 27.7% of the force to be reckoned with's creating a limit, which was sufficient to supply an expected 22.8% of worldwide power [3].

2 Literature review

Hydropower limit ascended by 3.6% to roughly 1055 GW, while different renewables all in all developed from about 18% to an expected aggregate drawing nearer 660 GW (REN21 2015). This was nearly trailed by the National Renewable Vitality Master Plan (NREMP) in 2005. NREMP imagined national reasonable

* Corresponding author: sdixit@ricssbe.edu.in

advancement through the usage of SPSI (Sambo 2009b), with the desire of a 10% RE entrance by 2020. Regardless of Nigeria's rich RE assets, the nation is still vigorously dependent on petroleum derivatives. The difficulties encompassing the usage of SPSI are ascribed to the low level of open mindfulness, costs, financing imperatives furthermore, powerless innovation scattering systems. Other factors incorporate an absence of shopper certainty due to poor item quality, an absence of sufficient approach, and an absence of gifted labour. As indicated by Okafor what's more, financial, social, specialized, what's more, political (counting instructive) obstructions must be overcome for the correct coordination of RE into the country's vitality blend. It has been contended that the acknowledgement of SPSI by the open is related to the general population's convictions, mentality also, the impression of such innovation. Open dispositions need to change to make significant improvements as far as the execution of SPSI doable. Open (social) acknowledgement abbreviates the time, what's more, their usage [4] [5].

It is thus that training on REs is respected as fundamental for the effective usage of SPSI and for open help. Instruction as an intense operator of social change not just brings issues to light about new improvements what's more, items; it likewise makes open trust in new items, what's more, prepares the general population to utilize them effectively [6] [7] recommends that the Nigerian government ought to sharpen the open, and also aid the production of business sectors for RE frameworks. And agrees by belligerence for the joining of RE training into the Nigerian instructive framework.

Table 1 issues/hurdles in solar energy implementation identified from the literature review and expert interview

Initial investment/high cost
Affordable/economical
Setting-up cost
Price to quality ratio/Profitability
Environment-friendly
Lack of policy
Lack of government initiative
Space requirement
Lack of knowledge
Reuse/recycle
Availability of raw material
Convenience of use
Uncertain power output
Lack of robust design
Automatic operation
Variation in electricity prices

2.1 Objectives of the study

I. To determine the factors influencing the companies to adopt sustainable supply chain management practices.
II. To prioritise the attributes on the basis of their relative importance index on various companies in the transition towards SSCM.

3 Research Methodology

For this research paper, a questionnaire survey method has been used to identify and analyse the hurdles/issues in the implementation of solar energy in India. For this, the structured questionnaire survey is shared with the professionals, the potential, and existing customers of solar energy. The received data were analyzed using Relative importance index (RII), relative importance index (Rii) was applied to prioritise the severity of the factors" [1] [2] [3] [4] and other techniques.

Quantitative research methodology is adopted for the study. The data is collected using structured questionnaire survey. Primary data has been collected for the study because of non-availability of secondary data on the said tittle. The convenience sampling technique is used to collect the data from the known and easily approachable respondents. The questionnaire is shared with 300 professionals working in different segments of the economy including but not limited to manufacturing, production, suppliers, customers, construction, energy, and others sectors. A total of 105 valid responses were received with a response rate of 35%.

The structured questionnaire is divided into three parts: the first portion has information about the topic and general study that is to be conducted, the second part is the questionnaire itself and the last portion consists the feedback and any suggestion for the study by the respondents. The respondents were asked to rate the questions on the Likert scale of 1-5 given against each question.

3.1 Reliability analysis

Dependability analysis is needed to check the consistency of the model after some time, and Cronbach's alpha test was the best way to check the reliability of the data collected through questionnaire [5]. The value of Cronbach's alpha for this study is 0.76 which is considered to be good for the study.

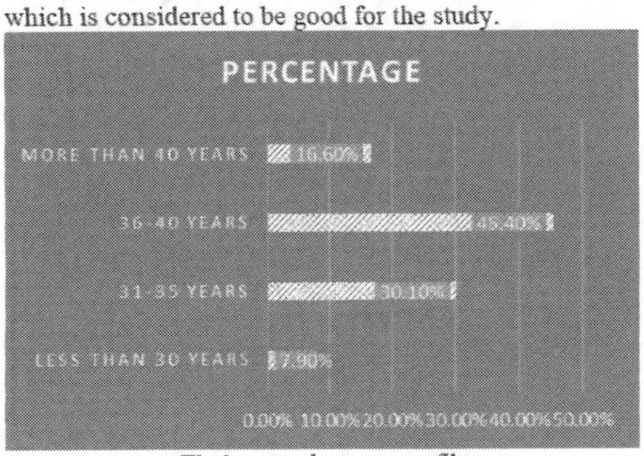

Fig 1 respondents age profile

Table 2 relative importance index ranking

Rank	*Total responses*	*Total score*	*RII*	*Attribute name*
1	105	467	0.89	Initial investment/high cost
2	105	444	0.85	Setting-up cost
3	105	426	0.69	Lack of government initiative
4	105	425	0.69	Lack of knowledge
5	105	413	0.69	Reuse/recycle
6	105	395	0.68	Convenience of use
7	105	393	0.68	Uncertain power output
8	105	392	0.67	Lack of robust design
9	105	391	0.67	Variation in electricity prices
10	105	385	0.67	Automatic operation
11	105	384	0.67	Availability of raw material
12	105	378	0.67	Space requirement
13	105	363	0.67	Affordable/economical
14	105	359	0.66	Price to quality ratio/Profitability
15	105	332	0.66	Environment friendly
16	105	327	0.66	Lack of policy

3.2 Relative importance index

The received responses by the respondents were summarized in an Excel data sheet and the data analysed using SPSS software. Relative importance indices (RII) is performed to determine the priority of the significant factors and then followed by Reliability analysis performed to check the consistency of the data received.

$$RII = \frac{\sum_{r=1}^{5} r * n_r}{5N} \quad (1)$$

r is the rating on a Likert scale (1-5) as for the impact on construction efficiency for a specific element influencing construction profitability, nr is the number of respondents providing a specific Likert scale rating r, N is the aggregate number of respondents to a specific question [6][4], [16], [17].

The value given to the Likert scale rating is as follows:

1. No effect (or no opinion)
2. Less effect (no or minimal effect)
3. Minor effect (minor problem)
4. Serious effect (medium problem)
5. Very serious effect (major problem

Table 3 Relative importance index

3.3 Most significant issues and hurdles

The significant issues and hurdles in the implementation of solar energy in Indian context are convenience of use, uncertain power output, lack of robust design, variation in electricity prices, and automation in operation having a relative importance index value by analyzing the received data is 68%, 67%, 67%, 67% and 67% respectively.

3.4 Significant issues and hurdles

The significant benefits through the transition towards SSCM are environmental benefits, competitive advantages, reduced carbon footprints, inventory control and helps in circular economy having a relative importance index value of 0.68, 0.68, 0.67, 0.67, and 0.67 respectively.

4 Discussion and conclusion

The study identified the current scenario of solar energy and the trends in the world, and in India. The contribution of India to solar energy production is 9 GW and it is in the 7 position. There is three type of establishment in solar energy are a government project, private projects, and individual household solar plants installation. There is a huge potential in all the three segments as mentioned above for the implementation of solar energy.

The main issues and hurdles in the expansion of solar plants were identified and analysed using RII and the top issues are mostly related to cost associated with the implementation of the plants. There are few schemes that provide subsidies on the installation of solar plants, but there is a loophole in these schemes i.e. the schemes provide subsidies amount on the purchase of solar panels and storage batteries only if they were purchased from the government agencies. And the item rate of these agencies is comparatively high to the private and locally available solar panels and batteries in the market. It is recommended to provide subsidies amount irrespective of the vendor or the market to all the traders working in the solar industry. The study provides an insight into the barriers in the functioning and expansion of solar energy in India.

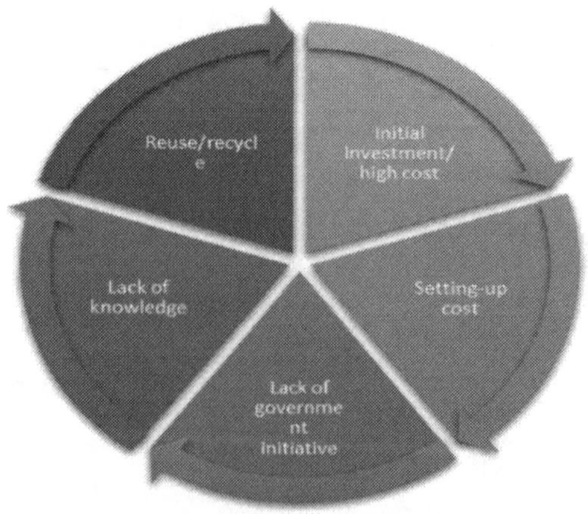

Fig 2 Top five issues/hurdles in the implementation of solar energy

5 Limitations of the study

The study is conducted using structured questionnaire survey and collection of primary data. The received responses are 105 considered for this study is comparatively small to generalise the findings to the larger scale [19]. It is recommended to conduct a similar kind of study in different regions of the country to have a better understanding of the issues and the hurdles in the implementation of solar energy as an alternative to the conventional energy sources in the Indian context

References

[1] IRENA, *Remap 2030: Renewable Energy Prospects for India*, no. May. 2017.

[2] R. Singh, B. B. Krishna, J. Kumar, and T. Bhaskar, "*Bioresour. Technol.*, vol. 199, pp. 398–407, 2016.

[3] M. S. Bhaskar, R. M. Kulkarni, S. Padmanaban, P. Siano, and F. Blaabjerg, *EEEIC 2016 - Int. Conf. Environ. Electr. Eng.*, pp. 1–6, 2016.

[4] I. A. Urzúa, J. C. Olmedo, and E. E. Sauma, *Renew. Sustain. Energy Rev.*, vol. 60, pp. 810–821, 2016.

[5] A. Sao, S. Singh, S. Dixit, A. K. Pandey, and S. Singh, *Int. J. Mech. Eng. Technol.*, vol. 8, no. 10, 2017.

[6] J. Cloke, A. Mohr, and E. Brown, *Energy Res. Soc. Sci.*, no. June, pp. 0–1, 2017.

[7] A. K. Pandey, S. Dixit, S. Bansal, S. Saproo, and S. N. Mandal, *Int. J. Civ. Eng. Technol.*, vol. 8, no. 10, 2017.

[8] C. Rodríguez-Monroy, G. Mármol-Acitores, and G. Nilsson-Cifuentes, *Renew. Sustain. Energy Rev.*, vol. 81, no. June 2017, pp. 937–945, 2018.

[9] S. Dixit, A. K. Pandey, S. N. Mandal, and S. Bansal, *Int. J. Civ. Eng. Technol.*, vol. 8, no. 6, 2017.

[10] C. Fogarassy and S. Istv, The ... Reykjaví k University Renewable Energy Summer Course Circular Economy Extracurricular Reykjaví k , 2017," no. August, 2017.

[11] A. Tapaninen, M. Seppänen, and S. Mäkinen, "Characteristics of innovation in adopting a renewable residential energy system," *J. Syst. Inf. Technol.*, vol. 11, no. 4, pp. 347–366, 2009.

[12] V. Khare, S. Nema, and P. Baredar, *Renew. Sustain. Energy Rev.*, vol. 58, pp. 23–33, 2016.

[13] A. Basu, B. Kumar, and T. K. Rana, pp. 52–55, 2017.

[14] L. Kong, Z. Li, L. Liang, and J. Xie, *Ind. Manag. Data Syst.*, vol. 117, no. 6, pp. 1145–1165, 2017.

[15] S. Dixit, S. N. Mandal, A. Sawhney, and S. Singh, *Int. J. Civ. Eng. Technol.*, vol. 8, no. 8, pp. 623–636, 2017.

[16] D. Castro-Lacouture and K. O. Roper, *Facilities*, vol. 27, no. 5/6, pp. 173–186, 2009.

[17] S. Griffiths, *Energy Transitions*, vol. 1, no. 1, p. 3, 2017.

[18] R. N. Wojuola and B. P. Alant, *African J. Sci. Technol. Innov. Dev.*, vol. 9, no. 4, pp. 399–409, 2017.

[19] S. Singh, A. Bala, S. Dixit, and D. Varshney, *Int. J. Civ. Eng. Technol.*, vol. 9, no. 1, 2018.

Effect of Blade Profiles on the performance of Bidirectional Wave Energy Turbine

P.Madhan Kumar[1], *Abdus* Samad[1,*]

[1] Wave Energy and Fluids Engineering Laboratory, Department of Ocean Engineering, Indian Institute of Technology Madras, Chennai 600036, India

Abstract. To fulfill the ever growing demands of world energy consumption, the wave energy should be extracted economically. The oscillating water column is most commonly used to extract energy from waves. It consists of a chamber in which waves drives the entrapped air column to rotate the Wells turbine. The Wells turbine is a self-rectifying low-pressure axial reaction turbine with 90° stagger angle. These turbines consist of symmetrical airfoil profile to achieve unidirectional rotation for the bi-directional airflow. The turbine performance predominantly depends on the aerodynamic characteristics of the airfoil profile used. In this study, the performance of Wells turbine with various symmetrical airfoil profiles was analysed using ANSYS CFX 14.5. The CFD analysis was performed by solving three dimensional steady Reynolds averaged Navier-Stokes equation with k-ω SST turbulence closure model. The reference geometry has NACA0015 as blade profile and the CFD results were compared with the experimental values. The performance characteristics of the new airfoil profiles were compared with the reference case to analyse the suitability of airfoils in wave energy extraction. The NACA0021 airfoil profile showed better performance in the post-stall regime compared to the NACA0015 and the S1046 airfoil profiles.

Keywords: Wave energy, Wells turbine, Airfoil profiles, CFD

1 Introduction

The Wells turbine is used to extract the energy from the bidirectional airflow inside the oscillating water column (OWC). It consists of a symmetrical airfoil with 90° stagger angle. It provides unidirectional torque for the oscillating flow without any guidevanes, which makes the power take-off system simple [1]. On the contrary, the Wells turbine has limitations such as narrow operating range, noisy operation, and poor starting characteristics. The performance of the turbomachines primarily depends on the airfoil profile used. There were many studies done on the effect of different symmetrical and non-symmetrical airfoil profiles on the performance of Wells turbine. The airfoils with thicker profiles improve starting characteristics and performance of the Wells turbine [2,3]. The stall characteristics of Wells turbine can be improved by optimization of airfoil profile. Webster et al. [4] optimized the NACA0015 airfoil profile and reported better stall characteristics. Setoguchi et al. [5] investigated the performance characteristics of four different airfoil profiles and concluded that NACA four-digit airfoil profiles with 20% thickness ratio are suitable for better performance of Wells turbine. Takao et al. [6] compared the performance characteristics of airfoils with sweep. They studied the following airfoil profiles NACA0015, NACA0020, CA9, and HSIM 15-262123-1576 and reported that NACA0015 is the desirable airfoil profile to achieve better performance. The airfoil profile CA9 with solidity 0.64 is suitable to achieve better performance in the real sea conditions [7]. Takao et al. [8] investigated airfoil profiles NACA0015, NACA0020 and modified Eppler472 and reported that modified Eppler472 airfoil profile is superior to the other profiles analysed. Mohamed et al. [9] proposed the use of non-symmetrical airfoil profiles in the Wells turbine. The optimization of airfoil profile in Wells turbine is also investigated by many researchers [9-11]. From the above studies, it is evident that the performance of Wells turbine is highly influenced by the nature of airfoil profile used. To analyse the influence of airfoil profiles on the Wells turbine performance three airfoil profiles are used in this study. Based on the previous studies airfoils NACA0015 and NACA0021 are chosen as the suitable candidates for the blade profile. Mohamed [12] evaluated the effects of various symmetrical and non-symmetrical airfoil profiles on the performance of Darrieus turbine. He reported that S1046 airfoil profile shown better aerodynamic characteristics and improved efficiency compared to other profiles. Hence S1046 airfoil profile is also included in this numerical study. It is symmetrical in nature and suitable for Wells turbine application. The performance characteristics of the

* Corresponding author: samad@iitm.ac.in

airfoil profiles are numerically analysed and compared with the reference case. The fluid dynamics is also explained in detail with the help of post-processed figures.

2 Description of rotor geometry

The Wells turbine geometry used in the work of Torresi et al. [13] is taken as the reference case for this study. It consists of NACA0015 as the blade profile. The specifications of the reference geometry are provided in Table 1.

Table 1. Reference geometry specifications

Blade profile	NACA0015
Number of blades	8
Chord length	125 mm
Tip diameter	600 mm
Hub diameter	400 mm
Mean radius	250 mm
Solidity	0.64
Tip clearance	1.25 mm (1% Chord length)
Rotational speed	2000 rpm

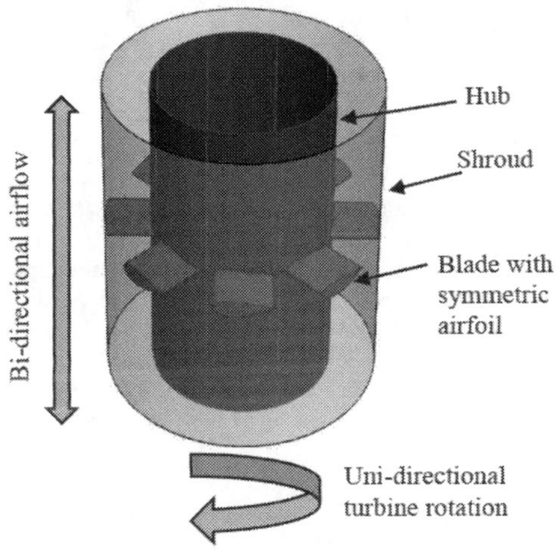

Fig. 1. Schematic of Wells turbine

The schematic of Wells turbine is illustrated in Figure 1. The modified geometries are created by replacing the NACA0015 airfoil profile with other airfoil profiles and all other design parameters are preserved same as the reference turbine. The description of the airfoil profiles used in this study is provided in Table 2 and the comparison of airfoil profiles used is shown in Fig 2.

Table 2. Details of airfoil profiles

Blade profile	Maximum thickness (%C)	Maximum thickness location (%C)
NACA0015	15	30
NACA0021	21	30
S1046	17	30.8

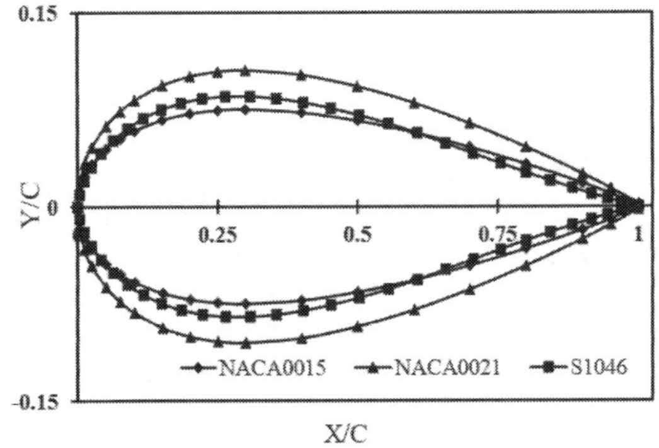

Fig. 2. Comparison of airfoil profiles

3 Numerical formulation

The computational domain for numerical analyses is shown in Fig 3. To minimize the computational time, a single blade with rotational periodicity is chosen as the computational domain. The upstream and the downstream length is fixed as 4C and 6C respectively. The computational domain was discretized with unstructured tetrahedral elements using ICEM CFD. To capture the near wall flow physics, 20 layers of prism elements were generated around the blades with a stretching ratio of 1.2.

The surface mesh of the computational domain is presented in Fig 4. The numerical analyses were done by solving incompressible steady three dimensional Reynolds averaged Navier-Stokes (RANS) equation using coupled solver ANSYS CFX 14.5. It solves the aerodynamic equations for (u,v,w,p) as a single system [14]. The rotation of turbine is realized by implementing the rotating reference method. The boundary conditions are applied in CFX-Pre. Air at 25°C is taken as the working fluid and the compressibility effects are neglected. Uniform velocity and zero pressure gradient boundary conditions are given on the inlet and outlet respectively. The blade and hub are imposed with no-slip boundary condition whereas the shroud is kept as counter-rotating no-slip wall. The lateral faces of the computational domain is taken as periodic faces to account for the rotational periodicity. The turbulence intensity at the inlet is fixed as 5% (medium intensity) and k-ω SST with automatic wall function is chosen as the turbulence closure model. It is suitable for flows with severe adverse pressure gradient and separation [15]. It activates k-ε model away from the wall and k-ω model closer to the wall, thereby incorporating the merits of both turbulence models. A high-resolution advection scheme is employed for spatial discretization to ensure second-order accuracy. The numerical simulations are run up to 2000 iterations and the residual convergence criteria value is set to 1e-5 to ensure convergence.

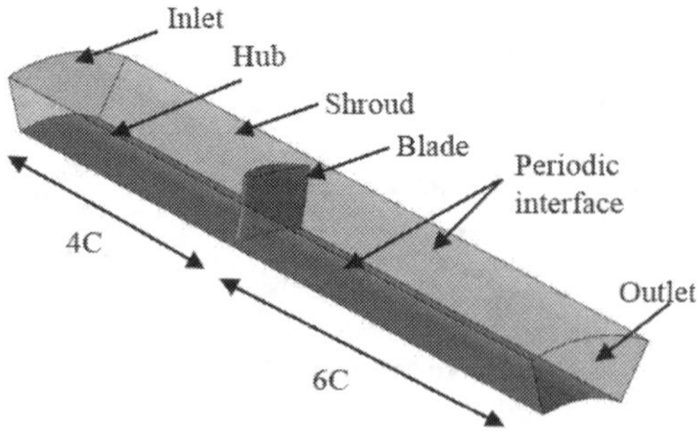

Fig. 3. Schematic of Wells turbine

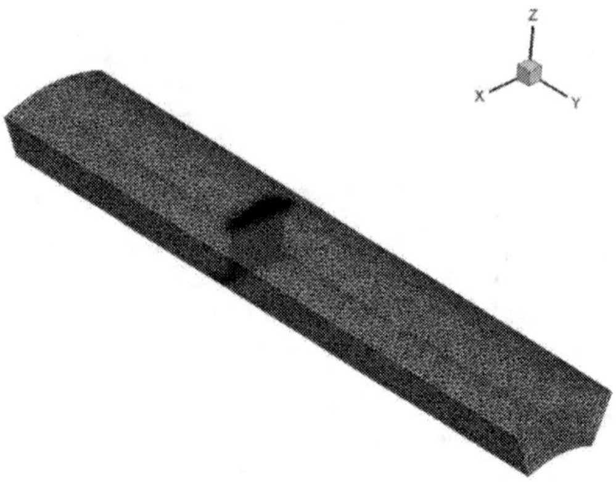

Fig. 4. Surface mesh of the computational domain

The performance characteristics of the Wells turbine is analysed by the following non-dimensional parameters.

Torque coefficient

$$T^* = \frac{T}{\rho\omega^3 R_{tip}^5} \quad (1)$$

Stagnation pressure drop coefficient

$$\Delta p^* = \frac{\Delta p_o}{\rho\omega^2 R_{tip}^2} \quad (2)$$

Efficiency

$$\eta = \frac{T\omega}{Q\Delta p_o} \quad (3)$$

Flow coefficient

$$U^* = \frac{U_\infty}{U_{tip}} \quad (4)$$

Where T is the torque, ρ is the density of air, ω is the angular velocity of the turbine, R_{tip} is the tip radius, Δp_o is the stagnation pressure drop, Q is the discharge, U_∞ is the free stream inlet velocity and U_{tip} is the blade tip velocity. The performance characteristics of the Wells turbine is obtained by plotting the above non-dimensional parameters against the flow coefficient (FC) U^*.

4 Results and discussion

4.1 Grid independence study

The grid independence study is performed to select the mesh with optimum resolution. It is essential to perform this study as the computational time will increase with an increase in a number of grid elements. To perform this study meshes with three resolutions such as coarse, medium and fine is selected and the efficiency is compared for the entire flow range. The results of all

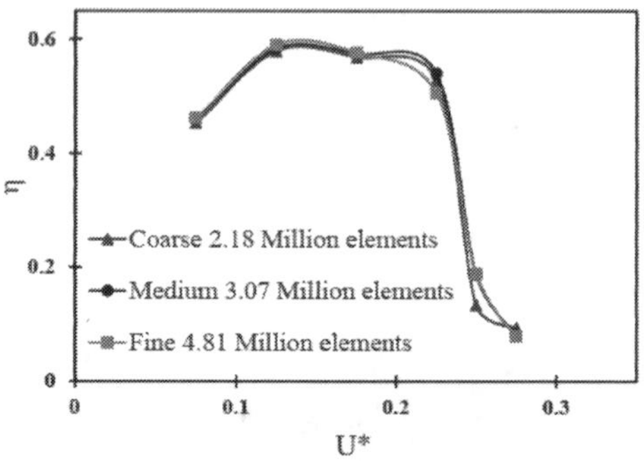

Fig. 5. Grid independence study

grids are almost same and the maximum deviation between the medium and fine grid is 6.8% which is acceptable. Hence the medium grid with 3 million elements is retained for numerical simulations throughout this study. The plot of grid independency study is shown in Fig 5.

4.2 CFD Validation

To ensure the numerical accuracy of the simulations performed the present numerical results are validated with the experimental results of Curran and Gato [16] and the numerical results of [13,17,18,19]. The CFD validation of the numerical results is presented in Fig 6. The performance characteristics of the Wells turbine for the entire flow range is plotted and verified. From the figure, it is evident the present numerical results follow the same trend as the existing CFD results except for the work of Shaaban and Abdel Hafiz [18]. In their work, they reported that the steady RANS model failed to predict the stall phenomenon of the Wells turbine.

4.3 Effect of Airfoil Profiles

The present work investigates the effect of three different airfoil profiles on the performance of Wells turbine. The study has been done for the entire flow range, including the post-stall regime. The comparison of the performance characteristics of the Wells turbine for different airfoil profiles is illustrated in Fig 7. It can be observed that all three airfoil profiles undergo stall after U^*=0.225, which is denoted by a rapid decrease in torque and efficiency due to flow separation.

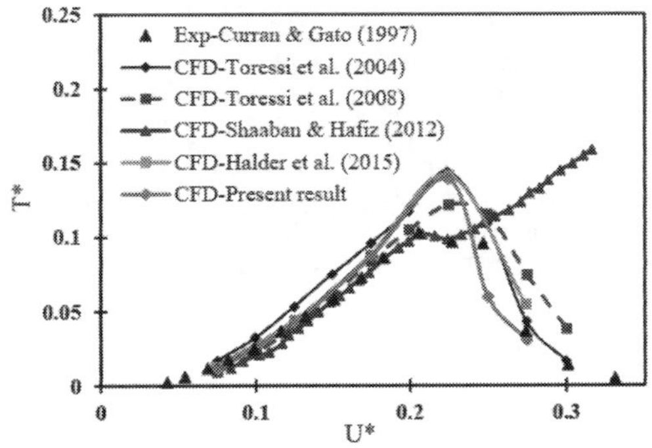

a) Torque coefficient

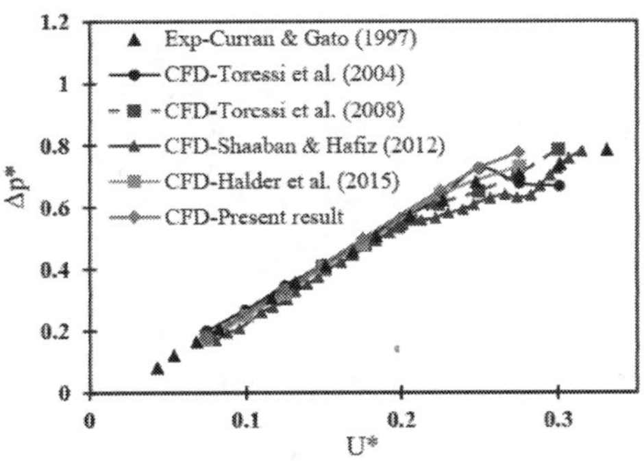

b) Pressure coefficient

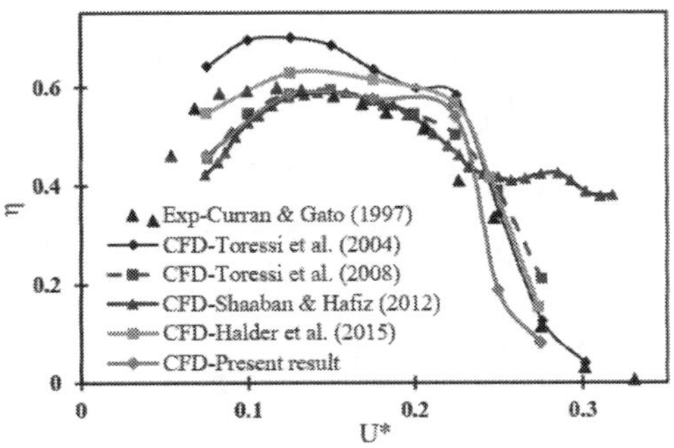

c) Efficiency

Fig. 6. Validation of numerical results

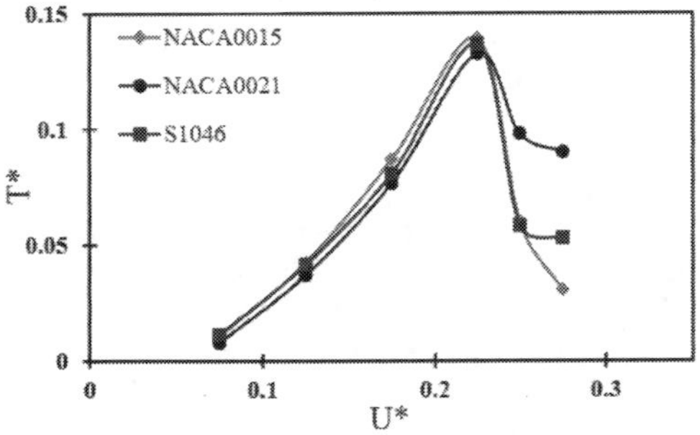

a) Torque coefficient

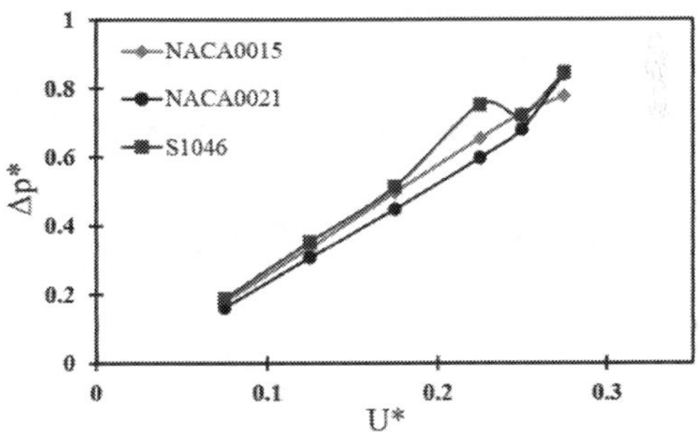

b) Pressure coefficient

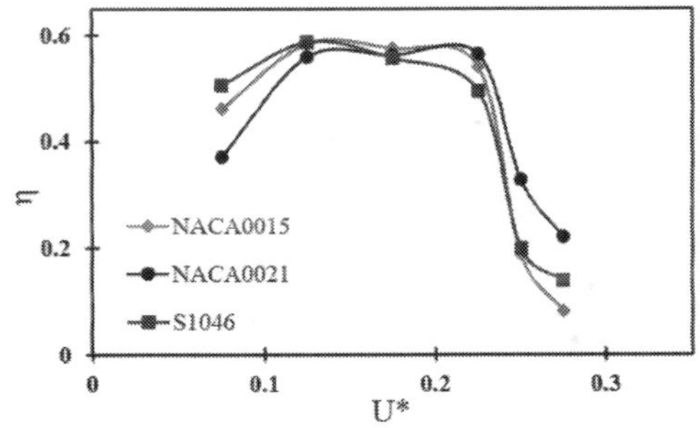

c) Efficiency

Fig. 7. Comparison of various airfoil profiles

From Fig. 7a, it is evident that the maximum torque coefficient is obtained for NACA0015 profile followed by S1046 and NACA0021. The NACA0021 airfoil profile performs better in the post-stall regime and the drop in torque and efficiency is not severe as compared to the other airfoil profiles. The pressure drop of the NACA0021 airfoil profiles is lower than the NACA0015 and S1046 airfoil profiles (Fig. 7b). The peak efficiency is obtained for the S1046 airfoil profile at U^*=0.125, however, the difference between the maximum efficiency of S1046 and NACA0015 is not sign9ificant. At low FC (U^*=0.075), it can be noticed that the efficiency of S1046 is significantly higher than the other airfoil profiles. The efficiency of NACA0021 is better in the post-stall regime compared to the S1046 and NACA0015 airfoil profiles. The static pressure distribution on the blade suction surface of the different airfoil profiles is presented in Fig 8.

At U^*=0.075, the pressure distribution is almost same for all the airfoil profiles. With the increase in FC, the incidence is also increased. At U^*=0.225, the suction pressure is minimum at the leading edge (LE) for all the cases. This pressure difference between the suction and the pressure side creates the lift force. At FC U^*=0.225, the peak torque is obtained for all the cases (Fig 7a). At U^*=0.275, a large low-pressure region is noticed on the on the suction side (SS) of the blade. This can be attributed to the high incidence of the flow at high FCs. The low-pressure region on the blade SS implicates adverse pressure gradient and it leads to flow separation. This can be corroborated by a sudden drop in torque and efficiency in Fig 7a and 7c. In order to analyse the stall phenomenon in detail, the surface streamlines on the blade suction surface are presented in Fig 9.

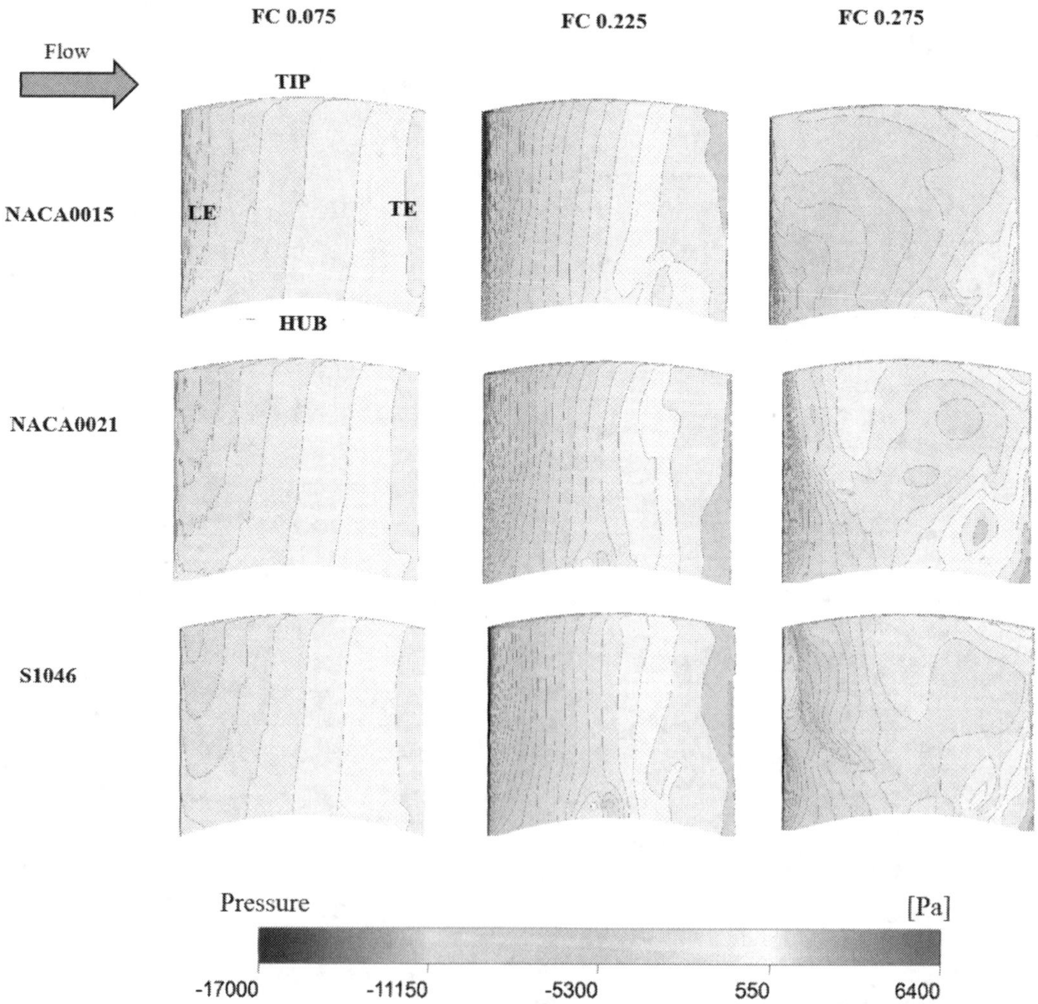

Fig. 8. Static pressure distribution on the blade suction surface

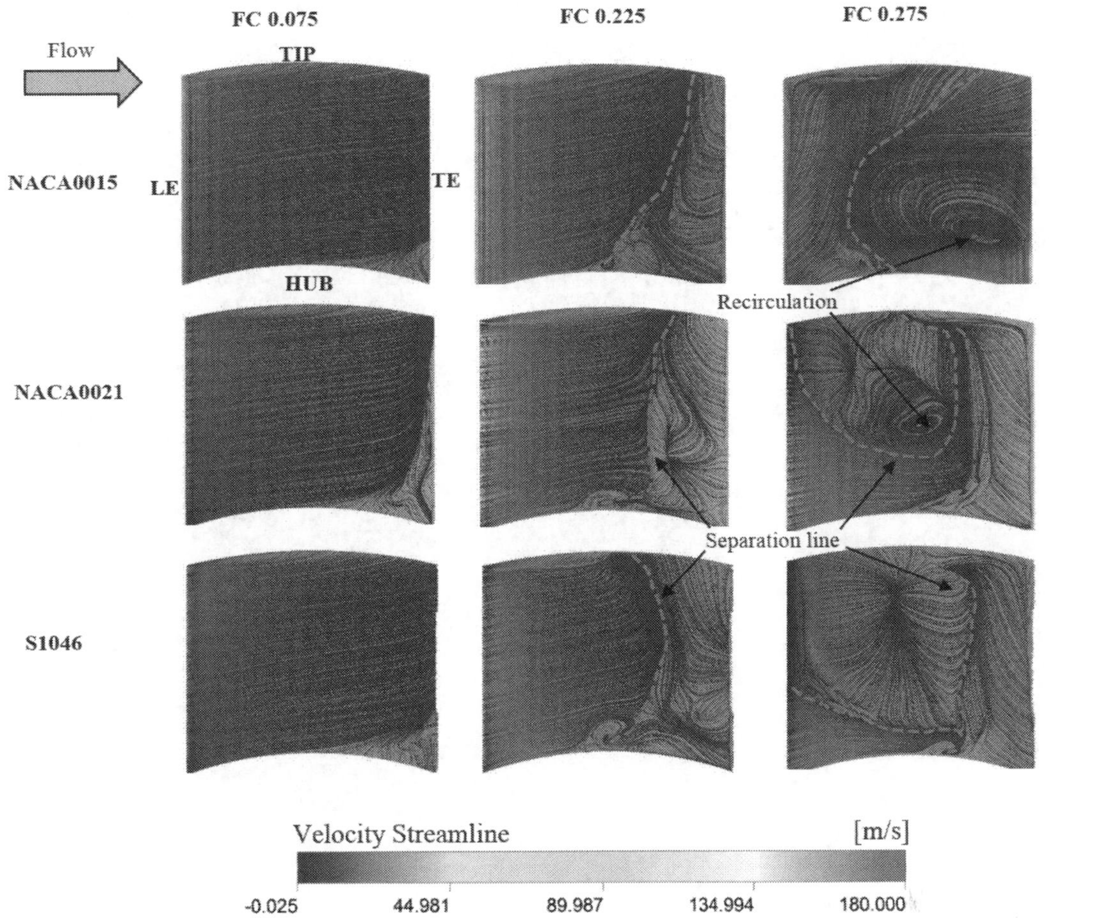

Fig. 9. Surface streamlines on the blade suction surface

At low FC (U^*=0.075), the flow is completely attached for all the cases. However, the hub separation is observed in all the three airfoil profiles. Albeit the flow is attached the torque produced is less because of the low incidence. At FC U^*=0.225, the peak torque is obtained for all the airfoil profiles and the flow separation region is shown in Fig 9. The area between the separation line and the trailing edge implicates the separated region of flow. In this case, it is evident that flow separation is confined closer to the trailing edge of the blade. With the increase in FCs, the separation line advances towards the leading edge and gradually stall occurs. It can be seen from Fig 9 at high FC (U^*=0.275), the separation line is further advanced towards the leading edge of the blade and the complete flow separation takes place for all the cases. It explains the sudden drop in torque and efficiency in the Fig 7a and 7c. At high FC, the effect of tip leakage flow is severe and it aids in the inception of stall phenomenon. The NACA0021 airfoil seems to have better performance in the post-stall regime compared to NACA0015 and S1046 airfoil profiles. The flow separation region of NACA0021 is relatively smaller than the other cases and it corroborates the higher torque and efficiency of NACA0021 in the post-stall regime from Fig 7a and 7c. In Fig 10, the streamlines on the blade midspan are presented. It can be witnessed from the figure that the flow is fully attached at low FC (U^*=0.075). As explained earlier, due to low incidence the flow is completely attached for all the airfoil profiles. At FC U^*=0.225, a recirculation zone is visible at the SS of the trailing edge for all cases of airfoil profiles. Since the flow separation is limited to the trailing edge region and due to high incidence, the peak torque is achieved at this FC. With further increase in FC (U^*=0.275), the separation line advances towards the SS of the leading edge and the stall occurs. The separated region of the NACA0015 airfoil profile is larger compared to NACA0021 and S1046 airfoil profiles. This can be attributed to the poor performance of NACA0015 in the post-stall regime. Moreover, the performance of NACA0021 airfoil is relatively better than the other airfoil profiles.

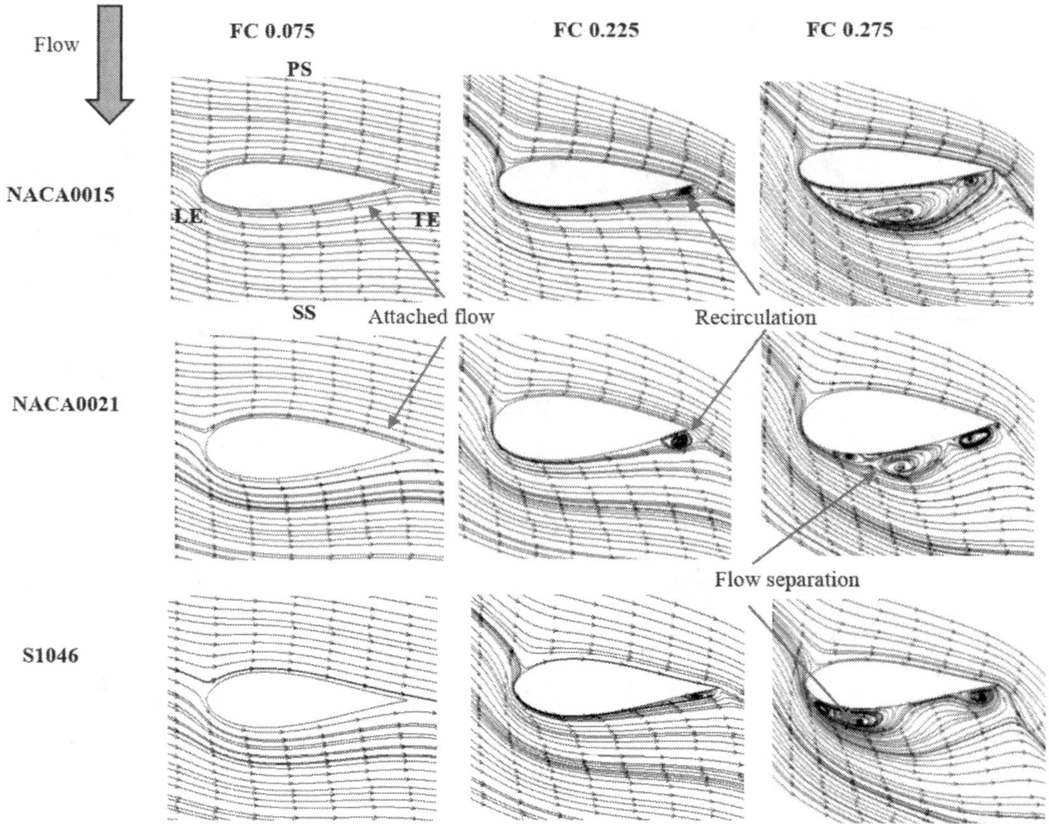

Fig. 10. Streamlines on the blade midspan

5 Conclusion

In this numerical study, the performance of Wells turbine with different airfoil profiles was investigated. Three symmetrical airfoil profiles such as NACA0015, NACA0021 and S1046 was selected for this study. The prominent conclusions drawn from this study were listed below.

- The performance of Wells turbine predominantly depends on the airfoil profile used.
- The maximum peak efficiency was achieved in the case of S1046 airfoil whereas the maximum peak torque was obtained in case of the NACA0015 airfoil.
- The NACA0021 airfoil showed better performance in the post-stall regime.
- Furthermore, optimization of these airfoil profiles is recommended to obtain an optimized airfoil profile with improved performance characteristics.

6 References

1. Raghunathan, S., C. P. Tan, and N. A. J. Wells. Journal of Energy **6**, 157-160. (1982)
2. Raghunathan, S., and C. P. Tan. *International journal of heat and fluid flow* **6**, no. 1, 17-22. (1985)
3. Raghunathan, S. *Progress in Aerospace Sciences* **31**, no. 4, 335-386. (1995)
4. Webster, M., and L. M. C. Gato. In *The Ninth International Offshore and Polar Engineering Conference*. International Society of Offshore and Polar Engineers, (1999).
5. Setoguchi, T., T. W. Kim, M. Takao, A. Thakker, and S. Raghunathan. *International journal of ambient energy* **25**, no. 3, 137-150. (2004)
6. Takao, Manabu, Ajit Thakker, Rahil Abdulhadi, and Toshiaki Setoguchi. *International Journal of Sustainable Energy* **25**, no. 01, 53-61. (2006)

7. Thakker, A., and R. Abdulhadi. *International Journal of Rotating Machinery*, (2007).
8. Takao, Manabu, Toshiaki Setoguchi, Shuichi Nagata, and Kazutaka Toyota. *OMAE Proceedings*, 625-632. (2008)
9. Mohamed, M. H., G. Janiga, and D. Thévenin. In *ASME Turbo Expo Conference,(GT2008-50815), Berlin, Germany*. (2008).
10. Mohamed, M. H., G. Janiga, E. Pap, and D. Thévenin. *Energy* **36**, no. 1, 438-446. (2011)
11. Gato, L. M. C., and J. C. C. Henriques. *ASME-PUBLICATIONS-FED*, **238**, 623-630. (1996)
12. Mohamed, M. H. *Energy* **47**, no. 1, 522-530. (2012)
13. Torresi, M., S. M. Camporeale, P. D. Strippoli, and G. Pascazio. *Renewable Energy* **33**, no. 4 (2008): 735-747.
14. CFX, ANSYS. "Solver Theory Guide. Ansys." *Inc., Canonsburg, PA*. (2011)
15. Menter, Florian R. *AIAA journal* **32**, no. 8, 1598-1605. (1994)
16. Curran, R., and L. M. C. Gato. *Proceedings of the Institution of Mechanical Engineers, Part A: Journal of Power and Energy* **211**, no. 2, 133-145. (1997)
17. Torresi, M., S. M. Camporeale, G. Pascazio, and B. Fortunato. In *59 Congresso ATI, Genova, Italy*. (2004)
18. Shaaban, S., and A. Abdel Hafiz. *Energy Conversion and Management* **61**, 51-58. (2012)
19. Halder, Paresh, Abdus Samad, Jin-Hyuk Kim, and Young-Seok Choi. *Energy* **86**, 219-231. (2015)

Review of Energy Status and Associated Conservational Issues in Botswana

A. Agarwal[1]*, *O.M.* Seretse[2], *M.T.* Letsatsi[3], *E.* Dintwa[4]
[1, 2, 4] Department of Mechanical Engineering, University of Botswana, Gaborone. Botswana
[3] Department of Industrial design and technology, University of Botswana, Gaborone. Botswana

Abstract. This paper discusses energy status of Botswana and the environmental concerns associated with the production of energy. It was offered in response to the growing concerns in industry about the need for the conservation of energy and for combating the increasing environmental degradation. Accomplishing vitality proficiency and diminishing ecological contamination are progressively essential parts of expert designing. The paper is meant to address top to bottom comprehension of basic issues of energy thrift and environmental consequences of irrational use of energy resources in industry. There is need for up to date technical knowledge and skills required for achieving the better management of energy, designing of energy-efficient systems and processes and the reduction and control of pollution cost-effectively. This information can successfully be coordinated to different segments of the economy for application and help the divisions in enhancing their aggressiveness despite lessening assets, likely considerable increment in unit vitality costs and the urgent requirement to comply with the increasingly-restrictive pollution-control standards.

1 Introduction

Energy is an essential element of modern societies. It is the forerunner for maintaining financial advancement and enhancing ways of life. Without a doubt present day vitality administrations are perceived as basic to enhance personal satisfaction and for supportable improvement as a rule. However the discerning and monetary utilization of vitality, with minimal harm to the earth is imperative for the future welfare of our planet. Providing energy for eradicating poverty has been a daunting task that has not been fully addressed. Efficient, affordable, adequate and reliable modern energy services are recognized as essential to improve quality of life and for sustainable development in general. The advancement of Botswana's economy is generally reliant on the abuse of assets, for example, minerals, water, vitality, natural assets, go assets and agro-biodiversity. Key financial parts, for example, tourism, mining, agribusiness are asset subordinate, while the employments of the lion's share, particularly in the country regions, are to a great extent gotten from characteristic assets. The accessibility and profitability of these assets, including woods, water, land and untamed life is vigorously affected by climatic conditions.[1] The digging division represents around 50 percent of the request, the business part around 20 percent, and the private segment around 25 percent. In the vicinity of 2004 and 2007, country access to power multiplied to 44 percent, however shy of the 60 percent focus under the National Development Plan. The Government's "Vision" goes for 100 percent provincial access to help the more extensive advancement objectives of access to instruction and wellbeing, and in addition business openings, to the country and the burdened populace. The extending vitality emergency over the Southern Africa sub-area is a noteworthy hindrance to Botswana's monetary development designs, represents a risk to solidness, and requires a noteworthy deliberate exertion at the national and territorial levels to address the vitality challenge. Botswana, similar to a few different nations in the Southern Africa sub-region, have up to this point depended on reasonable, plenteous, and solid power from South Africa.

2 Status of Energy Sources in Botswana

In 2008, Botswana imported around 2440 GWh (67 percent of its energy necessities) from Eskom, the national electric utility of South Africa, while its own little 25-year-old coal control plant (Morupule A, 4 x 33 MW) gave around 22 percent. The sub-locale, including outstandingly South Africa and different neighbors to Botswana, have been encountering extreme deficiencies of energy since the finish of 2007 because of high development and slacking interests in new limit. South Africa had begun stack shedding discontinuously since December 2007, a condition that is foreseen to compound through the medium term until the point when plentiful new age limit is fabricated and charged. Botswana is likewise encountering power outages and is thinking about load- Shedding because of its noteworthy reliance on Eskom, which has diminished its supply responsibilities from current levels to zero of every 2013 onwards under another business assertion. There is developing social and political weight in South Africa to stop fares to different nations.

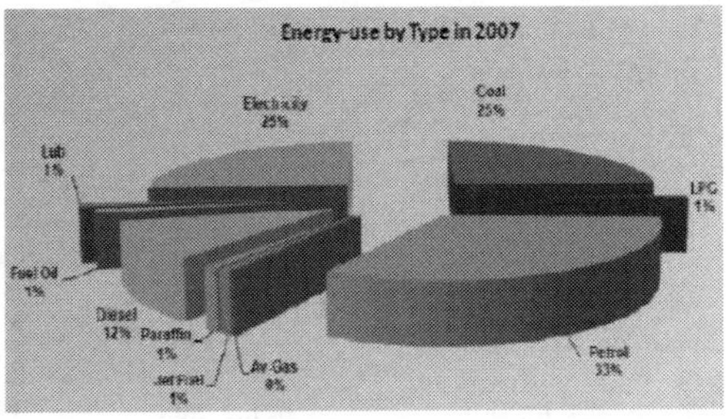

Fig. 1: A depiction of utilization in 2007 demonstrating the distinctive vitality sources, estimated in huge amounts of Oil Equivalent – TOE (Energy Affairs Department, 2007).

Figure 1 shows that, petrol is the most consumed energy source (33%), followed by coal and electricity (25% each), diesel (12%) and other sources at about 1% each including solar. Energy security has developed as the significant national basic for the legislature of Botswana. There is profound worry in the legislature that the energy circumstance may prompt an emergency of certainty and political shakiness, which the nation can't manage. The administration's technique to react to the test is thorough, tending to the short-to medium-term and additionally long haul issues in vitality. The principle components of the administration's methodology are abridged under three subjects: (I) vitality security angles, including vitality preservation and proficiency, judicious improvement of local vitality assets, pulling in private part, and so on.; (ii) regular assets and protections perspectives to advance dependable utilize and way to deal with natural and social effects; and (iii) more extensive advancement viewpoints to address national, territorial, and worldwide concerns (e.g., environmental change).[2]

2.1 Coal

Botswana is endowed with a vast amount of coal resources. Investigations into the availability of coal deposits started prior to the independence of the country. It is currently known that coal deposits are well distributed along the eastern part of the country. The main coal deposits are located in Morupule, Moijabana Letlhakeng, Ncojane, Dukwi, Mmamantswe, Mmamabula, Serule, Dutlwe, Foley and Bobonong. Coal resources in these deposits are estimated to be in the region of 212 billion tonnes of which 3.34 billion tones are measured. Botswana coal is a low rank bituminous containing slate and shale. It is characterized by high ash, high sulphur, low caloric value, high friability and low volatile properties. Coal production commenced in 1973 at the Morupule coal deposit. Morupule colliery, which is the only coal mine in the country, is located 270 km north of the capital city, Gaborone. The mine is connected to the national railway line and the major road net-work for ease of coal distribution countrywide. Annual coal production at Morupule has steadily increased from 20,299 metric tons in 1973 to 921,000 metric tonnes in 1995/96.[3] In mid-1996 it dropped to 800 000 metric tons due to the de-commissioning of Selebi Phikwe power station and again there has been a steady increase to 928 000 metric tons in 1998 and thereafter remained almost constant [4].

The mine has a production capacity of slightly over one million tonnes per annum. The urge for the country to have the security of coal supply for power generation, smelting of copper and nickel and of late, the production of soda ash at Sua-Pan has led to this steady increase in coal production. These major industrial sub-sectors of the economy consume close to 90% of the total production. In view of the abundant coal reserves, a study was conducted in 1984 with the assistance of Federal Republic of Germanys Technical Cooperation Agency (GTZ) with the purpose of investigating ways in which the use of coal could be broadened. One of the recommendations was the establishment of the Expanded Coal Utilization Project (ECUP), which was commissioned in 1987 with the overall objectives of: Reducing the depletion of the natural vegetation by relieving the use of fuel wood with coal where viable.

Reducing coal imports. Replacement of other energy sources like liquefied petroleum gas, where economically viable. To achieve these objectives, ECUP was charged with the responsibility of promoting coal utilization in industry, government institutions and households. ECUPs tasks also entailed the identification of coal using appliances that could burn on the locally produced coal safely, efficiently and in an environmentally friendly manner. The Project, through its mindfulness programs on the requirement for coal usage has empowered government foundations and ventures to set out on the utilization of coal for cooking, water warming and steam creation for various applications. Privately created coal is utilized as a part of the generation of meat items, material, cleanser, cowhide, blocks, and solid items, among others. Coal imports continue to flow into the country at a reduced level. 22,000 tonnes of coal were imported in 1981, while only 10,000 tonnes of coal were imported in 1994.[4]

Due to the quality of Botswana coal, a beneficiation

study has been undertaken that is geared at upgrading the quality of local coal to a level where coal imports could be substituted by locally produced coal. It is expected that the household sector would also benefit from the use of beneficiated coal briquettes which have less environmental hazards. Further Beneficiation or coal cleaning helps in the removal of inorganic sulphur. This is the cheapest way of reducing sulphur dioxide emissions.

2.2 Electricity- The Botswana Power Corporation

The bulk of the electricity used in Botswana is generated locally by the Botswana power Corporation (BPC). The rest is supplied by imports through cross border connections, the Department of Electrical and Mechanical Services (DEMS) and from private generation (generation by some industries for their own use).BPC is a parastatal under the Ministry of Mineral, Energy and Water Resources (MMEWR), was established in 1970. At its establishment, the national grid consisted of two parts, namely, the northern grid (electricity generation at Selibe Phikwe) and the southern grid (electricity generation in Gaborone). Prior to 1975, only major urban towns were electrified. Major villages and small towns had their own power stations and internal reticulations. Examples of the latter include Palapye, Serowe and Mahalapye which were electrified in 1975 through SIDA financial assistance.In order to supply electricity beyond the above areas, the Morupule Power Station was constructed and commissioned in 1985. The total installed generation capacity is 132 MW. In addition, the northern and southern grid were connected to Morupule Power Station and the Gaborone Power station was later decommissioned. BPC operated both the Morupule and Selibe-Phikwe and the latter was decommissioned in 1996. The Selibe-Phikwe plant was coal fired and Morupule plant is also coal fired, the coal being availed by Morupule colliery. The total coal consumption of the two plants for the 1993/94 financial year was 603050 tonnes which was 69.6% of total production of the colliery [3]. The national grid is mostly concentrated along the railway line where the majority of the people live. Most of the major villages and a few other locations more than 100 km from the railway line have also been connected to the national grid. The national framework is relied upon to grow further to the remote regions with the presentation of Rural Electrification Program which plans to charge 14 towns consistently. The administration chose to quicken this program by charging 70 towns for a long time from September 1999 to September 2001. The BPC used to work various diesel-fuelled power control stations (called Central Power Supply Stations, (CPSS)) in areas which by prudence of their separation from the national lattice, were not yet associated. This wellspring of power contributed under 2% to yearly aggregate power supply[5].However due to Botswana government efforts of extending the grid to as many parts of the country as possible most of the locations have finally been connected to the grid. BPC has since decommissioned all the CPSSs in Botswana. The Department of Building and Engineering Services (DBES) of the Ministry of Works, Transport and Communication has installed diesel generators in public institutions/offices which are located in villages not connected to the grid. The total installed generation capacity amounts to approximately 20MW and contributes less than 2% of the total electricity supply [8].The generation facilities operated by DBES usually supply only one government institution. On average 70% of the total electricity supplied in Botswana is imported [8]. Imported electricity is mainly from South Africa and is supplied through 3 x 132 KV lines each of which has a 75MW gross capacity. There are also some cross border connections from Namibia which account for approximately 25 MW gross capacity [8].

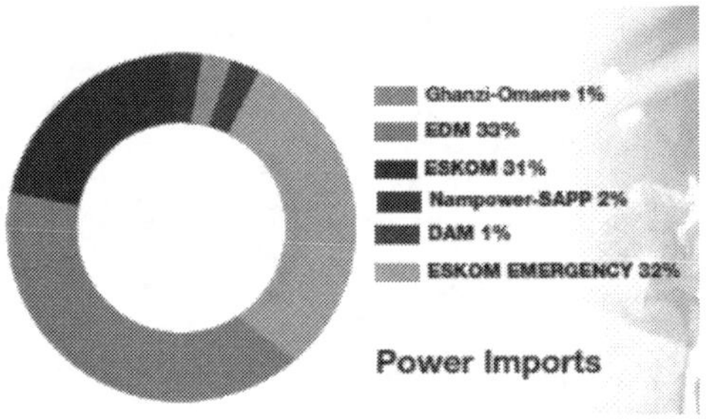

Fig. 2- Power import during 2015-16 [5]

Figure 2 shows the power imported during 2015-16. Hence the total electricity importation capacity from South Africa and Namibia has a firm capacity of 250MW. A further diversification of imported electricity supply to Botswana was achieved in 1991 through interconnection to Zambia by means of a 222 kV line. Although the use of this line is currently minimal, it is envisaged to provide 30MW of firm capacity in future. Botswana also has cross-border interconnections with Namibia. All interconnections are aimed at supplying electricity to remote villages adjacent to respective borders. Although most parts of Botswana have access to electricity, uptake of electricity varies greatly with settlement size and income group. In rural areas, small towns and low income areas of the larger settlements, connection rates are typically below 5%. [5] Households in such locations generally find connection costs unaffordable. In urban areas electricity uptake is about 26%. Overall, electricity uptake in Botswana is only about 10% of all the housing units due to high connection rates and tariffs [5].

On the demand side, Botswana's electricity sector shows a unique feature. Around 70% of its consumption are accounted for by only three consumers, i.e., the three largest mines (the Jwaneng and Orapa diamond mines and the copper-nickel mine -Bamangwato Concession Limited (BCL). This has an impact on BPCs overall load factor and also determines the peak load of the entire system. This could easily escalate into a big problem should one of these mines encounter some financial difficulties and close down. For example, the closing down of BCL which is currently subsidized by government and happens to be the leading consumer among the above three industries, could lead to the creation of excess capacity which would lower the gross profit margin of BPC and therefore create problems for both BPC and its other clients.

Botswana power get to achieved 71% of the populace in urban regions, while in rustic zones is as yet restricted to 24%, albeit expanding. The national general power get to rate is 53%. The Global Tracking Framework evaluates the national access to clean cooking arrangements at 62%. Strategy destinations with respect to vitality get to are guaranteeing enhanced security of supply and unwavering quality of vitality supply to all parts of the economy. Government has set national power get to focus of 82 % by 2016 and 100% by 2030. Botswana depends mostly on power, coal, fuel wood and oil for its vitality requests. Add up to vitality supply for the nation in 2008 was 1604.10 toe with oil being the real provider took after by coal (29%) and Electricity (17%).

Sustainable power source assets for Botswana are sun powered, wind, and different types of bioenergy that incorporate biofuels and biomass squanders .Botswana has no enduring waterways and extensive dams and in this way has no potential for hydro-control. With respect to, Botswana has coordinate typical light (DNI) of 3000kwh/m2/year, which is among the most noteworthy on the planet.

The region flanking Botswana, South Africa and Namibia is among the territories with the most astounding DNI. It is evaluated that utilizing under 1% of the nation territory, Botswana could meet its present power utilization. At a national level, a Rapid Assessment/Gap Analysis of the vitality area was delivered and following a demand dated January 2016, the nation will set out during the time spent building up its national SE4All Activity Agenda and Investment Prospectus with the specialized help of the SE4All Africa Hub.[6]

2.3 Fuel-Wood

While the resource potential, supply and demand figures of other energy sources are comparatively well documented, information on fuel-wood is fragmentary and insufficient for proper sub sector planning.

The total national growing stock of woody biomass has been estimated at 1400 million tonnes with an annual increment of 50 million tonnes. Regeneration capacity ranges from 0.3 t/ha/year to 2.1 t/ha/year and has a national average value of 0.93 t/ha/year.[7] Given an estimated standing stock density ranging from 3.6 tonnes/ha to 48.0 tonnes/ha, the regeneration rate is rather low. However, since the national fuel-wood demand side was estimated at slightly over 1.6 million tonnes in the 1993 energy balance, the supply of fuel-wood is sustainable if proper policies regarding the harvesting of fuel-wood and which give due regard to sustainability of the environment are put in force. Fuel wood accounts for 48 percent of total energy used and almost 100 percent of domestic energy in rural areas[8].

Fuel-wood is mainly used for cooking, water heating, and space heating and in some cases, lighting. The methods, patterns and effect of fuel-wood harvesting and collecting vary according to location and the demand for various energy sources in the specific location. Although population growth within the context of dynamic economic growth would not lead to a worsening of the fuel-wood situation, it has led to the expansion of a poor underclass and to an increase in the rural-urban migration rate which in turn has exerted a constant increase in the demand for fuel-wood.

Given the slow economic growth in rural areas, fuel-wood will remain the staple fuel for many years to come unless programmes that will make alternative sources of energy both available and affordable to the population in the low and middle income group are put in place. Given that contribution to fuel-wood supply from afforestation is small (small scale plantation forestry initiated in the 1970s under the Rural Afforestation Programme has stagnated at 650ha), the only solution to the unsustainable use of fuel-wood is inter fuel substitution. In urban areas and large villages, this will be a switch to paraffin, gas or electricity. The switch may be more difficult for the rural population which live in remote areas with neither the national grid nor proper distribution networks for the other fuels.

2.4 New and Renewable Sources of Energy

Renewable sources of energy can be defined as those sources that are inexhaustible. In Botswana these sources incorporate sun based, wind, bio-gas and to some degree, biomass vitality. With an end goal to expand the utilization of New and Renewable Sources of Energy which are right now utilized as a part of inconsequential sums, the Botswana government through applicable research foundations has done different exercises and pilot ventures. Nearby vitality assets thought to be in plenitude in Botswana incorporate coal (200 billion tons), daylight (3,200 hrs at 21MJ/m2), biogas (2.2 million steers, 3 kg excrement/LSU/day) and fuel wood (200 tons/annum).

2.5 Solar Energy

Botswana receives over 3200 hours of sunshine a year with completely cloudy days few and far between [9]. Average daily radiation on a horizontal surface is 21MJ/M2. These conditions are favorable for solar energy applications. Currently solar energy is mainly used for water heating (mainly in urban areas), and lighting and refrigeration in homes, rural schools and clinics. Other applications, though on a very small scale, include solar desalination, telecommunications, signals in Botswana Railways, telemetry by North South Carrier, rural libraries, security lighting at cordon fence barrier gates and water pumping by Water Affairs etc. In 1992, as a pilot project, the Botswana government through Energy Affairs Division electrified 42 private houses and a clinic using solar energy. The purpose of the project was to assess the technical, social and economic viability of the system in a village environment.

The results of an evaluation carried out in 1994 showed that the installations provided tangible benefits to the community, are technically viable and socially acceptable to the users. The results and recommendation of the study led to the introduction of the National Photo-Voltaic Rural Electrification (NPVRE) scheme. This scheme was financed by the Government and its implementation was done through Rural Industries Innovation Centre (RIIC). The scheme was targeted at semi urban areas and villages. The main objectives of the scheme was to disburse loans at prime rate of interest to rural households for the purpose of installing PV lighting systems. Promote PV lighting systems of various sizes.

Strive to improve quality of life of the rural people. In order to promote good quality and safe installations the EAD together with local solar energy equipment suppliers developed guidelines for installation of domestic solar water-heating systems. The guidelines were used as a working document so as to, ensure that the installations provide value for money for those who buy the systems. To complement the guidelines, the EAD also offered on annual basis training on maintenance of solar water heaters to central government (DBES) and council employees who are responsible for maintenance of solar water heaters. The above effort together with more extension work, were intended to encourage increased use of technically and economically viable renewable energy technologies.

2.6 Wind Energy

The moderately low normal breeze speeds (2 - 3 m/s) winning in Botswana confine wind vitality applications to water pumping. It is assessed that there are more than 200 windmills in activity, for the most part situated in the southern piece of the nation.

2.7 Biogas

Despite Botswana's relatively large cattle herd, the use of cow-dung to generate biogas has not taken off as initially expected. This is mainly because the grazing and animal husbandry practice used in the country of allowing the cattle to roam the veld looking for pastures is not conducive to efficient utilization of cow-dung as it (the dung) is deposited over wide areas, making its collection difficult.

2.8 Petroleum products

All of Botswana's petroleum requirements are imported in refined form, mainly from the Republic of South Africa (RSA) by local divisions of several international oil companies. Such companies include Shell Oil Botswana (Pty) Ltd, Total Botswana (Pty) Ltd B.P. Botswana (Pty) Ltd, Engen Botswana (Pty) Ltd, and Caltex Oil Botswana (Pty) Ltd. The Ministry of Mineral, Energy and Water Resources (MMEWR) through its Petroleum Management Unit (PMU) is responsible for monitoring of petroleum supply, pricing, contingency planning and management of strategic reserves. The actual procurement is carried out by the oil companies which also operate the strategic reserves under the supervision of the government. The real oil based commodities imported into the nation are recorded underneath: - There are no known oil holds in Botswana.

2.8.1 Petrol

Petrol is the petroleum product most used by the final consumer. It is mainly used in the transport sector. Petrol is used to fuel equipment in other sectors e.g. Agriculture, construction, residential and government. Government regulates the prices of petrol by periodically fixing base prices. However, petrol prices differ from place to place due to transport cost.

2.8.2 Diesel

The contribution of diesel to the total amount of the petroleum products used in the economy comes second after petrol [1]. It is also mainly used in the transport sector, although substantial amounts of it are used for electricity generation and for operating equipment in the sectors indicated under petrol. Pricing system for diesel is similar to that of petrol.

2.8.3 Paraffin

Paraffin accounts for about 3% of all the petroleum products used [4]. It is mainly used in households for lighting. Its use as a cooking fuel is on the increase, and the trend is expected to continue as fuel wood scarcity increases. Pricing system for paraffin is similar to that of petrol and diesel.

2.8.4 Liquefied Petroleum Gas (LPG)

LPG ranks fourth (after paraffin) in share of total petroleum imports[4]. LPG is mainly used for cooking and in a few cases for lighting as well. Its use is widely spread in the residential sector partly due to convenience and the high cost of electricity. The Government does not control LPG price. In addition to the oil companies, Botswana Oxygen and individual vendors also participate in the procurement and sale of LPG. As a result, its price varies widely even in one location.

2.8.5 Fuel Oil, Aviation Gas and Jet AI

Fuel oil is mainly used in industry while Aviation Gas and Jet A1 are used in the Aviation sub-sector. The contribution of the three fuels to total energy consumption is rising.

3 Environmental Issues Associated with the Use of Energy Sources

Nowadays issues of Environmental concern can neither be referred to as national or regional, they are a Global concern. The world's atmosphere is anticipated to change since human exercises are modifying the substance piece of the environment through the development of ozone depleting substances - essentially carbon dioxide, methane, and nitrogen oxides. The warmth catching property of these gases is undisputed. In spite of the fact that vulnerability exists about precisely how earth's atmosphere reacts to these gases, worldwide temperatures are rising. Vitality from the sun drives the world's climate and atmosphere, and warms the world's surface; thus, the earth emanates vitality once again into space. Air ozone depleting substances (water vapor, carbon dioxide and different gases) trap a portion of the active vitality; holding heat to some degree like the glass boards of a green house.
Without this regular nursery impact, temperatures would be much lower than they are currently, and life as known today would not be conceivable. Rather on account of ozone depleting substances, the world's normal temperature is a more affable 16oC. Likewise a gas like carbon dioxide is critical for living things that is vegetation and creatures. Trees require carbon dioxide for their photosynthesis procedure and thusly give oxygen to breath process in creatures. In any case, issues may emerge when the air centralization of ozone depleting substances increments.

Why are greenhouse gas absorptions increasing?

Researchers for the most part trust that the burning of petroleum derivatives and other human exercises are the essential purpose behind the expanded centralization of carbon dioxide. Plant breath and the decay of natural issue discharge in excess of 10 times the CO2 discharged by human exercises, yet these discharges have dependably been in adjust with the carbon dioxide consumed by plant photosynthesis.

What has changed in the last couple of hundred years is the extra arrival of carbon dioxide by human exercises?

Energy consumed to run autos and trucks, warm homes and organizations, and power manufacturing plants is in charge of around 80% of society's carbon dioxide discharges, around 25% methane outflows and around 20% of worldwide nitrous oxide emanations [9]. Expanded farming, deforestation, modern generation and mining additionally contribute a noteworthy offer of emanations.

3.1 Environmental impact as a result of locally used fossil fuels

Combustion of fossil fuels is a chemical process that results with its constituents which are elements and compounds, chemically combining with oxygen and each other to form a mixture of solids and gaseous compounds. The combustion products of a fossil fuel depends primarily upon:

a) The nature of the equipment that is used for its combustion process.
b) Constituents of the fuel.
c) The oxygen (air) that is made available for the process.
d) Proportional mixing of air and fuel (fuel-air ratio and Turbulence).
e) The time it takes for the process to complete.
f) The temperature at which the process takes place.

3.1.1 Nitrous oxide and other nitrogen oxides (NOX)

This is formed by a chemical combination of nitrogen and oxygen in the air for combustion. Atmospheric air contains only about 21% oxygen and about 70% nitrogen, the rest being other inert gases. Nitrogen is inert at temperatures below 1000 °C, above this it does chemically react with oxygen to form a variety of oxides which are all toxic and not environment friendly. This is evident in all internal combustion (IC) as they burn fuel at very high temperatures to produce tremendous power at very high combustion efficiency.

3.1.2 Ash

This is an incombustible combustion residue material containing mineral compounds. It is the main constituent of Morupule coal at about 20% [3]. Water vapour (H2O) - This is formed due to the chemical combination of Hydrogen (H) in a fuel and oxygen for combustion.

3.1.3 Fossil fuels, Combustion products

Carbon dioxide (CO2) - This is an odorless, nontoxic and invisible gas. It is framed by the concoction mix of carbon and oxygen. It is an indication of good or finish ignition and assumes a noteworthy part in vegetation. Be that as it may it adds to a dangerous atmospheric deviation.

3.1.4 Carbon monoxide (CO)

This is an odorless, invisible and very toxic gas. It is formed by the chemical combination of carbon and less oxygen. It forms due to inadequate oxygen for a combustion process. It is a result of inefficient and incomplete combustion.

3.1.5 Sulphur dioxide (SO2)

This is a toxic smelly invisible gas formed by the chemical combination of sulphur (mainly in coal) and oxygen in the air for combustion. It depends on the sulphur content of the fuel. It can combine with, atmospheric water vapour to form acid rain and water vapour in flue gases to form light sulphuric acid which is very corrosive.

3.1.6 Soot

These are tiny micro particles of unburnt carbon. They are a sign of incomplete combustion and inefficient combustion process.

3.1.7 Smog

This is a combination of Hydrocarbons {mainly Methane (CH4)} in a mixture form that escapes unburnt from a fuel. The escape normally happens at the initial ignition of a fuel. So far in Botswana there are a couple of segregated neighborhood natural effect related with the above items, because of the generation of vitality and other warm procedures like purifying of copper-nickel metal [9]. Further information on the emissions can be obtained from the Air Pollution Control section of the Department of Mines.

3.2 Environmental impact as a result of fuel wood use locally

Regions with low population density and high tree re-growth and woodland density (e.g. the north-western part of Botswana), have abundant availability of fuel-wood. On the other hand, in areas of higher population density and low regeneration rates (e.g. eastern Botswana and around towns and large villages), evidence of localized over-exploitation can be seen easily. This problem is caused by the non-sustainable harvesting of fuel-wood for use as an energy source and for timber extraction, and by overgrazing and land clearing for agriculture development.

3.3 Environmental impact as a result of electricity use locally

Apart from being an end product of fossil fuels, invisible and silent killer, there are no known local environmental impacts associated with electricity. Environmental impacts as a result of new and renewable energy use locally. There are no known local impacts associated with the use of new and renewable sources of energy. In fact, these are the most save and environmentally friendly energy sources. The high cost of the technology is the one that hampers the local and worldwide use of it.

4 Conclusion

Global environmental issues are addressed yearly in different forums in different countries around the world. All of these are aimed at reducing the emissions of greenhouse gases and eradication of ozone depleting chemicals. Various conventions (Rio, Montreal, Kyoto and so on.), of which Botswana is gathering to, have been marked all went for worldwide ecological concern. Studies directed as of late, especially under the Intergovernmental Panel on Climate Change (IPCC), have examined the connection between ozone depleting substances and worldwide environmental change. An accord has developed, prescribing that such discharges should be lessened altogether worldwide to battle the risk of fast and genuine climatic change. Most as of late the reception of the Framework Convention on Climate Change (FCCC) built up that emanations of ozone harming substances ought to be constrained or diminished through an organized and purposeful universal exertion [6]. It is indispensable that countries should formulate energy climate change policies laden with incentives for renewable energy and fuel efficient technologies. This can benefit the environment and create a lot of jobs.

References

1. "Environment and Energy," 2017. [Online]. Available: http://www.bw.undp.org/content/botswana/en/home/ourwork/environmentandenergy/overview.html. [Accessed: 25-Jan-2018].
2. "Botswana - energy sector overview." [Online]. Available: http://www.mmegi.bw/index.php?sid=1&aid=1253&dir=2012/June/Friday22/. [Accessed: 22-Jun-2012].
3. "Morupule Colliery monthly data compilation, Ultimate and Proximate Analysis."
4. "Energy Statistical Bulleting," *Energy affairs Division, Ministry of Minerals, Energy and Water resources*.

5. BPC, "Botswana Power Corporation Annual Reports."
6. "Botswana-At a glance." [Online]. Available: https://www.se4all-africa.org/se4all-in-africa/country-data/botswana/.
7. "Initial national Communication to the United nation Framework Convention on Climate Change."
8. M. Moyo, S.; O'Keefe, P.; Sill, *The Southern African environment: profiles of the SADC countries*. Earthscan Publications Ltd.
9. "Environmental Statistical Data,Central Statistics Office, Ministry of Finance and Development Planning," Gaborone, Botswana.
10. Energy Affairs Department. (2007). Energy Statistical Bulletin - 2007. Gaborone: Energy Affairs Department.

Design, Modelling and Experimental Investigation of an Economic Domestic STHW System Using T*Sol® Simulation in Botswana

O.M. Seretse[1]*, *A.* Agarwal[2], *M.T.* Letsatsi[3], *O.M.* Moloko[4], *M.S.* Batlhalefi[5]

[1-5] Department of Mechanical Engineering, University of Botswana, Gaborone. Botswana
[3] Department of Industrial design and technology, University of Botswana, Gaborone. Botswana

Abstract- The study describes a research and development directed at examining research and design a SHWS, and test a flat-plate solar collector capable of collection efficiency in excess of 50 percent at an inlet fluid temperature 22℃. The model was analyzed using theoretical and mathematical analysis and the T*SOL® simulation programme for solar thermal heating systems using radiation data to site. The average efficiency of mathematically designed system was higher as 70% than the designed value efficiency of 20. The effectiveness of the collector was seen to be higher toward the start of the test and reductions consistently as the test advances with time. The economic analysis of the whole system desired was done with an initial total investment of US$991.10. The economic performance of the system gives a positive net present value of US$4662.13with a payback period of 0.7 years. At a steady radiation, the higher the heat absorbed, the higher the efficiency of the collector. The use of the flat plate solar collector for domestic and commercial heating (for eight person bread and breakfast) purposes in Gaborone is feasible since reasonable temperatures (outlet from the collector,t_o) were realized.

Keywords- Solar energy, solar radiation, modelling, T*SOL® simulation, solar collector

1 Introduction-

Energy plays the most essential role in the economic growth, progress, and development, as well as security and poverty eradication of any nation. Future economic growth crucially depends on the long-term availability of energy from sources that are affordable, accessible, and environmentally friendly. Due to the fast increase in the world population, the need for the energy increases rapidly. Most of the total consumed primary energy comes from the burning of fossil fuels as is the case of Botswana. The combustion of fossil fuels for energy has caused a significant increase in greenhouse gas emissions to the atmosphere, which is widely believed to be the primary cause of global warming. This has resulted in changes in sea levels, ecosystems and weather events, which have directly, affect people's health and way of life, as well as cause irreversible losses to plant and animal species [1].With that been the case, the use of renewable energy sources will be a good way to cope with energy and environmental problems. Renewable energy sources can easily replace fossil-fuels in the near future especially in stationary plants. Depleting reserves of fossil-fuels and environmental problems make it necessary to use the reserves more carefully [2]. Solar energy has great potential in space heating for buildings, water heating etc. owing to its low-grade energy characteristics and is the most important alternative to fossil fuels. Solar systems with feasible design and installation have very short payback periods and meet the energy demand very effectively.Solar water heating is an environmentally-friendly solar thermal system. An active solar heating system consists of solar collectors, a storage tank, a fluid circulation system between collector and tank, an auxiliary heating system, and a control system to regulate the overall operation [3]. Most solar collectors for buildings are either flat-plate or concentrating types. Both installations requires long pipe lengths to receive water from the mains supply, have it delivered to the collectors and then to the outlet points. This raises the cost of the system, and sometimes affects the aesthetics of the building. In addition, there exist technical problems such as low efficiency, high heat loss, and poor solar energy harvesting capability. The key to guaranteeing an efficient operation of the system and achieving maximum benefits to users' lies in how well it is designed [3]. Every location on earth receives sunlight at least part of the year. The amount of solar radiation that reaches any one "spot" on the earth's Surface varies according to geographic location, time of the day, season, local landscape and local weather. The shape of the earth makes the sun to strike its surface at different angles ranging from 0° (just above the horizon) to 90° (directly overhead) [4]. The average rate of solar energy (or solar flux) as described by Wallace [5], arriving at the stratosphere before suffering any losses is 1.37 kW/m^2 (the solar constant). The surface area of the earth over which this flux is averaged over time is $4\pi r^2$. After averaging the sun oriented steady, and subtracting the scattered and consumed motion by the climate and mists, the last motion striking the world's surface is 174.7 W/m^2.

* Corresponding author: seretseom@mopipi.ub.bw

2 Literature Review-

Solar water heating systems depend on; the type of solar collector used, the weather conditions, the hot water demand and the temperature of the water heated can vary from tepid to nearly boiling[6]. Most solar systems are meant to furnish 20 to 85% of the annual demand for hot water, the remainder being met by conventional heating sources, which either raise the temperature of the water further or provide hot water when the solar water heating system cannot meet demand (e.g. at night)[6]. Solar hot water systems can be used wherever moderately hot water is required. Solar water heating technology has been embraced by a number of developing countries with both strong solar resources and costly or unreliable conventional energy supplies[6]. The dynamic frameworks are more unpredictable and costly and for the most part have higher efficiencies, their qualities being 35%– 80% higher than those of the inactive frameworks. Dynamic sun powered water radiators depend on electric pumps, valves, and controllers to flow water, or other warmth exchange liquids (as a rule a propylene-glycol blend) through the authorities [3]. Passive direct systems are also used in warmer climates.

The special normal for these frameworks is that they don't utilize pumps or other electrical segments, in this way giving a straightforward and solid framework. In an audit of SHWS for residential and industrial applications did by [7] and [8] SHWS were gathered into two general classifications (passive and active), each of them working in either direct or indirect mode. They reported their performances, uses and applications, and factors considered for their selection. The most well-known detached frameworks are the thermosyphon and fundamental authority stockpiling frameworks. Roundabout thermosyphon frameworks could likewise be utilized as a part of colder atmospheres [3]. Integral-collector capacity frameworks comprise of at least one stockpiling tanks set in a protected box with a coated side confronting the sun. The thermosyphon framework utilizes a level plate gatherer and a different stockpiling tank that must be found higher than the collector. The collector is similar to those used in active systems.

The first published analysis of thermosyphon solar water heater trail was by[8]. There have been numerous different distributions on the examination of these frameworks yet they are altogether in light of the first definition. For checking the hypothetical outcomes, he tried two thermosyphon frameworks with various qualities and the outcomes adjusted well to those anticipated [9], did the investigation of the warm execution of the thermosyphon water radiator unit to demonstrate its relevance in Bahrain. Their results show that the system is quite suitable for application in Bahrain weather conditions. [10] Made some fabrications and experimented on a single-phase open thermosyphon to utilize solar energy for water heating. Their exploratory work was finished utilizing protected water tank and protected associating funnel between the tank and the level plate authority. From their gathered information, execution parameters, for example, momentary gatherer.

3 Scenario in Botswana: Problem Statement-

Through observation and documented records, it was found out that most of the commercial and industrial hot water demands in Botswana are met mainly by the use of electric heaters. The issues related with the utilization of these electric warmers are: the rising vitality cost, ecological concerns, and the draining idea of the present essential vitality sources. These issues can be taken care of by removing a portion of the vitality interest for boiling water purposes from power in areas where the high temp water is promptly required, for example, in healing centers, inns and schools by utilizing sunlight based water warming frameworks.

On that point, the investigators proposed to design a thermosyphon SHWS for 8- person bed and breakfast to try to curl the problem of using electric heaters, as it will mean reduced load on the nation power grid. Other inline objectives are researching and designing (using software and mathematical models) a suitable solar thermal hot water heating system for eight person bed and breakfast. Identifying a solution for heating water to solve the power supply crisis for eight person bed and breakfast in Botswana by constructing a typical flat plate solar collector for the designed solar water heating system using ASTM standard and testing the fabricated flat plate collector for performance and efficiency.

4 Methodology

The territories of intrigue incorporates the comprehension of the different kinds of sun based gatherers normally utilized, sun powered boiling water estimating and utilization propensities particularly in motels and visitor houses or overnight boardinghouse places, standards of financial assessment while taking buy choices of sun powered water radiators, computation of yearly vitality and cost funds and in addition the payback time of sunlight based water warmers. The method for sizing a solar thermal system for 8 person bed and breakfast is on peak hot water demand with additional capacity to provide potential for future expansion and safety margins. Sizing a solar water heating system is by determining the collector area that will meet the heating load, depending upon the insolation level and in some cases, the collector area that will give maximum life cycle solar savings (minimize life cycle cost) of SWH system[3].

SWH framework estimating likewise includes the assurance of the required stockpiling tank ability to give a vitality

* Corresponding author: seretseom@mopipi.ub.bw

support between times of low insolation levels, for example, 30 night and overcast days, and times of high insolation levels. Different segments, for example, the sensors and mechanical help structures can be chosen as long as they are not out of extent since they are heartless to the warm execution of the SWH framework. According to the National Energy Renewable Laboratory (NREL) of the US Department of Energy, a workshop on solar thermal technology and applications by Roger Taylor in June 2006 provided the following values for hot water usage in various types of buildings [11].

Various mathematical models were investigated, and the following method below by [12] was adopted for the studies. This model was adopted for this study because it goes beyond the energy balance equations and adds heat transfer equations of collector's components. Hypothetical conditions will be utilized for assessing the warm effectiveness of the authority, outlet water temperature and different warmth misfortunes from the framework. The manufacturing simplicity and absence of moving parts make it an interesting technological solution for production of hot water in eight person bed and breakfast.

5 Useful Heat Output of a non-concentrating Solar Collector

If the rate of solar energy arriving on the collector surface area(A), m^2, perpendicular to the line of the sun is I the solar radiation, in W/m^2, then the amount of solar energy being received by the collector is:

$$Q_i = I \cdot A \qquad (1)$$

However, parts of this radiation are reflected back to the sky, absorbed by the glazing and the rest is transmitted through the glazing, reaching the absorbed plate as short wave radiation. Incorporating the rates of transmissions of the cover and the absorption rate of the absorber,

$$Q_i = I \cdot \tau \cdot \alpha \cdot A \qquad (2)$$

As the collector absorbs heat, its temperature becomes higher than its surroundings and a temperature gradient is created. This causes loss of heat to the atmosphere through convection and radiation. The heat loss rate Q_o is dependent on the collector overall heat transfer coefficient U_L, and the temperature difference between the collector t_c and ambience t_a.

$$Q_o = U_L \cdot A(t_c - t_a) \qquad (3)$$

$$U_L = \frac{1}{\frac{1}{h_n} + \frac{L_n}{k_n}} \qquad (4)$$

* Corresponding author: seretseom@mopipi.ub.bw

Thus, the rate of useful energy extracted by the collector Q_u, expressed as a rate of heat extraction under steady state conditions-

$$Q_u = Q_i - Q_o = I \cdot \tau \cdot \alpha \cdot A - U_L \cdot A(t_c - t_a) = A[I \cdot \tau \cdot \alpha - U_L(t_c - t_a)] \qquad (5)$$

The useful energy extraction rate may also be expressed as the amount of heat carried away in the HTF passed through it:

$$Q_u = \dot{m} \cdot c_p(t_o - t_i) \qquad (6)$$

Where, $\dot{m}$ is the mass flow rate, c_p is the specific heat, and t_o & t_i are the outlet and inlet temperatures of the HTF, respectively. A collector heat removal factor F_R is employed as this quantity. The F_R is the ratio of heat delivered to the HTF and heat in the condition that the overall collector plate temperature equals the inlet fluid temperature.

$$F_R = \frac{\dot{m} \cdot c_p(t_o - t_i)}{A[I \cdot \tau \cdot \alpha - U_L(t_i - t_a)]} \qquad (7)$$

Introducing the collector heat removal factor to Eqn. 7 forms the widely used relationship for measuring collector energy gain known as the *"Hottel-Whillier-Bliss equation."*

$$Q_u = F_R \cdot A[I \cdot \tau \cdot \alpha - U_L(t_i - t_a)] \qquad (8)$$

The performance of a solar collector could be measured using the collector efficiency η.

$$\eta = \frac{Q_u}{AI} = \frac{F_R \cdot A[I \cdot \tau \cdot \alpha - U_L(t_i - t_a)]}{AI} = F_R\left[\tau \cdot \alpha_{(1)} U_L\left(\frac{t_i - t_a}{I}\right)\right] \qquad (9)$$

6 Characteristics & simulation programming

We led an examination that included sun oriented accessibility, water use and vitality utilization information gathering and investigation. Online solar calculators were used to aid the experiment carried out on solar radiation to be used to determine the tilt angle. The warm execution and monetary investigation of the solar water heating system was carried out in this research using the T*SOL® programming software. The programme uses the online methodology employed to compute total global annual irradiation for the climate location, fraction of diffuse component of total global annual irradiation, mean outside temperature, the collector surface area irradiation, required size of collector area, thermal performance of the system in terms of the energy produced by the collectors and collector loop, installed collector power, CO2 emissions avoided, solar fraction, electricity savings and system efficiency.

The programme then performs the economic analysis of the system in terms of life cycle savings, total running cost, net present value, payback period and the cost of solar energy. In order to gather information on solar irradiance an instrument used to measure solar irradiance named pyranometer, is used to verify data obtained from department of metrological services and position the flat plate collector. T*SOL® software program which is an entire reproduction program that enables an expert to measure and recreate sun powered warm frameworks of various classes, for example, household boiling water, process warming, air authorities, space-warming, swimming pool, and support tank frameworks would be utilized to estimate the proposed sunlight based water warming framework for the clinic. Other than mimicking, the warm and monetary examination and reports would be created for the proposed sun powered water warming framework using T*SOL®.

7 Design and analysis

Solar irradiation data was collected by performing an experiment at University of Botswana, explore on contemporary sunlight based water warming innovations, and a monetary examination to plan a fitting sun powered water warming framework. Keeping in mind the end goal to get the most vitality from the sun oriented boards, it was necessary to let the panels face to the solar radiation directly for most of the time as recommended by Close[8]

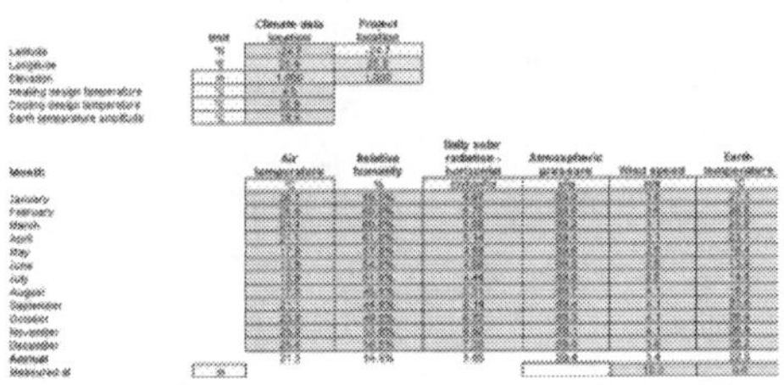

Fig. 1: Monthly average daily global solar irradiation data for Gaborone Source: RET Screen International climate data for Gaborone, 2012.[13]

Figure 1 shows the monthly average daily global solar irradiation data for Gaborone .To calculate these design parameters and evaluate the sensitivity and effect of some design parameters on the performance of thermosyphon solar water heating (TSWH) system, the design method from [14] was adopted.

Table 1: Hot water demand for different users at a hot temperature of 50°C

		Low demand (litres)	**Medium demand (litres)**	**High demand (litres)**
Residential buildings	per person and day	30	50	60
Sport facilities	per shower	20	30	50
Accommodation	per bed	20	40	60

10 rooms (8 beds) B= 8	Hot water demand per person (DHW): Medium demand
Assuming 100% occupancy (O) = 8 people	Hot water demand kitchen: 160 litres/day(HWDK)

Cold water: 22°C	Hot water (storage): 55°C	Hot water at shower: 50°C

*Average solar radiation: May to July (For high solar fractions: 75-80%)

The solar heating system is expected to heat up the cold water from mains supply line (from the water utilities cooperation) at 22°C to 55°C. When dealing with the calculation aspect, the energy demand can be thought as the equal value of heating the cold supply water from the water supply company from 22°C to 35°C in the ideal situation (neglect the energy losses in the system).
$Q = cm\Delta T$ Where:$c = 4200J/kg \cdot$°C; $m = \rho V$;
$V = 0.6m^3 per\ day$; $\rho = 1000kg/m^3$;$\Delta T = 55$°C $-$ 22°C $=$ 33°C
Therefore $Q = 4200 \times 0.6 \times 1000 \times 33 = 83160000J = 23.1kWh(per\ day) \approx 23kWh(per\ day)$
From the table it can be seen that the collector and the storage tank were sized properly as it can be seen from the table above. The calculated values correspond to the values in the table.

Table 2: Dimensioning of domestic hot water solar systems for southern African Conditions

Daily hot water demand [litres]	**Solar storage capacity [litres]**	**Solar storage capacity [litres] Collector area* SV [m2]**	**Collector area* SC [m²]**
50	50-75	1.0 – 1.5	0.9 – 1.3
100	100-150	2.0 – 3.0	1.5 – 2.5
200	200-300	3.5 – 4.5	3.0 – 4.0
300	300-450	4.5 – 6.0	4.0 – 5.0
500	500-750	7.5 - 10	6.0 – 8.5
1000	1000-1200	14 - 20	11 – 16
2000	2000-2500	30 - 40	24 – 32
4000	4000-5000	60 - 80	50 – 70
10000	10000-12000	140 - 200	110 – 160

*) depending on the required solar fraction (SV ... coating of solar varnish) (SC-selective coating) Source: [14]

* Corresponding author: seretseom@mopipi.ub.bw

https://doi.org/10.1051/matecconf/201817206004

8 Energy Calculations

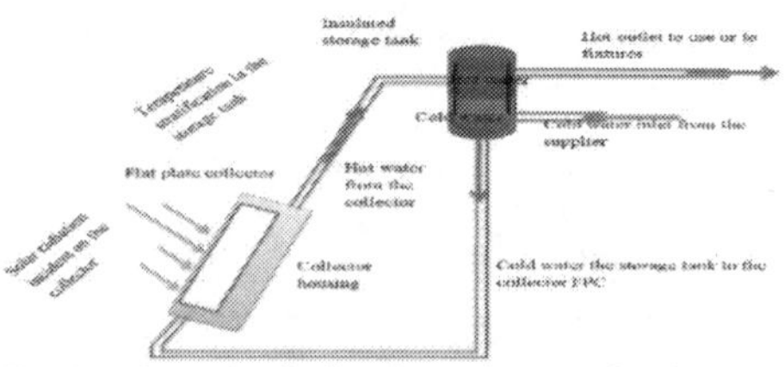

Fig. 2: Schematic of a flat plate solar water heating system

As per the figure 2 process, the water enters the solar flat panel at an average supply temperature$T_i = 27°C$. It is desired that after one pass through the collector, the water will exit at$T_o = 32°C$. Calculating for the mass flow rate of the fluid in the solar panel required to achieve this, the useful energy gained, and the efficiency of the collector. The amount of solar radiation in Botswana received by the flat-plate collector-

$$Q_i = I_C A = 469.9 \times (1.75 \times 1.17) = 0.96212025\ kW$$

The actual heat input into the flat-plate panel - $Q_{i,act} = \tau\alpha Q_i = 0.84 \times 0.97 \times 0.96212025 = 0.783935579\ kW$ and, the heat lost by the collector to the surroundings:

$$Q_o = U_c A(T_c - T_a)$$
$$= 6.844616513 \times (1.75 \times 1.17)(50 - 30)$$
$$= 0.2802870462\ kW$$

Using energy balance and assuming that $F_R = 1$:

$$\dot{m} C_P (T_o - T_i) = Q_i - Q_o$$
$$\dot{m} = \frac{Q_i - Q_o}{C_P(T_o - T_i)} = \frac{0.96212025 - 0.2802870462}{4.188(32 - 28)}$$
$$= \mathbf{0.0407016\ kg/s}$$
$$= \mathbf{2.442096003 l/min}$$

Therefore, in order to raise the temperature of water by 4 °C after every cycle, the HTF will have to flow $\mathbf{2.442096003 l/min}$.

- The useful energy gained per second from the flat-plate collector:

$$Q_u = \dot{m} C_P (T_o - T_i) = 0.0407016 \times 4.188 \times 4$$
$$= \mathbf{0.681833203\ kW}$$

- The efficiency of the collector $\eta = \frac{Q_u}{AI} = \frac{0.681833203}{0.96212025} = 70.86777388\%$

Solar water heating system for 8 person bed & breakfast has been analyzed using T*SOL® software programme by providing technical and economic input parameters described on table6. The detail procedure for the analysis is provided in appendix B. The flat plate collector (type kf) is supplied by Solahart industries Pty Ltd and the other one supplied by Standard (type standard flat plate collector) were analyzed.

The collector performance parameters and their technical specifications are given in the Table 3 below.

Table 3: Collector performance parameters and technical specifications

Performance parameters	Type of collector	
	Flat plate collector (kf)	**Flat plate collector (standard)**
Gross Area (Ag)	$8.2m^2$	$8.2m^2$
Aperture(active)Area(Ap)	$2.05m^2$	$2.05m^2$
Conversion factor (ηo)	79.9 %	78 %
Linear heat loss coefficient (a1)	$3.80(W/m^2 . K)$	$3.8(W/m^2K)$
Quadratic heat loss coefficient (a2)	$0.018(\frac{W}{m^2K^2})$	$0.03(W/m^2K^2)$
Collector fluid water Specific heat capacity (Cp)	4200 J/kg-°C	4200 J/kg-°C
Nominal flow rate per collector	0.030 kg/s	0.05 kg/s

*Source: T*SOL® software programme catalogue data.

9 Results and Discussions

From the two graphs obtained from experimental results, it can be seen that from graph 1 as the incident radiation experiences progressive increase there is a corresponding increase in the useful energy gained by the black body shown in graph 2 by rise in temperature. In relation to our project, this energy will be responsible for driving by convection buoyancy the system fluid flow, depending on the thermal properties of water.

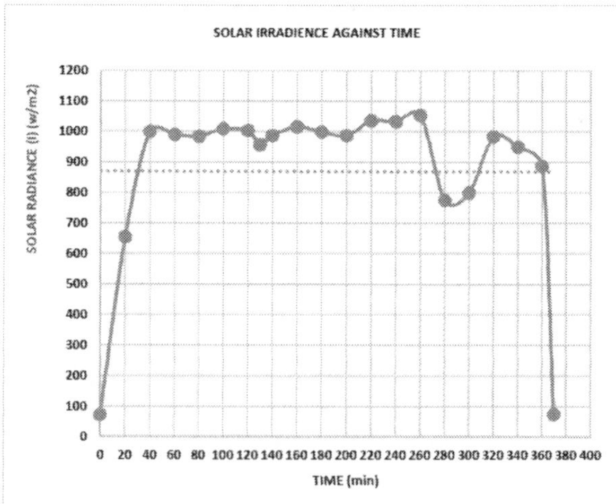

Fig.3: solar irradiance against time

* Corresponding author: seretseom@mopipi.ub.bw

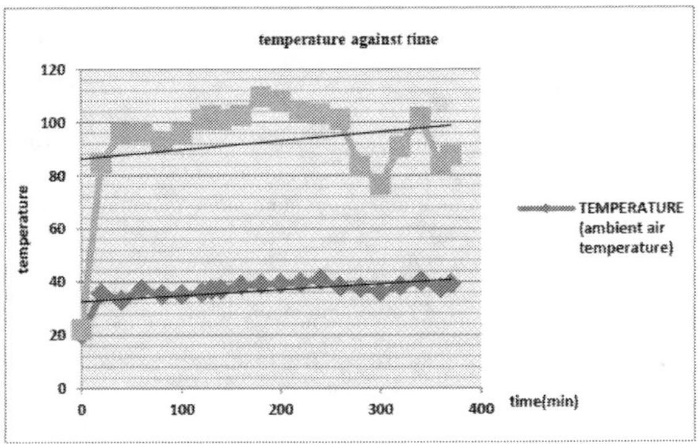

Fig.4. Temperature against time

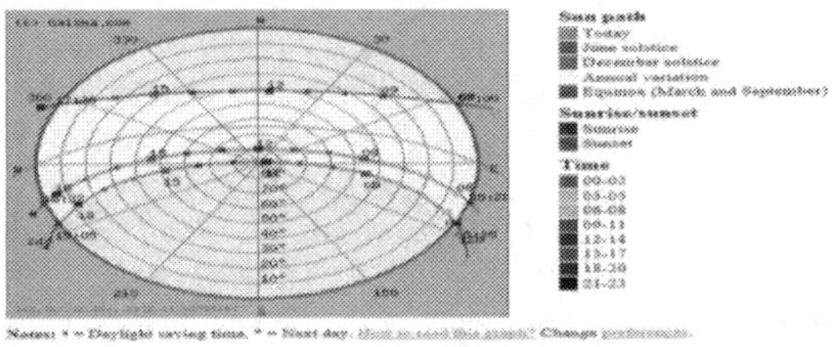

Fig. 5. Gaborone, Botswana-Sun Path Diagram
Source: http://www.gaisma.com/en/location/gaborone.html

It can be seen from figure 5 that the sun rises from the South-East (azimuth = 105°) in Gaborone at 05:28. Sunset happens at 18:32 when the sun is in the South-West (azimuth = 255°). On that day the elevation angle is approximately 80 degrees at noon. The insolation increases from 7.00 CAT until it reaches its maximum value (1000, 894, 640, 430 W/m2) at 12.00 CAT for the day of the experiment, summer, autumn/spring and winter seasons, then it begins to decrease. The useful energy follows the same pattern on figure 4 and auxiliary heat would be needed on winter and part of autumn/spring seasons' days. A solar-collection approach was to alter the solar collector at different tilt angles and azimuth rotation angle for the full duration of its lifetime. The geographic coordinates of Gaborone on [15] are: Latitude: -24.62821°, Longitude: +25.923147100°, Altitude: 1014m.

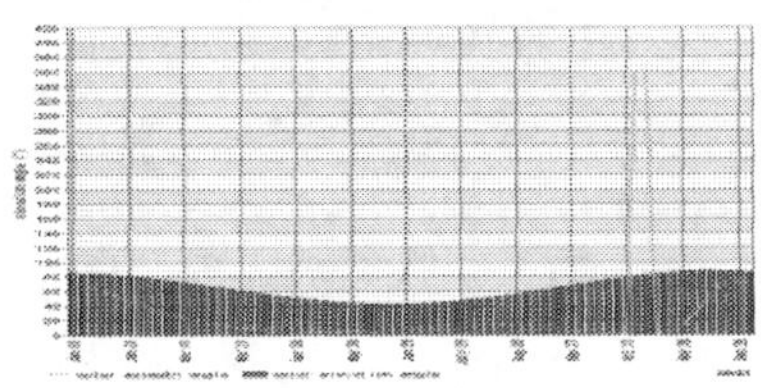

Fig. 6: Solar azimuth and elevation angle

Figure 6 shows the Gaborone solar azimuth and elevation angle for the whole year 2016, [15]

* Corresponding author: seretseom@mopipi.ub.bw

Therefore from the graph the mean solar elevation angle of 2016 Gaborone is 64.52°≈65° and the mean solar azimuth is 28.03° ≈ 28°. The solar panel will face true north because the installation site will be in Botswana which is in the southern hemisphere.

10 SWH Simulation Results

The detail procedure for the analysis is provided in appendix A. The figure beneath demonstrates the rundowns of the yearly execution of the sun based water warming framework in light of the two proposed sunlight based gatherers from the T*SOL® software programme in terms of the collector array and thermal performance. Therefore, from the summaries of the results in figure below, any of the two collectors can be used.

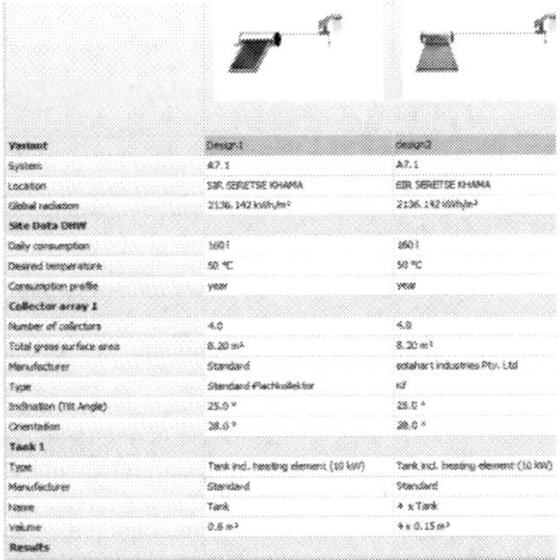

Variant	Design1	design2
System	A7.1	A7.1
Location	SIR. SERETSE KHAMA	SIR. SERETSE KHAMA
Global radiation	2136.142 kWh/m²	2136.142 kWh/m²
Site Data DHW		
Daily consumption	160 l	160 l
Desired temperature	50 °C	50 °C
Consumption profile	year	year
Collector array 1		
Number of collectors	4.0	4.0
Total gross surface area	8.20 m²	8.20 m²
Manufacturer	Standard	solahart industries Pty. Ltd
Type	Standard-Flachkollektor	kf
Inclination (Tilt Angle)	25.0 °	25.0 °
Orientation	28.0 °	28.0 °
Tank 1		
Type	Tank incl. heating element (10 kW)	Tank incl. heating element (10 kW)
Manufacturer	Standard	Standard
Name	Tank	4 x Tank
Volume	0.6 m³	4 x 0.15 m³
Results		

(a)

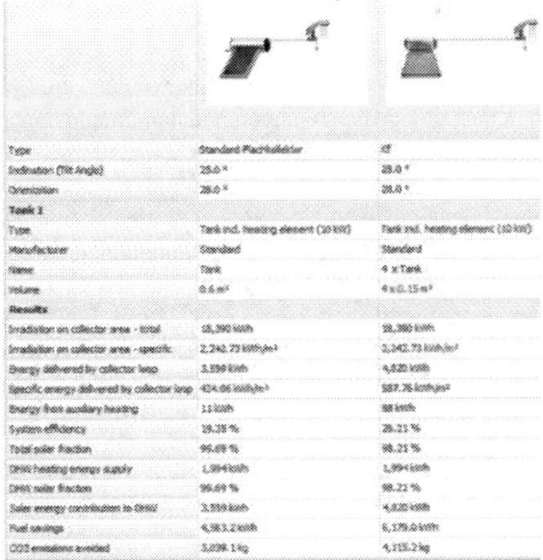

Type	Standard-Flachkollektor	kf
Inclination (Tilt Angle)	25.0 °	28.0 °
Orientation	28.0 °	28.0 °
Tank 1		
Type	Tank incl. heating element (10 kW)	Tank incl. heating element (10 kW)
Manufacturer	Standard	Standard
Name	Tank	4 x Tank
Volume	0.6 m³	4 x 0.15 m³
Results		
Irradiation on collector area - total	18,390 kWh	18,390 kWh
Irradiation on collector area - specific	2,242.73 kWh/m²	2,242.73 kWh/m²
Energy delivered by collector loop	3,359 kWh	4,820 kWh
Specific energy delivered by collector loop	424.06 kWh/m²	587.76 kWh/m²
Energy from auxiliary heating	11 kWh	88 kWh
System efficiency	19.28 %	26.21 %
Total solar fraction	99.49 %	98.21 %
DHW heating energy supply	1,994 kWh	1,994 kWh
DHW solar fraction	99.49 %	98.21 %
Solar energy contribution to DHW	3,359 kWh	4,820 kWh
Fuel savings	4,983.2 kWh	6,179.0 kWh
CO2 emissions avoided	3,038.1 kg	4,115.2 kg

(b)

Fig. 7. (a & b) Summary of the SWH System Annual Performance

https://doi.org/10.1051/matecconf/201817206004

11 Thermal performance analysis

The thermal performance of the SWH system based on the two collector types is given in terms of the fuel savings, solar fraction, solar contribution to domestic hot water heating load (E- Solar-DHW), energy required for auxiliary heating (E- Aux) and CO2 emissions avoided and system efficiency as shown in figure below.

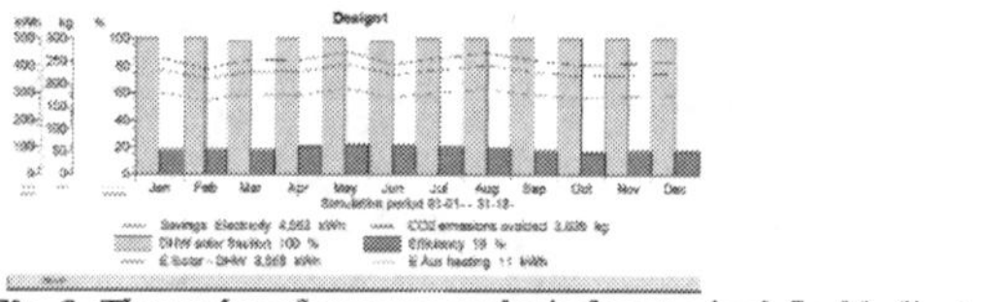

Fig. 8. Thermal performance analysis for standard-flachkollector

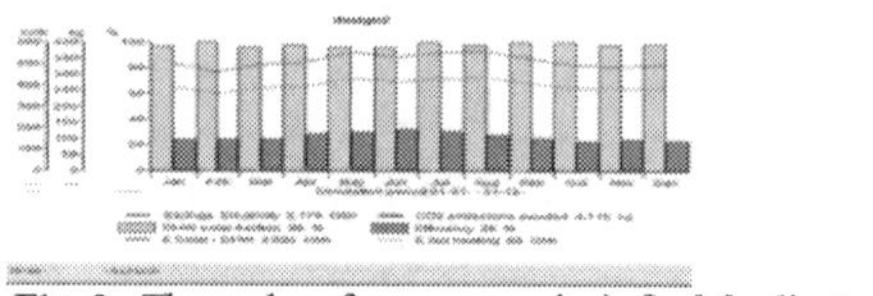

Fig. 9. Thermal performance analysis for kf collector

In terms of the annual collector performance analysis, the kf collector has lower solar fraction but shows increase of value in terms of fuel cost savings (saving electricity), solar contribution to annual heating load and the amount of CO2 emissions avoided more than that of the standard-flachkollektor as shown in Figures 8 and 9respectively.

12 Financial analysis

The figure 10 shows an annual loan instalment of BWP 5000 (taking if US$1=BWP1) payable to the financial institution that granted the loan for a period of 5 years. The life cycle cost of the system (comprising an initial investment and total operating cost) amounts to BWP 10000 given a positive net present value of BWP 47,040 indicating the economic viability of the solar water heating system as proposed for the 8 person bed and breakfast. The proposed system, also from the economic analysis, has a payback period of 0.7 years.

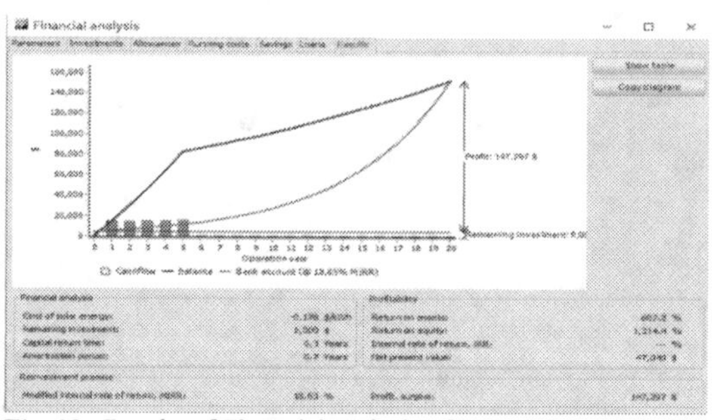

Fig. 10. Results of Financial analysis, If US$1=BWP1

13 Energy analysis of solar water heater

Ambient air Temperature at the start of the experiment =19.6°; Inlet temperature T_0 is assumed to be the same throughout the whole experiment =25°; Pressure of the system = 4 bars and the Flow rate was measured to be $0.013\,{}^{kg}/_{s}$. In table 4 below it should be noted that the results were recorded instantaneously when the water from the valve was opened immediately. The water was supplied directly from the water mains/stand pipe. The temperatures were measured in 30 minutes interval from 1035hrs to 1605hrs. The inlet temperatures were higher than the theoretical ones because the temperatures were measured instantaneously after a period of 30 minutes, from there the valve of the stand pipe was closed until the next reading was collected. Because of this the collector had enough time to overheat on its own without receiving water.

Table 4: Time taken for experiment and inlet and outlet temperatures recorded

Time	Inlet temp.(t_i)	Outlet temp.(t_o)	Temp. gain (ΔT) = ((t_o) − (t_i))
1035	25	40	15
1105	25	45	20
1135	25	63	38
1205	25	66.7	41.7
1235	25	85	60
1305	25	87.3	62.3
1335	25	76.4	51.4
1405	25	75	50
1435	25	50.2	25.2
1505	25	68.9	43.9
1535	25	59.1	34.1
1605	25	55	30
Avg temp	25	64.3	39.3

* Corresponding author: seretseom@mopipi.ub.bw

The highest outlet temperature of 87.3 °C was recorded at 1305hrs, this is the point at which the sun is directly in line with the collector. A large temperature drop was observed at 1435hrs, this was because of the cloud clover which was experienced around that time. The lowest temperature outlet was 40°C which were experienced at the beginning of the experiment.

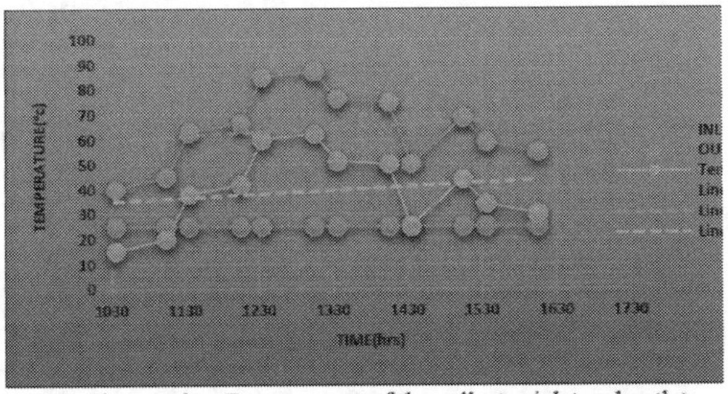

Fig.11. Time against Temperatures of the collector inlet and outlet

From figure 11 it can be seen that the temperatures were gradually increasing until at 1305hrs where the graph was at its pick point. From there the graph gradually decreases until at the end of the experiment. An unusual decrement was experienced at 1435hrs due to the above explained reason of cloud cover. The figure 12 below shows the relationship between time of the day versus Useful Energy Extraction Rate(Q_u) and Collector Efficiency (η) (%).Flow rate was measured to be = $\mathbf{0.077}\ ^{\boldsymbol{kg}}/_{\boldsymbol{s}}$

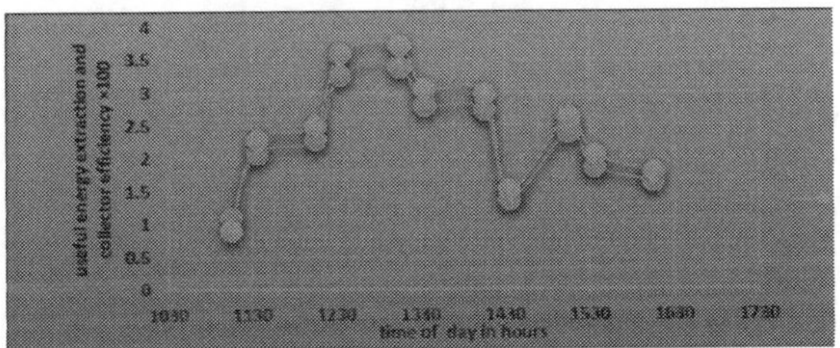

Fig. 12. Solar water heater experiment 2 results and analysis

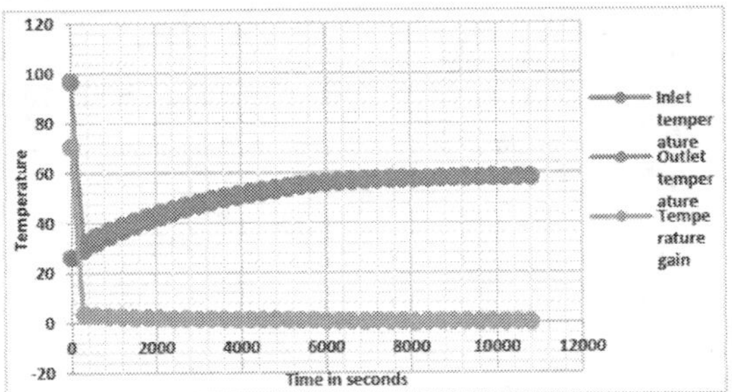

Fig.13. Relationship of temperature inlet, outlet and gain plotted against time

From figure 13 above it can be seen that the inlet and outlet temperatures were showing a direct relationship as compared to the temperature gain. Thereby, decreasing the heat absorbed by the collector. This is further shown by the graph which was plotted by the data harvester itself.

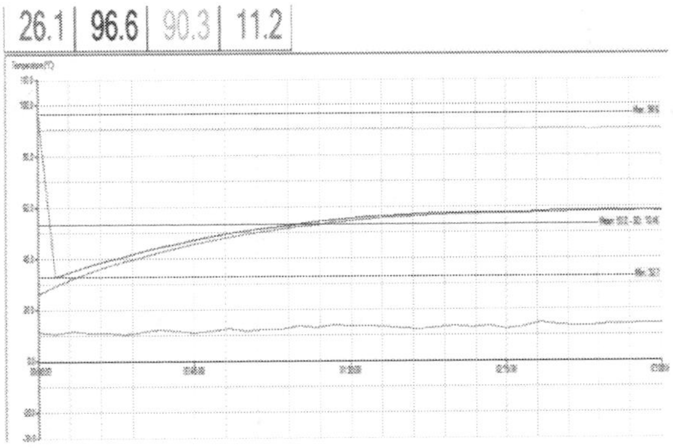

Fig. 14. Relationship of time of the day with mean temperature difference

The figure 14 above shows the relationship of time of the day with mean temperature difference by the circulating fluid red, the graph follows a trend of direct proportionality.

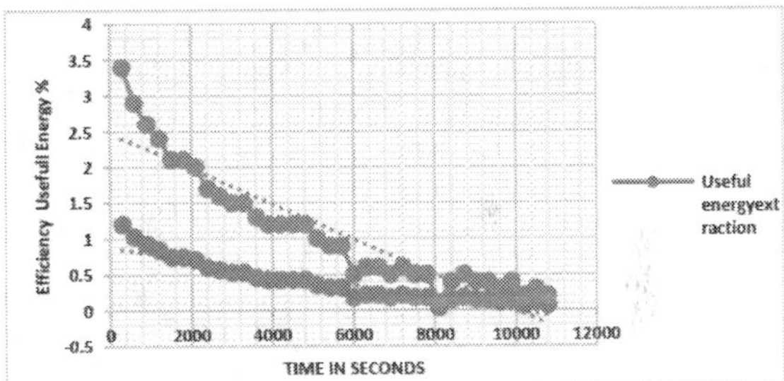

Fig. 15. The Efficiency Graph

Figure 15 above show the relationship of efficiency with heat absorbed by the circulating fluid. For consistent episode vitality, the proficiency of the level plate authority relies upon the warmth consumed by the circling water. The warmth consumed by the gatherer relies upon such a large number of elements, for example, air mass factor. Figure 16 below shows the schematic of proposed model.

$$\boldsymbol{Energy\ saved} = Q_u - W_{pump} = 0.3905 - 0.37$$
$$= 0.0488\ kW = \mathbf{20.5\ W}$$

* Corresponding author: seretseom@mopipi.ub.bw

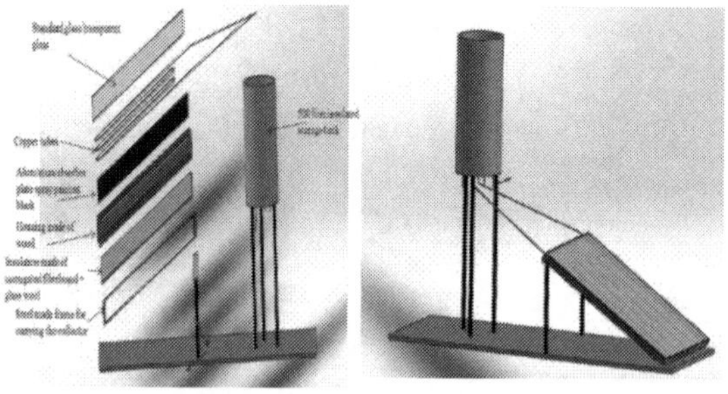

Fig. 16. Schematic of proposed model

14 Conclusion

The research was analyzed using theoretical and mathematical analysis and the T*SOL® simulation programme for solar thermal heating systems. The metrological solar data of Gaborone was processed using software's such as RET-Screen to obtain the monthly average daily solar radiation of Gaborone for recommended average day of the months. The month of April with the worst amount of the average daily solar radiation was considered as the design month and solar radiation and weather data of this month was used as input data for system design and parametric studies. System efficiency was 19.8% using T-Sol and of the mathematical modeling was 70%.

The average efficiency of mathematically designed system was higher than the designed value efficiency. The economic performance of the system has proved that the proposed solar water heating system is feasible because it gives a positive net present value of BWP 47,040 with a payback period of 0.7 years. However, bank negotiations to reduce the interest rate to 8 percent or less would ensure a payback period of 5 months or less. Since the prices of solar water heating system components are dynamic, using the cost of the system from T-sol item database and making presumptions concerning increment in vitality and running expense may digress from the genuine financial situations of the investigation that has been done in this examination. In conclusion, at a steady radiation, the higher the heat absorbed, the higher the efficiency of the collector.

The use of the flat plate solar collector for domestic and commercial heating (for eight person bread and breakfast) purposes in Gaborone, Botswana is feasible since reasonable temperatures (outlet from the collector,(t_o) were realized.

Reference

1. AEE – Institute for Sustainable Technologies, "Dimensioning and Design of Solar Thermal Systems."
2. IEA, *World Energy Outlook 2014*. (2014).
3. J. a. Duffie, W. a. Beckman, and W. M. Worek, *Solar Engineering of Thermal Processes, 4nd ed.*, vol. 116. (2003).
4. K. Chuawittayawuth and S. Kumar, "Experimental investigation of temperature and flow distribution in a thermosyphon solar water heating system," *Renew. Energy*, vol. 26, no. 3, pp. 431–448, (2002).
5. P. V. H. John M.Wallace, *Atmospheric Science: An Introductory Survey*, no. November. Canada: Elsevier Inc., (1999).
6. H. Garg, *Solar Water Heating Systems*. Springer Science & Business Media., (2012).
7. M. P. Malkin, S. A. Klein, J. A. Duffie, and A. B. Copsey, "A Design Method for Thermosyphon Solar Domestic Hot Water Systems," *J. Sol. Energy Eng.*, vol. 109, no. 2, pp. 150–155, May (1987).
8. D. J. Close, "The performance of solar water heaters with natural circulation," *Sol. Energy*, vol. 6, no. 1, pp. 33–40, (1962).
9. A. A. Karaghouli and W. E. Alnaser, "Experimental study on thermosyphon solar water heater in Bahrain," *Renew. Energy*, vol. 24, no. 3–4, pp. 389–396, (2001).
10. S. Sadhishkumar and T. Balusamy, "Performance improvement in solar water heating systems - A review," *Renew. Sustain. Energy Rev.*, vol. 37, pp. 191–198, 2014.
11. T. Roger, "Solar thermal technology and applications," 2006, p. 7.
12. G. N. Tiwari, A. Tiwari, and Shyam, *Handbook of Solar Energy*. 2016.
13. RETscreen, "Retscreen international climate data for gaborone, 2012." [Online]. Available: https://eosweb.larc.nasa.gov/sse/RETScreen/.
14. W. Weiss, "SOUTHERN AFRICAN SOLAR THERMAL TRAINING AND DEMONSTRATION INITIATIVE," *AEE - Institute for Sustainable Technologies*, 2009.
15. Keisan, "KEISAN ONLINE CALCULATOR," 2016. [Online]. Available: http://keisan.casio.com. [Accessed: 21-Sep-2016].

* Corresponding author: seretseom@mopipi.ub.bw

MATEC Web of Conferences **172**, 06005 (2018) https://doi.org/10.1051/matecconf/201817206005
ICDAMS 2018

Performance evaluation of a 15 kW_{th} biomass gasifier in downdraft and updraft operating modes: an experimental study

K. Shanmuganandam[1,*], J.Anichai[2] and V. Jayakumar[1]

[1]Department of Mechanical Engineering, Saveetha School of Engineering, Saveetha Institute of Medical and Technical Sciences, Chennai- 602105, Tamilnadu, INDIA.
[2] Department of Piping Engineering, Saipem India Projects Ltd, ENI Group, Chennai 600034, Tamilnadu, INDIA.

Abstract: Biomass gasifiers are equipment's that can generate producer gas which is a renewable, alternate source of energy that can be employed for power generation and thermal applications. In this experimental study the gasifier is tested in updraft and downdraft modes and performance analysis was carried out. For both the studies, casuarina wood and ambient air were used as feed stock and gasification agent respectively. From the experimental analysis it was inferred that the performance of the biomass gasifier was higher in downdraft mode than updraft mode.

1 Introduction

Amongst the viable alternative technologies available for power generation, biomass gasification technology is a highly viable option that can cater to the requirements of current energy scenario [1-3]. Biomass gasification is a thermochemical process by which the raw wood is converted into a burnable gas termed as producer gas. Furthermore this process is a highly suitable technology for decentralized locations such as villages that couldn't be connected by normal electrical grids [4-6]. Moreover this technology operates on feed material such as wood stock that is available in the vicinity of the villages. This technology can be operated by employing available atmospheric air as the gasification agent. These gasifiers capacity can range from few kW to multiple MWs. Adoption of this technology can reduce the incurring running cost, can create job opportunities in villages, spur the economic activities in villages directly and indirectly, apart from providing electrical power to villages. Based on the economic point of view these gasifiers can be fabricated at cheaper cost, rather than coal based centralized power plants that require quantum of expenditure.

In addition the gasifiers create lesser pollution in contrast to the centralized power plants. On the nut shell this gasification technology is a boon to the remote locations in the country that are still to be electrified.This gas can be used for thermal applications or can be used for operating internal combustion (IC) engines or can be used for power generation [7-9]. The producer gas consists of CO (18-22)%, CO_2 (9-12)%, H_2 (13-19) %, CH_4 (1-5)%, N_2 (45-55)%.

Based on the movement of fuel, air and producer gas the gasifiers are named as updraft, downdraft and cross draft gasifiers. In downdraft gasifiers both the feed material and ambient air moves from top of gasifier to its bottom and the producer gas exits from bottom of the gasifier. In updraft gasifier the wood moves from top to bottom, but the air and producer gas moves from bottom of gasifier to its top. The downdraft gasifiers finds application in power generation as it contains least tar. The updraft gasifier is employed for thermal applications as it contains sizeable quantity of tar [10-12].

In this experimental study an attempt is made to study the performance of 15 kW_{th} biomass gasifier in updraft and downdraft modes.

* Corresponding author: shanmugam_anandam@yahoo.com

https://doi.org/10.1051/mateconf/201817206005

2. Experimental and Instrumentation setup

The experimental setup comprises of biomass gasifier comprising of top and bottom shell. The shell is made of mild steel material of 5 mm thick and is coated with refractory cement for 1 cm to repel the heat to the interiors of the gasifier. The gasifier has a throat at the middle region. The throat region consists of convergent – divergent region and the meeting region of both is termed as throat. The grate is located below the throat. The grate is the region where the biomass is combusted. The ash collection tray is available below the grate by which the formed ash is removed periodically.

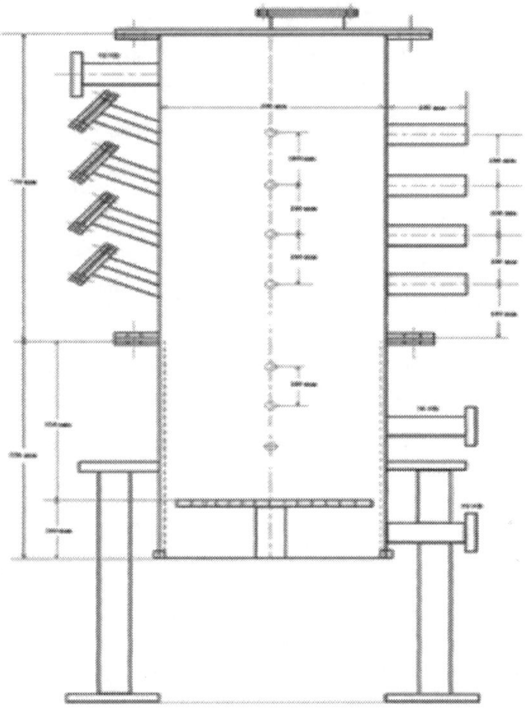

Fig. 1(a). Schematic of the updraft gasifier

The gasifier is located on stand so that it can be operated comfortably. The biomass inlet port is located at the top of the gasifier, through which the biomass is fed into the gasifier. The air is supplied into the gasifier by employing a suitable centrifugal blower ahead of the gasifier. The air supplied by the gasifier is controlled by a gate valve. The producer gas formed in the gasifier is flared through a firing port or flare duct. The updraft and downdraft gasifier diagram is depicted vide Figure 1(a) and (b).

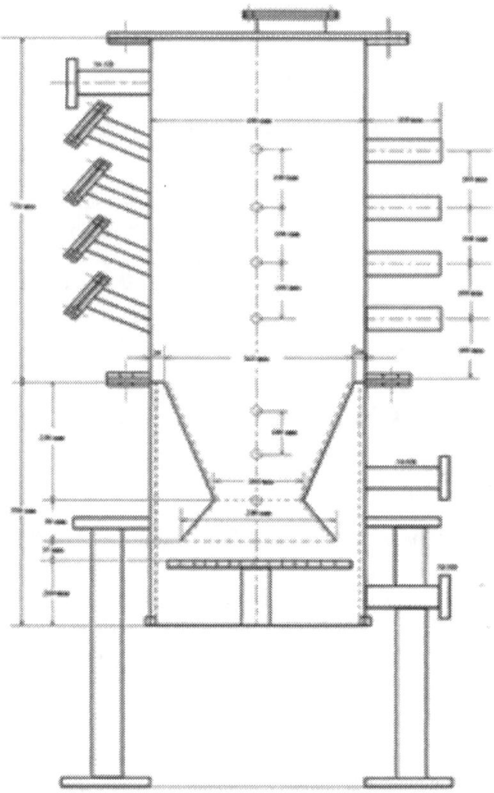

Fig. 1(b). Schematic of the downdraft gasifier

The instrumentation employed in the gasifier is useful to measure the critical operating parameters such as temperature, pressure, air flow rate, producer gas flow rate, producer gas composition and tar formed in the gasifier. The temperature formed in the gasifier is detected by deploying "K" type thermocouples. The pressure in the gasifier is measured by employing the "U" tube manometers. The air flow rate is measured using a orifice meter. The producer gas flow rate is measured using the venturimeter. The producer gas composition is measured using a suitable gas analyzer. A tar sampling set up is used for sampling the tar in the producer gas. The collected samples are then analyzed to estimate the tar formed during the gasification process.

3 Results and discussion

The feed stock used in this experimentation is casuarina wood. Figure 2 depicts the photograph of casuarina wood feed stock.

Fig. 2. Photograph of Casuarina wood feed stock

The characterization findings of this feed stock are tabulated in Table 1.The moisture content of 12.5% is in agreement of the range denoted by Peter et al (2002) ie lesser than 30 %. The required ash content should be lesser than 5%. The tabulated value is 2% which lies in this range. The volatile content of 67.5% indicates the possibility for formation of more tar during gasification process. That indicates the need for gas cleaning technologies such as physical, thermal method or catalytic tar mitigation methods. The gasifier is operated first in downdraft and then in updraft mode and the key findings are tabulated in Table 2.

Table 1. Characterization of feed stock

Feed material	Proximate analysis % by weight				Ultimate analysis % by weight			
	Ash	FC	VM	M	C	H	O	N
Casuarina wood	2	18	67.5	12.5	47.5	5.9	42.5	2.1

Table 2. Gasification parameters at Updraft and downdraft operating modes

Operating modes	Average Values of producer gas composition					Average Values of CV (MJ/Nm^3)
	CH_4 (% Vol.)	CO (% Vol.)	H_2 (% Vol.)	CO_2 (% Vol.)	N_2 (% Vol.)	
Updraft	1.58	10.42	10.62	12.56	64.82	3.03
Downdraft throat	1.45	14.70	13.82	11.22	58.79	3.87

The temperature distribution in the gasifier is depicted vide Figure 3.

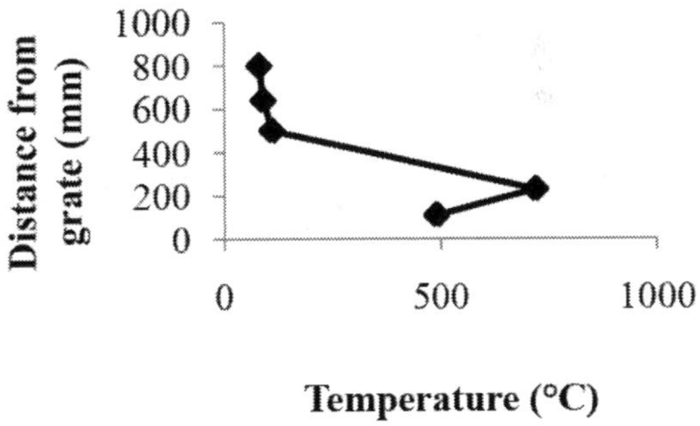

Fig. 3. Temperature distribution across the gasifier

The temperature distribution curve indicates a temperature of 750°C in the throat region. At this existing temperature condition the gasification parameters for updraft and downdraft modes are analyzed as indicated in Table 2. From the Table it is inferred that the downdraft mode of operation yields better performance than updraft mode, due to the reason that the downdraft mode has generated a producer gas with a calorific value of 3.87 MJ/Nm^3, whereas the updraft mode has generated a lesser caloric value producer gas of 3.03 MJ/Nm^3. This is due to the generation of 13.82 vol % of H_2 in downdraft mode which is higher than 10.62 vol% in updraft mode. Similarly another important component which contributed to the higher calorific value was the formation of 14.7 vol% of CO which was higher than 10.42vol% in updraft mode. On the contrary the CH4 content during downdraft mode was 1.45 vol% which was lesser than 1.58 vol% in updraft mode. The CO_2 content and N_2 content in downdraft mode is lower than updraft mode.

The better performance of downdraft mode of operation than updraft mode has generated higher caloric value of gas in downdraft mode predominantly due to larger cracking of tar in downdraft mode, based on the fact that due to presence of narrowed throat region, higher temperature has occurred, enhancing water - gas, water – shift, Boudard and tar reforming reactions which has cracked the hydrocarbons in the tar into Hydrogen and carbon mon Oxide gas. Furthermore the cracking of CH_4 has increased the formation of CO and H_2 gas composition which is evident from Table 2.On the contrary the temperature distribution in a updraft gasifier will be lesser due to the non availability of throat, which has resulted in lesser cracking of tar. This has resulted in formation of lesser vol% of CO and H_2 and lesser calorific gas.

4 Conclusions

From the above experimentation the following conclusions were derived.

1. The performance of the downdraft gasifier is better than updraft gasifier.
2. The calorific value of producer gas from a downdraft gasifier is higher than updraft mode gasifier.
3. The Vol% of CO and H_2 in downdraft mode was higher than updraft mode
4. The higher performance of downdraft gasifier is due to the reason that higher temperature in the throat region has favored the water gas, water shift, boudard and tar cracking reactions that has increased the formation more vol% of CO and H_2 and higher caloric value of gas than updraft mode of operation.

References

1. SA. Channiwala, 'On biomass gasification process and technological development some analytical and experimental investigations', Ph. D Thesis, IIT Delhi (1989)
2. M.J.Groenveld, Co-Current down Draft Gasifier, Ph.D Thesis, Twente University of Technology, The Netherlands (1980)
3. I.R.JanVenselaar, Design rules for down draft wood gasifiers a short review, prepared by project JTA-9A - Research Developmentat the Institute Teknologi Bandung, Indonesia August (1982)
4. Peter McKendry, J.Biores Tech. **83**, 55-63 (2002)
5. Rajiv Varshney, JL. Bhagoria, CR. Mehta, J. Engg, Sci and Manag. Edu. **3**, 33-40 (2010)
6. Prabir Basu , *Biomass gasification and pyrolysis – practical design book* (Academic press, 2010)
7. V.P. Rathod, JK. Rathnadhariya, S.A. Channiwala, Proceedings of the Int Conf on Mech Eng (ICME 2003) 26-28 , Dhaka, Bangladesh (2003)
8. Sanjeev H Kulkarni, TR.Anil, Int. J. Eng. Res. **3**(6), 384-389 (2014)
9. H. Susanto, Beenackers, Fuel. **75** (11), 1339-1347 (1996)
10. Yi Chen, Yong-hao Luo, Wen-Guang Wu, Yi Su 2009, Energy Fuels. **23**, 4659-4667 (2009)
11. S. Monteiro Nunes, N. Paterson, AA. Herold, DR. Dugwell and R. Kandiyoti, Energy & Fuels. **22**, 1955-1964 (2008)

A PMV-PPD model based study of thermal comfort in Low-Income Group house in Chhattisgarh

Nisha Netam[1,*], *Shubhashis* Sanyal[1], and *Shubhankar* Bhowmick[1]

[1]Department of Mechanical Engineering, National Institute of Technology Raipur, Chhattisgarh, 492010, India

Abstract. People tend to maintain symmetry between comfort and economy while choosing essential commodities needed in their life. Families buy a house which may offer comfort condition, but at minimum in term expenses of energy throughout a life. Thus, it is most important to erect a house to provide comfortable condition and moderate the lifetime expenditure by saving energy consumption. Sensation of thermal comfort varies from people to people, even in an identical environment. To minimize the consumption of energy of building, cost of consumed energy and to provide a comfortable house, thermal comfort analysis in indoor environments have attracted many researchers. Fanger's Predicted Mean Vote (PMV) - Predicted Percentage of Dissatisfied (PPD) model is widely accepted theory for assessment of building indoor thermal conditions. In the present work, thermal comfort of an LIG house in Chhattisgarh region of India has been analyzed based on PMV-PPD method for months representing three different seasons in a year i.e. May, September and December representing summer, post monsoon and winter respectively. Cooling, heating and actual energy load of LIG house has been calculated and reported for the above mentioned months. From analysis it is concluded that inhabitants are comfortable only during the winter.

Keywords: Thermal comfort, PMV-PPD model, Cooling and heating load, Thermal sensation.

1 Introduction

People tend to maintain symmetry between comfort and economy while choosing essential commodities needed in their life. Families buy or construct a house which may deliver comfort; however, these comfort conditions are mostly attained at the cost of high energy consumption. Thus, it is important to construct a house to provide conducive living conditions and also reduce the lifetime expenditure by saving energy consumption.

A (thermal) comfort perception varies between people to people, even in identical environment. To minimize the building energy consumption, cost of consumption and a comfortable house, thermal comfort analysis of the indoors in buildings have attracted many researchers. Fanger's Predicted Mean Vote (PMV)- Percentage of Dissatisfied (PPD) model is widely used theory for assessment of indoor thermal ambiance in buildings [1].The PMV index suggested by Fanger collects the response of inhabitants exposed to the climate spaces as per the seven-point thermal sensation scale of ASHRAE. A mean vote (MV) is estimated by calculating the mean of feeling/sensation provided by all the inhabitants at that condition. The PPD shows the percentage of the inhabitants sensing discomfort in that particular environment using the seven-point thermal sensation scale of ASHRAE (-3 to +3) as shown in table 1.

[2] Investigated the thermal environment and comfort related answers of occupants of Hyderabad. The survey was reported for 100 subjects, providing 3962 datasets and was analyzed for different categories of people based on: tenure, gender, age and economy. It is found that thermal reception of women and old persons were higher. Presently, two different methods for thermal comfort analysis are adopted by researchers: the rational approach and the adaptive approach reported by [3]. The former approach uses the collected data from climate chamber observations to support its theory and the latter one uses field data of people who reside in building. [4] Conducted a questionnaire survey based on the subjective responses to report thermal comfort of both the modern and traditional buildings. It is found that traditional residences in Kerala has comfortable indoors in all seasons because of its unique design, materials and unique methods adopted for construction.

[5] Reported a thermal comfort and people's behaviour inspection, in the living rooms of Kanto, Japan for hot and humid season, to report the need of indoor air circulation in the building to attain thermal comfort. [6] Reported an embodied energy analysis of mud house under the Delhi climate, and concluded that mud house has little embodied energy as compared to RCC building. Cooling/heating load and actual energy analysis has also reported for mud houses. Similar analysis has been reported for houses at different locations of the country. [7] Proposed the use of simulation tools for the building energy analysis such as Energy plus, e QUEST, TRNSYS and DOE-2.1E.

* Corresponding author: nishane.me@nitrr.ac.in

Table 1 ASHRAE thermal sensation scale

Value	Sensation
+3	Hot
+2	Warm
+1	Slightly warm
0	Neutral
-1	Slightly Cool
-2	Cool
-3	Cold

Literature survey reports that thermal comfort analysis of habitats at specific locations around the world has been conducted and highlighted the need of thermal comfort analysis for different regions to re-define the building design regulations. From the review it is concluded that very little work has been reported for India. Review also points towards the scarcity of such investigations for the region of Chhattisgarh.

Although the city realizes different season throughout a year, the general year long climate is hot and dry. The present study reported for summer (May) and Post monsoon months (September) and winter (December) in the year 2016.

2 Methods

2.1 Measurement of indoor and outdoor data

The outdoor environment data of temperature and humidity were taken from IMD metrological website. For collecting the indoor data LIG house has been chosen in Kabir Nagar, Raipur (21.25°N, 81.63°E), Chhattisgarh (Fig 1). Data logger with Wi-Fi enabled humidity and temperature sensors has been used to collect the data of room air humidity and temperature at three different season of the year (Fig. 2).

2.2 Method to calculate Indoor thermal comfort indexes

Fanger's approach is used for thermal comfort analysis which is based on field study data; it gives realistic and actual information of thermal environment. The interior thermal comfort indexes are calculated by the model proposed by Fanger.

PMV can be calculated by the following equation [8]:

$$PMV = (0.303e^{-0.036M} + 0.028)\{(M-W) - 3.96E^{-8} f_{c1}[(t_{c1}+273)^4 - (\bar{t}_r+273)^4] - f_{c1}h_c(t_{c1}-t_a)] - 3.05\,[5.733 - 0.007(M-W) - P_a] - 0.42[(M-W) - 58.15] - 0.0173M(5.867 - P_a) - 0.0014M(34 - t_a)\} \quad (1)$$

Surface temperature of clothing (t_{cl}) :

$$t_{cl} = \{35.7 - 0.028(M-W)\} - 3.96*10^{-8} I_{cl} *0.155* f_{c1}[(t_{cl}+273)^4 - (t_r+273)^4] - I_{cl}*0.155* f_{cl}h_c(t_{cl}-t_a) \quad (2)$$

Clothing factor (f_{cl})

$$f_{cl} = \begin{cases} 1.0 + 0.2I_{cl} & I_{cl} < 0.5clo \\ 1.05 + 0.1I_{cl} & I_{cl} > 0.5clo \end{cases} \quad (3)$$

Clothing thermal insulation (R_{cl}):

$$R_{cl} = 0.155I_{cl} \quad (4)$$

Convective heat transfer coefficient (h_c):

$$hc = 12.1V^{1/2} \quad (5)$$

PPD can be calculated by the following equation:

$$PPD = 100 - 95e^{[-(0.3353PMV^4 + 0.2179PMV^2)]} \quad (6)$$

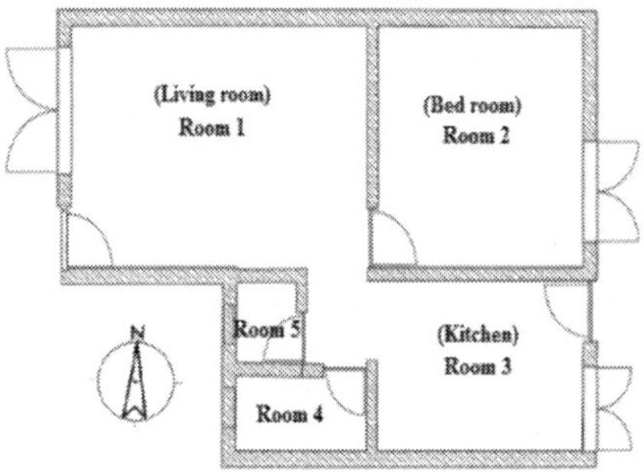

Fig. 1 LIG house plan. Living room (3.65 X3.05), bedroom (3.05 X 3.05), Kitchen (3.05 X 2.13), room 4 (2.13 X 1.21), room 5 (1.21 X 1.21). (All dimensions are in m)

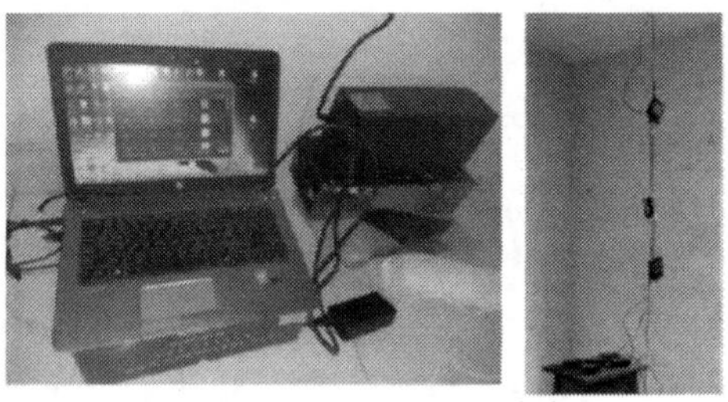

Fig. 2 Experimentation setup Data logger

2.3 Method to calculate cooling load and actual energy

Cooling load is the heat gain in the closed space that is needed to be removed, to maintain a steady indoor temperature. As Raipur experiences dry and hot climate mostly around the year, quantification of the annual energy use in a building, thus, requires determination of cooling loads of building. Cooling and heating load can be calculated by the determination of heat gains or heat losses of an indoor space respectively. The additional heat sources are occupants, equipment and appliances present in the house, and infiltration.

Cooling load/ Heating load $= Q_{wall} + Q_{roof} + Q_{door} + Q_{window} + Q_{\inf iltration}$ (7)

$$Q_{wall,roof,door} = UA(T_{sol} - T_0) \quad (8)$$

$$Q_{win} = [AU(T_{sol} - T_0) + A\tau I] \quad (9)$$

$$Q_{\inf iltration} = \frac{\rho V C_a N(T_0 - T_{amb})}{3600} \quad (10)$$

Where, Tsol is sol air temperature:

$$T_{sol} = \left(\frac{\alpha I}{h_o} + T_{amb} - \frac{\varepsilon \Delta R}{h_o} \right) \quad (11)$$

For horizontal and vertical surface, $\varepsilon\Delta R = 60\ W/m^2$, and 0 respectively. T_0 is the desired room temperature. (T_0=23°C)

The actual energy load of room of LIG house is estimated from the equation proposed by [6] Actual energy load depends on the room air temperature and size of the room. Actual energy load of house:

$$Q_{\text{Actual load}} = \left(M_a C_a\right)_1 X \left(T_d - T_r\right) X \frac{1}{\left(3600 sh^{-1} X 1000\right)} \quad (12)$$

In eq. (12) the chosen room air temperature is T_d (=20 °C, throughout the year for the house).

3 Results and discussion

Thermal comfort analysis for selected LIG house has been performed for three different season winter, post monsoon and summer in the month of December, September and May, respectively. The indoor temperature and ambient temperature measured at a particular day on May 20, September 2 and December 22, is plotted in Fig. 3.

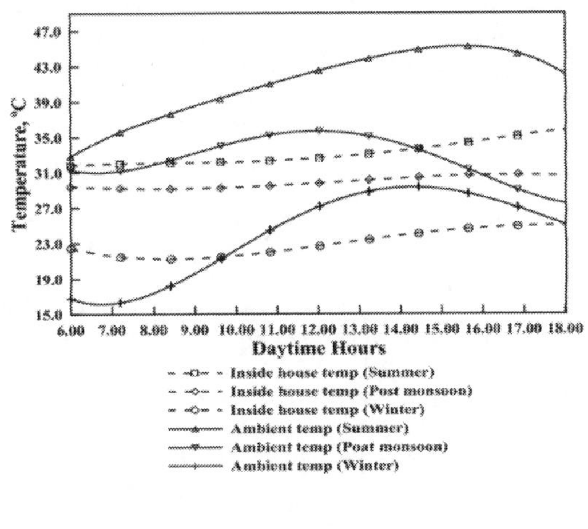

Fig.3 Ambient temperature and indoor temperature for three different season of the year in LIG house.

3.1 Thermal comfort at different seasons

Using correlation proposed by Fanger's (Eq (1) and (6)) PMV and PPD calculated for the time period of 6:00 hours to 18:00 hours as shown in Figs 4 and 5. In summer indoor temperature of house throughout a day is observed above 33°C and calculated PMV value is greater than +3 throughout a day. Recommended thermal comfort PMV should be in the range of ±0.5 and PPD<10%. In summer occupant feels highly uncomfortable inside the LIG house with PPD value of 100%. In Post monsoon indoor temperature of house throughout a day is between 29 °C to 30.4 °C and calculated PMV value between +1.8 to +2.4 throughout a day. The occupant feels warm inside the LIG house with PPD value of nearly100%. In winter indoor temperature of house throughout a day is between 21 °C to 24.5 °C and calculated PMV value is between -0.06 to +1.0 throughout a day. Based on seven points thermal sensation the occupant feels neutral and slightly warm during daytime inside the LIG house with PPD value of 1.5 % to 15.5%.

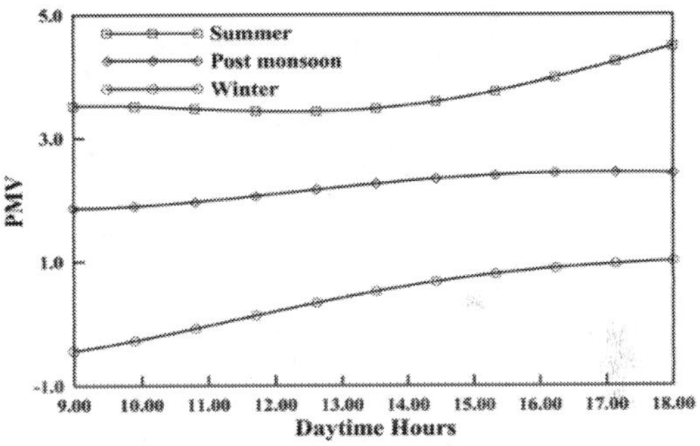

Fig. 4 Comparison on daytime PMV values for the day of three different seasons in LIG house

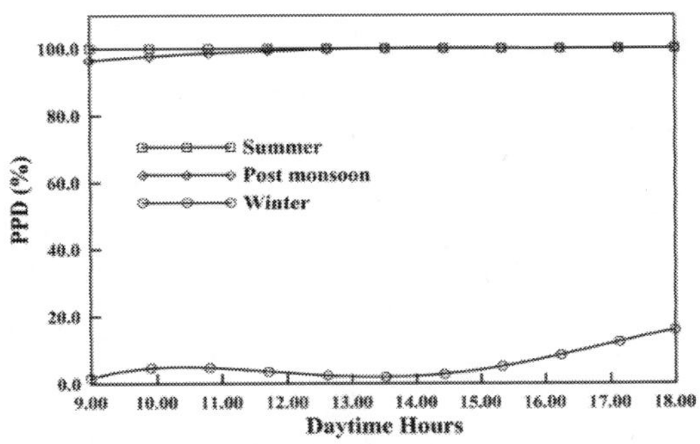

Fig. 5 Daytime PPD variation for the day of three different seasons in LIG house

3.2 Cooling and heating load calculation

Cooling and heating load has been calculated for May, September and December, which are the month of three different season summer, post monsoon and winter respectively. Maximum cooling load for the month of May is 702.5 kWh/month and followed by September is 505.5 kWh/month. Heating load for the month of December is 169.5 kWh/month as shown in Fig 6.
Actual energy load of each room of LIG house is calculated from the Eq. (12), hourly basis the total actual energy load of LIG house is calculated as shown in Fig. 7.

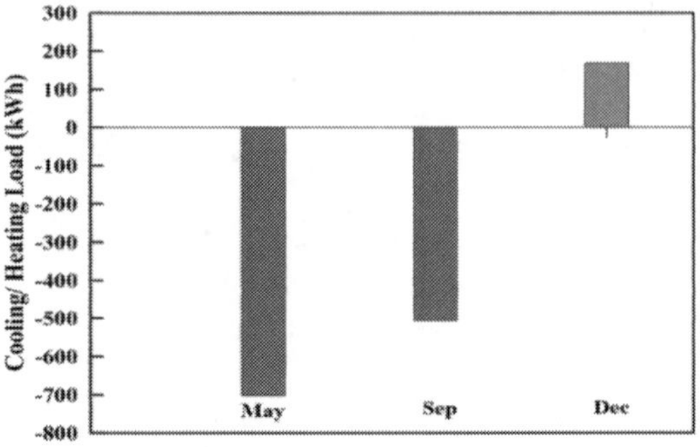

Fig. 6 Cooling and heating load analysis of LIG house.

3.3 Cooling and heating load calculation with modification on envelope

To study cooling/heating load of the house under different house envelope materials, house is then modelled by adding RCC roof with green roof and tiles roof. Similar comparison is performed where house is modelled by replacing burnt bricks of external wall with the brick wall with insulation material layer (U=0.903) and wall made with the ECBC suggested material (U=0.440). Figs. 8 and 9 show the cooling/ heating load for different wall and roof materials.

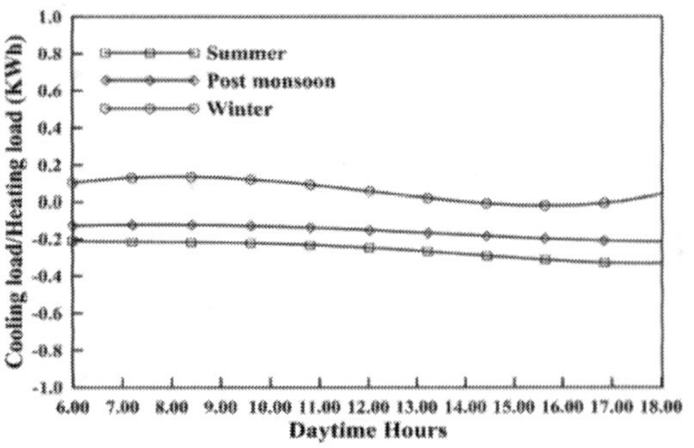

Fig. 7 Daytime actual energy load of LIG house for three different seasons.

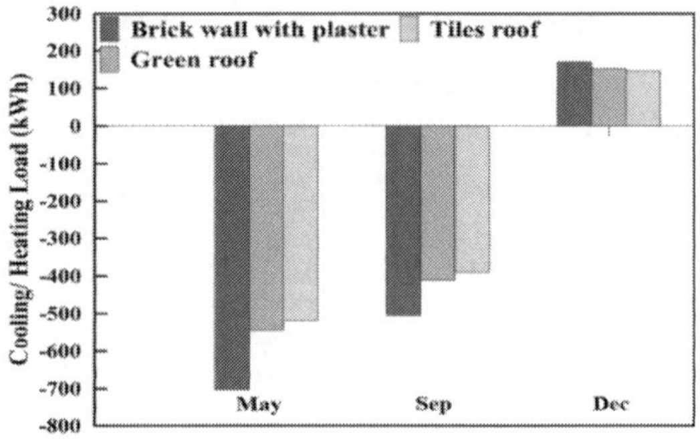

Fig. 8 Comparison of cooling and heating load analysis of LIG house for different roof materials.

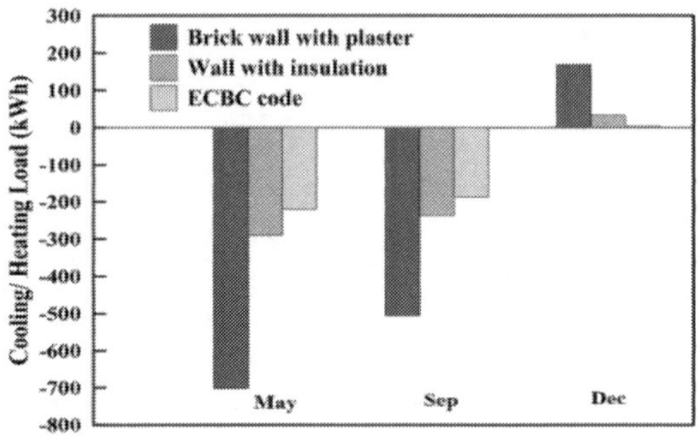

Fig. 9 Comparison of cooling and heating load analysis of LIG house for different wall materials.

4 Conclusions

In order to identify the comfort condition and to investigate discomfort in the LIG house at three different season of the year, data has been collected experimentally. Based on the collected data thermal comfort analysis has been investigated. The recommended PMV and PPD values are $-0.5 < PMV < +0.5$ and $PPD < 10\%$ for thermal comfort. Results shows that in summer, the level discomfort is very high and occupants feel the heat. During post monsoon, the occupant feels warm. In winter, the occupant feels neutral and slightly warm inside the LIG based on thermal sensation scale. Inhabitants are comfortable only during the winter season inside LIG house which remains for a period of for two months.

The comfort condition of house can be improved by:

- Replacing RCC roof of LIG house, with green roof and concrete tile roof reduced the heating/cooling load by 22.5% and 26.2 % respectively.
- Replacing burnt bricks external wall, with the brick wall with insulation material layer and wall made with the ECBC suggested material

reduced the heating/cooling load by 58% and 68% respectively.

Nomenclature

A	Area (m^2)
C_a	Specific heat of air (J/KgK)
e	Euler's number (2.718)
fcl	Clothing factor
hc	Convective heat transfer coefficient
ho	Outside convective heat transfer coefficient
I	Total solar Radiation (W/m^2)
Icl	Clothing insulation (clo)
M	Metabolic rate (W/m^2) 115 for all scenarios
Ma	Air mass (kg)
N	Air change rate (h^{-1})
pa	Vapor pressure of air [kPa]
Q	Heat transfer (J)
Rcl	Clothing thermal insulation
ta	Air temperature (°C)
tcl	Surface temperature of clothing (°C)
tr	Mean radiant temperature (°C)
Tr	Room temperature (K)
Tsol	Sol-air temperature (K)
τ	Transmissivity
U	Overall heat transfer coefficient (W/m^2K)
V	Air velocity (m/s)

References

1. P.O. Fanger, *Thermal Comfort—Analysis and Applications in Environmental Engineering,* McGraw-Hill Book Company, New York, NY, 1972.
2. M. Indraganti, K. D. Rao Effect of age, gender, economic group and tenure on thermal comfort: A field study in residential buildings in hot and dry climate with seasonal variations. Energy and Buildings, **42**(3), 273–281 (2010)..
3. N. Djongyang, , R. Tchinda, & D. Njomo,. Thermal comfort: A review paper. Renewable and Sustainable Energy Reviews, **14**(9), 2626–2640 (2010).
4. A. S. Dili, M. A. Naseer, & T. Z. Varghese, Thermal comfort study of Kerala traditional residential buildings based on questionnaire survey among occupants of traditional and modern buildings. Energy & Buildings, **42**(11), 2139–2150 (2010)..
5. H. Rijal, Investigation of Comfort Temperature and Occupant Behavior in Japanese Houses during the Hot and Humid Season. Buildings, **4**(3), 437–452 (2014).
6. A. Chel, G. N. Tiwari, Thermal performance and embodied energy analysis of a passive house - Case study of vault roof mud-house in India. Applied Energy, **86**(10), 1956–1969 (2009).
7. T. Ramesh, R. Prakash, & K. K. Shukla, Life cycle energy analysis of buildings: An overview. Energy and Buildings, **42**(10), 1592–1600 (2010).
8. K. Horikiri, Y. Yao, & J. Yao, Numerical optimisation of thermal comfort improvement for indoor environment with occupants and furniture. Energy & Buildings, **88**, 303–315 (2015).

AUTHOR INDEX

AUTHOR INDEX

EDP Sciences – Web of Conferences
17, Avenue du Hoggar
Parc d'Activité de Courtabouf
BP 112
F-91944 Les Ulis Cedex A
France

ISBN 978-1-5108-6467-2

CURRAN ASSOCIATES INC.
proceedings.com

International Cooperation for Education About Standardization (ICES 2018)

Strengthening Industry and Engineering, Science, and Management Education Through Standardization Learning

Joint International Conference with 5th Annual Conference on Industrial and System Engineering and World Standard Cooperation Academic Day

SHS Web of Conferences Volume 49 (2018)

Yogyakarta, Indonesia
3 – 5 July 2018

edp sciences